The Evolution of Population Biology

This is the third of three volumes published by Cambridge University Press in honor of Richard Lewontin. The first volume, *Evolutionary Genetics from Molecules to Morphology*, honors Lewontin's more technical contributions to population and evolutionary genetics, and the second volume, *Thinking about Evolution: Historical, Philosophical, and Political Perspectives*, honors Lewontin's contributions to the history and philosophy of biology and to the controversial field of sociobiology. This volume honors his contributions to population biology: the nexus between population genetics and ecology.

This unique collection of essays deals with the foundation and historical development of population biology, and its relationship to population genetics and population ecology on one hand and to the rapidly growing fields of molecular quantitative genetics, genomics, and bioinformatics on the other. Such an interdisciplinary treatment of population biology has never been attempted before. The volume is set in a historical context, but it has an up-to-date coverage of material in various related fields. The areas covered are the foundation of population biology, life history evolution and demography, density- and frequency-dependent selection, recent advances in quantitative genetics and bioinformatics, evolutionary case history of model organisms focusing on polymorphisms and selection, mating system evolution and evolution in the hybrid zones, and applied population biology including conservation, infectious diseases, and human diversity.

The volume brings out the central role of population biology in all aspects of its connection to population genetics and population ecology and it is a must for all graduate students and researchers in population genetics and ecology.

RAMA S. SINGH is a Professor in the Department of Biology at McMaster University.

MARCY K. UYENOYAMA is a Professor in the Department of Biology at Duke University.

The Evolution of Population Biology

Edited by

RAMA S. SINGH
McMaster University

MARCY K. UYENOYAMA
Duke University

CAMBRIDGE UNIVERSITY PRESS
Cambridge, New York, Melbourne, Madrid, Cape Town, Singapore, São Paulo, Delhi

Cambridge University Press
The Edinburgh Building, Cambridge CB2 8RU, UK

Published in the United States of America by Cambridge University Press, New York

www.cambridge.org
Information on this title: www.cambridge.org/9780521112116

First published 2004
This digitally printed version 2009

A catalogue record for this publication is available from the British Library

Library of Congress Cataloguing in Publication data

The evolution of population biology / edited by Rama Singh, Marcy Uyenoyama.
p. cm.
Includes bibliographical references and index.
ISBN 0 521 81437 5
1. Population biology. I. Singh, Rama S. (Rama Shankar), 1945–
II. Uyenoyama, Marcy, 1953–
QH352.E96 2003
577.8′8 – dc21 2003044035

ISBN 978-0-521-81437-9 hardback
ISBN 978-0-521-11211-6 paperback

Contents

Contributors

Braswell, W. E., Department of Biology, New Mexico State University, Las Cruces, NM 88003

Britch, S. C., Department of Biology, New Mexico State University, Las Cruces, NM 88003

Capy, P., Laboratoire Populations, Génétique et Evolution, Centre National de la Recherche Scientifique, 91198 Gif-sur-Yvette, France

Cariou, M.-L., Laboratoire Populations, Génétique et Evolution, Centre National de la Recherche Scientifique, 91198 Gif-sur-Yvette, France

Castillo-Davis, C. I., Department of Organismic and Evolutionary Biology, Harvard University, Cambridge, MA 02138

Cavalieri, D., Harvard Center for Genomics Research, Harvard University, Cambridge, MA 02138

Cavalli-Sforza, L. L., Department of Genetics, Stanford University, Stanford, CA 94305-5120

Charlesworth, B., Institute of Cell, Animal and Population Biology, University of Edinburgh, Ashworth, King's Buildings, West Mains Road, Edinburgh EH9 3JN, UK

Christiansen, F. B., Department of Genetics and Ecology, University of Aarhus, Build. 540 DK-8000 Aarhus C, Denmark.

Clarke, B., Institute of Genetics, University of Nottingham, Queen's Medical Centre, Clifton Boulevard, Nottingham NG7 2UH, UK

Cochran, G. M., Department of Anthropology, University of Utah, Salt Lake City, UT 84112

David, J. R., Laboratoire Populations, Génétique et Evolution, Centre National de la Recherche Scientifique, 91198 Gif-sur-Yvette, France

Ewald, P., Department of Biology, University of Louisville, Louisville, KY 40292

Gillespie, J., Section of Evolution and Ecology, University of California, Davis, CA 95616-8755

Golding, G. B., Department of Biology, McMaster University, 1280 Main Street West, Hamilton, Ontario, Canada L8S 4K1

Hartl, D., Department of Organismic and Evolutionary Biology, Harvard University, Cambridge, MA 02138

Hedrick, P., Department of Biology, Arizona State University, Tempe, AZ 85287

Holmes, E., Department of Zoology, University of Oxford, South Parks Road, Oxford OX1 3PS, UK

Howard, D. J., Department of Biology, New Mexico State University, Las Cruces, NM 88003

Joly, D., Laboratoire Populations, Génétique et Evolution, Centre National de la Recherche Scientifique, 91198 Gif-sur-Yvette, France

Lachaise, D., Laboratoire Populations, Génétique et Evolution, Centre National de la Recherche Scientifique, 91198 Gif-sur-Yvette, France

Lemeunier, F., Laboratoire Populations, Génétique et Evolution, Centre National de la Recherche Scientifique, 91198 Gif-sur-Yvette, France

Levins, R., Department of Population Sciences, School of Public Health, Harvard University, 667 Huntington Avenue, Boston, MA 02115

Lewontin, R. C., Museum of Comparative Zoology, Harvard University, Cambridge, MA 02138

Mackay, T. F., Department of Genetics, North Carolina State University, Raleigh, NC 27695-7614

Marshall, J. L., Department of Biology, The University of Texas at Arlington, Box 19498, 501 South Nedderman Dr., Arlington, TX 76019-0498

Meiklejohn, C. D., Department of Organismic and Evolutionary Biology, Harvard University, Cambridge, MA 02138

Morton, R., Department of Biology, McMaster University, 1280 Main Street West, Hamilton, Ontario, Canada L8S 4K1

Provine, W. B., Department of Ecology and Evolutionary Biology, E139 Corson Hall, Cornell University, Ithaca, NY 14853

Ranz, J. M., Department of Organismic and Evolutionary Biology, Harvard University, Cambridge, MA 02138

Singh, R. S., Department of Biology, McMaster University, 1280 Main Street West, Hamilton, Ontario, Canada L8S 4K1

Takebayashi, N., Department of Biology, Box 90338, Duke University, Durham, NC 27708-0338

Taylor, P., Program on Critical and Creative Thinking, Graduate College of Education, University of Massachusetts, Boston, MA 02125

Townsend, J. P., Department of Organismic and Evolutionary Biology, Harvard University, Cambridge, MA 02138

Tuljapurkar, S., Mountain View Research, 2251 Grant Road, Los Altos, CA 94204

Uyenoyama, M. K., Department of Biology, Box 90338, Duke University, Durham, NC 27708-0338

Wakeley, J., Department of Organismic and Evolutionary Biology, Harvard University, Cambridge, MA 02138

Watt, W. B., Department of Biological Sciences, Stanford University, Stanford, CA 94305

Publications of R. C. Lewontin

1. 1952. An elementary text on evolution. Review of *A Textbook of Evolution*, by E. O. Dodson. *Evolution* 6(2):247–8.
2. 1953. The effect of compensation on populations subject to natural selection. *Am. Nat.* 87:375–81.
3. 1954. Review of *Problems of Life*. *Am. J. Sci.* 252:123–4.
4. 1954. Review of *Biology and Language*. *Am. J. Sci.* 252:124–6.
5. 1954. Familial occurrence of migraine headache: A study of heredity (with H. Goodell and H. G. Wolff). *A.M.A. Archives of Neurology and Psychiatry* 72:325–34.
6. 1955. The effects of population density and composition on viability in *Drosophila melanogaster*. *Evolution* 9:27–41.
7. 1956. Estimation of the number of different classes in a population (with T. Prout). *Biometrics* 12:211–23.
8. 1956. Studies on homeostasis and heterozygosity. I. General considerations. Abdominal bristle number in second chromosome homozygotes of *Drosophila melanogaster*. *Am. Nat.* 90:237–55.
9. 1956. A reply to Professor Dempster's comments on homeostasis (Letters to the Editors). *Am. Nat.* 90:386–8.
10. 1957. The adaptations of populations to varying environments. *Cold Spring Harbor Symp. Quant. Biol.* 22:395–408.
11. 1958. A general method for investigating the equilibrium of gene frequency in a population. *Genetics* 43:419–34.
12. 1958. Studies of heterozygosity and homeostasis. II. Loss of heterosis in a constant environment. *Evolution* 12:494–503.
13. 1959. On the anomalous response of *Drosophila pseudoobscura* to light. *Am. Nat.* 93:321–8.
14. 1959. The goodness-of-fit test for detecting natural selection (with C. C. Cockerham). *Evolution* 13:561–4.
15. 1960. *Quantitative Zoology*, 2nd edn. (with G. G. Simpson and A. Roe). New York: Harcourt Brace.

16. 1960. Interaction between inversion polymorphism of two chromosome pairs in the grasshopper *Moraba scurra. Evolution* 14:116–29.
17. 1960. The evolutionary dynamics of a polymorphism in the house mouse (with L. C. Dunn). *Genetics* 45:705–22.
18. 1960. The evolutionary dynamics of complex polymorphisms (with K. Kojima). *Evolution* 14:458–72.
19. 1960. Review of *Introduction to Quantitative Genetics* by D. S. Falconer. *Am. Sci.* 48:274–6A.
20. 1961. Evolution and the theory of games. *J. Theor. Biol.* 1:382–403.
21. 1961. Review of *Biochemical Genetics. Am. Sci.* 49:190A.
22. 1962. Review of *Introduction to the Mathematical Theory of Genetic Linkage* by N. T. J. Bailey. *Am. Sci.* 50:320–2A.
23. 1962. Review of *An Outline of Chemical Genetics* by B. Strauss. *Human Biol.* 34:235–6.
24. 1962. Interdeme selection controlling a polymorphism in the house mouse. *Am. Nat.* 96:65–78.
25. 1963. Interaction of genotypes determining viability in *Drosophila busckii* (with Y. Matsou). *Proc. Natl Acad. Sci. USA* 49:270–8.
26. 1963. Relative fitness of geographic races of *Drosophila serrata* (with L. C Birch, T. Dobzhansky and P. O. Elliott). *Evolution* 17:72–83.
27. 1963. Models, mathematics, and metaphors. *Synthese* 15:222–44.
28. 1963. Cytogenetics of the grasshopper *Moraba scurra.* VII. Geographic variation of adaptive properties of inversions (with M. J. D. White and L. E. Andrew). *Evolution* 17:147–62.
29. 1964. The interaction of selection and linkage. I. General considerations: Heterotic models. *Genetics* 49:49–67.
30. 1964. A molecular messiah: the new gospel of genetics? Essay review of *The Mechanics of Inheritance* by F. Stahl. *Science* 145:525.
31. 1964. The role of linkage in natural selection. In *Genetics Today*, pp. 517–25. Proc. XI Int. Cong. of Genetics, The Hague, September 1964. Vol. 2, ed. S. J. Geerts. Oxford: Pergamon Press.
32. 1964. The interaction of selection and linkage. II. Optimum models. *Genetics* 50:757–82.
33. 1964. Review of *Elizabethan Acting.* The Seventeenth Century News.
34. 1964. The capacity for increase in chromosomally polymorphic and monomorphic populations of *Drosophila pseudoobscura* (with T. Dobzhansky and O. Pavlovsky). *Heredity* 19:597–614.
35. 1965. Selection in and of populations. In *Ideas in Modern Biology*, pp. 299–311. Proc. XVI Int. Cong. of Zool. Vol. 6, ed. J. A. Moore. Garden City, NY: Natural History Press.
36. 1965. The robustness of homogeneity tests in 2 × *N* tables (with J. Felsenstein). *Biometrics* 21:19–33.

37. 1965. Selection for colonizing ability. In *The Genetics of Colonizing Species*, ed. H. Baker, pp. 77–94, New York: Academic Press.
38. 1965. Review of *The Effects of Inbreeding on Japanese Children. Science* 150:332–3.
39. 1965. Review of *Stochastic Models in Medicine and Biology. Am. Sci.* 53: 254–5A.
40. 1966. Adaptation and natural selection (essay review). *Science* 152: 338–9.
41. 1966. Is nature probable or capricious? *BioScience* 16:25–7.
42. 1966. Differences in bristle-making abilities in scute and wild-type *Drosophila melanogaster* (with S. S. Young). *Genet. Res.* 7:295–301.
43. 1966. On the measurement of relative variability. *Syst. Zool.* 15:141–2.
44. 1966. Stable equilibria under optimizing selection (with M. Singh). *Proc. Natl Acad. Sci. USA* 56:1345–8.
45. 1966. A molecular approach to the study of genic heterozygosity in natural populations. I. The number of alleles at different loci in *Drosophila pseudoobscura* (with J. L. Hubby). *Genetics* 54:577–94.
46. 1966. A molecular approach to the study of genic heterozygosity in natural populations. II. Amount of variation and degree of heterozygosity in the natural populations of *Drosophila pseudoobscura* (with J. L. Hubby). *Genetics* 54:595–609.
47. 1966. Hybridization as a source of variation for adaptation to new environments (with L. C. Birch). *Evolution* 20:315–36.
48. 1966. Review of *The Theory of Inbreeding. Science* 150:1800–1.
49. 1967. The genetics of complex systems. *Proc. V Berkeley Symp. on Math. Stat. Prob.* Berkeley: University of Calif. Press, Vol. IV, pp. 439–55.
50. 1967. The interaction of selection and linkage. III. Synergistic effect of blocks of genes (with P. Hull). *Der Zuchter* 37:93–8.
51. 1967. The principle of historicity in evolution. In *Mathematical Challenges of the Neo-Darwinian Theory of Evolution*, ed. P. S. Moorhead and M. M. Kaplan. Wistar Symp. Monograph No. 5:81–94. Philadelphia, PA: Wistar Institute Press.
52. 1967. An estimate of average heterozygosity in man. *Am. J. Hum. Genet.* 19:681–5.
53. 1967. Population genetics. *Annu. Rev. Genet.* 1:37–70.
54. 1968. A molecular approach to the study of genic heterozygosity in natural populations. III. Direct evidence of coadaptation in gene arrangements of *Drosophila* (with S. Prakash). *Proc. Natl. Acad. Sci. USA* 59:398–405.
55. 1968. A note on evolution and changes in the quantity of genetic information (with C. H. Waddington). In *Towards a Theoretical Biology*, ed. C. H. Waddington, Vol. I, pp. 109–10. Chicago: Aldine Publ.
56. 1968. Essay review of *Phage and the Origins of Molecular Biology*, ed. J. Cairns *et al.* (Cold Spring Harbor Lab. of Quant. Biol.,

Cold Spring Harbor, NY, 1966, xii + 340). *J. Hist. Biol.* 1(1): 155–61.

57. 1968. The concept of evolution. In article "Evolution". In *International Encyclopedia of the Social Sciences*, pp. 202–10. New York: Macmillan and The Free Press.
58. 1968. The effect of differential viability on the population dynamics of *t* alleles in the house mouse. *Evolution* 22:262–73.
59. 1968. Selective mating, assortative mating, and inbreeding: definitions and implications (with D. Kirk and J. Crow). *Eugenics Quart.* 15:141–3.
60. 1969. The bases of conflict in biological explanation. *J. Hist. Biol.* 2(1):35–45.
61. 1969. On population growth in a randomly varying environment (with D. Cohen). *Proc. Natl Acad. Sci. USA* 62(4):1056–60.
62. 1969. A molecular approach to the study of genic heterozygosity in natural populations. IV. Patterns of genic variation in central, marginal and isolated populations of *Drosophila pseudoobscura* (with S. Prakash and J. L. Hubby). *Genetics* 61:841–58.
63. 1969. The meaning of stability. (Reprinted from *Diversity and Stability in Ecological Systems.*) *Brookhaven Symp. in Biol.* 22:13–24.
64. 1970. On the irrelevance of genes. In *Towards a Theoretical Biology*, ed. C. H. Waddington, Vol. 3: Drafts, pp. 63–72. Edinburgh: Edinburgh University Press.
65. 1970. Race and intelligence. *Bull. Atom. Sci.* 26(Mar):2–8.
66. 1970. Further remarks on race and the genetics of intelligence. *Bull. Atom. Sci.* 26(May):23–5.
67. 1970. Genetic variation in the horseshoe crab (*Limulus polyphemus*), a phylogenetic "relic" (with R. K. Selander, S. Y. Yang and W. E. Johnson). *Evolution* 24:402–14.
68. 1970. Is the gene the unit of selection? *Genetics* 65:707–34 (with I. Franklin).
69. 1970. The units of selection. *Annu. Rev. Ecol. Syst.* 1:1–18.
70. 1971. Genes in populations – end of the beginning (with J. F. Crow and M. Kimura). Review of *An Introduction to Population Genetics Theory. Quart. Rev. Biol.* 46:66–7.
71. 1971. Evolutionary significance of linkage and epistasis (with K. Kojima). In *Biomathematics*, Vol. I: *Mathematical Topics in Population Genetics*, ed. K. Kojima pp. 367–88. Berlin: Springer-Verlag.
72. 1971. The effect of genetic linkage on the mean fitness of a population. *Proc. Natl Acad. Sci. USA* 68:984–6.
73. 1971. The Yahoos ride again. *Evolution* 25:442.
74. 1971. Science and ethics. *BioScience* 21:799.
75. 1971. A molecular approach to the study of genic heterozygosity in natural populations. V. Further direct evidence of coadaptation

in inversions of *Drosophila* (with S. Prakash). *Genetics* 69: 405–8.

76. 1972. Testing the theory of natural selection. *Nature* 236:181–2.
77. 1972. The apportionment of human diversity. *Evol. Biol.* 6:381–98.
78. 1972. Comparative evolution at the levels of molecules, organisms and populations (with G. L. Stebbins). *Proc. VI Berkeley Symp. on Math. Stat. Prob.* Berkeley: University of Calif. Press, Vol. V, pp. 23–42.
79. 1972. The apportionment of human diversity. In *Evolutionary Biology*, Vol. 6, ed. T. Dobzhansky, M. K. Hecht and W. C. Steere, pp. 381–98. New York: Appleton-Century-Crofts.
80. 1973. Distribution of gene frequency as a test of the theory of the selective neutrality of polymorphism (with J. Krakauer). *Genetics* 74:175–95.
81. 1974. Population genetics. *Annu. Rev. Genet.* 7:1–17.
82. 1974. Molecular heterosis for heat-sensitive enzyme alleles (with R. S. Singh and J. L. Hubby). *Proc. Natl. Acad. Sci. USA* 71:1808–10.
83. 1974. *The Genetic Basis of Evolutionary Change.* New York: Columbia University Press. 346 pp.
84. 1974. Annotation: The analysis of variance and the analysis of causes. *Am. J. Hum. Genet.* 26:400–11.
85. 1974. Darwin and Mendel – the materialist revolution. In *The Heritage of Copernicus: Theories "More Pleasing to the Mind"*, ed. J. Neyman, pp. 166–83. Cambridge, MA: MIT Press.
86. 1975. The problem of genetic diversity. *Harvey Lecture Series* 70:1–20.
87. 1975. Selection in complex genetic systems. III. An effect of allele multiplicity with two loci (with M. W. Feldman, I. R. Franklin and F. B. Christiansen). *Genetics* 79:333–47.
88. 1975. Review of *The Modern Concept of Nature* by H. J. Muller. *Soc. Biol.* 22:96–8.
89. 1975. The heritability hang-up (with M. W. Feldman). *Science* 190: 1163–8.
90. 1975. Genetic aspects of intelligence. *Annu. Rev. Genet.* 9:382–405.
91. 1976. Review of *Race Differences in Intelligence* by J. C. Loehlin, G. Lindzey and J. N. Spuhler. *Am. J. Hum. Genet.* 28:92–7.
92. 1976. *Adattamento genetico. Enciclopedia del Novecento* (Rome) 1:61–8.
93. 1976. Genetic heterogeneity within electrophoretic "alleles" of xanthine dehydrogenase in *Drosophila pseudoobscura* (with R. S. Singh and A. A. Felton). *Genetics* 84:609–29.
94. 1976. Race and intelligence. In *The IQ Controversy*, ed. N. J. Block and G. Dworkin, pp. 78–92. New York: Pantheon Books.
95. 1976. Further remarks on race and the genetics of intelligence. In *The IQ Controversy*, ed. N. J. Block and G. Dworkin, pp. 107–12. New York: Pantheon Books.

96. 1976. The analysis of variance and the analysis of causes. In *The IQ Controversy*, ed. N. J. Block and G. Dworkin, pp. 179–93. New York: Pantheon Books.
97. 1976. The problem of Lysenkoism (with R. Levins). In *The Radicalisation of Science*, ed. H. Rose and S. Rose, pp. 32–64. London: Macmillan.
98. 1976. The fallacy of biological determinism. *The Sciences* 16:6–10.
99. 1976. Sociobiology – a caricature of Darwinism. *Phil. Sci. Assoc.* 2:22–31.
100. 1977. The relevance of molecular biology to plant and animal breeding. In *Proc. Int. Conf. on Quant. Gen.*, ed. E. Pollak, pp. 55–62. Ames: Iowa State University Press.
101. 1977. Population genetics. *Proc. 5th Int. Cong. of Hum. Gen., Excerpta Medica Int. Cong.* Series No. 441, pp. 13–8.
102. 1977. *Adattamento. Enciclopedia Einaudi* (Turin) I:198–214.
103. 1977. Biological determinism as a social weapon. In *Biology as a Social Weapon*, pp. 6–18. Minneapolis, MN: Burgess.
104. 1977. Caricature of Darwinism. Review of *The Selfish Gene*, by R. Dawkins. *Nature* 266:283–4.
105. 1978. Heterosis as an explanation for large amounts of genic polymorphism (with L. R. Ginzburg and S. D. Tuljapurkar). *Genetics* 88:149–70.
106. 1978. Fitness, survival and optimality. In *Analysis of Ecological Systems*, ed. D. J. Horn *et al.* Columbus: Ohio State University Press.
107. 1978. The extent of genetic variation at a highly polymorphic esterase locus in *Drosophila pseudoobscura* (with J. Coyne and A. A. Felton). *Proc. Natl Acad. Sci. USA* 75:5090.
108. 1978. *Evoluzione. Enciclopedia Einaudi* (Turin) V:995–1051.
109. 1978. Adaptation. *Sci. Am.* 239(3):212–28.
110. 1978. *Geneticheskie Osnovi Evolutsil.* Moscow: Mir Publishers (Translation: *The Genetic Basis of Evolutionary Change*, 1974).
111. 1979. *La Base Genética de la Evolución.* Ediciones Omega, S.A., Casanove, 220 Barcelona, Spain. 328 pps. (Translation: *The Genetic Basis of Evolutionary Change*, 1974).
112. 1979. Sociobiology as an adaptationist program. *Behav. Sci.* 24:5–14.
113. 1979. Theodosius Dobzhansky. (Biographical article) In *International Encyclopedia of the Social Sciences.* New York: The Free Press, Macmillan.
114. 1979. Single- and multiple-locus measures of genetic distance between groups. *Am. Nat.* 112(988):1138–9.
115. 1979. The genetics of electrophoretic variation (with J. A. Coyne and W. F. Eanes). *Genetics* 92:353–61.

116. 1979. The spandrels of San Marco and the Panglossian paradigm: a critique of the adaptationist programme (with S. J. Gould). *Proc. R. Soc. Lond. Ser. B.* 205:581–98.
117. 1979. The sensitivity of gel electrophoresis as a detector of genetic variation (with J. A. M. Ramshaw and J. A. Coyne). *Genetics* 93: 1019–37.
118. 1979. *Mutazione/selezione. Enciclopedia Einaudi* (Turin) IX:647–95.
119. 1980. Dialectics and reductionism in ecology (with R. Levins). *Synthese* 43:47–78.
120. 1980. Sociobiology: another biological determinism. *Int. J. Health Sci.* 10(3):347–63.
121. 1980. Economics down on the farm. Review of *Farm and Food Policy: Issues of the 1980's* by D. Paarlberg. *Nature* 287:661–2.
122. 1980. The political economy of food and agriculture (World Agricultural Research Project). *Int. J. Health Sci.* 10:161–70.
123. 1981. *An Introduction to Genetic Analysis*, 2nd edition (with D. T. Suzuki, and A. J. F. Griffiths). San Francisco, CA: W. H. Freeman. 711 pp.
124. 1981. Evolution/creation debate: a time for truth. *BioScience* 31:559.
125. 1981. Sleight of hand. Review of *Genes, Mind and Culture*, by C. J. Lumsden and E. O. Wilson. *The Sciences* 21:23–6.
126. 1981. Review of *The Mismeasure of Man* by S. J. Gould. *NY Rev. of Books* 28(16).
127. 1981. Gene flow and the geographical distribution of a molecular polymorphism in *Drosophila pseudoobscura* (with J. S. Jones, S. H. Bryant, J. A. Moore and T. Prout). *Genetics* 198:157–78.
128. 1981. Theoretical population genetics in the evolutionary synthesis. In *The Evolutionary Synthesis*, ed. E. Mayr and W. Province. Cambridge, MA: Harvard University Press.
129. 1981. *Dobzhansky's Genetics of Natural Populations I-XLIII* (ed. with J. A. Moore, W. B. Provine and B. Wallace). New York: Columbia University Press.
130. 1981. L'évolution. *La Pensée* 223:16–24.
131. 1982. Artifact, cause and genic selection (with E. Sober). *Phil. Sci.* 49:157–80.
132. 1982. A study of reaction norms in natural populations of *Drosophila pseudoobscura* (with A. Gupta). *Evolution* 36:934–8.
133. 1982. Review of *Matter, Life and Generation* by S. A. Roe. *The Sciences*, December.
134. 1982. Prospectives, perspectives, and retrospectives. Review of *Perspectives on Evolution* by R. Milkman. *Paleobiology* 8(3).
135. 1982. Organism and environment. In *Learning, Development and Culture: Essays in Evolutionary Epistemology*, ed. H. Plotkin. Chichester: Wiley.

136. 1982. Review. *Evolution and the Theory of Games* by J. M. Smith, Cambridge University Press. *Nature* 300:113–14.

137. 1982. *Human Diversity*. Redding, CT: Scientific American and W. H. Freeman.

138. 1982. Elementary errors about evolution. Review of *Intentional Systems in Cognitive Ethology: The 'Panglossian Paradigm'* defined by D. Dennett. *Behav. and Brain Sci.* 6:367.

139. 1983. The corpse in the elevator. Reviews of *Against Biological Determinism* and *Toward a Liberating Biology* by The Dialectics of Biology Group. *NY Rev. of Books* 29(21&22).

140. 1983. Biological determinism. In *The Tanner Lectures on Human Values* Vol. IV, pp. 147–83. Salt Lake City: University of Utah Press.

141. 1983. Science as a social weapon. In *Occasional Papers I*, pp. 13–29. Amherst: Instit. for Advanced Study in the Humanities, University of Massachusetts.

142. 1983. Review of books by J. Miller & B. Van Loon, J. M. Smith, B. G. Gale, J. Gribben & J. Cherfas, N. Eldrige & I. Tattersall, D. Futuyma, P. Kitcher & M. Ruse. *NY Rev. of Books* 30(10).

143. 1983. The organism as the subject and object of evolution. *Scientia* 188:65–82.

144. 1983. Gene, organism and environment. In *Evolution from Molecules to Men*, ed. D. S. Bendall, pp. 273–85. Cambridge: Cambridge University Press.

145. 1983. Introduction. In *Scientists Confront Creationism*, ed. L. R. Godfrey and J. R. Coles. New York: W. W. Norton.

146. 1983. Discussion: Reply to Rosenberg on genic selectionism (with E. Sober). *Phil. Sci.* 50: 648–50.

147. 1983. *Il Gene e La Sua Mente. Biologia, Ideologia e Natura Umana.* Milan: Mondadari. (Translation: *Not in Our Genes*, 1984).

148. 1984. Detecting population differences in quantitative characters as opposed to gene frequencies. *Am. Nat.* 123(1):115–24.

149. 1984. *Not in Our Genes: Biology, Ideology and Human Nature* (with S. Rose and L. Kamin). New York: Pantheon Press. 322 pp.

150. 1984. Review of *Women in Science: Portraits of a World in Transition* by V. Gornick. *NY Rev. of Books* 31(6).

151. 1984. *Le déterminisme biologique comme arme social.* In *Les Enjeux du Progrès*, ed. A. Cambrosio and R. Duchesne, pp. 233–41. Sainte-Foy: Presses de l'Université du Québec.

152. 1984. *La Diversité des Hommes.* Paris: L'Univers. de la Science. 179 pp. (Translation: *Human Diversity*, 1982).

153. 1985. Nearly identical allelic distributions of xanthine dehydrogenase in two populations of *Drosophila pseudoobscura* (with T. P. Keith, L. D. Brooks, J. C. Martinez-Cruzado and D. L. Rigbny.) *Mol. Biol. Evol.* 2:206–216.

154. 1985. Population genetics. *Annu. Rev. Genet.* 19:81–102.
155. 1985. Population genetics. In *Evolution, Essays in Honor of John Maynard Smith*, ed. J. Greenwood and M. Slatkins, pp. 3–18. Cambridge: Cambridge University Press.
156. 1985. *The Dialectical Biologist* (with R. Levins). Cambridge, MA: Harvard University Press. 303 pps.
157. 1985. Review of books by R. W. Clark, C. Darwin, V. Orel, L. J. Jordanova and J. P. Changeux. *NY Rev. of Books* 32(15).
158. 1985. This week's citation classic (with J. L. Hubby). *Current Contents/Life Sciences* 43:16.
159. 1985. *Menselijke Verscheidenheld. Natuur en Technick.* Maastricht, Brussels. 178 pp. (Translation: *Human Diversity*, 1984).
160. 1985. *Nous Ne Sommes Pas Programmé. Génétique, Hérédité, Idéologie.* Editions La Décca verte. (Translation: *Not in Our Genes*, 1984).
161. 1986. Technology, research, and the penetration of capital: The case of agriculture (with J. P. Berlan). *Monthly Rev.* 38:21–34.
162. 1986. The political economy of hybrid corn (with J. P. Berlan). *Monthly Rev.* 38:35–47.
163. 1986. Review of *In the Name of Eugenics*, by D. J. Kevles. *Rev. Symp. Isis* 77(2):314–7.
164. 1986. *Education and Class* (with M. Schiff). Oxford: Oxford University Press. 243 pp.
165. 1986. A comment on the comments of Rogers and Felsenstein. *Am. Nat.* 127(5):733–4.
166. 1986. *An Introduction to Genetic Analysis*, 3rd edition (with D. Suzuki, A. Griffith, and J. Miller). New York: W. H. Freeman. 612 pp.
167. 1986. Breeder's rights and patenting of life forms (with J. P. Berlan). *Nature* 322:785–8.
168. 1986. How important is genetics for an understanding of evolution? In *Science as a Way of Knowing*, ed. J. A. Moore, Vol. III, pp. 811–20. Thousand Oaks, CA: Am. Soc. of Zoologists.
169. 1986. *Inte i Vara Gener: Biologi, Ideologi och Människans natur.* Göteborg: Bokskogen. (Translation: *Not in Our Genes*, 1984.)
170. 1987. The shape of optimality. In *The Latest on the Best: Essays on Evolution and Optimality*, ed. J. Dupré, pp. 151–9. Cambridge, MA: MIT Press.
171. 1987. Sequence of the structural gene for xanthine dehydrogenase (*rosy* locus) in *Drosophila melanogaster* (with T. P. Keith, M. A. Riley, M. Kreitman, D. Curtis and G. Chambers). *Genetics* 116: 64–73.
172. 1987. Polymorphism and heterosis: old wine in new bottles and *vice versa*. *J. Hist. Biol.* 20:337–49.
173. 1987. Are the races different? In *Anti-Racist Teaching*, ed. D. Gill and L. Levidor, pp. 198–207. London: Free Association Books.

174. 1987. *L'Evolution du vivant: enjeux idéologiques.* In *Les Scientifiques Parlent,* ed. A. Jacquard, pp. 54–73. Paris: Hachette.

175. 1988. A general asymptotic property of two-locus selection models (with M. W. Feldman). *Theor. Pop. Biol.* 34(2):177–93.

176. 1988. On measures of gametic disequilibrium. *Genetics* 120:849–52.

177. 1988. Aspects of wholes and parts in population biology (with R. Levins). In *Evolution of Social Behavior and Integrative Levels,* ed. G. Greenberg and E. Tobach, pp. 31–52. Mahwah, NJ: Lawrence Erlbaum Associates.

178. 1988. *La paradoja de la adaptación biológica.* In *Polémicas Contemporáneas en Evolución,* ed. A. O. Franco, pp. 57–65. Mexico City: AGT Editor, SA.

179. 1989. Inferring the number of evolutionary events from DNA coding sequence differences. *Mol. Biol. Evol.* 6(1):15–32.

180. 1989. Review of *Controlling Life: Jacques Loeb and the Engineering Ideal in Biology* by P. Pauly, and *Topobiology: An Introduction to Molecular Embryology* by G. Edelman. *NY Rev. of Books* 36(7).

181. 1989. DNA sequence polymorphism. In *Essays in Honor of Alan Robertson,* ed. W. G. Hill, and T. F. C. Mackey, pp. 33–37. Edinburgh: University of Edinburgh Press.

182. 1989. On the characterization of density and resource availability (with R. Levins). *Am. Nat.* 134:513–24.

183. 1989. Distinguishing the forces controlling genetic variation at the *Xdh* locus in *Drosophila pseudoobscura* (with M. A. Riley, and M. E. Hallas). *Genetics* 123:359–69.

184. 1989. Review of *Evolutionary Genetics* by J. Maynard Smith. *Nature* 339:107 (May 11).

185. 1989. *An Introduction to Genetic Analysis,* 4th edition (with D. T. Suzuki, A. J. F. Griffiths, and A. J. H. Miller). New York: W. H. Freeman 768 pp.

186. 1990. The political economy of agricultural research: the case of hybrid corn. In *Agroecology,* ed. C. R. Carroll, J. H. Vandermeer and P. Rosset, pp. 613–26. New York: McGraw-Hill.

187. 1990. Review of *Wonderful Life: The Burgess Shale and the Nature of History* by S. J. Gould. *NY Rev. of Books* 37(10).

188. 1990. The evolution of cognition. In *Thinking, an Invitation to Cognitive Sciences,* Vol. 3, ed. D. N. Osherson and E. E. Smith, pp. 229–240. Cambridge, MA: MIT Press.

189. 1991. *How Much Did the Brain Have to Change for Speech?* (Pinker and Bloom Commentary) San Diego, CA: Academic Press.

190. 1991. Review of *The Structure and Confirmation of Evolution Theory* by E. A. Lloyd, Greenwood Press, NY, 1988. *Biol. and Phil.* 6: 461–6.

191. 1991. Foreword. In *Organism and the Origins of Self,* ed. A. I. Tauber, pp. xiii–xix. Boston, MA: Kluwer Academic.

192. 1991. Facts and the factitious in natural science. *Critical Inquiry* 18:140–53.

193. 1991. *Biology as Ideology: The Doctrine of DNA*. Toronto, Ontario: Stoddart. 100 pp.
Also published as:
Biology as Ideology. The Doctrine of DNA. New York: Harper Collins. 100 pp. 1993.
The Doctrine of DNA. Biology as Ideology. London: Penguin Books. 1298 pp. 1993.
Biologia come Ideologia: La dottrina del DNA. Torino, Italy: Bollati Boringhieri. 98 pp. 1994.
De DNA Doctrine. Amsterdam: Uitgevery Bert Bakker. 179 pp.
Biology as Ideology. In *The Dancer and the Dance,* ed. E. Sagarra and M. Sagarra, pp. 54–64. Trinity Jameson Quartercentenary Symposium.

194. 1991. Population genetics. *Encycl. Hum. Biol.* 6:107–15.

195. 1991. Perspectives: 25 years ago in genetics: electrophoresis in the development of evolutionary genetics: milestone or millstone? *Genetics* 128:657–62.

196. 1991. Population genetic problems in forensic DNA typing (with D. L. Hartl). *Science* 254:1745–50.

197. 1992. Genotype and phenotype. In *Keywords in Evolutionary Biology,* ed. E. F. Keller and E. A. Lloyd, pp. 137–44. Cambridge, MA: Harvard University Press.

198. 1992. Biology. In *Academic Press Dictionary of Science and Technology,* ed. C. Morris, p. 261. San Diego, CA: Academic Press.

199. 1992. The dream of the human genome. *NY Rev. Books* 39 (10):31–40.

200. 1992. *Polemiche sul genoma umano, I. La Rivista dei Libri,* Oct. 7–10.

201. 1992. *Polemiche sul genoma umano, II. La Rivista dei Libri,* Nov. 6–9.

202. 1992. The dimensions of selection (with P. Godfrey-Smith). *J. Phil. Sci.* 60:373–95.

203. 1992. Open peer commentary: gene talk on target. *Soc. Epist.* 6(2):179–81.

204. 1992. Letter to the Editor: Which population? *Am. J. Hum. Genet.* 52:205.

205. 1992. Inside and Outside: Genetics, Environment and Organism. *Heinz Werner Lecture Series,* Vol. XX. Worcester, MA: Clark University Press. 48 pp.

206. 1992. DNA data banking and the public interest (with N. L. Wilker, S. Stawksi, and P. R. Billings). In *DNA on Trial,* ed. P. R. Billings, pp. 141–149. Cold Spring Harbor, NY: Cold Spring Harbor Press.

207. 1993. Correlation between relatives for colorectal cancer mortality in familial adenomatous polyposis (with S. Presciuttini, L. Bertario, P. Saia, and C. Rossetti). *Ann. Hum. Genet.* 57:105–15.

208. 1993. Letters (with D. L. Hartl). *Science* 260:473–474.
209. 1993. *Risposta a Sgaramella* (Reply to Sgaramella). *La Rivista dei Libri* (Florence) 3(7/8):43–4.
210. 1993. *Chelovechskaya Individulnost.* Moscow: Progress Publishers. (Translation: *Human Diversity*, 1982).
211. 1993. *An Introduction to Genetic Analysis*, 5th edition (with A. J. F. Griffiths, J. H. Miller, D. T. Suzuki, and W. M. Gelbart). New York: W. H. Freeman. 840 pp.
212. 1994. Comment: the use of DNA profiles in forensic context. In *DNA Fingerprinting: a Review of the Controversy* by K. Roeder. *Stat. Sci.* 9(2):259–262.
213. 1994. Women *versus* the biologists. *NY Rev. of Books* 41(7):31–5.
214. 1994. Correspondence. Forensic DNA typing dispute. *Nature* 372:398.
215. 1994. Letters. DNA fingerprinting (with D. L. Hartl). *Science* 266:201.
216. 1994. Here we go again. Neri e bianchi per mi pari sono. *La Repubblica* (Milan), Oct. 25, pp. 24–5.
217. 1994. Response to Goldberg and Hardy. *NY Rev. of Books.*
218. 1994. Holism and reductionism in ecology (with R. Levins). *CNS* 5(4): 33–40.
219. 1994. Facts and the factitious in natural sciences. In *Questions of Evidence*, ed. J. Chandler, A. I. Davidson and H. Harootunian, pp. 478–91. Chicago: University of Chicago Press.
220. 1994. A rejoinder to William Wimsatt. In *Questions of Evidence*, ed. J. Chandler, A. I. Davidson and H. Harootunian, pp. 504–509. Chicago: University of Chicago Press.
221. 1994. Letters. Response to Lander and Budowle. *Nature* 372:398.
222. 1995. *Human Diversity.* New York: Scientific American Library. 156 pp.
223. 1995. *Promesses, promesses. Génétique et Evolution* 11:II–V.
224. 1995. Genes, environment and organisms. In *Hidden Histories of Science*, ed. R. B. Silvers, pp. 115–139. New York NY Rev. of Books Collections.
225. 1995. The detection of linkage disequilibrium in molecular sequence data. *Genetics* 140:377–88.
226. 1995. *A la recherche du temps perdu. Configurations* 2:257–65.
227. 1995. Sex, lies and social science. *NY Rev. of Books* 42(7):24–9.
228. 1995. *Il potere del progetto* (The power of the project). *SFERA Magazine* (Rome) 43:11–34.
229. 1995. Theodosius Dobzhansky – a theoretician without tools. In *Genetics of Natural Populations: The Continuing Importance of Theodosius Dobzhansky*, ed. L. Levine, pp. 87–101. New York: Columbia University Press.
230. 1995. The dream of the human genome. In *Politics and the Human Body*, ed. J. B. Elshtain and J. T. Lloyd, pp. 41–66. Nashville, TN: Vanderbilt University Press.

231. 1995. IV. Primate models of human traits. In *Aping Science. A Critical Analysis of Research on the Yerkes Regional Primate Research Center*, ed. B. P. Reines, pp. 17–35. New York: Committee on Animal Models in Biomedical Research.

232. 1996. What does electrophoretic variation tell us about protein variation? (with A. Barbadilla, and L. M. King). *Mol. Biol. Evol.* 13(2):427–32.

233. 1996. *La recherche du temps perdu*. In *Science Wars*, ed. A. Ross, pp. 293–301. Durham, NC: Duke University Press.

234. 1996. Of genes and genitals. In *Transitions: An International Review*, pp. 178–93. Cambridge, MA: Transition Publishers.

235. 1996. The end of natural history? (with R. Levins). *CNS* 7(1):1–4.

236. 1996. Pitfalls of genetic testing (with R. Hubbard). *NEJM (Sounding Board)* 334(18):1192–4.

237. 1996. Authors' reply, (with R. Hubbard). *NEJM (Correspondence)* 335:1236–7.

238. 1996. In defense of science. *Society* 33(4):30–1.

239. 1996. Evolution as engineering. In *Integrative Approaches to Molecular Biology*, ed. J. Collado, T. Smith and B. Magasnik, pp. 1–10. Cambridge, MA: MIT Press.

240. 1996. Primate models of human traits. In *Monkeying with Public Health*, ed. B. Reines and S. Kaufman, pp. 17–35. Gays Mills, WI: One Voice.

241. 1996. Indiana Jones meets King Kong. Review of three books by M. Crichton. *NY Rev. of Books* 43(4).

242. 1996. Letter (Zola biography review). *NY Rev. of Books* 43(9).

243. 1996. Evolution and religion. Am. Jewish Congress.

244. 1996. Letter (On Horton's essay on genetics and homosexuality). *NY Rev. of Books.*

245. 1996. Population genetic issues in the forensic use of DNA. In *The West Companion to Scientific Evidence*, ed. D. Faigman, pp. 673–96. St Paul, MN: West Publishing Co.

246. 1996. Detecting heterogeneity of substitutions along DNA and protein sequences (with P. J. E. Goss). *Genetics* 143:589–602.

247. 1996. False dichotomies (with R. Levins). *CNS* 7(3):27–30.

248. 1996. The return of old diseases and the appearance of new ones (with R. Levins). *CNS* 7(2):103–7.

249. 1996. *An Introduction to Genetic Analysis*, 6th edition (with A. J. F. Griffiths, J. H. Miller, D. T. Suzuki, and W. M. Gelbart). New York: W. H. Freeman. 915 pp.

250. 1997. Billions and billions of demons. Review of *The Demon Haunted World: Science as a Candle in the Dark* by C. Sagan. *NY Rev. of Books* 44(1).

251. 1997. Nucleotide variation and conservation at the dpp locus, a gene controlling early development (with B. Richter, M. Long, and E. Nitasaka). *Genetics* 145(2):311–23.
252. 1997. A scientist meets a visionary. In *Bucky's 100.* In press.
253. 1997. Population genetics. In *Encyclopedia of Human Biology*, 2nd edition. New York: Academic Press.
254. 1997. Genetics, plant breeding and patents: conceptual contradictions and practical problems in protecting biological innovation (with M. de Miranda Santos). *Plant Genetic Resources Newsletter* 112:1–8.
255. 1997. The evolution of cognition: questions we will never answer. In *Invitation to Cognitive Science*, Vol. 4. gen. ed. D. N. Osherson, ed. D. Scarborough and S. Sternberg, pp. 107–12. Cambridge, MA: MIT Press.
256. 1997. Dobzhansky's "Genetics and the Origin of Species": Is it still relevant? *Genetics* 146:351–5.
257. 1997. The Cold War and the transformation of the Academy. In *The Cold War and the University*, Vol. 1, pp. 1–34. New York: The New Press.
258. 1997. The biological and the social (with R. Levins). *CNS* 8(3):89–92.
259. 1997. Chance and necessity (with R. Levins). *CNS* 8(2):95–98.
260. 1997. Confusion over cloning. Review of *Cloning Human Beings: Report and Recommendations of the National Bioethics Advisory Commission* (June, 1997). *NY Rev. of Books* 44(16).
261. 1997. *Biologie als Ideologie: Ursache und Wirkung bei der Tuberkulose, den menschlichen Genen und in der Landwirtschaft. Streitbarer Materialismus* 21:111–28.
262. 1997. A question of biology: are the races different? In *Beyond Heroes and Holidays*, ed. E. Lee, D. Menkart and M. Okazawa-Rey. Washington DC: Network of Educators on the Americas.
263. 1998. Review of *"Unto Others" The Evolution and Psychology of Unselfish Behavior* by E. Sober and D. S. Wilson, Harvard University Press. *NY Rev. of Books* 45(16):59–63.
264. 1998. *Gene, organismo, e ambiente. I rapporti causa-effetto in biologia.* Rome: Guis. Laterza and Figli Spa. 97 pp.
265. 1998. The maturing of capitalist agriculture: farmer as Proletarian. *Monthly Rev.* 50(3):72–85.
266. 1998. Survival of the nicest? *NY Rev. of Books* 45(16):59–63.
267. 1998. Foreword (with R. Levins) In *Building a New Biocultural Synthesis*, ed. A. H. Goodman and T. L. Leatherman, pp. xi–xv. Ann Arbor: University of Michigan Press.
268. 1998. The confusion over cloning. In *Flesh of My Flesh. The Ethics of Cloning Humans*, ed. G. E. Pense, pp. 129–39. Lanham, MD: Rowman and Littlefield Publishers.

269. 1998. How different are natural and social science? (with R. Levins). *CNS* 9(1):85–89.
270. 1998. Does anything new ever happen? (with R. Levins). *CNS* 9(2): 53–6.
271. 1998. Life on other worlds (with R. Levins). *CNS* 9(4):39–42.
272. 1998. *Las bases del conflicto en la explicación biológica.* In *Historia y Explicación en Biologia* ed. A. Barahorio and S. Martínez, pp. 96–106. Mexico City: Universidad Nacional Autónoma de México, Fondo de Cultura Económica.
273. 1998. *Realidades y ficciones en las ciencias naturales.* In *Historia y Expilicación en Biologia* ed. A. Barahorio and S. Martínez, pp. 107–22. Mexico City: Universidad Nacional Autónoma de México, Fondo de Cultura Económica.
274. 1998. The evolution of cognition: questions we will never answer. In *An Invitational to Cognitive Science: Methods, Models, and Conceptual Issues,* Vol. 4. general ed. D. N. Osherson, ed. D. Scarborough and S. Sternberg, pp.107–32. Cambridge, MA: MIT Press.
275. 1999. *Modern Genetic Analysis* (with A. J. F. Griffiths, W. M. Gelbart, and J. H. Miller). New York: W. H. Freeman. 675 pp.
276. 1999. Are we programmed? (with R. Levins). *CNS* 10(2):71–5.
277. 1999. Does Culture Evolve? (with J. Fracchia). In *History and Theory.* Theme Issue 38, *The Return of Science: Evolutionary Ideas and History,* pp. 52–78, Middletown, CT: Wesleyan University.
278. 1999. Locating regions of differential variability in DNA and protein sequences (with H. Tang). *Genetics* 153:485–95.
279. 1999. The problem with an evolutionary answer. Review of *The Dark Side of Man: Tracing the Origin of Male Violence* by M. P. Gighlieri. Perseus Books. *Nature* 400:728–9.
280. 2000. Foreword. In *The Ontogeny of Information,* ed. S. Oyama, pp. vii–xv. Durham, NC: Duke University Press. 273 pps.
281. 2000. Let the numbers speak (with R. Levins). *CNS* (11)2:63–7.
282. 2000. The politics of averages (with R. Levins). *CNS* 11(2):111–14.
283. 2000. Computing the organism. End paper. *Nat. Hist.* 4:94–5.
284. 2000. What do population geneticists know and how do they know it? In *Biology and Epistemology,* ed. R. Creath and J. Maienschen, pp. 191–214. New York: Cambridge University Press.
285. 2000. The problems of population genetics. In *Evolutionary Genetics: From Molecules to Morphology,* ed. R. S. Singh and C. B. Krimbas, pp. 5–23. Cambridge, UK: Cambridge University Press.
286. 2000. *It Ain't Necessarily So: The Dream of the Human Genome and Other Illusions.* NY: New York Review of Books 330 pp.
287. 2000. *The Triple Helix. Gene, Organism, and Environment.* Cambridge, MA: Harvard University Press. 192 pp.

288. 2000. *An Introduction to Genetic Analysis,* 7th edition (with A. J. F. Griffiths, J. H. Miller, D. T. Suzuki and W. M. Gelbart). New York: W. H. Freeman. 860 pp.

289. 2000. The maturing of capitalist agriculture, the farmer as proletarian. In *Hungry for Profit,* ed. F. Magdoff, F. H. Buttel and J. B. Foster, pp. 93–106. New York: Monthly Review Press.

290. 2000. Genetic diversity in human populations. In *Humankind Emerging,* ed. B. G. Campbell and J. D. Loy. Boston, MA: Longman Publishers. 679 pp.

291. 2000. Schmalhausen's Law (with R. Levins). *CNS* 11(4):103–8.

292. 2000. Cloning and the fallacy of biological determinism. In *Human Cloning,* ed. B. MacKinnon, pp. 37–49. Chicago: University of Illinois Press. 171 pp.

293. 2001. Dobzhansky, Theodosius. In *Encyclopedia of Genetics,* ed. S. Brenner and J. H. Miller, pp. 557–558. London: Academic Press.

294. 2001. Preface. In *The Character Concept in Evolutionary Biology,* G. P. Wagner, pp. i–xxiii. San Diego, CA: Academic Press. 622 pp.

295. 2001. Genotype and phenotype. In *International Encyclopedia of the Social and Behavioral Sciences,* ed. N. J. Smelser and P. D. Baltes. Oxford: Elsevier Science, Pergamon.

296. 2001. Natural history and formalism in evolutionary genetics (pp. 7–20) and Interview of R. C. Lewontin (with D. B. Paul, J. Beatty and C. B. Krimas) (pp. 21–61). In *Thinking About Evolution: Historical, Philosophical, and Political Perspectives,* ed. R. S. Singh, C. B. Krimbas, D. B. Paul and J. Beatty. Cambridge: Cambridge University Press. 606 pp.

297. 2001. In the beginning was the word. Review of *Who Wrote the Book of Life? A History of the Genetic Code* by L. E. Kay. Stanford, CA: Stanford University Press. 470 pp. Science 291:1263–4.

298. 2001. Genes in the food. Review. *NY Rev. of Books* 48(10):81–4.

299. 2001. Gene, organism and environment. In *Cycles of Contingency,* ed. S. Oyama, P. Griffiths and R. Gray, pp. 54–66. Cambridge, MA: MIT Press. 377 pp.

300. 2001. After the genome, what then? *NY Rev. of Books* 48(12):36–7, July 19.

301. 2001. *El sueña del genoma humano y otras ilusiones.* Barcelona, Spain: Ediciones Paidó Ibé, S.A. 206 pp. (Translation: *It Ain't Necessarily So,* 2000).

302. 2002. *Il sogna del genoma umano, e altre illusione della scienza.* Rome, Italy: Editori Laterza. 210 pp. (Translation: *It Ain't Necessarily So,* 2000).

303. 2002. The politics of science. Review. *NY Rev. of Books* 49(2):28–31.

304. 2002. *Die Dreifachhelix. Gen, Organismus und Umwelt,* Heidelberg: Springer-Verlag. (Translation: *The Triple Helix*).

Preface

Scientists earn their reputation by making special contributions in a variety of ways. Some become known for a discovery that revolutionizes their science. Others are respected as intellectual leaders for significant contributions leading to sustained progress in their field. Still others become known for providing guidance, opportunity, and uniquely inspiring rapport with a large number of graduate students, writers, and research colleagues. A rare few do all the above, and remarkably enough still find time to deal with the broader issues of epistemology, philosophy, history, and sociology of science. Richard Lewontin is one of these rare scientists.

If we are to attach a major discovery or a conceptual breakthrough to Lewontin's name (like Haldane's cost of natural selection, Fisher's fundamental theorem of natural selection, Wright's shifting-balance theory, or Maynard Smith's game theory applications), then the successful completion of the genetic variation research program of the Chetverikov–Dobzhansky school will be known as the outstanding highlight of Lewontin's career. Dobzhansky and his students and collaborators pursued the twin problems of the amount and the adaptive role of genetic variation for nearly 25 years without a satisfactory solution. All estimates of genetic variation were indirect or inadequate as there was no reductionist research program that could allow the study of genetic variation at the level of the gene. Lewontin's pioneering success in the application of protein electrophoresis to the problem of genetic variation changed the scene radically. The estimation of electrophoretic variation was direct and more useful than anyone had expected. The technique also removed the experimental limitations imposed by genetic incompatibility among species and allowed reliable comparisons of genetic variation among populations and species without any need to make genetic crosses. The impact and the anticipation of the avalanche of future results from the use of electrophoresis were discussed in his well-known book, *The Genetic Basis of Evolutionary Change* (1974). This book sets out the problem of population genetics in a rationally constructed historical context and is required reading for all aspiring population geneticists.

Evolutionary research requires broad interest and versatility in modeling experimental design, statistics, field biology, and much more. Such breadth allowed Lewontin to be successful, time and again, in designing new experimental systems or suggesting key concepts to answer old questions or pursue new ones. Lewontin became interested in the uniqueness of the phenotype– and the genotype–environment interactions inspired mainly by the Russian biologist I. Schmalhausen's book *Factors of Evolution.* His doctoral thesis studied fitness as a function of genotype frequency and density and showed that "viability of a genotype is a function of the other genotypes which coexist with it, the result of any particular combination not being predictable on the basis of the viabilities of the coexisting genotypes when tested in isolation." This was followed by studies of interlocus epistatic interactions in fitnesses and the evolution of naturally occurring inversion polymorphism in *Drosophila.* His mathematical work on linkage disequilibrium provided a new direction for research and results from a series of papers on multilocus fitness effects anticipated discussion on the units of selection. His experimental work on norms of reaction in *Drosophila* was exemplary in exposing the problem of the genetic determination and led to a new appreciation of genotype–environment interaction and phenotypic plasticity. He pointed out the importance of developmental time in fitness, something which is usually forgotten when describing fitness components. His 1972 paper on "Apportionment of human diversity," pointing out that any genetic difference between races has to be compared with genetic variation within population and races, is a landmark in human genetics and evolution. More recently his laboratory has been a major center for studies of DNA sequence variation. Lewontin has provided training and guidance to a large number of graduate students and postdoctoral fellows. The number is well over one hundred! Many more have worked in Lewontin's laboratory but have not necessarily coauthored publications with him.

But what makes Lewontin known more in the wider circle of evolutionary biology and in science in general is his role as a critic of how science is done, on the one hand, and his passionate engagements with the issues of science and society, on the other. He has made important contributions and has influenced research workers in the history and philosophy of science and in areas of science and society such as agriculture, social health problems, bioethics, and genetics, and IQ. If you drop Lewontin's name in any group of biologists, an animated discussion is sure to follow! These discussions are not about science but about its relevance and applications to human affairs. His concern about social issues springs directly from his unique perspective of evolutionary biology. Lewontin's research program may be reductionist but he is not. He has encouraged and challenged evolutionary biologists to find the most desirable combination of Platonic and Aristotelian traditions in studying nature. Accordingly the mathematical rigor of early population biology must be extended to accommodate interactive, hierarchical, probabilistic, and historical factors as learned empirically in the field. To him "Context and interaction are of the essence" (Lewontin 1974, p. 318), whether one is talking about interactions between hierarchical levels, between organisms and the environment, or between causes and effects. A reductionist approach to science does not necessitate a reductionist view of the world. No level of analysis is specially privileged for a general understanding of causality. Genetic and environmental effects are interdependent and the phenotypic variance cannot be partitioned into fixed components. Organisms do not fit in preexisting ecological niches but create their own niches. History and contingencies are so important in evolution that looking for adaptive explanations for all organismic traits undermines the role of natural history. These ideas essentially follow from his belief that relationships between organisms and their environments, and likewise, those between groups and hierarchical levels, are governed by forces so weak that the outcomes are neither fixed nor predetermined.

John Maynard Smith has written (first volume of this series, pages 628–640) that "Richard Lewontin has contributed to science not only by his own work on evolutionary theory and molecular variation and by his influence on the many young scientists who have worked with him but also by asking us to think about the relationships between the science we do and the world we do it in." While you may not agree with Lewontin on all issues (he would be surprised if you did!) one thing is sure – Lewontin has been a colorful personality who has made evolutionary biology rigorous and interesting at the same time. We affectionately dedicate this volume to him.

We sincerely thank Subodh Jain for his encouragement and valuable contribution in the early planning of this volume. At Cambridge University Press, we express our sincere thanks to Ellen Carlin for her enthusiastic support and early work on this project and to Maria Murphy for her supervision in the completion of this project. Thanks are also due to Aaron Thomson, McMaster University, who did the maddening job of checking up references and preparing the manuscripts for final submission.

Introduction

This series provides a forum for the review and discussion of some facets of the extraordinary intellectual contributions of Richard C. Lewontin. Previous volumes have addressed evolutionary genetics and the philosophy and politics of evolutionary biology. This volume invites consideration of the Lewontin/Levins vision of an integrated population biology.

R. C. Lewontin and Richard Levins outline the conceptual framework of population biology. They provide examples of the methodological developments the establishment of this field demands and of the kinds of insights it would inspire. Both, however, decry that this vision remains largely unfulfilled, nearly half a century after its explicit formulation as a scientific objective.

One advantage of starting with such a dyspeptic assessment is that each subsequent chapter can only contribute positive evidence that the precepts of population biology have become an inextricable part of modern evolutionary thought. In each of the several constituent fields of population biology, it has been the focused study of particular forces in isolation from the global network that has permitted major advances. This approach has entailed reduction of complex processes into more basic component mechanisms within a given level of biological organization and the substitution of simplified surrogates that represent lower levels in their entirety. The Lewontin/Levins vision of population biology challenges workers to transcend this very productive research strategy: to confront the considerable residual complexity that separates a biological system at a given level from its crude effigy assembled from independent building blocks and, even more difficult, to address interactions between levels. While acknowledging that this grand synthesis has not yet been achieved, contributors to this volume offer abundant evidence that the vision of an integrated population biology has enriched and deepened its component fields.

The most significant development since the 1950s in evolutionary genetics and ecology (and indeed in all of biology) has of course been access to molecular genetic information. Trudy F. C. Mackay provides a comprehensive review of the tremendous increase in resolution of the genetic basis of morphological traits, from QTL to gene to particular nucleotides. Daniel L.

Hartl *et al.* illustrate the characterization of the genomic response of an individual organism to its environment. G. Brian Golding describes methods for characterizing entire populations with respect to genomic properties.

However complex is the mapping between genotype and phenotype in the laboratory, it becomes incalculably more complex within the natural context of evolutionary change. Ward B. Watt presents case studies detailing adaptation to strong environmental challenges through the structural modification of key catalytic enzymes. Bryan Clarke calls for the analysis of biotic and abiotic components of the environment that may serve as sources of balancing selection for the maintenance of nonsynonymous polymorphism. Complementing the study of adaptive polymorphism, John H. Gillespie addresses the process of adaptive substitution in response to changing selection pressures.

In addition to greatly increasing resolution of the genetic basis of evolutionary change, the analysis of molecular-level variation now permits detailed reconstruction of the demographic context of evolution. L. Luca Cavalli-Sforza recounts the development of key concepts and methods that have endowed population biology with the historical dimension fundamental to evolutionary analysis. John Wakeley presents new methods for the inference of history from neutral variation. Daniel Lachaise *et al.* describe the various demographic contexts in which closely related species of *Drosophila* have diverged from a common ancestor.

A progressive broadening of perspective has enriched population genetics, with evolving units increasingly characterized as a series of nested, interconnected networks rather than as a set of independent, homogeneous gene pools. Rama Singh and Richard Morton use the framework of Sewall Wright's shifting balance theory to explore how interactions among multiple levels, from ontogeny to population structure, collaborate in the maintenance and origin of adaptation. Edward C. Holmes explores whether the entire gene pool of RNA viruses evolves as a unit, according to its own set of rules for the generation of variation and for selection, or whether processes at the level of individual viruses must also be considered. Daniel J. Howard *et al.* present a historical review of the study of hybrid zones, arguing that hybrid zones themselves constitute an evolutionary force that can act to facilitate, maintain, or limit the process of adaptation.

Recognition that genetic and ecological change may not only evolve towards different ends but actually contravene one another has become integral to evolutionary biology. A fundamental of this view is evolutionary conflict among organisms and among units of selection. Paul W. Ewald and Gregory M. Cochran review some key insights these concepts have permitted into the coevolution of host and pathogen. Marcy K. Uyenoyama and Naoki Takebayashi address the evolution of self-incompatibility, a mechanism that promotes outbreeding in flowering plants but only at the cost of postponing the expression of inbreeding depression and altering the course of evolution of the mating specificities themselves.

Optimization principles and other key generalizations are fundamental to the analysis of the evolutionary process at any given level of organization. While the importance of such insights is beyond question, the development of an integrated population biology challenges even these scaffolds from which the study of each successive level of organization is built. To what extent do maximization of population size or rate of growth, for example, determine the course of evolutionary change at the population level? Freddy Bugge Christiansen's examination of the extent to which evolutionary ecology can rely on implicit characterizations of evolutionary processes at lower levels serves to define and delimit the sphere of influence of key generalizations. Within the context of conservation biology, for which the consequences of a less than comprehensive understanding of the implications of intervention policies are immediate and global, Philip Hedrick addresses the effect of genetic composition of a species on its evolutionary potential and resistance to extinction. In his review of life-history theory, Brian Charlesworth explores the conditions under which principles developed for models that ignore age structure can serve as guides to the evolution of age-dependent fertility and mortality schedules. Shripad Tuljapurkar documents recent rapid declines in human mortality, explaining the increasing importance of an understanding of the evolution of mortality schedules and the nature of evolutionary constraints on lifespan.

While changes in mortality schedules have been attributed to environmental changes alone, the determinants of cultural or social differences among human populations have historically been regarded as genetic to the virtual exclusion of other factors. William B. Provine richly illustrates how prevailing belief systems have predisposed scientists, no less than other members of society, towards genetic determinism of social traits, in the absence and sometimes even contradiction of direct evidence. Peter Taylor describes analytical methods developed from conceptual frameworks that admit broader perspectives of the ontogeny of cultural phenotypes.

Have the efforts of innumerable scientists who view themselves as population biologists succeeded in realizing the Lewontin/Levins vision of a population biology that integrates interactions at all levels of organization? While the answer is clearly negative, it must be qualified by noting that the development of the field was not so much abandoned as postponed. Genetically based parameters represent keystones of population biology and ecological genetics, fields bridging population genetics and ecology. It is only now, after the pervasion of the molecular revolution throughout evolutionary biology, that the central concerns of population biology, especially interactions among levels of organization, can be explored in depth. This collection of essays offers various perspectives on the profound conceptual and methodological transformations that have brought population biology to the threshold of full realization.

PART I

HISTORICAL FOUNDATIONS AND PERSPECTIVES

1

Building a science of population biology

RICHARD C. LEWONTIN
Museum of Comparative Zoology, Harvard University, Cambridge

There has long been a distinction made by biologists between those phenomena and explanations that are at the level of individual organisms and their constituent parts, and those that are at the level of populations. Developments in individual-level biology are almost entirely motivated by prior empirical discoveries, despite efforts, largely unsuccessful, to create a mathematical basis for molecular, cellular, and physiological events. In contrast, the investigation of population level phenomena has been almost entirely theory driven. Models have been created of population-level phenomena which are then represented as mathematical structures, with specific functional forms, variables, and parameters. Empirical work is, at the very least, informed by these models or, more often, is designed to measure the variables and estimate the parameters of a specific model, or to test whether a particular model is an adequate representation of the natural process. That does not mean that the structures of models are not reciprocally informed by empirical findings. In some cases discoveries of new phenomena at the individual level may require the enrichment of the variety of population models, as for example when the discovery of non-Mendelian segregation patterns such as t-alleles in mice (Dunn 1957) and segregation distorters in *Drosophila* (Sandler *et al.* 1959) required the inclusion of the segregation ratio in gamete pools of heterozygotes as a parameter to be empirically determined. In other cases phenomena that appear in population experiments require the enrichment of standard models, as for example the inclusion of density dependence and composition dependence in models of natural selection (Lewontin 1955). But these enrichments of models are not typical and usually become part of the standard corpus only when the need in a specific case seems compelling.

Because the investigation of population-level phenomena is so organized by specific models, a contemplation of the bulk of these models quickly reveals a characteristic of "population biology" as a science – its nonexistence. There are essentially two sets of problems that are represented at the population

The Evolution of Population Biology, ed. R. S. Singh and M. K. Uyenoyama. Published by Cambridge University Press.

level. One is population and evolutionary genetics, whose problematic is the rate and direction of genetic change within populations and the genetic divergence between populations consequent on the phenomena of genetic segregation, mating pattern, mutation, migration, stochastic events, and natural selection. The second is population ecology, which is concerned with changes in population size and age distribution within a population as a consequence of interactions of organisms with the physical environment, with individuals of their own species, and with organisms of other species. For most of the history of their study population genetics and population ecology have been carried out independently of each other and with a curiously complementary and non-overlapping structure that has minimized the degree to which they have been melded into a coherent science of "population biology." It is important not to confound with a general science of population biology the kind of evolutionary study embodied in "ecological genetics," whose purpose is to describe and measure in a natural population the actual patterns of mating, migration, and reproduction of different genotypes and to provide a physiological, behavioral, and ecological basis for understanding the operation of natural selection in a specific case (e.g., Grant and Grant 2002). Ecological genetics, in this sense, is the attempt to map the abstract quantities of population genetics onto concrete biological processes.

Beginning in the late 1950s and continuing during the 1960s there was a self-conscious movement among population ecologists and evolutionists to create a coherent science of population biology. Among the manifestations of this movement were the appearance of university courses in "population biology," of training programs in population biology, of a few textbooks on population biology that attempted to bring ecology and evolution together (e.g., MacArthur and Connell 1966), of symposia on population biology (e.g., Lewontin 1968), and of the creation of population biology programs in granting agencies. An important outcome of this movement was the creation of a body of theoretical and empirical research that brought together concepts in population ecology and population genetics. These included a theory of the evolution of ecological niches (MacArthur and Levins 1967) and the successful development of a dynamical model of natural selection in continuously breeding species, a problem that even R. A. Fisher had failed to solve correctly (see the chapter by Charlesworth in this volume for a history of this development). Despite its early successes, however, this movement failed to become the general model for work in population biology, and the communities of population ecologists and population geneticists have remained largely independent in their work.

First, the theories operate in different dynamical state spaces. The basic state space in which population genetic changes are modeled is one of the relative frequency of different genetic variants. The laws of transformation in that space are framed entirely in terms of the changes over time in the frequencies of genotypes that are induced by mating, mutation, migration, natural selection, and stochastic sampling. Absolute numbers of individuals appear as a formal parameter, N, the effective population size, but this is a statistical abstraction used in evaluating the size of stochastic effects, rather than an actual census number of individuals. In contrast, in population ecology the basic dynamic state space is not that of relative frequency but of absolute numbers of individuals or, more rarely, of biomass or total energy flux. For special purposes such absolute numbers can be reduced to relative proportions of different classes, as for example changes in the age distribution within a population, but the basic dynamic model is one of numbers rather than frequency.

Second, the bases on which the laws of transformation are derived are of a different kind of generality. The rigidity and near-universality of the mechanism of passage of DNA between generations in sexually or asexually reproducing populations provide an unchallenged skeletal framework which is then the basis for further elaboration of simple perturbing "forces" like mutation or migration or differential reproduction of different genotypes. For sexually reproducing populations modeling begins with Mendel's principle of segregation. The famous Hardy–Weinberg proportions of genotypes (which appear explicitly even in the analysis of nonrandom mating) are nothing but the quantitative expression of the consequences of segregation of alleles at meiosis. Even the law of transformation of allelic frequencies by mutation, which appears to be framed entirely in allele frequency terms, with no reference to the frequencies of diploid genotypes, depends for its validity on the phenomenon of equal segregation of alternative alleles in the gametes produced by heterozygotes. Thus, the laws of transformation in population genetics appear as universals operating in an abstract space of relative frequencies, the contingencies of environment being effective only as determinants of specific parameter values. As we will see, this appearance of universality can be maintained only by a commitment to a form of natural selection that is seldom realized in actual biology and to a form of life history, discrete, nonoverlapping generations, that does not apply to a large fraction of the living world.

Population ecological models are, from their foundation, more contingent in form. In contrast to the case of population genetics, there is no well-established universal mechanism that produces an unchallenged basic model and mathematical formulation. Certainly it is the case that organisms are produced by parent organisms, so that the absolute growth rate in numbers of a population must, all other things being equal, increase linearly as the population size increases. But it is also universal that organisms need external finite resources for their reproduction, so there is a countervailing decrease in the rate of population growth with increasing numbers as the competition for limiting resources grows greater. The usual assumption is that the rate of growth in numbers decreases linearly as the population size increases so an equilibrium size is reached at the "carrying capacity" of the environment in which the population lives. But no universal phenomenon of biology forces us to assume that the decrease of growth rate with the competition from increasing numbers is linear. Moreover, some resources that are ultimately consumed are themselves produced by the very species that is consuming them (worker ants build nests, farm, and forage, increasing the resources for the colony as a whole), further complicating the relationship between population growth rate and numbers. At least one important school of population ecology has denied that populations are generally found at or near the carrying capacity of the environment (Andrewartha and Birch 1954).

A further level of contingency in the laws of population growth arises when the interactions with other species are considered. These may be competitors or predators, decreasing the growth rate of a given species, or they may be resources for the species. The dynamic laws of community ecology must then consider the simultaneous differential equations of population growth for multiple interacting species. The usual models, which make each species growth rate a simple linear function of the abundances of each of the other species with which it interacts, are arbitrary simplifications of what may turn out to be a rather messy multispecies interaction.

From the standpoint of creating a coherent population biology, the most important feature of most models in population ecology is the mirror image of that in population genetics: the failure to include the dynamics of its complementary phenomenology. In their classical form neither demography nor community ecology included a consideration of genetic heterogeneity within populations, treating the species demography typologically. It follows that the dynamical changes that are occurring in the biological properties of the species as a result of changes in genetic composition of the population are not taken into account. In many cases this genetic heterogeneity can be safely ignored because the processes of genetic evolution are slow compared with demographic processes, but they cannot be ignored for species in newly disturbed and disrupted habitats or in species suffering drastic reductions in population numbers.

1.1 Natural selection and demography

A great irony in the separate histories of evolutionary genetics and demography is that Fisher's original development of a genetical theory of natural selection was explicitly demographic. The *Genetical Theory of Natural Selection* (1930) is an attempt to derive the dynamics of natural selection from the theory of population growth. Fisher postulated a species whose reproduction is continuous in time and which is at a stable age distribution and growing at an equilibrium rate *m*, that is the root of the Euler equation

$$1 = \int e^{-mx} l_x b_x dx,$$

where l_x is the probability of living to age x, and b_x is the probability of producing an offspring in the age interval x to $x + dx$. He then made a fatal error. He supposed that if a population grows in number at a rate, m, given by its schedule of births and deaths, then a genotype within a population will grow at a rate m_i (its Malthusian parameter) given by its own particular schedule of births and deaths. If different genotypes have different reproductive properties then they will change in frequency within the population because each will grow at its own rate m_i. But Fisher forgot about sex. He confused the rate of reproduction *of* a genotype with the rate of reproduction *by* a genotype. In a sexually reproducing population genotypes do not reproduce themselves. Heterozygotes and homozygotes each produce homozygotes and heterozygotes as a result of mating and segregation, so that the Malthusian parameter is not the rate of increase of a genotype, unlike in the case of reproduction of a species or of an asexually reproduced genotype.

The problem of how to predict genotypic frequency changes in the case of continuously breeding populations with differential reproductive schedules remained unsolved for 40 years until the work of Charlesworth (1970) laid a correct foundation for this problem. The important lesson learned from this solution is that there is no parameter that is a combination of the mortality and fertility schedules that correspond to the notion of fitness, a notion that is built into the standard model of natural selection for species with discrete generations. The situation is made worse by the demonstration by Charlesworth and Geisel (1972) that the relative l_x and b_x schedules do not, in fact, contain sufficient information by themselves to determine which genotype will increase and which decrease, because the same genotype that increases in a growing population may decrease in a shrinking population. If the population is growing, then early reproduction is advantageous, but if the population is shrinking a genotype that postpones its reproduction is at an advantage. But the dependence of evolutionary change on a more detailed set of demographic information does not lead to the conclusion that evolutionary predictions are not possible from information on fecundity and longevity. Rather, it tells us that there is no simple parameter, "fitness," that is adequate and that detailed demographic information is relevant. This result goes to the heart of all heuristic discussions of differential fitness. If two genotypes differ in their l_x and b_x schedules, which is more fit than the other? There is no general way to answer that question. It all depends.

A second major misunderstanding of classical population genetics in its intersection with demography was a connection made between changes in the mean genotypic fitness in a population and demographic properties of the population as a whole summed up in the notion of "population fitness." Both Fisher (1930) and Wright (1931) showed that if genotypes segregating in a population had fixed fitness values, then the frequencies of the genotypes would change under natural selection so as to increase the frequency-weighted mean of the genotypic fitnesses. As a consequence the "mean fitness" would eventually reach a maximum, either when the population reached a stable intermediate polymorphism as a result of higher fitness of heterozygotes or, in the absence of heterosis, when it became monomorphic for the more fit homozygote. The conclusion flows directly from Fisher's Fundamental Theorem and from the algebraic form of Wright's equations for gene frequency change. This result was widely interpreted (and is still so interpreted in heuristic discussions of evolution) as meaning that populations would become larger, more resistant to extinction, or more resource efficient as a result of evolutionary changes. That is, the mean fitness *in* a population, a well-defined internal statistical property of populations calculated only from relative genotypic frequencies and dimensionless relative genotypic fitnesses, was conflated with a demographic property of the population as a whole, the fitness *of* a population.

A number of experimental papers appeared in the late 1950s and the 1960s documenting the larger size or biomass or intrinsic rate of increase, *m*, of stably polymorphic populations as compared with populations monomorphic for one of the alleles (e.g., Carson 1958, Beardmore *et al.* 1960, Dobzhansky *et al.* 1964). But such results cannot be general. The distinction made by Wallace (1968) between "hard" and "soft" selection makes this clear. If the biology of selection is such that selective deaths do not relieve the competitive deaths that would otherwise occur, for instance selective deaths of reproducing adults in which resources for adults are not in short supply (hard selection), then those selective deaths will indeed result in a reduction in population size. But if the selective deaths are among a cohort, say larvae, that is competing for resources in short supply so that selective death is an alternative to nonselective death, then the population size of adults will be unaffected (soft selection). Indeed, selection favoring greater offspring production can actually lead to a *decrease* in population size and even extinction. A genotype that doubles the fecundity of a holometabolous insect will spread quickly by natural selection, doubling the size of the population of newly hatched larvae. But if the increase in abundance of larvae should cause a potential bird predator to switch its search image, then the population could be drastically reduced. In judging the effect of selection on demography there is no way to avoid the particularities of biology.

1.2 The determination of components of fitness

The problems in the use of fitness in population genetic formulations are not restricted to continuously breeding populations, but arise even in the simple case of nonoverlapping generations. The standard model of population genetics assigns a fixed relative fitness value to each genotype in the population. But the evidence is abundant that fitnesses are often, if not usually, variable from causes internal to the population demography. Fitnesses are dependent on population density, on the relative frequency of genotypes in the population, as for example in cases of selection for mimicry, and on the mixture of interacting genotypes. All three of these dependencies appear even in simple larval competition experiments in *Drosophila* (Lewontin 1955). In such experiments (1) selection intensities are greatest at both the lowest and highest densities and almost absent at intermediate densities; (2) relative viabilities vary with the frequency of the competing genotypes; and (3) the ordering of relative survivorship among multiple genotypes cannot be judged by their survivorship in isolation or in pairwise competitions. This last result is particularly important. Competing genotypes can play a game of "scissors–paper–stone" in which genotype A is superior in competition with B and B is superior to C but C is superior to A, because in each competitive interaction a different set of attributes is involved: A is stronger than B, B is faster moving than C and C is more aggressive than A. A lack of transitivity in fitness disturbs not only the predictions of gene frequency analysis but of the current applications of game theory to evolutionary prediction, which depend on the transitivity of utilities. The theoretical effects of allowing density and frequency dependence into models of gene frequency change are considerable (see the chapters by Christiansen and Clarke in this volume) and include the possiblity, with rather simple models of frequency dependence, that selection leads to a *minimization* of mean fitness at equilibrium (Lewontin 1958).

1.3 The evolution of population ecological processes

As described earlier in this chapter, the classic models of population growth and codetermination of numbers in interacting species have been constructed from simple generalizations of the way in which it is supposed that organisms interact. Organisms produce organisms at a rate determined by the environment, but they also compete for resources in short supply. The simplest mathematical expression of these relations is the logistic equation for the changes in numbers with time

$$dN/dt = rN(1 - N/K) \tag{1.1}$$

where r is the "intrinsic" rate of increase expressing the reproductive force of an individual and K is "carrying capacity" of the environment, the maximum to which the population can grow because of the competition for a limiting resource. This particular form of the growth equation is motivated by the notion of a carrying capacity of the environment and a linear decrease in effective reproductive rate, r_{eff}, as that carrying capacity is approached, where

$$r_{\text{eff}} = r(1 - N/K).$$

This formulation gives an apparent independent reality to two parameters, the intrinsic rate of increase, r, a biological property of the reproductive force of species that is independent of numbers, and the carrying capacity, K, a property of the demand of the species for units of a fixed resource necessary to life. Both are contingent on the particular fixed external environment. This form of the basic equation has led to an evolutionary theory of species growth in which natural selection operates to augment the intrinsic rate of increase, r, when the species is low in density and growing in numbers, and to increase the carrying capacity, K, when it is near its maximum density. These evolutionary changes are presumed to be independent and a species is either "r-selected" for characters that produce rapid growth when resources are abundant or "K-selected" for characters that give it the greatest equilibrium numbers.

But there are other ways in which the original simple biological claims can be cashed out. One is to use a pairwise "collision" model which supposes that organisms have a force of reproduction, r, as before, and that individuals meet each other in a pairwise struggle for a necessary resource in short supply. This description leads to a different model form

$$dN/dt = rN - aN^2, \tag{1.2}$$

which makes explicit that the reduction in growth is proportional to the square of numbers on the analogy of reacting molecules meeting each other in a medium. But Equations 1.1 and 1.2 are equivalent if we set $a = r/K$. In Equation 1.2 the carrying capacity does not appear explicitly even though both equations are of identical functional form but with different parameterizations, so the justification for a reified carrying capacity that can be selected for seems to disappear.

Yet a third way to express the basic biological claim is to relax the claim that the reduction in growth rate from competition is linear, or a consequence of pairwise collisions, and to allow that it is of an unknown form that may vary from species to species. We can still write an expression for the growth rate using the principle that any function can be expanded in a power series. If we make the not unreasonable claim that higher order interactions are negligible in their effect, we can stop at the second order term, which captures the density effect, and say that

$$dN/dt = aN + bN^2$$

without committing ourselves to any biological interpretation of either a or b, in which case we can make no evolutionary argument at all. The advantage of this approach is that we can add yet more terms in higher order if we want to claim that there are both competitive and productive consequences of increasing numbers on growth rate.

The standard way to add the complication of interacting species is to combine the r, k reification for the intraspecies portion and the collision model for the interspecies effect to produce the classic Lotka–Volterra model of community ecology for n species

$$dN_i/dt = r_i(1 - N_i/K_i) + \sum \alpha_{ij} N_i N_j$$

where $(i, j = 1 \text{ to } n)$.

It is then possible to make models of the evolution of the rate of increase, of the carrying capacity, and of the pairwise species interaction coefficients α_{ij}.

The movement to produce a unified population biology made a great deal of use of the logistic and Lotka–Volterra apparatus and the "community matrix" of α_{ij} coefficients in its efforts to make community ecology evolutionary and to provide some generalized overview of what sorts of communities would be stable (see, for example, MacArthur 1968, Levins 1968, May 1973). This effort has largely ceased. Partly this has resulted from the realization that the models are both overly specific and arbitrary in their mathematical form so that they may not catch the important reality of the interactions. Some indication of their inadequacy was seen in the generally successful attempt by Vandermeer (1969) to test the validity of the approach. Using a system of four ciliates in laboratory culture, he estimated the values of r and K separately for each species grown in isolation and the values of the α_{ij} by all the pairwise competition experiments. He then asked how well these estimated parameter values predicted the outcome of the four-way competition. They were quite successful in predicting the order of equilibrium densities of the four species including the extinction of one, but one of the remaining species reached a considerably higher equilibrium density than was predicted. His general conclusion was that higher order interactions were not generally important in this simple system (much to his expressed disappointment!), but that they were significant for one of the species. In the absence of many such experiments we are left in an uncertain state.

A more recent approach to the problem of the uncertain generality of functionally specific models has been to make inferences from the qualitative structural properties of the network of feedback loops among interacting species and to explore how sensitive those properties are to variations in the postulated mathematical forms that model interactions (see the chapter by Levins in this volume). In the absence of detailed knowledge of the biology of specific cases these inferences may be the best that can be done and for many purposes they may be quite good enough.

The other major challenge to the models of population ecology comes from a rejection of the assumption of a causal asymmetry between the properties of organisms and the properties of their environment. The standard view of the relation between organisms and their environments is that the development of the organism is a consequence of an interaction between its genotype and its developmental and metabolic environment, but that the environment is a consequence of autonomous forces, independent of the organism. The differential equations representing the short-term and evolutionary change in organisms (O) and their environments (E) are then of the form

$$dO/dt = \mathrm{f}(O, E),$$

$$dE/dt = \mathrm{g}(E).$$

That is, organisms and their populations evolve as a consequence of their state and the state of the environment, but environments change independently of organisms. Yet it has become increasingly clear that the environments of organisms do not preexist the organisms (although a physical world certainly exists independent of their existence), but are created and modified by the organisms themselves. The environment of an organism consists of aspects of the external world that are made relevant to the organism juxtaposed in certain functionally relevant ways by the activities of the organism (Lewontin 1983). As a result organisms and their environments *coevolve* and the appropriate coupled differential equations, showing both organism and environment as both causes and effects, are

$$dO/dt = \mathrm{f}(O, E),$$
$$dE/dt = \mathrm{g}(O, E).$$

Of course, population ecologists have always known this in the special case of prey and predator where each is the environment of the other and the prey–predator equations are coupled. Nevertheless, the notion of the autonomous ecological niche that sets the parameters of population growth, but is not itself evolving as a consequence of that growth, is the usual model. Only recently has the notion of niche construction been included in the formal development of population biology (Laland *et al.* 1996).

1.4 What is to be done?

The attempts within population genetics to find a general quantity, fitness, that can be a parameter of a simple but general equation for genetic change, to find a general principle of maximization, to assert a simple relationship between average fitness and population demographic characteristics, like the attempts within population ecology to find simple generalized equations of demography and community ecology from which general demographic and evolutionary predictions can be made, flow from a view of science that is inappropriate for population biology. We are all educated to prize a model of science that comes from physics, and in particular the physics of immense bodies and submicroscopic particles. To be the model of a population biologist, in this view, is to produce the general equation, or matrix or inequality or extremal principle, that can be succinctly expressed as a closed relation. But biology, and especially population biology, is neither particle physics nor solar system astronomy. Organisms and their populations are a nexus of a large number of individually weakly determining interacting forces. One consequence is that different cases have different dynamics and that simple general functional forms may miss the important action (see Taylor 2001 for a more articulated explication of this view).

While a general theory may be more "beautiful," science is not primarily an aesthetic endeavor but an attempt to find methods to deal adequately with the material world. To construct a successful population biology we need to learn the lesson of statistics. For most of its modern history, statistics was a branch of applied mathematics in which the ideal was to find a closed form for the distribution of an estimator or a test statistic. In order to carry out that program, however, it was necessary to make assumptions in each case without any knowledge of how robust the test distributions were to deviations from those assumptions, because no mathematical solutions to the distribution problems could be found if the assumptions were relaxed. So, up until the second half of the twentieth century *t*-tests, *F*-tests, analyses of variance, and large numbers of *ad hoc* test statistics were performed without knowing whether they were robust to the assumption of normality. One simply crossed one's fingers and repeated the Central Limit Theorem when questioned. Then, with the invention of high-speed digital computers and the development of Monte Carlo methods of generating replicated data sets from arbitrary distributions, we learned that the *t*-test is robust, but that the *F*-test for the equality of two sample variances is not. More generally, the invention of simple test statistics of dubious properties has been replaced by computer simulations of sampling processes, permutation tests, and other stochastic procedures. Far from being diminished by the application of Monte Carlo simulations, statistics now has a claim to our confidence that did not exist before. In place of a general closed-form result, we have substituted a general methodology.

Population biology is concerned with the outcome of the interaction of an immense variety of biological processes involving individual behaviors and physiologies that are contingent on variations in genetics and on environment in its broadest sense. Rather than dispensing with the biological details in the interest of a simplified general theory, we need to explore the properties of many specific models of biological interaction, chiefly by computational procedures, but also by using results in such fields as topology, graph theory, and qualitative behavior of differential equations. The diversification of models and their means of investigation does not mean that every case is unique and that no generalizations are possible. Such generalizations, however, will turn out not to be of universal application but will be properties that apply to local regions of the model space. The task of population modeling is then to map these regions and find their properties. A great deal of this work can be done by well-designed exploratory computer simulation and computation, producing observed regularities for classes of models, regularities that can then be confirmed and extended by analytic methods. For example, this approach has proved quite fruitful in exploring the effects of linkage and natural selection in multilocus genetic systems, a problem that is much simpler than those generated by the evolution of community structure, yet that is already beyond the range of the usual equation-solving techniques (see, for example,

Lewontin and Feldman 1988). In this sense, theoretical work in population biology becomes an *inductive* science, producing observed generalities from a collection of carefully controlled and related trial structures.

When we consider the experimental work necessary to the building of a successful population biology, it becomes immediately clear that directions of vital importance have yet to be explored. Up until the present, empirical population biology that pretends to any generality has chiefly consisted of observations of stability or change in the genetic, demographic, or community structure of natural or experimental populations. The immediate use of these observations has been, for the most part, to try to fit the observed population characteristics into some simplified model of dynamics with the ultimate object of estimating parameters or testing hypotheses about the adequacy of particular models. Is natural selection operating on some genetic polymorphism in nature (e.g., Dobzhansky and Levene 1948)? Is the Lotka–Volterra formulation adequate to explain community structure (Vandermeer 1969)? Do changes in genetic composition of a population cause predictable changes in population demography (Carson 1958)? Is there some characteristic of the environment that can explain differences in species diversity in some group of organisms (MacArthur and MacArthur 1961)?

All of these questions have in common that they take some population or community property as the thing to be explained, and some general property of population models as the putative explanation. What is lacking is a large body of experimental observation on the constituent biological relations that would generate the basic form of the models in the first place. If it is true that the effective environment of an organism is created and modified by the organism and its conspecific neighbors in a population, then the evolution of a population causes evolution of the environment. There is a large literature in population genetics showing that the direction and speed of genetic changes in a population depend on the environment. There is, however, virtually no body of work on the reciprocal causal pathway, measuring and characterizing the changes in environment that are induced by changes in the genetic composition of the population and the way in which such environmental changes will affect the selection of genotypes. The related question of the commonness and form of frequency-dependent and density-dependent changes in fitness also provides a very large field for investigation. It is not sufficient simply to document examples of such dependence. The interactions must be understood in sufficient detail to generate predictions of their population effects. In population ecology the entire question of the opposing effects of facilitation and competition for resources on reproductive rates needs to be investigated so that realistic models of the dependency of population growth rates on density can be constructed.

In summary, population biology can only be built by breaking down the distinction between the study of population processes and the study of individual properties. Phenomena at the level of populations are the consequences of properties of individuals and, in turn, those properties are influenced by the individual's immersion in an external world that is in large part the consequence of the accumulated activity and interactions of other individuals. Population-level phenomena summarized in fitness, growth rates, population size, stability, and interspecific interaction rates are the outcome of material interactions of living organisms. There can be no successful population biology that is not, at root, biological.

REFERENCES

Andrewartha, H. G., and Birch, L. C. (1954). *The Distribution and Abundance of Animals.* Chicago: University of Chicago Press.

Beardmore, J. A., Dobzhansky, Th., and Pavlovsky, O. A. (1960). An attempt to compare the fitness of polymorphic and monomorphic populations of *Drosophila pseudoobscura. Heredity* 14: 19–33.

Carson, H. L. (1958). Increase in fitness in experimental populations resulting from heterosis. *Proc. Natl. Acad. Sci. U.S.* 44: 1136–41.

Charlesworth, B. (1970). Selection in populations with overlapping generations. I. The use of Malthusian parameters in population genetics. *Theor. Pop. Biol.* 1: 352–70.

Charlesworth, B., and Geisel, J. T. (1972). Selection in populations with overlapping generations. II. Relations between gene frequencies and demographic variables. *Am. Nat.* 106: 388–401.

Dobzhansky, Th., and Levene, H. (1948). Genetics of natural populations. XVII. Proof of the operation of natural selection in wild populations of *Drosophila pseudoobscura. Genetics* 33: 537–47.

Dobzhansky, Th., Lewontin, R. C., and Pavlovsky, O. A. (1964). The capacity for increase in chromosomally polymorphic and monomorphic populations of *Drosophila pseudoobscura. Heredity* 19: 597–614.

Dunn, L. C. (1957). Evidence of evolutionary forces leading to the spread of lethal genes in wild populations of house mice. *Proc. Natl. Acad Sci. U.S.* 43: 158–63.

Fisher, R. A. (1930). *The Genetical Theory of Natural Selection.* London: Oxford University Press.

Grant, P. R., and Grant, B. R. (2002) Unpredictable evolution in a 30-year study of Darwin's finches. *Science* 296: 707–11.

Laland, K. N., Odling-Smee, F. J., and Feldman, M. W. (1996). On the evolutionary consequences of niche construction. *J. Evol. Biol.* 9: 293–316.

Levins, R. (1968). *Evolution in Changing Environments.* Princeton, NJ: Princeton University Press.

Lewontin, R. C. (1955). The effect of population density and composition on viability in *Drosophila melanogaster. Evolution* 9: 27–41.

Lewontin, R. C. (1958). A general method for investigating the equilibrium of gene frequency in a population. *Genetics* 43: 419–34.

Lewontin, R. C. (ed.) (1968) *Population Biology and Evolution.* Syracuse, NY: Syracuse University Press.

Lewontin, R. C., and Feldman, M. W. (1988). A general asymptotic property of two-locus selection models. *Theor. Pop. Biol.* 34: 177–93.
MacArthur, R. H. (1968). The theory of the niche. In R. C.Lewontin (ed.) *Population Biology and Evolution.* Syracuse, NY: Syracuse University Press.
MacArthur, R. H., and Connell, J. H. (1966). *The Biology of Populations.* New York: John Wiley.
MacArthur, R. H., and Levins, R. (1967). The limiting similarity, convergence and divergence of coexisting species. *Am. Nat.* 101: 377–85.
MacArthur, R. H., and MacArthur, J. W. (1961). On bird species diversity. *Ecology* 42: 594–8.
May, R. (1973). *Stability and Complexity in Model Ecosystems.* Princeton, NJ: Princeton University Press.
Sandler, L., Hiraizumi, Y., and Sandler, I. (1959). Meiotic drive in natural populations of *Drosophila melanogaster.* I. The cytogenetic basis of segregation distortion. *Genetics* 44: 233–50.
Taylor, P. J. (2001). From natural selection to natural construction to disciplining unruly complexity. The challenge of integrating ecological dynamics into evolutionary theory. In R. S. Singh, C. B. Krimbas, D. R. Paul, and J. Beatty (eds.). *Thinking About Evolution.* Cambridge, U.K.: Cambridge University Press.
Vandermeer, J. H. (1969). The competitive structure of communities: an experimental approach with protozoa. *Ecology* 50: 362–71.
Wallace, B. (1968). Polymorphism, population size and genetic load. In R. C. Lewontin (ed.) *Population Biology and Evolution.* Syracuse, NY: Syracuse University Press.
Wright, S. (1931) Evolution in Mendelian populations. *Genetics* 16: 97–159.

2

Toward a population biology, still

RICHARD LEVINS
School of Public Health, Harvard University, Boston, and Institute of Ecology and Systematics, Havana

Dick Lewontin's Triple Helix updates a conceptual framework that has been evolving since at least the 1950s for seeing biology as the joint action of the genetic system, the organism in its development and physiology, and the organism in its environment. But this program for an integrated population biology remains an aspiration that has not been carried out in practice. Instead, we see a gross imbalance among these components and a continued separation of the disciplines. There has always been a much finer sophistication of our understanding of genetic variation than of the environment. Some of the reasons for this have been discussed in previous volumes of this series: the gene-centered reductionist view of evolution located all the richness of evolution in the genes, so that the nuances of genetic structure had priority. The mechanisms of Mendelian genetics allowed for the formulation of precise dynamic models, while statistical theory encouraged attempts to measure the relevant parameters even when unrealistic assumptions were needed. This has led to highly simplified and casual assumptions about environments being reflected in the "just so" stories of sociobiology, where it was enough to declare something to be a trait that it would be advantageous to consider as genetically determined and whose evolution had been explained. It also made it possible for psychologists to talk easily about twins raised in "similar" or "different" environments without examining the dimensions of similarity and difference. Meanwhile, much of the work in population genetics has been aimed at answering the questions of population genetics in the narrow sense, such as estimating selection pressure or effective population size. The demand for precision has tended to overwhelm the criteria of realism.

But a successful study of evolution requires the recognition of the complexity not only of the genotype but also of the environment and of the whole organism in its development and its physiological flux. Of course all scientific work requires simplifying assumptions, and it is no great contribution to point them out. The exercise of scientific judgment enters in the decision as

The Evolution of Population Biology, ed. R. S. Singh and M. K. Uyenoyama. Published by Cambridge University Press.

to whether these simplifications have done their job and are now oversimplifications, obfuscating more than they reveal.

After Schmalhausen's *Factors of Evolution* was translated into English in 1949, there was a flurry of interest in phenotypes in their environments. His idea of stabilizing selection focused attention toward homeostasis, phenotypic plasticity or variance, the distinction between familiar environments in which phenotypes are the result of selection and new environments in which the response of the organism is random, not random with respect to the developmental biology of the organism but with respect to its contribution to fitness. Waddington's (1957) genetic assimilation and canalized development and Michael Lerner's (1954) genetic homeostasis continued in this tradition.

There was a beginning interest in the systematic study of environment in order to account for the surprisingly high degree of genetic polymorphism in populations (Mayr 1942). The Dobzhansky school had emphasized balanced selection based on the superior homeostasis of the heterozygote (Dobzhansky 1951). But this was too demanding a condition to explain all the observations, and the search was undertaken for less stringent conditions than heterosis that might maintain polymorphism in populations. Ludwig (1950), Li (1955) and Levene (1953), working independently, showed that when organisms occupy more than one "niche" (that is, when population density is regulated within more than one subpopulation), then the conditions for stable polymorphism are less stringent than in a uniform environment. The superiority of the harmonic mean of the fitnesses of the heterozygote rather than the arithmetic mean was shown to be sufficient. Since the harmonic mean is closer to the minimum fitness than to the mean fitness, a reduced variance of fitness over a range of environments gives a greater harmonic mean. Now this was linked to heterogeneous environments in the determination of polymorphism with the proposition: since heterozygotes are better buffered in the face of varying conditions, polymorphism will be greater in more heterogeneous environments. Pierre Dansereau proposed a way of measuring the heterogeneity of the environment for *Drosophila*, and field work supported the general proposition that heterogeneous environments permit greater genetic diversity. Other work modeled temporal variation in the environment, and examined the role of linkage disequilibrium in maintaining balanced selection or allowing for hitch-hiking by less favored alleles linked to a strongly favored one. Frequency dependence, density dependence, multiple alleles, segregation distortion, gametic selection, inbreeding and other complications of the original single-locus, two-allele Mendelian model, and other explorations extended the conditions for polymorphism. The coupling of life table parameters to genetic parameters showed connections of evolution to demography.

The Waddington–Dobzhansky–Mayr program never reached fruition. The dramatic breakthroughs of the molecular geneticists starting with the genetic code encouraged a reductionism in biology that satisfied the prevailing philosophical biases of the scientific community toward "fundamental" controlling factors. As applications of molecular approaches appeared, this direction increasingly coincided with economic interests since single factor explanation makes it possible to seek patented, marketable commodities. The preferred strategies of pesticides in pest control, antibiotics in treatment of infectious diseases and vaccines in prevention of epidemics favored the search for specific genetic causes of traits and a simplification of causal models.

The questions population geneticists asked remained those of accounting for genetic polymorphism in populations and estimating selection coefficients or effective population size. They were still within classical population genetics rather than part of a population biology that would unite the various perspectives on biology above the level of the individual. Meanwhile, in population ecology, work on species coexistence in communities proceeded independently of the study of conditions for genetic polymorphism even when the methods were similar: simultaneous equations for interacting variables and a focus on the approach to steady states.

An integrated population biology must not only seek answers in the interactions of these separated domains but must also take up new questions specific to that integration. Thus we cannot limit ourselves to looking to ecology to offer answers to the questions posed by population genetics but must also ask the questions of ecology and development. For instance, does genetic variability contribute to survival of populations in changing environments? Does gene frequency track changing conditions? How much selection separates the adaptations of particular local populations to local conditions from those of other populations of the same species in different environments? What is required for a microorganism to change hosts or mode of transmission or clinical picture? Why are some taxa evolutionarily and ecologically conservative, but occasionally produce versatile subtaxa that bud off species into all sorts of environments beyond the usual ecological range of the larger group? What is the role of environmental history in microevolution? Why do genetic differences in the kinetics of intermediate metabolism usually show little fitness difference? Does genetic diversity in a population make it less susceptible to competitive displacement by invading species? What are the physiological and genetics bases for the common distinction between "*r*"- and "*K*"-selected species and the eventual abandonment of this dichotomy? Why is it so difficult to find evidence of selection in populations and yet easy to observe rapid changes when microorganisms respond to antibiotics or plant viruses change hosts?

I think that biology above the cellular level needs integration among the components of the Triple Helix more than exquisite refinements of precision within the fields that are still too separate. In what follows I look briefly at some of the links among these three areas.

2.1 The genotype and the whole organism

We need to adopt a developmental biology that sees the parts of the organism as environments to each other with reciprocal feedbacks that remove the notion of one link being primary, and the internal and external environments as providing developmental "information" rather than merely reading a genetic "program." David Moore's book *The Dependent Gene: the Fallacy of Nature versus Nurture* (2002) emphasizes what we all sort of know but seldom act on in our theories: development takes place through the reciprocal feedbacks among the elements of a developing organism in which each part is "environment" to the others. Traits are formed in the course of development by these influences. It is the whole genome/organism/environment system that develops.

Those aspects of the external or internal environment that are always present in the lives of mammals, birds and other groups can be thought of as an external yolk for the embryo. They are taken as givens, as the context for selection in the evolution of internal developmental mechanisms and as part of the developing system. But they pass unnoticed because they do not give variance. Even the behavior of parents, so necessarily present for the development of the offspring, often remains invisible because it is nearly universal.

Organisms do not respond to their "environment" as a whole but to some aspect that serves as a signal. Like those environmental signals such as day length that trigger diapause in insects, there is no necessary relationship between the physical form of the signal and the response it evokes or the environmental features for which it is adaptive. Thus Moore notes that the sound of a duckling's own peeping in the egg rather than prenatal exposure to the call of female mallards is the signal that attunes it to responding later to its mother's call. The cat's visual cortex does not develop in relation to what the cat sees but to binocular vision *per se*. It is therefore easy to miss the environmental determination of organ formation or behavioral patterns and maintain that these are "hard-wired" in the genome.

Further, the mutual actions within the system take place constantly during development, not once and for all, so that the organism is rolling along the channels of a Waddingtonian epigenetic landscape that is itself changing: the organism not only follows the pathways of the landscape but also transforms them. This requires a dynamic model in which the process depends on the positive and negative feedbacks among the elements of the system. In the case of the tent caterpillar, we have positive feedback: the healthier individuals seek habitats better exposed to sunlight that produce healthier offspring, and then their healthier offspring are better able to find favorable sites. But negative feedbacks also occur: higher temperature during insect development results in smaller individuals. If these then suffer greater heat and desiccation stress, they may seek out the cooler, moister places and produce somewhat larger offspring.

Among insects, when higher temperature is associated with a drier habitat, selection would act to favor larger body size whereas the direct environmental impact produces smaller individuals. Along this environmental gradient it is possible to observe very little phenotypic variation, but this apparent uniformity hides opposing genetic and environmental influences. But if it is the cooler habitats that are drier, selection for larger size and the direct impact of lower temperature act in the same direction, enhancing the steepness of the phenotypic gradient. The dual role of environment as altering development as well as selecting among the altered organisms, and the dual role of the organism in "responding" to its environment and selecting, transforming, and transducing it, make the triad of genome, organism, and environment the unit of evolution and the source of developmental and evolutionary "information."

2.1.1 The norm of reaction revisited

Against the facile genetic determinism of his day, Dobzhansky (1951) argued for the genome as a norm of reaction, a pattern of response to the environment. Thus we could raise different lines of *Drosophila* at a series of temperatures or transplant plants from different populations along an altitudinal gradient as in the classic Clausen *et al.* (1940) experiments and plot some measurable character as a function of the environmental factor. When this is done, several observations are usually apparent:

(a) The phenotypic expression of each genotype or variety varies along an environmental gradient.
(b) Some genotypes vary more than others in response to the same gradient.
(c) The norms of reaction cross each other so that no one genotype exceeds all the others in all environments, and performance in one environment does not predict performance in all environments.
(d) The within-genotype variance (or coefficient of variation, when the mean values are quite different) and the among-genotype variance also change with the environment, usually increasing in the extreme environments.

Some years ago a group of us studied the responses of plants to varying conditions.[1] A collection of corn hybrids and "acquisitions" from the International Center for the Improvement of Maize and Wheat (CIMMYT) and some common weeds were grown under a range of temperature, humidity and light conditions, after which the plants were either moved to other growth

[1] This work was done in the 1970s at the University of Wisconsin Biotron with support from the Ford Foundation. It was never published because for too many years I expected to be able to finish the project; however, funding never became available, the group dispersed and we all acquired other interests.

Table 2.1. *Coefficients of variation of daily growth in corn lines and the weed giant foxtail grown at different temperatures.*

Variety	11 °C	14 °C	28 °C	36 °C	40 °C
Hybrid 1	51	19	14	10	21
Hybrid 2	33	33	19	–	38
TL75B	30	31	23	34	50
TO75	39	32	24	45	82
Guat	48	28	23	37	37
PR	28	40	38	24	68
BA73	121	19	20	36	40
CCC	43	4	26	29	46
Giant foxtail	95	87	23	23	36

Table 2.2. *Coefficients of variation of daily growth of corn according to temperature of growth and temperature of pretreatment*

Pretreatment temperature	Test temperature 11 °C	28 °C	40 °C
11 °C	61	35	78
28 °C	413	145	261
40 °C	339	121	416

chambers or kept in the same environment. Plant growth was measured by increase in height per day. We found the following general results: growth was maximum at 28 °C and diminished at 11 °C and at 40 °C for all lines. The coefficient of variation both within and among lines of corn usually showed a U-shaped response with minimum variability at 28 °C. Table 2.1 shows the coefficients of variation in one series of a number of corn lines raised at different temperatures during successive periods of a few days. They all show the typical U-shaped pattern of increased variation at extreme cold or hot conditions. A similar result was obtained over day lengths.

When plants were moved among temperatures, both the growth and the variability depended on the pretreatment. The coefficients of variation for one series are shown in Table 2.2 (the diagonal is not consistent with Table 2.1 because it comes from a different series of experiments). For each pretreatment, the variability showed minima at the optimal temperature of 28 °C. The order in which the two temperatures were presented affected the outcome.

It is clear that as we make the environment more complex the norm of reaction increases into a higher dimension. Perhaps we have to delve further

Table 2.3. *Variability range/mean of health outcomes by "race" across states*

Outcome over states	White	Black	Asian	Native American
LBW	0.27	0.43	–	–
VLBW	0.40	0.70	–	–
IMR	0.56	0.58	–	–
AADR	0.30	0.74	0.75	1.67

LBW, low birth weight; VLBW, very low birth weight; IMR, infant mortality rate; AADR, age-adjusted death rate.

into the physiology of growth to identify the accumulation or breakdown of growth-promoting factors that are temperature sensitive. In different experiments there were positive or negative correlations between growth in consecutive periods. But the main point is the active construction of phenotype in the course of development and the significance of the variability as part of phenotype, as an indicator of canalization.

2.1.2 Schmalhausen's Law

The above results might be seen as examples of Schmalhausen's Law (Lewontin and Levins 2000):

An organism (or other system) at the boundary of its tolerance or in extreme or unusual conditions along any dimension of its requirements will be more sensitive to variation of the environment along any dimension. This will be seen as increased variability.

Schmalhausen's Law shows up in other situations as well:

(a) Blood pressure in a cohort of people shows increasing variance with age as the vicissitudes of living erode homeostasis. But this process is more rapid in Afro-Americans than among whites so that at comparable ages Afro-Americans show a greater variance. The difference corresponds roughly to ten years. Other health outcomes also show greater variability for Afro-Americans than for whites (Table 2.3).
(b) Across a sample of cities of a given size, the life expectancy of both men and women varies as a result of varying social, economic and physical conditions. But for all categories of city size the variability is greater among Afro-Americans than among whites (Table 2.4). Either the conditions of deprivation vary geographically or uniform deprivation makes people more vulnerable to small differences in other conditions that have less impact on whites.
(c) Populations of a species do not really occupy their whole geographic range. Rather they occur with more or less frequency in the environmental patches that are suitable for them. A simple model suggests that the proportion p of

Table 2.4. *Variability (range/mean) of age-adjusted death rates for Afro-American and white males and females*

Range	White male	Black male	White female	Black female
Urbanization classes	0.13	0.22	0.06	0.12
Core metropolitan	0.07	0.20	0.07	0.12
Small metropolitan	0.06	0.23	0.03	0.08
Nonmetropolitan city	0.17	0.40	0.11	0.17
Suburbs	0.08	0.14	0.06	0.08
Rural	0.22	0.25	0.15	–

possible sites that are occupied is given by

$$p = 1 - e/r$$

where e is the local extinction rate and r the rate of recolonization. Near the boundary of the range, e/r is close to 1 and p near 0. Suppose that $p = 0.01$ so that $e/r = 0.99$. Now increase r or decrease e by 1% so that $e/r = 0.98$. The prevalence becomes 0.02, a doubling in response to an environmental change that in practice would not be measurable. But in a more central part of the range where $p = 0.5$ and $e/r = 0.5$, a 1% change in e/r results in a 1% change in p. Therefore climatic or other environmental change will not result in a whole fauna moving north intact, but in big changes for marginal species and small ones for those that are more common. The same model applies to the prevalence of infectious disease.

(d) When a substance such as a toxic pollutant enters an organism and is removed by some enzymatic activity, it can reach an equilibrium level at

$$X = kI/(v - I)$$

where X is the steady state level, I the input rate, and k and v the kinetic constants of the system. The concave shape of X as a function of either I or v shows that when I is near v the sensitivity to changes in input is greater than when there is reserve detoxification capacity, and that variance in either v or I increases the average steady state load X. For any nonlinear response, the average effect over inputs is not the same as the effect at average input.

(e) When *Drosophila* from different populations but the same normal morphology are subjected to a heat shock at about 18 hours into pupal development, they produce abnormalities of the wing structure: extra vein segments, missing vein segments, holes in the blade, long familiar as Richard Goldschmidt's (1938) phenocopies. But each population has its own spectrum of abnormalities showing that they differed in the proportions of different genotypes that affected the norm of reaction. The treatment revealed latent genetic variation. A similar situation sometimes occurs when crop varieties are introduced into new habitats and "tend to vary."

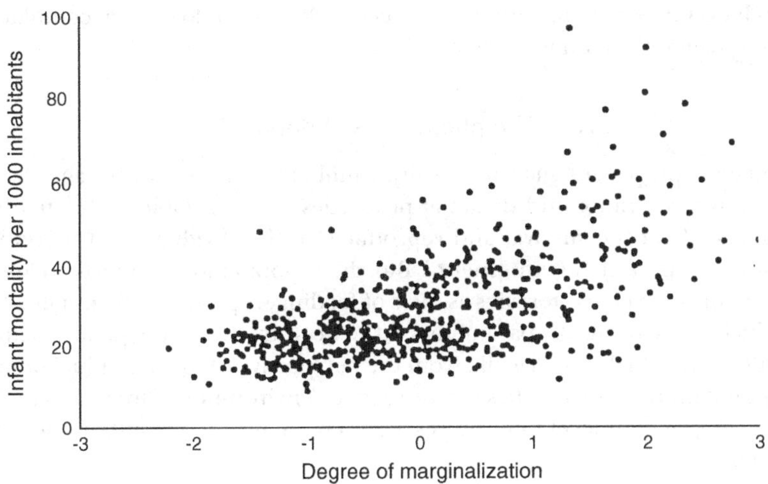

Figure 2.1. Schmalhausen's trumpet. The infant mortality rate for Mexican villages 1990–1996 is plotted against their degree of marginalization, a composite measure of indicators. Mortality increases with degree of marginalization as expected, but also spreads out more as outcomes are increasingly sensitive to local conditions. (Adapted from Lozano *et al.* 2001.)

The practical value of Schmalhausen's Law for population genetics is that it allows the intensity of selection to be negligible most of the time but latent phenotypic variance can appear and be available for sporadic selection.

If data are plotted as a scatter of unconnected points for some measurement along an environmental gradient, the scatter spreads out in the more severe environments, producing the pattern we have labeled Schmalhausen's trumpet. In Figure 2.1 we show infant mortality plotted against degree of marginalization of Mexican villages as reported by Lozano *et al.* (2001).

2.1.3 Shades of Lamarck

There are many examples of extragenetic heritability by way of the transmission of cytoplasmic ingredients, transplacental influences, habitat selection, learned behavior, and "cultural" continuity. While the impact of the environment on the organism is familiar autecology, and the impact of the organism on its environment is less often reported, the feedback between them is not commonly examined. The critical question here is the nature of the feedback between organism and environment: does the phenotype produced in a given environment increase or decrease that environmental effect? For instance, organisms that are more efficient feeders may be larger and therefore more capable of capturing prey, perpetuating their larger size in a positive feedback. But they can also make their environments less suitable for themselves by depleting resources or increasing contamination more rapidly in a negative feedback. Then an initial environmental impact will be reduced.

Thus it would be systems with positive feedback that would show accumulated pseudo-Lamarckian effects.

2.1.4 The phenotypes of populations

Population properties such as density, numbers, rate of growth, age distribution and sex ratio, and dynamic properties such as stable or fluctuating numbers, affect the survival and reproduction of individuals and therefore constitute a part of "environment." But these population properties clearly depend on such characteristics as rates of feeding, reproduction, and mortality which are sensitive to parameters influenced by the genotype and acting through individuals. Thus while from the perspective of the individual organism population characteristics can be seen as "environment," they can also be seen as population level phenotypes and may change as a result of selection and demography.

In the context of public health, age-specific mortality is a property of major concern. A curious observation is that while the death rates for those populations that are targets of racism are generally higher than for the more privileged populations, in the United States the age-specific mortality curves cross over late in life. Furthermore, the cross-over age has become later over time. An obvious explanation is some kind of selection throughout life so that at an advanced age only the hardiest are still around. Here selection only means differential survival and does not imply genetic differences or evolution.

However, the risk of dying is not a fixed trait but changes with circumstance. Suppose, for example, that a population can be divided into health classes that reflect all the accumulated circumstances of exposure and resistance. Suppose further that in each health class x there is a certain probability of dying, m_x, and probabilities p_{ij} transfer from health class j to health class i. We start with a population with a certain distribution across health classes, $P(x)$, probabilities of death from each class, and probabilities of transition from one risk class to another. For convenience, at a preliminary stage we assume transitions only to adjacent health classes. Since these models follow cohorts without recruitment, the population eventually dies out in all cases. We are interested both in the life expectancy for the whole population and the proportion of the population that dies from healthier risk classes, presumably with less lingering illness. "Policy" interventions are seen as changes in the mortalities from given classes and the transition probabilities among them.

Suppose first that people always remain in their original risk class from birth. Then, over time, selection dominates the process and in the end we have a population mostly in the relatively healthier class. The stronger the selection (the greater the mortality differential) the faster the change in cohort composition under selection. Then comparing two populations we would see a cross-over eventually as those who start out at higher risk are eliminated.

Suppose now that the probabilities of transition among risk classes are large. Then selection has only a slight effect on population composition. We might assume that transitions from one class to another are mostly to adjacent classes, and that the dominant direction is toward deteriorating health with age, but that they also allow significant risk reduction with acquisition of immunity, knowledge, and progressive social change. Mohtashemi and Levins (2002) studied trends in the cross-over age for Afro-Americans and Hispanics with whites.

Finally, suppose that at birth children begin life in risk classes influenced by the risk class of their parents or mother. Then we have a long-term model that can be adjusted for the effects of aging and changes in the parameters brought about by social and medical innovation.

2.1.4.1 Population/environment feedback

Here we look at one model in order to examine the tracking of the environment by the population. Later it will be expanded to consider selection directly. Suppose that a resource enters the system at some steady rate a. We could use a prey population as the resource, but here I would like to introduce the less familiar model in which the resource is not self-reproducing. It might be a photosynthate produced by a host plant which is then converted into insoluble elements of structure, such as reproductive structures or leaf surface, at a rate c. Then the equation for the resource R might be

$$dR/dt = a - R(px + c) \tag{2.1}$$

where a is the rate of photosynthesis, c the rate of removal, and p the rate of feeding by population x of herbivores. The equation for the herbivore is

$$dx/dt = x(pR - m) \tag{2.2}$$

where m is the mortality of x. We chose not to have x directly density dependent.

In the absence of x, R reaches an equilibrium at $R^* = a/c$, the ratio of photosynthesis to removal. The conversion of photosynthate is Rc, equal to a, so that c itself is selectively neutral since a lower c results in a greater R. Selection in the plant may act to increase the photosynthetic rate a. Note therefore that in the absence of a herbivore there is selection on a but in its presence selection depends only on the parameters of the herbivore equation. This is a consequence of the choice of model that excluded density dependence in the herbivore.

Now introduce the herbivore with population density x that consumes R at a rate p. Here R reaches an equilibrium at

$$R^* = m/p.$$

Thus R no longer depends on its own photosynthetic rate but on parameters of the herbivore. (However, m and p, although they appear as properties of the herbivore, may be influenced by selection in the plant: leaf structure may alter the ease of feeding, early leaf production may avoid the herbivore, or secondary compounds might increase mortality.)

The herbivore population reaches an equilibrium at

$$x^* = a/m - c/p.$$

(This corresponds to the carrying capacity K of logistic models. Thus K can depend on density-independent factors in its environment. When x is rare and R is close to a/c, then

$$dx/dt = mp(a/m - c/p)x \quad \text{or} \quad mpkx.$$

This is equivalent to the r_0 of logistic growth. Thus r_0 and K are not independent biological properties that could be selected separately. The reification of r_0 and K as distinct biological entities has done much mischief in ecology.)

The relative importance of selection in the herbivore for the different parameters can be found by differentiating the population size with respect to any of them. Some of the results are paradoxical. When mortality is high, selection to reduce mortality is relatively weak (proportional to $-a/m^2$) but becomes more important as mortality is reduced. The same occurs with feeding rate. When x^* is small so that a/m is close to c/p, small changes in any parameter can have large relative effects on x^*. The rate of photosynthesis is a parameter of the plant, but can be influenced also by the herbivore. For instance, the sorghum shoot fly kills the main stem of the sorghum plant. The plant responds by tillering, producing several secondary shoots. This increases the vegetative mass for photosynthesis, although it reduces yield if the sorghum is harvested mechanically.

But environments are not constant. Then instead of equilibrium we can find long-term average values of the parameters using the methods of time averaging. The basic idea is that the average value of a derivative is $1/t$ times the integral of the derivative up to time t, which is $[x(t) - x(0)]/t$ so that if $x(t)$ is bounded the average is less than $(\text{Max}(x) - \text{Min}(x))/t$ and as t increases this goes to zero. (A recent exposition of the method can be found in Rapport *et al.* 1998). We could of course let any of the parameters vary. Suppose that the parameters of the plant equation are all constant so that the system is driven from the herbivore end, by variation in the mortality of herbivores. Since the long-term average value of the derivative of a bounded variable is zero, we can average the right side of Equation 2.1 using the symbol E for average or expected value:

$$0 = a - E\{R\}(pE\{x\} + c) - p\text{Cov}(R, x)$$

or

$$a/E\{R\} = pE\{x\} + c + p\text{Cov}(R, x)/E\{R\}$$

where E is the expected value and Cov is the covariance. If we divide Equation 2.1 by R before averaging we have

$$(dR/dt)/R = a/R - (px + c).$$

Since the left side is the derivative of $\ln(R)$, which is also bounded, its expected value is zero so that

$$E\{a/R\} = pE\{x\} + c.$$

Since for a positive variable R, $1/E\{R\}$ is less than $E\{1/R\}$, $\text{Cov}(R, x)$ is negative.

Thus if the environmental parameter m, the herbivore mortality, varies while the other parameters are constant then there will be a negative correlation between the abundance of the herbivore and its resource. High populations of the herbivore would be hungry while sparse populations would be well fed. But if the input rate "a" or the conversion rate c varies and m remains constant then there is no correlation between R and x and hence between population density of the herbivore and its nutritional state. (Had we chosen to allow density dependence in the equation for x, there would have been a positive correlation between R and x, the average R would be reduced, and the dense herbivore populations would be well fed while sparse populations would be hungry, giving us an anti-Malthusian dynamic.) The average level of R is the average m/p and therefore not changed. These differences in the relation between abundance and nutritional status can also set the stage for the action of selection.

Finally we note that populations can be far from their steady state eventual behavior. In that case we need to look at the transient behavior. For instance, suppose that the system is driven by variation in the parameters a or c of Equation 2.1. Then we can look at average values for Equation 2.2. The average value of the derivative is

$$E\{dx/dt\} = (x(t) - x(0))/t$$

and

$$E\{(dx/dt)/x\} = \ln(x(t)/x(0))/t$$

so that

$$\text{Cov}(R, x) = [x(t) - x(0) - E\{x\}\ln(x(t)/x(0))]\,p/t.$$

Thus the covariance for a short period of observation depends on the direction and history of change. If population growth is exponential then $\text{Cov}(R, x) = 0$, but if it is slower at first and then accelerates, $E\{x\}$ is smaller and the covariance is positive. Thus the correlation between the resource and herbivore depends in part on the direction and history of change in the resource.

In studies of scale insects (*Lepidosaphes gloverii*) and their natural enemies on Cuban citrus in collaboration with Tamara Awerbuch, Sonja Sandberg and Cuban colleagues at the Institute for Fruit Research (in press), we found a positive correlation over time between the scale insect population and its natural enemies, a predator (a wasp) and two or three infective fungi, but a negative correlation in space. We interpret this to mean that temporal change is driven by factors entering the system from below, from the trees affecting the herbivore and passing on the effect in the same direction to the enemies. (In fact, the scale population builds up after the spring flush of leaves.) But spatial differences among trees and parts of trees act directly on the enemies of the herbivore and transmit their effect in the opposite direction to their prey. These and other statistical relations in communities are useful in the diagnostics of dynamics even when not all variables have been observed.

2.1.5 The role of history

Science operates on the assumption of a Markovian world in which the present state of the system fully determines the next state or at least its probability distribution. Thus if we had the pair of Equations 2.1 and 2.2 and the initial conditions $R(0)$ and $x(0)$ the system is fully described and can be solved for future times. The trouble is that usually we do not have a full description of a system. It may be that we can observe the herbivore x but not measure the photosynthate, or we monitor the state of the plant and know that there is herbivore damage but do not know the identity of the herbivore. Then we can eliminate one of the variables from the pair of equations and get a second order equation. Solving Equation 2.2 for R we get

$$R = ((dx/dt)/x + m)/p$$

so that

$$dR/dt = (d^2x/dt^2/x - (dx/dt)^2/x^2)/p$$

Substituting R in Equation 2.1 we have

$$(d^2x/dt^2)/x - (dx/dt)^2/x^2 = a - [(dx/dt)/x + m](px + c)/p. \quad (2.3)$$

This is a second order equation in x. That is, it takes into account not only the present population growth but also the change in growth rate, the difference between present and past growth rates. If we had used a discrete model we would have had an equation in x_{n+1} as a function of both x_n and x_{n+1}. Thus the history is necessary because we can never fully characterize the present. A two-variable system of equations in the present can be equivalent to a single-variable equation of higher order, including the past.

If we solved Equation 2.3, we could estimate the parameters from data. The parameters a and c refer to the missing variable while p depends on their interaction and m refers to the herbivore. We could have carried out the reverse process, solving Equation 2.1 for x as a function of R. Then we would have a second order equation in R that would reveal parameters of R and also of x. Knowing the feeding rate p and the mortality m of the herbivore might help in its identification.

2.2 A biological definition of environment

The characterization of "environment" as a multidimensional pattern of influences on development, behavior, and survival is a precondition for modeling the effects of environmental change on ecological and genetic ensembles. Attempts to anticipate the effects of climate change have concentrated on the meteorological side of the changes and correlations between weather anomalies and crop yields or disease outbreaks. However, the full richness of the characterization of environment has not been used. A taxonomy of environments would include various factors, discussed below.

2.2.1 Patchiness

Does the organism face a heterogeneous spatial pattern as coarse or fine grained? In a coarse-grained environment the organism lives in one of several alternative patches, or else lives in several of them for long enough so that survival depends on survival in each patch. In a fine-grained environment there is relatively rapid movement among patches, or conditions shift rapidly enough (compared with the metabolic rate of the organism) so that conditions can be averaged. In the case of plant pathosystems such as those associated with the gemini viruses propagated by whiteflies, we have to distinguish different crops as patches that succeed each other in time and show a spatial mosaic. A crop might be a suitable feeding host for the whitefly but not a reproductive host, or it may be a reproductive host, or it may be neither and serve as the space over which the flies have to move to get to one of the other types. In addition, each crop is characterized by whether or not it is susceptible to the virus. Thus we have a pattern of at least six patch types, and for each we have the relation between numbers of infected and noninfected flies that enter and leave.

2.2.2 Variance of life factors

Two environments may be very different for us but very similar in their life table parameters for the vector or pathogen. The crucial question here is how different they are compared with the tolerance of the vectors for the differences, how much an observed difference of X °C means to the organisms.

Here we have to use biologically relevant aspects of physical features and then consider how these may affect the direction and intensity of natural selection. In all discussion of "the environment" it is necessary first to translate physical conditions into ecologically meaningful ones.

2.2.3 Temperature

Consider, for example, "temperature." Temperature can be measured precisely in the field. In a patchy and changing environment, the exact site for measurement has to be specified since distances of centimeters can be associated with temperature differences as great as the average difference between sites several degrees of latitude apart. The smaller the organism, the more important these microsite differences are, but even for large insects and rodents the spatial patchiness cannot be ignored.

Once temperature is measured, it has to be understood biologically in its different significances for different processes:

(a) It determines the rate of development of organisms, and hence the generation time and rate of increase.
(b) Therefore, elevated temperatures produce smaller insects. These in turn are less tolerant of desiccating environments, cannot fly as far, and therefore a deme may fragment into much more local patches.
(c) Smaller body size usually implies lower fertility.
(d) At sufficiently high temperatures, proteins coagulate and death is almost instantaneous. But at somewhat lower temperatures there are thresholds for specific activities. The maximum temperature at which a whole generation can survive and reproduce is not the same as the temperature that allows flight or foraging.
(e) Temperature affects the relative success of different species. For instance, we observed that in direct sunlight on a tropical island, the ant *Brachymyrmex heeri* could collect food at a bait, while the more aggressive *Pheidole megacephala* could not. When a shadow fell on the food supply, *Pheidole* displaced *Brachymyrmex* within 20–30 minutes. If the sunlight returned, *Pheidole* left the bait very soon. Therefore if shadow and sunlight alternate rapidly, say faster than 20 minutes, *Brachymyrmex* got most of the food but if the alternations were on a time scale of hours the food would be more evenly divided. In the open vegetation where we did the study, shadows were cast by the small leaves of lianas and the alternations were rapid, but in a more wooded location a shadow or sun fleck lasted longer, and on cloudy days the alternations depended on the size of clouds and the wind speed.
(f) Temperature changes can break the synchrony between a predator and its prey or a herbivore and its food source. Here even very small differences can have large effects. The development time is given by days = required degree

days/(actual temperature – threshold temperature). When species are close to their threshold temperatures a single degree difference can double or halve development time, while when they are further from threshold the changes are more modest. Thus suppose that a prey species needs to accumulate 50 degree days and has a threshold of 14 °C. At 15 °C it will take 50 days. A parasitoid may require 100 degree days but with a threshold of 13 °C will also emerge in 50 days. Now change the temperature to 16 °C. The first species accumulates two degree days per day and emerges in 25 days while the second accumulates three degree days per day and finishes development in 33 days. Thus synchrony is broken. The same phenomenon occurs when two species use different microhabitats. Suppose that their thresholds are the same but the temperatures they are exposed to differ by 1 °C. A uniform change in the temperature in all habitats will also break synchrony. When parasites invade unfamiliar hosts, the chances of having synchronized development are small even if the invader can succeed in entering the host body. In contrast, unsuccessful parasites might succeed when a climate change accidentally provides just the right temperature for synchronized development.

(g) Temperature as part of climate affects the length of the growing season for plants and also therefore the possibilities for agriculture. Summer temperature can determine the survival of cohorts of pine at the edge of the tundra. On a microscale, temperature can determine whether the dewdrops last long enough on a leaf for fungal spores to germinate and send out hyphae that infect the plant.

The effective environment that a species is exposed to depends also on its behavior. For instance we found that in the presence of the fire ant *Solenopsis geminata*, the more timid *S. globularia* was displaced from the shaded baits into the more exposed sites of bare rock where the fire ant could not follow as readily. Thus in islands where the two species coexisted, *S. globularia* was exposed to higher temperatures and showed higher heat tolerance. Thus weather box temperature is only a rough indicator of the environment experienced by the organism. Not only do organisms select their own environments, but they do so under the influence of conflicting demands. A place that has abundant food may be a dangerous place. In fact it seems to be widely the case that a good place to feed is also a good place to be eaten. Thus benthic organisms such as arrow worms rise to the surface to feed on plankton and then go quickly back to the bottom, birds nest above where they feed, and mice come to the surface to feed and return underground.

2.2.4 Opposing demands

Pathogens face the opposing demands of finding a rich nutritional environment, safety from the body's defenses, and an exit to other hosts. The blood is a rich habitat. It is brimming with useful molecules, but it is also the most dangerous place, where the immune system is most effective. The central nervous system is also rich nutritionally and a safer place to hide, but with no good exit except when the host dies. The skin provides an easier exit but has poor circulation and is therefore not as accessible to the immune system; it is also poorer nutritionally. The intensity of selection on these three opposing demands will depend on the past history of the organism, how much time is available for reproduction, and transmission. Paul Ewald's work has concentrated on examining the opposition between "virulence" (which he identifies with within-host fitness) and transmission. It can also escape from the opposition itself by making the blood a safer place, either by changing the antigenic properties rapidly, as the trypanosomes do, or by destroying the immune system as in AIDS.

Although organisms transform their environments, this does not always make life easier for them. While adaptive responses to selection may ameliorate the environment for the organism, their direct impact (nutrient depletion, waste accumulation, attraction of enemies) can cause its deterioration. The net result is that a species is not necessarily most abundant or its growth most luxurious in habitats where it would be best when alone.

Because of the multidimensional nature of environmental requirements, I propose the hypothesis that *all organisms live in suboptimal environments for themselves.* Furthermore, since each "trait" can affect more than one fitness component, it is possible for the same trait to reflect a compromise among opposing demands or to be selected for only one contribution to survival while other traits are selected for their contributions to the remaining fitness components. This would allow similar organisms to adapt to similar conditions in quite different ways. A response to light intensity or duration might take place through leaf area or number of leaves, as in Susan Sultan's work; insect pests might be met by morphological barriers, phenology, or specific chemical defenses; *Drosophila* species might confront a geographic gradient in temperature through rapid physiological acclimation, permanent morphological differences, genetic differentiation in heat tolerance or behavior. This supports Antonovic's observation that there are often negative correlations among fitness components. It becomes important to know if selection and the direct impact of the environment change the phenotype in the same or opposite directions.

2.2.5 Predictability

All organisms live in uncertain environments. In preliminary work for examining the preparation for the resurgence of old diseases and the emergence of new diseases in human populations, we examined the modes of coping with uncertainty among other species. There are four principal modes and their combinations. The distinctions are not mutually exclusive.

(a) Detection of a relevant environmental change, with response. In order to be effective, the detection has to be accurate and the response rapid enough compared with the duration of the environmental condition.
(b) Prediction. Prediction requires first that there be correlations between some signal and the relevant environmental condition. The signal can be the same environmental factor, in which case it must be autocorrelated, or some other predictor such as day length for later temperature and food supply. There are short-term and long-term predictors, and organisms can even detect the first derivative of the environmental factor. This happens because different physiological conditions change at different rates. Thus in studies of malnutrition, the height for age of children is an integral of food intake during their whole previous life, while weight for height reflects recent food intake (over months or weeks). Therefore we can distinguish all combinations of long-term and recent malnutrition. The statistical structure of the environment makes the prediction possible, but for each organism only some "filters" of environmental history are available. The evolution of predictive processes would seem to be an important target for research. The response then depends on the available repertoire of physiological and behavioral changes and the coupling of the signal to the appropriate response. The study of such couplings would be analogous to conditioning experiments in psychology. Since responses are likely to involve packages of processes that are tightly bound to each other but only loosely connected to the signals, we should be able to explore the versatility of prediction/response systems in different groups and the resolving power of the genome to detect environmental differences. One of many failed experiments of my student days (for dramatic technical reasons) was an attempt to select *Drosophila* for large size at 16 °C and small size at 20 °C and the opposite, large size at 20 °C and small size at 16 °C.
(c) Horizontal resistance. Plant breeders distinguish vertical from horizontal resistance of crops to disease. Vertical resistance is usually mediated by single genes, confers complete protection only against a specific genotype of pathogen, and is of short duration before the pathogen adapts. Horizontal resistance is polygenic, confers resistance to a broad spectrum of pathogens, and resistance is incomplete and usually long lasting. In single species, a combination of phenology or behavior, morphology, and chemical defenses contribute to pathogen resistance. At a population level, factors which slow down the feeding by vectors, increase the incubation period, reduce the sensitive period of the crop to damage, or reduce vector reproduction, contribute to a generalized horizontal crop resistance. Elements such as biodiversity usually offer horizontal resistance to ecosystems.
(d) Prevention. Here the organism reaches out into the environment to avoid confronting the stressors.
(e) Mixed strategy. Since no defense works all the time, most species meet uncertainty through a mixture of different strategies. Which ones actually are used depends on the past history and evolutionary repertoire of the species.

2.3 Selection in variable environments

Environmental experience is transmissible across generations in a number of ways: as cytoplasmic conditions, nutrients, parental care, and behavior. The formalization of this relation for evolution and ecology is difficult. While we have equations that could in principle represent changing gene frequencies, once we allow that the conditions of selection are in part determined by the organism we can only suggest plausible models or derive particular models empirically for special circumstances. But qualitative modeling can capture some essential features. Here we ask the single question, can selection track a varying environment?

First we return to the model of Equations 2.1 and 2.2 in which a resource R is consumed by a herbivore. But now consider two variants of the herbivore (competing species, haploid genotypes for microbial herbivores, or with an adjustment of the constants, selection without dominance). Then for the resource R and consumers x and y we have

$$dR/dt = a - R(p_1 x + p_2 y + c)$$

where p_1 and p_2 are the feeding rates of x and y, and for the consumers

$$dx/dt = x(p_1 R - m_1) \tag{2.3}$$

and

$$dy/dt = y(p_2 R - m_2)$$

for mortalities m_1 and m_2. We can track the frequency z of x,

$$z = x/(x + y)$$

so that

$$dz/dt = z(1 - z)[(p_1 - p_2)R - (m_1 - m_2)]. \tag{2.4}$$

Note that if $p_1 > p_2$ but $m_1 > m_2$ then z will increase when R is sufficiently abundant but decrease when R is low.

Now the resource equation becomes

$$dR/dt = a - R[Np_2 + c + (p_1 - p_2)Nz]$$

where N is the total population size, $x + y$. If N varies with changing proportions of x and y there can be no polymorphism. Equations 2.3 and 2.4 can only be solved when x or y is zero. The variant for which $R^* = m_i/p_i$ is smaller completely displaces the other. For the convenience of this model, however, we fix N, perhaps by fixing the number of available nesting sites. Then the persistence of an intermediate value of z would be shown if z increases near 0 and decreases near 1. Near $z = 0$, R equilibrates at $R = a/(p_2 N + c)$ and

$$dz/dt = z(1 - z)[(p_1 - p_2)a/(p_2 N + c) - (m_1 - m_2)]$$

so that z would increase if

$$a/(p_2N+c) > (m_1-m_2)/(p_1-p_2).$$

Similarly, near $z=1$, z decreases if

$$a/(p_1N+c) < (m_1-m_2)/(p_1-p_2)$$

or finally

$$p_1N > a(p_1-p_2)/(m_1-m_2) > p_2N.$$

What would happen then is that when R is high enough selection would favor the better feeder even though it has higher mortality, while when R is sufficiently small selection would favor the variant with lower mortality even though it is a poorer feeder. Nonetheless, the response to selection does not improve fitness. If the system is being driven from below (that is, if either of the parameters a or c vary with the external environment), then Equation 2.4 can be averaged as follows. Divide the equation by $1-z$. Then

$$(dz/dt)/(1-z) = z[(p_1-p_2)R-(m_1-m_2)]. \tag{2.5}$$

The left-hand side is the derivative of $-\ln(1-z)$ so that if z is bounded away from 0 and 1 it has an expected value of zero. The right side has the expected value

$$E(z)E[(p_1-p_2)R-(m_1-m_2)]+(p_1-p_2)\text{Cov}(z, R)=0.$$

If we now divide Equation 2.5 by z and again average, we get

$$E[(p_1-p_2)R-(m_1-m_2)]=0$$

so that $\text{Cov}(z, R)=0$ and we find there is no correlation between z and R. If the system is driven from above, with variation in m_1 or m_2, then there is a negative correlation between R and z, so that responding to selection results in higher frequencies of the good feeder with high mortality when food is scarce. Responding to selection here is deleterious. Each allele is most abundant when conditions are most unfavorable for it.

Finally, for completeness we show one example where the tracking of the environment might be successful. Suppose that at any one time there is an optimal gene frequency that depends on the environment in such a way that

$$dz/dt = z(1-z)[s_1+s_2-z]$$

where s_1 is responsive to z while s_2 is either constant or varies due to exogenous factors. If environmental variation enters the system by way of an equation for s_1, then there will be a positive correlation between z and s_1, between the actual and optimal gene frequencies. But if variation enters the system through changes in a or c then the correlation depends on whether s is increased or decreased by higher values of z (that is, if the feedback between z and s is positive or negative) and whether s is self-damped. If the feedback is negative, say

$$ds/dt = a - s(z + c),$$

then the correlation will be negative and tracking the environment will be harmful, whereas with a positive feedback such as

$$ds/dt = az - cs$$

the correlation will be positive and tracking successful.

The object of the exercise is to show that natural selection does not necessarily track a changing environment. Whether it does or not depends on the feedback between gene frequency and the environment. And even if it tracks the environment this does not necessarily improve fitness since the maintenance of genetic variance itself detracts from the population ever being all of some optimal phenotype for any environment.

2.3.1 Selection of complex physiological systems

The physiological responses of organisms to fluctuating conditions constitute the homeostatic system. They can be visualized as a metabolic network in which the genes are expressed as kinetic constants of enzymatic reactions. The structure and dynamics of this network provide the linkage between the organismic physiology and population genetics by identifying the direction and magnitude of the response to selection. Then it becomes obvious how the intensity of selection varies depending on the rest of the network, bearing in mind that most molecules have multiple effects on vital processes, that control in metabolic networks is diffuse, with positive and negative feedbacks amplifying or buffering the impacts of parameter (genetic) change according to circumstance, and that genes interact at the fitness level according to how the processes they influence are sequential, parallel, competitive, supplementary or antagonistic. Therefore a given environmental change percolates through the network, being amplified along some pathways, buffered along others, and even inverted along some pathways according to the layout of the feedbacks. In general, where we have studied a process in sufficient detail such as the regulation of glucose in humans, we find diffuse control, a many-to-many relationship in which each relevant trait is controlled by many molecules, but each molecule has multiple effects. Since any one of them may have a dominant influence at any one time, we can see why it is that so many studies reach conclusions about the very strong effect of the molecule studied on some medical outcome, until heart disease, cancer, diabetes, and other pathologies are "explained" many times over and yet leave much of the variation still unaccounted for. This also helps to explain why some traits seem to be stable in evolution while others are labile and vary readily from deme to deme or in time. An evolutionary explanation of a trait must consider not only why it is advantageous but also how come it is responsive when other equally advantageous traits do not seem to respond to the new circumstances, and why the change occurred in some species and not others.

Much of the interest in genome/environment interaction has focused on development. But the most active interface between the organism and the environment it constructs is the domain that Schank and Wimsatt (2001) designate as peripheral, the domain of shifting concentrations of small molecules. There are a large number of molecules that play quite diverse roles in different contexts within the organism. Glutamine or its derivatives appear as part of protein structure, as a precursor for glutathione in the detoxification pathways of the liver, among the antioxidants that protect the tissues and enzymes against free radicals, as a precursor of the inhibitory neurotransmitter GABA, and as glutamic acid as an excitatory neurotransmitter. It can also be used as an energy source in the citric acid cycle. Tryptophan is a precursor of serotonin but also of niacin. Homocysteine seems to be involved as an antioxidant in the arteries and an inhibitor of collagen formation, and is associated with increased risk of cardiovascular disease, cancer, and general mortality. Estrogens show up not only in the reproductive processes but also in bone maintenance, cardiac health, and psychological states.

Thus the semi-independence that seems to apply at earlier levels in the pathways from genes to "phenotype" is not apparent here. The concentration of any molecule depends on its rate of synthesis and its rate of breakdown. Specific enzymes are involved in these processes but so are nutrients and molecules that may interact with the molecules in question. The available antioxidants are reduced by free radicals, and so on. The complex network of intermediate metabolism guarantees that the effects on fitness of any gene that participates in the transformation of these metabolites will depend on the levels of other metabolites, which in turn depend on external conditions and many genes. For instance, phase I detoxification by the 50 or so P-450 enzymes in the liver removes potential toxins but produces intermediates that are themselves toxic and are removed by phase II pathways. Therefore while these enzymes are necessary, if they are too powerful compared with the phase II pathways they can become lethal. This supports the idea that much selection is toward some optimal value which itself varies.

Consider a network of interacting genetic loci and suppose that there is no frequency-dependent selection, inbreeding or linkage. Then Sewall Wright's (1931) equation for selection on gene frequency p_i is

$$dp_i/dt = p_i(1 - p_i)\partial W/\partial p_i$$

where the adaptive value W may be a function of all the gene frequencies and will depend on environmental conditions. At equilibrium, when $\partial W/\partial p_i = 0$, the matrix of first partials determines local stability and resistance to parameter change. Each element of the matrix

$$a_{ij} = p_i(1 - p_i)\partial^2 W/\partial p_i \partial p_j + (1 - 2p_i)\partial W/\partial p_j. \quad (2.6)$$

The second term is zero at equilibrium. The first term reflects epistatic interactions or if $i = j$, it reflects dominance and heterosis. Note therefore that $a_{ij} = a_{ji}$, i.e., the matrix is symmetrical. The diagonal elements $a_{ii} = 0$ if fitness is a linear function of gene frequency: that is, without dominance or heterosis. Since W is more likely to be at a peak than a trough in the fitness landscape, the second derivative is negative. (Mutation away from the allele in question adds another negative term.) The sharper the peak of W the more negative is the second derivative. That is, if fitness is very sensitive to the frequency of this gene then the diagonal term will be strongly negative. A large value of $p_i(1 - p_i)$ (that is, a high degree of polymorphism) also increases this negative effect.

Since bacteria are all haploid, dominance does not enter and the a_{ii} would be zero. But bacteria interact in ways that can be frequency dependent. Suppose for instance that a mutation blocks the synthesis of some necessary nutrient. If such a mutant is uncommon in a population it may be able to absorb the nutrient that leaks from the majority of cells that do synthesize it, while being spared the energy and material expenditure of doing the synthesis itself. Therefore it will have an advantage when rare that decreases as it becomes more common. This kind of frequency dependence makes the diagonal element of the matrix $a_{ii} < 0$.

Where there is dominance, frequency dependence or other effects, the first term in Equation 2.6 is different from zero. Then the factor $p(1 - p)$ enters: diagonal elements are small for genes near fixation and more important as the equilibrium contributes more heterogeneity.

The matrix a_{ij} can be represented by a signed digraph in which the vertices are gene frequencies, and the sides are interactions. This allows us to examine local stability properties from the matrix and also the effect of a change of any parameter on the ensemble of equilibrium gene frequencies.

The question we ask is, how will a change in some parameter that appears in one of the equations for a gene frequency change the frequencies of all the genes in the network? That is, we look for $\partial p_i/\partial C$ where C is a parameter that appears in the equation for one or more gene frequencies.

The formal result is

$$\partial p_i/\partial C = \sum (\partial f_j/\partial C)\ P_{ij} F\,(\text{complement of the path})/F\,(\text{whole network})$$

where: the numerator is the sum of the initial change in the equations where the parameter C enters, times the sum of the products of the a_{ij} along each path from variable j to variable i, times the gain or feedback of the complementary subsystem, that is $(-1)^k$ times the determinant of order k of that part of the network which is not on the path. The numerator can therefore be different for each path and for each parameter. The denominator is the

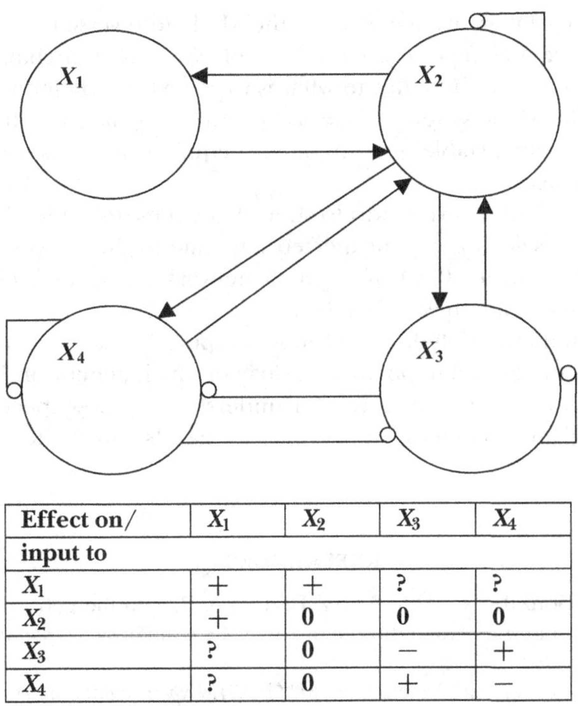

Effect on/ input to	X_1	X_2	X_3	X_4
X_1	+	+	?	?
X_2	+	0	0	0
X_3	?	0	−	+
X_4	?	0	+	−

Figure 2.2. A hypothetical gene frequency interaction network and its response to parameter change.

feedback of the whole network, $(-1)^n$ times the determinant of the whole network. Therefore this term is common to all $\partial p_i/\partial C$ and is a measure of resistance to selection of the whole population. We suggest that this term is different for different species and as a measure of resistance would be a useful object of investigation.

Figure 2.2 shows a hypothetical network of interacting gene frequencies. Suppose that an environmental parameter changes. Changes that enter the system through a parameter of X_1 affect all gene frequencies but are ambiguous without some quantification since the short path $X_1 \to X_2 \to X_3$ is multiplied by the feedback of X_4 and is therefore positive while the longer path $X_1 \to X_2 \to X_4 \to X_3$ is negative. Similarly for the two paths from X_1 to X_4. The direct path between X_3 and X_4 or from X_4 to X_3 has as its complement only the positive loop X_1, X_2 since X_1 is not self-damped because there is no dominance or frequency dependence. Therefore a parameter change arising in X_2 will have no effect on the frequencies of X_2, X_3, or X_4. In each case the path has in its complement the zero feedback of X_1. Only X_1 will change. Similarly, parameter changes arising in X_2 or X_3 will not change X_2. Pathways between X_3 and X_4 have either zero complement if they pass through X_2 or else have the positive complement from the positive loop (X_1, X_2) which is then divided by a negative feedback of the whole to reverse the effect of the path. Therefore a change in a parameter of X_3 or X_4 will change each of these in the opposite direction to what is expected by common sense. The feedback of the whole system is $F_4 = a_{12}a_{21}(a_{44}a_{33} - a_{34}a_{43})$ which we assume to be negative for a stable gene frequency equilibrium. It is a measure of general resistance.

The object of this exercise was to show that genes imbedded in networks can respond to selection in unexpected ways due to the network structure, and to introduce the feedback of the whole network as a measure of resistance that could become an object of study.

All of the simple models discussed above support the claim that it is possible to develop an integrated population biology in which genetic and ecological concerns are merged into a search for understanding how species respond to fluctuating and long-term environmental trends which they themselves influence.

REFERENCES

Awerbuch, T., Gonzàlez, C., Hernàndez, D., Levins, R., Sandberg, S., Sibat, R., and Tapia, J. L. (in press). The natural control of the scale insect *Lepidosaphes gloverii* on Cuban citrus.

Clausen, J. D., Keck, D., and Heisey, W. M. (1940). *Experimental Studies on the Nature of Species*, I. Washington DC: Carnegie Institute Wash. Pbl. 520.

Dobzhansky, T. (1951). *Genetics and the Origin of Species*, 3rd edition, New York: Columbia University Press.

Goldschmidt, R. B. (1938). *Physiological Genetics.* New York: McGraw-Hill.

Lerner, M. (1954). *Genetic Homeostasis.* Edinburgh: Oliver & Boyd.

Levene, H. (1953). Genetic equilibrium when more than one niche is available. *Am. Nat.* 87:331–3.

Levins, R. (1998). Qualitative methods for understanding, prediction and intervention in complex systems. In D. Rapport *et al.* (eds) *Ecosystem Health: Principles and Practice.* Malden, MA: Blackwell Science.

Lewontin, R., and Levins, R. (2000). Schmalhausen's Law. *Capitalism, Nature, Socialism* 11(4):103–8.

Li, C. C. (1955). The stability of an equilibrium and the average fitness of a population. *Am. Nat.* 89:281–96.

Lozano, R., Zurita, B., Franco, F., Ramírez, T., Hernández, P., and Torres, J. L. (2001). Mexico: marginality, need, and resource allocation at the county level. In T. Evans, M. Whitehead, F. Diderichsen, and A. Bhuiya (eds) *Challenging Inequities in Health: From Ethics to Action*, chapter 19. Oxford: Oxford University Press.

Ludwig, W. (1950). Zur Theorie der Konkurrenz. Klatt-Festschrift. *Neue Ergeb. Probleme Zool.* pp. 516–37.

Mayr, E. (1942). *Systematics and the Origin of Species.* New York: Columbia University Press.

Mohtashemi, M., and Levins, R. (2002). Qualitative analysis of all-cause black–white mortality crossover. *Bull. Math. Biol.* 64:147–73.

Moore, D. (2002). *The Dependent Gene: the Fallacy of Nature vs. Nurture.* New York: W. H. Freeman.

Rapport, D. *et al.* (eds). *Ecosystem Health: Principles and Practice.* Malden, MA: Blackwell Science.

Schank, J. C., and Wimsatt, W. C. (2002). Evolvability: adaptation and modularity. In R. S. Singh, C. B. Krimbas, D. B. Paul, and J. Beatty (eds) *Thinking about Evolution: Historical, Philosophical, and Political Perspectives,* pp. 322–35. Cambridge: Cambridge University Press.

Schmalhausen, I. I. (1949). *Factors of Evolution.* Philadelphia: Blakiston.

Waddington, C. H. (1957). *The Strategy of the Genes.* New York: Macmillan.

Wright, S. (1931). Evolution in Mendelian populations. *Genetics* 16:97–159.

PART II

GENOTYPES TO PHENOTYPES: NEW GENETIC AND BIOINFORMATIC ADVANCES

3

Genetic dissection of quantitative traits

TRUDY F. C. MACKAY
Department of Genetics, North Carolina State University, Raleigh

It is clear that descriptions of the genetic variation in populations are the fundamental observations on which evolutionary theory depends. Such observations must be both dynamically and empirically sufficient if they are to provide the basis for evolutionary explanations and predictions.... (W)e see that a sufficient description of variation is necessarily a description of the statistical distribution of genotypes in a population, together with the phenotypic manifestation of those genotypes over the range of environments encountered by the population. The description must be genotypic because the underlying dynamical theory of evolution is based on Mendelian genetics. But the description must also specify the relations between genotype and phenotype, partly because it is the phenotype that determines the breeding system and the action of natural selection, but also because it is the evolution of the phenotype that interests us. Population geneticists, in their enthusiasm to deal with the changes in genotype frequencies that underlie evolutionary changes, have often forgotten that what are ultimately to be explained are the myriad and subtle changes in size, shape, behavior, and interactions with other species that constitute the real stuff of evolution.

(Lewontin 1974, p. 19)

Evolutionarily significant genetic variation is then, almost by definition, variation that is manifest in subtle differences between individuals, often so subtle as to be completely overwhelmed by effects of other genes or of the environment... We see here the fundamental contradiction inherent in the study of the genetics of evolution. On the one hand the Mendelian genetic system dictates the frequencies of genotypes as the appropriate genetic description of a population. The enumeration of these genotypes requires that the effect of an allelic substitution be so large as to make possible the unambiguous assignment of individuals to genotypes. On the other hand, the substance of evolutionary change at the phenotypic level is precisely in those characters for which individual gene substitutions make only slight differences as compared with variation produced by the genetic background and the environment. What we can measure

The Evolution of Population Biology, ed. R. S. Singh and M. K. Uyenoyama. Published by Cambridge University Press.

is by definition uninteresting and what we are interested in is by definition unmeasurable.

(Lewontin 1974, pp. 22–3)

3.1 Introduction

What an irresistible challenge! Unfortunately, responses to the challenge were largely polarized along the very Mendelist ("genotype up") vs. Biometrician ("phenotype down") approaches that need to be bridged if we are to understand the genetic basis of adaptation. In the 30 years since the publication of Lewontin's classic *The Genetic Basis of Evolutionary Change* (1974), we have seen great advances in our understanding of the patterning of molecular genetic variation within and between species, and concomitant development of theory with which to test departures of observed deposition of molecular variation from that predicted from a balance between neutral mutation and drift in finite populations. However, "phenotype" in these studies is redefined as the amino acid sequence of proteins, enabling the unambiguous enumeration of alleles at different loci at the level of individuals, but bypassing the fundamental issue of the connection between genotype and organismal phenotype for traits that are "the real stuff of evolution." Nevertheless, the same 30 years brought an explosion of studies applying quantitative genetic methods to a wide variety of phenotypes of evolutionary importance – but "genotype" in these studies is at the level of genetic variance components, not allele frequencies at loci affecting variation in the trait. Thus, the need to integrate the disciplines traditionally known as population and quantitative genetics remains as pressing today as it has in the past.

Why has progress towards this integration been so sluggish? This is apparent when we examine what we need to know before understanding the genetic basis of any adaptive trait (Robertson 1967, Mackay 2001):

1. the numbers and identities of all genes (quantitative trait loci, or QTL) in the developmental, physiological and/or biochemical pathway leading to the trait phenotype;
2. the mutation rates at these loci;
3. the numbers and identities of the subset of QTL that are responsible for variation in the trait within populations, between populations and between species;
4. the homozygous and heterozygous effects of new mutations and segregating alleles on the trait;
5. all two-way and higher order epistatic interaction effects;
6. the pleiotropic effects on other quantitative traits, most importantly reproductive fitness;
7. the extent to which additive, dominance, epistatic, and pleiotropic effects vary between the sexes, and in a range of ecologically relevant environments;

8. the molecular polymorphism(s) that functionally define(s) QTL alleles (the QTN, or quantitative trait nucleotides);
9. the molecular mechanism causing the variation in trait phenotype;
10. QTL allele frequencies.

Until recently, we have been stymied by steps 1 and 3 – the necessity to understand quantitative traits at the level of individual QTL. Once the genes affecting variation in adaptative traits are identified, application of molecular population genetics methods is straightforward. Here I shall review recent progress that has been made towards these goals, problems to be overcome, and future prospects.

3.2 Identifying genes affecting quantitative traits

3.2.1 Mutagenesis

The first task on our list of requirements – to identify all genes that could conceivably affect the trait(s) in which we are interested – is traditionally addressed by mutagenesis. However, most mutation screens to date have concentrated on unraveling the genetic basis of fundamental cellular and developmental processes, and mutations disrupting such processes are often recessive lethal or have very large phenotypic effects. Annotation of completed eukaryotic genome sequences (Goffeau *et al.* 1996, The *C. elegans* Sequencing Consortium 1998, Adams *et al.* 2000, The Arabidopsis Genome Initiative 2000, International Human Genome Sequencing Consortium 2001, Venter *et al.* 2001) have revealed how few loci have been characterized genetically (i.e., by analysis of mutations) even in model organisms, and how many predicted genes exist with unknown functions.

Could it be that part of the explanation for this discrepancy lies in the bias of mutagenesis screens towards detecting mutations with large qualitative effects that completely destroy the gene product (null alleles)? On the one hand, there may be loci for which the mutational spectrum is constrained to alleles with subtle, quantitative effects, as might be expected for genes in a pathway exhibiting functional redundancy. Support for the existence of such a class of genes comes from an experiment (Winzeler *et al.* 1999), in which 6925 yeast strains were constructed, each containing a precise deletion of one of 2026 open reading frames (about a third of the yeast genome). Only 17% of the deleted open reading frames were essential for viability under standard conditions, and 40% showed quantitative growth defects. On the other hand, many of the loci identified and characterized on the basis of null mutations may have pleiotropic effects on quantitative traits that can only be perceived if the trait is assessed on heterozygotes or on mutant alleles that more subtly impair the functioning of the gene (hypomorphic alleles). Indeed, one of the earliest and surprising results of the genetic analysis of artificial selection lines in *Drosophila* was that selection response (and limits to selection response) were often attributable to homozygous lethal genes with quantitative heterozygous effects on the selected trait (Clayton and Robertson 1957, Frankham *et al.* 1968).

These considerations motivate the need to examine subtle, quantitative effects of mutations on the morphological, physiological, and behavioral traits in which we are interested. Such experiments are more difficult to execute than screens for mutations with effects that can be scored unambiguously on single individuals. It is necessary to perform the quantitative mutation screen in a homozygous genetic background, since the magnitude of the effects we wish to detect are of the same order as naturally segregating variation. Further, mutations with quantitative effects that are sensitive to environmental variation can only be detected if multiple individuals bearing the same mutation are evaluated for the trait phenotype. This requires that stocks are established for each mutation prior to phenotypic assessment. Mobilizing transposons to generate insertional mutations in an inbred background satisfies these criteria. Further, the transposons act as molecular tags, enabling the cloning of the affected QTL (Doebley *et al.* 1997, zur Lage *et al.* 1997, Lai *et al.* 1998).

Only a few direct screens for mutations with quantitative phenotypic effects have been performed to date. Genetic variation for mouse body weight was significantly increased in lines harboring multiple retrovirus insertions relative to their co-isogenic control lines (Keightley *et al.* 1993), but the genes responsible were not identified. In *Drosophila*, highly significant quantitative mutational variation was found for activities of enzymes involved in intermediary metabolism (Clark *et al.* 1995), sensory bristle number (Lyman *et al.* 1996) and olfactory behavior (Anholt *et al.* 1996) among single *P* transposable element insertions that were co-isogenic in a common inbred background. Insertion sites were not determined in the enzyme activity study, but statistical arguments suggested that the insertions were highly unlikely to be in enzyme-coding loci (Clark *et al.* 1995). Of the 50 insert lines with significant effects on bristle number, nine were hypomorphic mutations at loci known to affect nervous system development while the remaining 41 inserts did not map to cytogenetic regions containing loci with previously described effects on adult bristle number (Lyman *et al.* 1996). Most of the mutational variance for olfactory behavior was attributable to *P* element inserts in 14 novel *smell-impaired* (*smi*) loci (Anholt *et al.* 1996). Several of the *smi* loci disrupted by the *P* element insertions are predicted genes with hitherto unknown olfactory functions, as well as genes with more obvious roles in olfactory behavior (Anholt and Mackay 2001).

Screening for quantitative effects of induced mutations is a highly efficient method both for discovering new loci affecting quantitative traits and determining novel pleiotropic effects of known loci on these traits. There is clearly a need for more experiments of this sort; a view that has been explicitly advocated for genetic dissection of quantitative traits in the mouse

(Nadeau and Frankel 2000). Projects to induce targeted disruptions of all known and predicted genes in several model organisms (Spradling *et al.* 1999, Winzeler *et al.* 1999, DeAngelis *et al.* 2000, Fraser *et al.* 2000, Gönczy *et al.* 2000, Nolan *et al.* 2000) will provide valuable resources for these studies, although segregating genetic backgrounds will in many cases limit the power to identify mutations with small effects. Knowledge of mutations with subtle phenotypic effects will also help prioritize for further study candidate loci in regions to which naturally segregating QTL map (see below).

3.2.2 QTL mapping

Mutagenesis is essential to identify loci and pathways important for the normal expression of the adaptive phenotype of interest. However, most questions in evolutionary genetics are focused on the subset of these loci affecting *variation* of the phenotype within and between natural populations and species. This requires that we identify the QTL at which naturally occurring variation for the trait segregates.

Since the effects of individual QTL are too small to be tracked by segregation in pedigrees, QTL must be mapped by linkage disequilibrium (LD) with polymorphic markers that do exhibit Mendelian segregation. If a QTL is linked to a marker locus, there will be a difference in mean values of the quantitative trait among individuals with different genotypes at the marker locus. This simple principle was recognized nearly 80 years ago (Sax 1923); why, then, does the genetic basis of most quantitative traits remain obscure? The answer is that, until relatively recently, dense maps of polymorphic markers that did not affect the phenotype were simply not available in any species. The discovery of abundant molecular polymorphism and development of rapid methods for determining molecular marker genotypes, as well as the development of sophisticated statistical methods for localizing QTL relative to these markers, have opened the floodgates for mapping naturally segregating QTL (Mackay 2001).

LD between QTL and marker alleles can be generated experimentally by crossing lines with divergent gene frequencies at the QTL and markers, and also occurs in naturally outbreeding populations within families or extended pedigrees that are segregating for the QTL genotypes. Traditionally, *linkage* mapping of QTL relies on the LD between markers and trait values that occurs within mapping populations or families. However, LD also occurs in outbred natural populations in which QTL alleles are in drift-recombination equilibrium (Hill and Robertson 1968, Weir 1996), in populations resulting from an admixture between two populations with different gene frequencies for the marker and mean values of the trait (Falconer and Mackay 1996, Weir 1996, Hartl and Clark 1997), and in populations which have not reached drift-mutation equilibrium due to a recent mutation at the trait locus, or a recent founder event followed by population expansion (Weir 1996, Lynch and Walsh 1998). *Association* mapping of QTL relies on marker–trait linkage disequilibria in such populations.

The precision of mapping QTL depends on the scale of LD between the QTL and marker alleles, which in turn depends on the number of recombinations that have occurred between markers and trait loci. The extent of LD (in cM) is inversely proportional to the number of recombinations; therefore, QTL mapping is most precise where recombination is high. In principle, QTL could be mapped to the level of genetic locus in a single study using either linkage or association analyses, provided the numbers of individuals or families sampled and the density of polymorphic markers were both sufficiently large. However, such large-scale strategies are currently impractical, given the cost and time necessary to genotype markers on large numbers (say, of the order of 10^4) of individuals, although they may not be in the future. It is currently most cost-effective, in terms of marker genotyping, to adopt an iterative strategy to map QTL. An initial low-resolution genome scan to identify genomic regions containing QTL is performed using linkage analysis, followed by higher resolution confirmation studies of detected QTL, and culminating with recombination or association mapping to identify genes corresponding to the QTL.

3.2.2.1 Genome scans

Linkage mapping of QTL in organisms amenable to inbreeding begins by choosing parental inbred strains that are genetically variable for the trait of interest. A mapping population is then derived by back-crossing the F_1 progeny to one or both parents, mating the F_1 *inter se* to create an F_2 population, or constructing recombinant inbred lines (RIL) by breeding F_2 sublines to homozygosity. These methods are very efficient for detecting marker–trait associations, since crosses between inbred lines generate maximum LD between QTL and marker alleles, and ensure that only two QTL alleles segregate, with known linkage phase. Outbred populations pose additional challenges for QTL mapping, since one is restricted to obtaining information from existing families. Only parents that are heterozygous for both markers and linked QTL provide linkage information, and individuals may differ in QTL–marker linkage phase. Further, not all families will be segregating for the same QTL affecting the trait, and in the presence of genetic heterogeneity different families may show different associations.

Having generated (or collected) a mapping population and determined phenotypes for the trait and genotypes for the markers on all individuals in this population, all that remains is to determine whether QTL are in LD with the markers, and to estimate the QTL map positions and effects using an appropriate statistical method. There is no shortage of the latter from which to choose (reviewed by Lynch and Walsh 1998, Mackay 2001). Interval mapping methods, in which the QTL position is localized between flanking markers, are preferred over single marker analyses, since QTL map position and effect are not confounded with the former method, but are with the latter. Methods that take segregating markers affecting the trait into account at the same time as evaluating effects of a specific interval greatly improve the power and precision of mapping (e.g., Jansen and Stam 1994, Zeng 1994).

Two important statistical considerations transcend the experimental design and method of statistical analysis used to map QTL: power and significance threshold. Large sample sizes are required to reliably detect QTL (Falconer and Mackay 1996, Lynch and Walsh 1998). If power is low, not all QTL will be detected, leading to poor repeatability of results, and the effects of those that are detected can be overestimated (Beavis 1994). Second, genome scans entail multiple tests for marker–trait associations. To maintain the conventional experiment-wise significance level of 0.05, one needs to set a more stringent significance threshold for each test performed. Permutation (Churchill and Doerge 1994, Doerge and Churchill 1996) or other resampling methods (Zeng *et al.* 1999) are widely accepted as providing appropriate significance thresholds.

Genome scans for QTL have proliferated (reviewed by Lynch and Walsh 1998, Mackay 2001) since the first report of QTL localizations using a high-resolution molecular marker map (Paterson *et al.* 1988). A comprehensive review is not possible here due to space constraints. Suffice it to say that QTL have been mapped for traits of agronomic importance in crops and domestic animals, susceptibility to common human diseases, medically important and behavioral traits in model organisms, model quantitative traits in *Drosophila*, and morphological traits accompanying species divergence. It is important to recognize that the numbers of QTL mapped in these studies are in all cases minimum estimates of the total number of loci that potentially contribute to variation in the traits (Mackay 2001), for two reasons. First, larger sample sizes enable mapping QTL with smaller effects, and the increased number of recombinations enable the separation of linked QTL. Second, one can only map QTL at which different alleles are fixed in the two parent strains, which are a limited sample of the existing genetic variation.

Some confusion has occasionally arisen in the literature when QTL mapped by genome scans are referred to as "genes." This is, of course, not true. QTL mapping pinpoints the chromosomal regions containing one or more loci affecting the trait, and is a major step forward since it limits our search for the genes affecting our traits to particular regions of the genome. However, there is still much work to be done after the initial QTL mapping to narrow the contribution to the observed variation down to single genes. In *Drosophila*, the average size of intervals containing significant QTL from recent studies was 8.9 cM and 4459 kb, with a range of 0.1–44.7 cM and 98–19 284 kb. An "average" *Drosophila* gene is 8.8 kb. Therefore, an "average" QTL in these studies encompasses 507 genes, with a range from 11 to 2191 genes (Mackay 2001).

The large number of genes in intervals to which QTL map, limited genetic inferences that can be drawn from analysis of most mapping populations, and poor repeatability of many initial QTL mapping efforts that had low power due to small samples, has engendered some pessimism about the value of this approach (Nadeau and Frankel 2000). However, understanding the genetic basis of naturally occurring variation is too important a problem to ignore the challenge of mapping QTL to the level of genetic locus.

3.2.2.2 High-resolution mapping

The challenge posed for high-resolution QTL mapping is that individual QTL are expected to have small effects that are sensitive to the environment, and therefore the phenotype of a single individual is not a reliable indicator of the QTL genotype. High-resolution QTL mapping is thus contingent on increased recombination in the QTL interval, and accurately determining the QTL genotype (Darvasi 1998). Hunting for informative recombinants in the QTL interval is a straightforward exercise, needing only larger samples and/or more advanced generations. In organisms amenable to genetic manipulation, the effects of individual QTL can be magnified by constructing strains that are genetically identical except for a defined region surrounding the QTL. These strains are a permanent genetic resource from which multiple measurements of the trait phenotype can be obtained to increase the accuracy of determining the QTL genotype. The region surrounding the QTL can be a whole chromosome (chromosome substitution lines) or a smaller interval (variously called introgression lines, interval-specific congenic strains, and near-isoallelic lines (NIL)).

If constructing introgression lines is necessary for high-resolution mapping, why not dispense with the traditional low-resolution genome scan and develop methods that both map QTL and produce the first generation introgression lines simultaneously? In *Drosophila*, the availability of balancer chromosomes enables the rapid construction of chromosome substitution lines. This technique has been adopted to substitute single homozygous chromosomes from a high scoring strain into the homozygous genetic background of a low scoring strain, and to subsequently map QTL one chromosome at a time (Shrimpton and Robertson 1988, Long *et al.* 1995, Gurganus *et al.* 1999, Weber *et al.* 1999). Chromosome substitution strains have been constructed in mice (Nadeau *et al.* 2000), enabling chromosomal localization of QTL by screening just 20 lines. Populations of introgression lines, derived by selection on marker loci, have been used for mapping QTL for tomato fruit traits (Eshed and Zamir 1995) and morphological traits associated with divergence between two *Drosophila* species (Laurie *et al.* 1997). Alternatively, introgression can also be accomplished by selecting on the quantitative trait, rather than markers, to identify QTL of large effect while producing NILs for further analysis (Hill 1998). When combined with progeny testing of recombinant genotypes within intervals (Thoday 1961, Wolstenholme and Thoday 1963, Spickett and Thoday 1966, Shrimpton and Robertson 1988, Eshed and Zamir 1995), linked QTL with small effects within each interval can be individually identified.

Several examples of fine-scale recombination mapping of QTL include: (1) *Idd3*, a susceptibility allele for type 1 diabetes mellitus in the mouse, maps to a 145 kb interval containing a single known gene, *Il2* (Lyons *et al.* 2000). (2) *NIDDM1*, a susceptibility gene for human type 2 diabetes mellitus, was mapped by linkage analysis to a 5 cM region containing 7 known genes and 15 ESTs (Horigawa *et al.* 2000). (3) The tomato fruit weight QTL, *fw2.2*, was localized to a 1.6 cM interval containing four unique transcripts (Frary *et al.* 2000). (4) A tomato fruit-specific apoplastic invertase gene *Lin5* has been shown unambiguously to correspond to the *Brix9-2-5* QTL, which affects fruit glucose and fructose contents, since recombinants in a 484 base pair region within this gene cosegregate with the QTL phenotype (Fridman *et al.* 2000).

Drosophila geneticists have an additional tool for mapping QTL to sub-cM intervals: deficiency chromosomes. Deficiency complementation mapping (Mackay 2001) has been used to resolve QTL for *Drosophila* life span that were initially detected by recombination mapping (Nuzhdin *et al.* 1997) into multiple linked QTL, one of which mapped to a 50 kb interval containing one known and two predicted genes (Pasyukova *et al.* 2000). As for all complementation tests, failure to complement deficiencies could be attributable to epistatic, rather than allelic interactions. Thus, QTL inferred using this method must be subsequently confirmed using a different approach.

3.2.3 From QTL to gene

High-resolution mapping whittles down the genomic region containing our QTL to intervals containing a small number of genes. It is now necessary to determine which of these genes correspond to the QTL. This can be achieved by: (1) demonstrating cosegregation of intragenic recombinant genotypes in a candidate gene with the QTL phenotype; (2) functional complementation by transgenic rescue; (3) genetic complementation; and (4) association mapping. The first two methods constitute clear proof that the QTL corresponds to the candidate gene, while further corroborative evidence is required for the latter two methods. The list of QTL that have been mapped to the gene level is growing.

As noted above, Fridman *et al.* (2000) have clearly shown that the tomato apoplastic invertase gene is a fruit sugar content QTL, by ultra-high-resolution recombination mapping. Two further genes were shown to correspond to QTL by transgenic rescue: *ORFX* in tomato is the *fw2.2* tomato fruit weight QTL (Frary *et al.* 2000), and *Pla2g2a* (encoding a secretory phospholipase) corresponds to the mouse tumor suppressor gene, *Modifier-of-Min1* (Cormier *et al.* 1997). The first study was successful since the apoplastic invertase gene contained a recombination hot spot, and the latter two were facilitated by the large effects of the QTL, and their dominant gene action. In many cases genes corresponding to QTL will not be easy to identify by intragenic recombination and functional complementation. For example, if the effect of the QTL is not large, it will be necessary to construct multiple independent transgenic lines to account for position effects of the transgene on the quantitative trait.

Genetic complementation of QTL alleles to mutations of candidate genes is a rapid method for nominating candidate genes for further study. For example, a QTL explaining most of the difference in inflorescence morphology between maize and teosinte mapped to a region including the maize gene, *teosinte-branched1* (*tb1*) (Doebley and Stec 1991, 1993). A near-isoallelic line (NIL) containing the teosinte QTL in a maize background failed to complement the maize *tb1* mutant allele, suggesting that *tb1* is the QTL (Doebley *et al.* 1995). The maize *tb1* gene was subsequently cloned by transposon tagging, and expression patterns were consistent with the gene corresponding to the QTL (Doebley *et al.* 1997). Quantitative complementation tests showed failure of QTL alleles for *Drosophila* sensory bristle number to complement mutations at candidate genes affecting peripheral nervous system development (Long *et al.* 1996, Mackay and Fry 1996, Gurganus *et al.* 1999, Lyman and Mackay 1998, Lyman *et al.* 1999). Associations of molecular polymorphisms in *Delta* (Long *et al.* 1998) and *scabrous* (Lyman *et al.* 1999) with phenotypic variation in bristle number lend support to the contention that these genes are bristle number QTL.

For organisms in which controlled crosses can be made and for which genetic resources in the form of mutant stocks at loci in the region to which the QTL maps are available, systematic quantitative complementation to each locus in turn is an unbiased method for identifying loci with undescribed and unexpected pleiotropic effects on the trait, and assigning a phenotype to loci with unknown function. That is, in a rare role reversal, quantitative genetics can actually be a valuable gene discovery tool. Strains containing single gene knockouts of all known and predicted genes in model organisms (Spradling *et al.* 1999, Winzeler *et al.* 1999, DeAngelis *et al.* 2000, Fraser *et al.* 2000, Gönczy *et al.* 2000, Nolan *et al.* 2000) will be an exceptionally valuable resource for quantitative complementation tests in the future.

In organisms that are less genetically amenable, putative candidate genes corresponding to QTL can be mapped by LD between the trait phenotype and marker locus genotype in samples taken directly from the natural population (Risch and Merikangas 1996). When a new mutation occurs in a population at a locus affecting a quantitative trait, all other polymorphic alleles in that population will initially be in complete LD with the mutation. Over time, however, recombination between the mutant allele and the other loci will create the missing haplotypes and restore linkage equilibrium between the mutant allele at all but closely linked loci. The length of the genomic fragment surrounding the original mutation in which LD between the QTL and other loci still exists depends on the average amount of recombination per generation experienced by that region of the genome, the number of generations that have passed since the original mutation, and the population size (Hill and Robertson 1968, Falconer and Mackay 1996, Weir 1996, Hartl and Clark 1997). For old mutations in large equilibrium populations, strong LD is only expected to extend over distances of the order of kilobases or less. Larger tracts of LD are expected in expanding populations derived from a recent founder event or in population isolates with small effective size.

The principle of LD mapping is superficially simple: determine trait phenotypes on individuals sampled from a population varying genetically for the trait of interest, genotype molecular markers in the QTL interval, and determine whether differences in trait phenotypes are associated with marker genotypes for each marker in turn, correcting the threshold for statistical significance for multiple tests (Churchill and Doerge 1994, Doerge and Churchill 1996, Long *et al.* 1998, Lyman *et al.* 1999). Samples of at least 500 individuals are necessary to detect a gene contributing 5% of the total phenotypic variance (a large effect) with 80% power; larger samples are required for genes with smaller effects (Risch and Merikangas 1996, Long and Langley 1999, Luo *et al.* 2000). If the genotypes at all polymorphic sites in the region of interest are determined, one of them must correspond to the functional variant (the QTN).

This scale of genotyping is not feasible using current technology, however, but will change in the future as methods for rapid and cost-effective genome resequencing become available. The first complexity to be addressed, therefore, is the issue of how many markers should be genotyped. The estimate of the effect of a linked marker on the trait underestimates the effect of the polymorphism at the QTN by an amount that is proportional to the LD between the marker and QTN (Lai *et al.* 1994, Nielsen and Weir 1999). Therefore, markers should be spaced such that they are in LD with each other, otherwise marker–trait associations will be missed (Kruglyak 1999, Long *et al.* 1998, 2000, Long and Langley 1999). Unfortunately, however, the relationship between LD between marker loci and physical distance varies by at least an order of magnitude among different gene regions (Miyashita and Langley 1988, Aguadé *et al.* 1989, Miyashita 1990, Long and Langley 1999, Cardon and Bell 2001) and breaks down over short physical distances, with some closely linked sites in linkage equilibrium, and others in strong disequilibrium (Miyashita and Langley 1988, Miyashita 1990, Clark *et al.* 1998, Nickerson *et al.* 1998, Eaves *et al.* 2000, Taillon-Miller *et al.* 2000). This variation reflects differences in recombination rates (Aguadé *et al.* 1989, Begun and Aquadro 1992), natural selection, genetic drift, marker mutations, and other genetic processes (Weir and Hill 1986, Hill and Weir 1988, 1994, Weir 1996). Knowledge of the LD landscape in the region of interest for a small sample would assist the optimal spacing of marker loci.

The second complexity is that marker–trait associations in natural populations are not necessarily attributable to linkage, but can be caused by admixture between populations that have different gene frequencies at the marker loci and different values of the trait (Falconer and Mackay 1996, Hartl and Clark 1997). This problem can be alleviated by experimental designs that control for population structure and jointly test for linkage and association, such as the transmission-disequilibrium test (TDT) (Spielman *et al.* 1993). Methods to explicitly account for population structure using unlinked markers outside the QTL region are also promising (Thornsberry *et al.* 2001). Replication of the association in independent populations helps allay concerns that associations are spurious.

An example of the successful application of LD mapping is the positional cloning of the human calpain-10 (*CAPN10*) gene, a putative susceptibility gene for type 2 diabetes in the interval containing the *NIDDM1* QTL (Horigawa *et al.* 2000). Associations between 21 single nucleotide polymorphisms (SNPs) and diabetes susceptibility in the 5 cM region to which *NIDDM1* mapped narrowed the search to three genes in a 66 kb interval. Association tests for 63 additional SNPs revealed that only one SNP in *CAPN10* was both associated with the disease and evidence for linkage, implicating *CAPN10* as the disease susceptibility locus. Haplotypes at three polymorphic SNPs in *CAPN10* were associated with diabetes in the study population and two unrelated populations.

3.2.4 From gene to QTN

Integration of quantitative and population genetics requires that we estimate allele frequencies at genetic loci corresponding to QTL. To do this we need to be able to define functional alleles – the polymorphic sites (QTN) that cause the difference in the trait phenotype. This is not a trivial problem, given the high levels of intragenic polymorphism in most populations. For example, there are 42 nucleotide differences distinguishing the two parental alleles of the tomato fruit weight QTL, *ORFX* (Frary *et al.* 2000); 11 molecular variants in 484 base pairs differentiate the parental alleles of the tomato QTL affecting fruit sugar content (Fridman *et al.* 2000); 88 variant sites in 71 individuals for the human lipoprotein lipase gene (Nickerson *et al.* 1998); and at least 108 variant sites in *CAPN10* (Horigawa *et al.* 2000). Which of these sites are causally associated with the phenotypes of interest?

The same LD mapping methods used to focus in on a candidate gene can also be used to determine which of the polymorphic sites at the candidate gene is associated with the quantitative trait phenotype. Again, the optimal density of markers must be adjusted relative to the local scale of LD. In *Drosophila* regions of high polymorphism and recombination, the optimal spacing of markers could be as small as 200 base pairs (Long *et al.* 1998).

To date, no study of genotype–phenotype associations at candidate genes has utilized the optimal density of markers. Indeed, many studies in humans have been conducted with a single marker, leading to difficulty in replicating associations due to variation among populations in the degree of LD between the marker and the putative causal variant (Cardon and Bell 2001). The few studies in which multiple polymorphic markers within candidate genes have been examined for association with phenotypic variation reveal some interesting features. All kinds of molecular variation (transposable element insertions, small insertions/deletions and SNPs) at *Drosophila* bristle number candidate genes have been associated with quantitative variation in bristle number, and all significant associations have been for molecular polymorphisms in introns and noncoding flanking regions (Mackay and Langley 1990, Lai *et al.* 1994, Long *et al.* 1998, 2000, Lyman *et al.* 1999). Polymorphisms in noncoding regions of the *CAPN10* putative diabetes susceptibility locus in humans are associated with disease risk (Horigawa *et al.* 2000). It is thus possible that variation in regulatory sequences causes slight differences in message levels, timing and tissue-specificity of gene expression, and protein stability that lead to subtle quantitative differences in phenotype. The *CAPN10* story is unexpectedly complex. Two haplotypes of three SNPs in this gene have significantly increased risk of diabetes, but the at-risk genotype is a homozygote for one SNP and heterozygous for the others; homozygotes for either haplotype do not have increased risk (Horigawa *et al.* 2000). This suggests that at least two interacting risk factors within a single gene are required to affect susceptibility to diabetes.

Correlation is not causality, and the ultimate proof that a molecular variant is functionally associated with differences in phenotype will be biological, not statistical. One such functional test is to construct by *in vitro* mutagenesis alleles that differ for each of the putative QTN, separately and in combination, and to assess their phenotypic effects by germ-line transformation into a null mutant background (Stam and Laurie 1996). Mimicking complex effects such as those observed at *CAPN10* could prove to be challenging.

3.3 Properties of genes affecting quantitative traits

3.3.1 Additive effects

Is genetic variation for quantitative traits caused by a very large number of QTL with very small and equal allelic effects, as assumed by the "infinitesimal" model (Falconer and Mackay 1996, Lynch and Walsh 1998); or is the distribution of allelic effects exponential, with a few loci with large effects causing most of the variation in the traits, as proposed by Robertson (1967)? Observed distributions of QTL effects are in accord with Robertson's prediction (Mackay 2001), suggesting that much progress in understanding quantitative genetic variation can be made by a detailed study of the manageable number of "leading factors" affecting most traits. A caveat is that overestimation of effects of detected QTL could be an artifact of small sample size (Beavis 1994) or LD between QTL, low heritability and epistatic interactions between QTL (Brost *et al.* 2001). However, high-resolution recombination mapping (Doebley *et al.* 1997, Frary *et al.* 2000, Fridman *et al.* 2000) and transformation of cloned QTL

(Symula *et al.* 1999, Frary *et al.* 2000) prove the existence of at least some QTL with large effects, and significant associations of molecular polymorphisms at candidate genes with moderate to large QTL effects is consistent with an exponential distribution of effects.

3.3.2 Interaction effects

QTL effects typically vary according to the genetic, sexual and external environment. If the magnitude and/or the rank order of the differences in phenotypes associated with QTL genotypes also changes depending on the genetic or environmental context, then that QTL exhibits genotype-by-genotype interaction (epistasis), genotype-by-sex interaction (GSI) or genotype-by-environment interaction (GEI) (Mackay 2001). Extensive epistasis, GSI and GEI have practical and theoretical consequences. First, estimates of QTL positions and effects are relevant only to the sex and environment in which the phenotypes were assessed, and may not replicate across sexes and in different environments. Second, estimates of main QTL effects will be biased in the presence of epistasis. Third, genetic variation for quantitative traits can be maintained at loci exhibiting GSI and GEI (Levene 1953, Gillespie and Turelli 1989, Fry *et al.* 1996).

The power to detect epistasis between QTL in mapping populations is low. Even large mapping populations contain few individuals in the rarer two-locus genotype classes; segregation for other QTL can interfere with detecting epistasis between the pair of loci under consideration; and only strong interactions remain significant after appropriate adjustment of the significance threshold to account for the multiple statistical tests involved in searching for epistatic interactions. Therefore, it is not surprising that many studies report largely additive QTL effects or do not test for epistasis. On the other hand, strong interactions have been observed between QTL affecting *Drosophila* bristle number (Spickett and Thoday 1966, Shrimpton and Robertson 1988, Long *et al.* 1995); mouse body size (Routman and Cheverud 1997); grain yield in rice (Li *et al.* 1997); susceptibility to diabetes in humans (Cox *et al.* 1999); and longevity in *Drosophila* (Leips and Mackay 2000) and *C. elegans* (Shook and Johnson 1999).

However, genotype-specific QTL effects are common when more precise tests for interactions are performed. One such test is to introgress a mutant allele into several different wild-type genetic backgrounds. Typically, the expression of the mutant will be enhanced or suppressed, and the degree of dominance and pleiotropic effects on other traits can be modified (Polaczyk *et al.* 1998, Rutherford and Lindquist 1998, Gibson *et al.* 1999, Nadeau 2001). These results indicate a hidden reservoir of genetic variation that is only revealed in the mutant background. Specific tests for epistasis between QTL also reveal considerable interaction. Diallel crosses between co-isogenic NIL containing QTL for yield-related traits in tomato exhibited antagonistic epistasis (Eshed and Zamir 1996). Similarly, diallel crosses between co-isogenic *P* element insertions affecting odor-guided behavior in *Drosophila* could be organized in a network of enhancing and suppressing interactions (Fedorowicz *et al.* 1998). Tests for additive-by-additive epistasis require the construction of the four double homozygous genotypes at two bi-allelic loci, while estimating all classes of epistatic interactions involves synthesizing the nine two-locus genotypes. Such tests have revealed considerable epistasis between QTL affecting divergence in plant architecture between maize and teosinte (Lukens and Doebley 1999) and between *P* element insertions on metabolic activity in *Drosophila* (Clark and Wang 1997). Further, functional assays to identify QTN responsible for differences in protein concentration at the *Drosophila Adh* locus have revealed epistasis between QTN within this gene (Stam and Laurie 1996).

Genotype-by-sex interaction exists when the cross-sex genetic correlation of the trait is less than unity. When this occurs, there is genetic variation for sex dimorphism of the trait, giving the potential for evolution of sex dimorphism under natural or artificial selection. Sex-specific effects of genes affecting quantitative traits are surprisingly common, and have been observed in *Drosophila* for spontaneous and *P*-element-induced mutations, QTL and molecular polymorphisms in candidate genes for traits as diverse as sensory bristle number, olfactory behavior, longevity and enzyme activity (Mackay 2001). While most QTL for mouse body size affect both males and females, some affect only males or females (Vaughn *et al.* 1999). It is likely that this is a general phenomenon, and sex-specific effects of QTL affecting adaptive phenotypes in other organisms are to be expected.

QTL genotype by environment interaction is common, but by no means ubiquitous (Lynch and Walsh 1998). In maize, QTL effects for grain yield, ear height and plant height were largely independent of the environment, while QTL for days to tassel, grain moisture and ear number were highly environment-dependent (Cockerham and Zeng 1996). In *Drosophila*, GEI has been observed for QTL affecting sensory bristle number (Gurganus *et al.* 1998), life span (Leips and Mackay 2000; Vieira *et al.* 2000) and fitness (Fry *et al.* 1998). Some *Drosophila* QTL have even more complicated effects, with significant three-way interactions of QTL, sex and environment (Gurganus *et al.* 1998; Leips and Mackay 2000; Vieira *et al.* 2000), and epistatic interactions that are sex and environment specific (Leips and Mackay 2000).

3.3.3 Pleiotropy

The most important property of a QTL from an evolutionary perspective is its pleiotropic relationship to fitness. Fitness effects at all loci contributing to variance in a trait determine the magnitude of genetic variation segregating for the trait and relative contributions of additive, dominance and epistatic variance, and response of the trait to evolutionary forces of natural selection and inbreeding (Falconer and Mackay 1996). Since it is very difficult to determine the relationship of traits to fitness in nature (Kingsolver *et al.* 2001) and even in the laboratory (Nuzhdin *et al.* 1995), it follows that it will be exceedingly difficult, if not impossible, to measure empirically fitness effects of all loci affecting variation in a trait. Selection acting on any single locus affecting a quantitative trait is likely to be quite weak (Lewontin 1974, Kimura 1983, Kingsolver 2001), and one needs to consider the whole range of ecologically relevant environments. Yet this is exactly what is required, because the relationship of a trait to fitness could be caused by multiple mechanisms, with quite different consequences for the nature and magnitude of segregating variation. For example, optimal fitness for intermediate trait values could be caused by direct selection on the trait phenotype, deleterious pleiotropic fitness effects of alleles causing extreme phenotypes (Barton 1990, Keightley and Hill 1990, Kondrashov and Turelli 1992), or overdominance of alleles associated with intermediate trait values (Robertson 1967, Barton 1990). In the first and second cases, variation for the trait may be attributable to a balance between the input of new deleterious alleles by mutation and their elimination by natural selection, which occurs when mutant alleles are rare (Barton 1990). In the third case, alleles will be maintained at intermediate frequencies. To add to the complexity, it is likely that there is heterogeneity in causal relationships to fitness among loci affecting a trait. Both common (Long *et al.* 1998, 2000) and rare (Mackay and Langley 1990, Long *et al.* 2000) molecular polymorphisms at candidate bristle number QTL have been associated with variation in bristle number, even within a single gene (Long *et al.* 2000), indicating different selective effects of the underlying QTN.

Finally, we have come full circle. There is a rich body of population genetics theory for inferring the action of historical selection from data on DNA sequence variation (Hartl and Clark 1997). We are now in a position to apply these tests to sequences of cloned QTL to detect the signatures of purifying selection, selective sweeps, balancing selection and neutrally evolving polymorphisms (Wang *et al.* 1999).

3.4 Future prospects

Several landmark technological advances have punctuated the history of experimental evolutionary genetics and catalyzed bursts of progress: visualization of polymorphic proteins by gel electrophoresis, restriction fragment length polymorphism analysis of cloned genes, DNA sequencing, and the discovery of abundant polymorphic molecular markers with which to map QTL. Now we stand to benefit from further new technologies that continue to develop in the post-genome era. Targeted disruptions of all known and predicted genes in several model organisms (Spradling *et al.* 1999, Winzeler *et al.* 1999, DeAngelis *et al.* 2000, Fraser *et al.* 2000, Gönczy *et al.* 2000, Nolan *et al.* 2000) will provide new candidate genes for evaluation of mapped QTL, and will be valuable resources for quantitative complementation tests. The cost, efficiency, and accuracy of high-throughput genotyping is the current technical limitation to high-resolution QTL mapping. When the technology for rapidly and economically resequencing whole genomes is developed, one should be able to map QTL with high resolution by generating and phenotyping huge mapping populations, and genotyping the extremes. Similar considerations apply to the mapping of QTN within candidate genes. Combining analyses of whole genome variation in transcript levels using expression arrays (Lockhart and Barlow 2001) with information from QTL mapping should be useful for nominating candidate genes for further study. One can envision future tests for epistasis based on analyses of genome-wide changes in expression in response to NIL or single mutations affecting a common trait. Development of techniques for knocking in alternative QTL and QTN alleles at the endogenous sites, without altering genetic backgrounds, will provide the rigorous standard of proof for functional association of genotype and phenotype. We stand on the verge of being able to describe genetic variation in populations in terms of gene and genotype frequencies as well as phenotypic and fitness effects at loci affecting phenotypes that are the real stuff of evolution; finally, measuring the unmeasurable.

3.5 Acknowledgments

This work was supported by grants from the NIH, and is a publication of the W. M. Keck Center for Behavioral Biology.

REFERENCES

Adams, M. D., Celniker, S., Holt, R. A., Evans, C. A., Gocayne, J. D., *et al.* (2000). The genome sequence of *Drosophila melanogaster*. *Science* 287:2185–95.

Aguadé, M., Miyashita, N., and Langley, C. H. (1989). Reduced variation in the *yellow-achaete-scute* region in natural populations of *Drosophila melanogaster*. *Genetics* 122:607–15.

Anholt, R. R. H., and Mackay, T. F. C. (2001). The genetic architecture of odor-guided behavior in *Drosophila melanogaster*. *Behav. Genet.* 31:17–27.

Anholt, R. R. H., Lyman, R. F., and Mackay, T. F. C. (1996). Effects of single *P* element insertions on olfactory behavior in *Drosophila melanogaster*. *Genetics* 143:293–301.

Barton, N. H. (1990). Pleiotropic models of quantitative variation. *Genetics* 124:773–82.

Beavis, W. D. (1994). The power and deceit of QTL experiments: lessons from comparative QTL studies. In *49th Annual Corn and Sorghum Research Conference*, pp. 252–68. Washington DC: American Seed Trade Association.

Begun, D., and Aquadro, C. H. (1992). Levels of naturally occurring DNA polymorphism correlate with recombination rates in *D. melanogaster*. *Nature* 356:519–20.

Brost, B., de Vienne, D., Hospital, F., Moreau, L., and Dilmann, C. (2001). Genetic and nongenetic bases for the L-shaped distribution of quantitative trait loci effects. *Genetics* 157:1773–87.

Cardon, L. R., and Bell, J. I. (2001). Association study designs for complex diseases. *Nat. Rev. Genet.* 2:91–9.

Churchill, G. A., and Doerge, R. W. (1994). Empirical threshold values for quantitative trait mapping. *Genetics* 138:963–71.

Clark, A. G., and Wang, L. (1997). Epistasis in measured genotypes: *Drosophila P* element insertions. *Genetics* 147:157–63.

Clark, A. G., Wang, L., and Hulleberg, T. (1995). *P*-element-induced variation in metabolic regulation in *Drosophila. Genetics* 139:337–48.

Clark, A. G., Weiss, K. M., Nickerson, D. A., Taylor, S. L., Buchanan, A., *et al.* (1998). Haplotype structure and population genetic inferences from nucleotide sequence variation in human lipoprotein lipase. *Am. J. Hum. Genet.* 63:595–612.

Clayton, G. A., and Robertson, A. (1957). An experimental check on quantitative genetical theory. II. The long-term effects of selection. *J. Genet.* 55:152–70.

Cockerham, C. C., and Zeng, Z.-B. (1996). Design III with marker loci. *Genetics* 143:1437–56.

Cormier, R. T., Hong, K. H., Halberg, R. B., Hawkins, T. L., Richardson, P., *et al.* (1997). Secretory phospholipase Pla2g2a confers resistance to intestinal tumorigenesis. *Nat. Genet.* 17:88–91.

Cox, N. J., Frigge, M., Nicolae, D. L., Concannon, P., Hanis, C., *et al.* (1999). Loci on chromosomes 2 (*NIDDM1*) and 15 interact to increase susceptibility to diabetes in Mexican Americans. *Nat. Genet.* 21:213–15.

Darvasi, A. (1998). Experimental strategies for the genetic dissection of complex traits in animal models. *Nat. Genet.* 18:19–24.

DeAngelis, M. H., Flaswinkel, H., Fuchs, H., Rathkolb, B., Soewarto, D., *et al.* (2000). Genome-wide, large-scale production of mutant mice by ENU mutagenesis. *Nat. Genet.* 25:444–7.

Doebley, J., and Stec, A. (1991). Genetic analysis of the morphological differences between maize and teosinte. *Genetics* 129:285–95.

Doebley, J., and Stec, A. (1993). Inheritance of the morphological differences between maize and teosinte: comparison of results from two F_2 populations. *Genetics* 134:559–70.

Doebley, J., Stec, A., and Gustus, C. (1995). *teosinte branched1* and the origin of maize: evidence for epistasis and the evolution of dominance. *Genetics* 141:333–46.

Doebley, J., Stec, A., and Hibbard, L. (1997). The evolution of apical dominance in maize. *Nature* 386:485–8.

Doerge, R. W., and Churchill, G. A. (1996). Permutation tests for multiple loci affecting a quantitative character. *Genetics* 142:285–94.

Eaves, I. A., Merriman, T. R., Barber, R. A., Nutland, S., Tuomilehto-Wolf, E., *et al.* (2000). The genetically isolated populations of Finland and Sardinia may not be a panacea for linkage disequilibrium mapping of common human disease genes. *Nat. Genet.* 25:320–3.

Eshed, Y., and Zamir, D., (1995). An introgression line population of *Lycopersicon pennellii* in the cultivated tomato enables the identification and fine-mapping of yield-associated QTL. *Genetics* 141:1147–62.

Eshed, Y., and Zamir, D. (1996). Less-than-additive interactions of quantitative trait loci in tomato. *Genetics* 143:1807–17.

Falconer, D. S., and Mackay, T. F. C. (1996). *Introduction to Quantitative Genetics.* Harlow, Essex: Addison Wesley Longman. 464 pp.

Fedorowicz, G. M., Fry, J. D., Anholt, R. R. H., and Mackay, T. F. C. (1998). Epistatic interactions between *smell-impaired* loci in *Drosophila melanogaster. Genetics* 148:1885–91.

Frankham, R., Jones, L. P., and Barker, J. S. F. (1968). The effects of population size and selection intensity for a quantitative character in *Drosophila.* III. Analysis of the lines. *Genet. Res.* 12:267–83.

Frary, A., Nesbitt, T. C., Frary, A., Grandillo, S., van der Knaap, E., *et al.* (2000). *fw2.2*: a quantitative trait locus key to the evolution of tomato fruit size. *Science* 289: 85–8.

Fraser, A. G., Kamath, R. S., Zipperien, P., Martinez-Campos, M., Sohrmann, M., *et al.* (2000). Functional genomic analysis of *C. elegans* chromosome I by systematic RNA interference. *Nature* 408:325–30.

Fridman, E., Pleban, T., and Zamir, D. (2000). A recombination hotspot delimits a wild-species quantitative trait locus for tomato sugar content to 484 bp within an invertase gene. *Proc. Natl. Acad. Sci. USA* 97:4718–23.

Fry, J. D., Heinsohn, S. L., and Mackay, T. F. C. (1996). The contribution of new mutations to genotype–environment interaction for fitness in *Drosophila melanogaster. Evolution* 50:2316–27.

Fry, J. D., Nuzhdin, S. V., Pasyukova, E. G., and Mackay, T. F. C. (1998). QTL mapping of genotype–environment interaction for fitness in *Drosophila melanogaster. Genet. Res.* 71:133–41.

Gibson, G., Wemple, M. and van Helden, S. (1999). Potential variance affecting homeotic *Ultrabithorax* and *Antennapedia* phenotypes in *Drosophila melanogaster. Genetics* 151:1081–91.

Gillespie, J. H., and Turelli, M. (1989). Genotype–environment interactions and the maintenance of polygenic variation. *Genetics* 121:129–38.

Goffeau, A., Barrell, B. G., Bussey, H., Davis, R. W., Dujon, B., *et al.* (1996). Life with 6,000 genes. *Science* 274:546–67.

Gönczy, P., Echeverri, C., Oegema, K., Coulson, A., Jones, S. J. M., *et al.* (2000). Functional genomic analysis of cell division in *C. elegans* using RNAi of genes on chromosome III. *Nature* 408:331–6.

Gurganus, M. C., Fry, J. D., Nuzhdin, S. V., Pasyukova, E. G., Lyman, R. F., *et al.* (1998). Genotype–environment interaction at quantitative trait loci affecting sensory bristle number in *Drosophila melanogaster. Genetics* 149:1883–98.

Gurganus, M. C., Nuzhdin, S. V., Leips, J. W., and Mackay, T. F. C. (1999). High-resolution mapping of quantitative trait loci for sternopleural bristle number in *Drosophila melanogaster. Genetics* 152:1585–604.

Hartl, D. L., and Clark, A. G. (1997). *Principles of Population Genetics,* 3rd edition. Sunderland, MA: Sinauer. 542 pp.

Hill, W. G. (1998). Selection with recurrent backcrossing to develop congenic lines for quantitative trait loci analysis. *Genetics* 148:1341–52.

Hill, W. G., and Robertson, A. (1968). Linkage disequilibrium in finite populations. *Theor. Appl. Genet.* 38:226:31.

Hill, W. G., and Weir, B. S. (1988). Variances and covariances of squared linkage disequilibria in finite populations. *Theor. Popul. Biol.* 33:54–78.

Hill, W. G., and Weir, B. S. (1994). Maximum likelihood estimation of gene location by linkage disequilibrium. *Am. J. Hum. Genet.* 54:705–14.

Horigawa ,Y., Oda, N., Cox, N. J., Li, X., Orho-Melander, M., *et al.* (2000). Genetic variation in the gene encoding calpain-10 is associated with type 2 diabetes mellitus. *Nat. Genet.* 26:163–75.

International Human Genome Sequencing Consortium (2001). Initial sequencing and analysis of the human genome. *Nature* 409:860–921.

Jansen, R. C., and Stam, P. (1994). High resolution of quantitative traits into multiple loci via interval mapping. *Genetics* 136:1447–55.

Keightley, P. D., and Hill, W. G. (1990). Variation maintained in quantitative traits with mutation-selection balance: pleiotropic side-effects on fitness traits. *Proc. R. Soc. Lond. Ser. B* 242:95–100.

Keightley, P. D., Evans, M. J., and Hill, W. G. (1993). Effects of multiple retrovirus insertions on quantitative traits of mice. *Genetics* 135:1099–106.

Kimura, M. (1983). *The Neutral Theory of Molecular Evolution.* Cambridge: Cambridge University Press. 367 pp.

Kingsolver, J. G., Hoekstra, H. E., Hoekstra, J. M., Berrigan, D., Vignieri, S. N., *et al.* (2001). The strength of phenotypic selection in natural populations. *Am. Nat.* 157:245–61.

Kondrashov, A. S., and Turelli, M. (1992). Deleterious mutations, apparent stabilizing selection and the maintenance of quantitative variation. *Genetics* 132:603–18.

Kruglyak, L. (1999). Prospects for whole-genome linkage disequilibrium mapping of common disease genes. *Nat. Genet.* 22:139–44.

Lai, C., Lyman, R. F., Long, A. D., Langley, C. H., and Mackay, T. F. C. (1994). Naturally occurring variation in bristle number and DNA polymorphisms at the *scabrous* locus of *Drosophila melanogaster. Science* 266:1697–1702.

Lai, C., McMahon, R., Young, C., Mackay, T. F. C., and Langley, C. H. (1998). *que mao,* a *Drosophila* bristle locus, codes for geranylgeranyl pyrophosphate synthase. *Genetics* 149:1051–61.

Laurie, C. C., True, J. R., Liu, J., and Mercer, J. M. (1997). An introgression analysis of quantitative trait loci that contribute to a morphological difference between *Drosophila simulans* and *D. mauritiana. Genetics* 145:339–48.

Leips, J., and Mackay, T. F. C. (2000). Quantitative trait loci for life span in *Drosophila melanogaster*: interactions with genetic background and larval density. *Genetics* 155:1773–88.

Levene, H. (1953). Genetic equilibrium when more than one ecological niche is available. *Am. Nat.* 87:331–3.

Lewontin, R. C. 1974. *The Genetic Basis of Evolutionary Change.* New York: Columbia University Press. 346 pp.

Li, Z., Pinson, S. R., Park, W. D., Paterson, A. H., and Stansel, J. W. (1997). Epistasis for three grain yield components in rice (*Oryza sativa* L.). *Genetics* 145:453–65.

Lockhart, D. J., and Barlow, C. (2001). Expressing what's on your mind: DNA arrays and the brain. *Nat. Rev. Neuroscience* 2:63–8.

Long, A. D., and Langley, C. H. (1999). Power of association studies to detect the contribution of candidate genetic loci to complexly inherited phenotypes. *Genome Res.* 9:720–31.

Long, A. D., Lyman, R. F., Langley, C. H., and Mackay, T. F. C. (1998). Two sites in the *Delta* gene region contribute to naturally occurring variation in bristle number in *Drosophila melanogaster. Genetics* 149:999–1017.

Long, A. D., Lyman, R. F., Morgan, A. H., Langley, C. H., and Mackay, T. F. C. (2000). Both naturally occurring insertions of transposable elements and intermediate frequency polymorphisms at the *achaete-scute* complex are associated with variation in bristle number in *Drosophila melanogaster. Genetics* 154:1255–69.

Long, A. D., Mullaney, S. L., Mackay, T. F.C., and Langley, C. H. (1996). Genetic interactions between naturally occurring alleles at quantitative trait loci and mutant alleles at candidate loci affecting bristle number in *Drosophila melanogaster. Genetics* 114:1497–1510.

Long, A. D., Mullaney, S. L., Reid, L. A., Fry, J. D., Langley, C. H., *et al.* (1995). High resolution mapping of genetic factors affecting abdominal bristle number in *Drosophila melanogaster. Genetics* 139:1273–91.

Lukens, L. N., and Doebley, J. (1999). Epistatic and environmental interactions for quantitative trait loci involved in maize evolution. *Genet. Res.* 74:291–302.

Luo, Z. W., Tao, S. H., and Zeng, Z.-B. (2000). Inferring linkage disequilibrium between a polymorphic marker locus and a trait locus in natural populations. *Genetics* 156:457–67.

Lyman, R. F., and Mackay, T. F. C. (1998). Candidate quantitative trait loci and naturally occurring phenotypic variation for bristle number in *Drosophila melanogaster*: The *Delta-Hairless* gene region. *Genetics* 149:983–98.

Lyman, R. F., Lai, C., and Mackay, T. F. C. (1999). Linkage disequilibrium mapping of molecular polymorphisms at the *scabrous* locus associated with naturally occurring variation in bristle number in *Drosophila melanogaster. Gen. Res.* 74: 303–11.

Lyman, R. F., Lawrence, F., Nuzhdin, S. V., and Mackay, T. F. C. (1996). Effects of single *P* element insertions on bristle number and viability in *Drosophila melanogaster. Genetics* 143:277–92.

Lynch, M., and Walsh, B. (1998). *Genetics and Analysis of Quantitative Traits.* Sunderland, MA: Sinauer. 980 pp.

Lyons, P. A., Armitage, N., Argentina, F., Denny, P., Hill, N. J., *et al.* (2000). Congenic mapping of the type 1 diabetes locus, *Idd3*, to a 780-kb region of mouse chromosome 3: identification of a candidate segment of ancestral DNA by haplotype mapping. *Genome Res.* 10:446–53.

Mackay, T. F. C. (2001). The genetic architecture of quantitative traits. *Annu. Rev. Genet.* 35:303–9.

Mackay, T. F. C., and Fry, J. D. (1996). Polygenic mutation in *Drosophila melanogaster*: genetic interactions between selection lines and candidate quantitative trait loci. *Genetics* 144:671–88.

Mackay, T. F. C., and Langley, C. H. (1990). Molecular and phenotypic variation in the *achaete-scute* region of *Drosophila melanogaster. Nature* 348:64–6.

Miyashita, N. (1990). Molecular and phenotypic variation of the *Zw* locus region in *Drosophila melanogaster. Genetics* 125:407–19.

Miyashita, N., and Langley, C. H. (1988). Molecular and phenotypic variation of the *white* locus region in *Drosophila melanogaster. Genetics* 120:199–212.

Nadeau, J. H. (2001). Modifier genes in mice and humans. *Nat. Rev. Genet.* 2:165–74.

Nadeau, J. H., and Frankel, D. (2000). The roads from phenotypic variation to gene discovery: mutagenesis versus QTL. *Nat. Genet.* 25:381–4.

Nadeau, J. H., Singer, J. B., Matin, A., and Lander, E. S. (2000). Analysing complex genetic traits with chromosome substitution strains. *Nat. Genet.* 24:221–5.

Nickerson, D. A., Taylor, S. L., Weiss, K. M., Clark, A. G., Hutchinson, R. G., *et al.* (1998). DNA sequence diversity in a 9.7-kb region of the human lipoprotein lipase gene. *Nat. Genet.* 19:233–40.

Nielsen, D. M., and Weir, B. S. (1999). A classical setting for associations between markers and loci affecting quantitative traits. *Genet. Res.* 74:271–7.

Nolan, P. M., Peters, J., Strivens, M., Rogers, D., Hagan, J., *et al.* (2000). A systematic, genome-wide, phenotype-driven mutagenesis programme for gene function studies in the mouse. *Nat. Genet.* 25:440–3.

Nuzhdin, S. V., Fry, J. D., and Mackay, T. F. C. (1995). Polygenic mutation in *Drosophila melanogaster*: the causal relationship of bristle number to fitness. *Genetics* 139:861–72.

Nuzhdin, S. V., Pasyukova, E. G., Dilda, C. L., Zeng, Z.-B., and Mackay, T. F. C. (1997). Sex-specific quantitative trait loci affecting longevity in *Drosophila melanogaster*. *Proc. Natl Acad. Sci. USA* 94:9734–9.

Pasyukova, E. G., Vieira, C., and Mackay, T. F. C. (2000). Deficiency mapping of quantitative trait loci affecting longevity in *Drosophila melanogaster*. *Genetics* 156: 1129–46.

Paterson, A. H., Lander, E. S., Hewitt, J. D., Peterson, S., Lincoln, S. E., *et al.* (1988). Resolution of quantitative traits into Mendelian factors by using a complete linkage map of restriction fragment length polymorphisms. *Nature* 335:721–26.

Polaczyk, P. J., Gasperini, R., and Gibson, G. (1998). Naturally occurring genetic variation affects *Drosophila* photoreceptor determination. *Dev. Genes and Evol.* 207:462–70.

Risch, N., and Merikangas, K. (1996). The future of genetic studies of complex human diseases. *Science* 273:1516–17.

Robertson, A. (1967). The nature of quantitative genetic variation. In A. Brink (ed.) *Heritage from Mendel*, pp. 265–80. Madison WI: University of Wisconsin.

Routman, E. J., and Cheverud, J. M. (1997). Gene effects on a quantitative trait: two-locus epistatic effects measured at microsatellite markers and at estimated QTL. *Evolution* 51:1654–62.

Rutherford, L. S., and Lindquist, S. (1998). Hsp90 as a capacitor of morphological evolution. *Nature* 396:336–42.

Sax, K. (1923). The association of size differences with seed-coat pattern and pigmentation in *Phaseolus vulgaris*. *Genetics* 8:552–60.

Shook, D. R., and Johnson, T. E. (1999). Quantitative trait loci affecting survival and fertility-related traits in *Caenorhabditis elegans* show genotype–environment interactions, pleiotropy and epistasis. *Genetics* 153:1233–43.

Shrimpton, A. E., and Robertson, A. (1988). The isolation of polygenic factors controlling bristle score in *Drosophila melanogaster*. II. Distribution of third chromosome bristle effects within chromosome sections. *Genetics* 118:445–59.

Spickett, S. G., and Thoday, J. M. (1966). Regular responses to selection. 3. Interactions between located polygenes. *Genet. Res.* 7:96–121.

Spielman, R. S., McGinnis, R. E., and Ewens, W. J. (1993). Transmission test for linkage disequilibrium: the insulin gene region and insulin-dependent diabetes mellitus (DDM). *Am. J. Hum. Genet.* 52:506–16.

Spradling, A. C., Stern, D., Beaton, A., Rhem, E. J., Laverty, T., *et al.* (1999). The Berkeley *Drosophila* genome gene disruption project: single *P* element insertions mutating 25% of vital *Drosophila* genes. *Genetics* 153:135–77.

Stam, L. F., and Laurie, C. C. (1996). Molecular dissection of a major gene effect on a quantitative trait: the level of alcohol dehydrogenase expression in *Drosophila melanogaster*. *Genetics* 144:1559–64.

Symula, D. J., Frazer, K. A., Ueda, Y., Denefle, P., Stevens, M. E., *et al.* (1999). Functional screening of an asthma QTL in YAC transgenic mice. *Nat. Genet.* 23:241–4.

Taillon-Miller, P., Bauer-Sardiña, I., Saccone, N., Putzel, J., Laitinen, T., *et al.* (2000). Juxtaposed regions of extensive linkage disequilibrium in human Xq25 and Xq28. *Nat. Genet.* 25:324–8.

The Arabidopsis Genome Initiative (2000). Analysis of the genome sequence of the flowering plant *Arabidopsis thaliana*. *Nature* 408:796–815.

The *C. elegans* Sequencing Consortium (1998). Genome sequence of the nematode *C. elegans*: a platform for investigating biology. *Science* 282:2012–18.

Thoday, J. M. (1961). Location of polygenes. *Nature* 191:368–70.

Thornsberry, J. M., Goodman, M. M., Doebley, J., Kresovitch, S., Nielsen, D., *et al.* (2001). *Dwarf8* polymorphisms associate with variation in flowering time. *Nat. Genet.* 28:286–9.

Vaughn, T. T., Pletscher, S., Peripato, A., King-Ellison, K., Adams, E., *et al.* (1999). Mapping quantitative trait loci for murine growth: a closer look at genetic architecture. *Genet. Res.* 74:313–22.

Venter, J. C., Adams, M. D., Myers, E. W., Li, P. W., Mural, R. J., *et al.* (2001). The sequence of the human genome. *Science* 291:1304–51.

Vieira, C., Pasyukova, E. G., Zeng, S., Hackett, J. B., Lyman, R. F., *et al.* (2000). Genotype-environment interaction for quantitative trait loci affecting life span in *Drosophila melanogaster*. *Genetics* 154:213–27.

Wang, R.-L., Stec, A., Hey, J., Lukens, L., and Doebley, J. (1999). The limits of selection during maize domestication. *Nature* 398:236–9.

Weber, K., Eisman, R., Morey, L., Patty, A., Sparks, J., *et al.* (1999). An analysis of polygenes affecting wing shape on chromosome 3 in *Drosophila melanogaster*. *Genetics* 153:773–86.

Weir, B. S. (1996). *Genetic Data Analysis II*. Sunderland, MA: Sinauer. 445 pp.

Weir, B. S., and Hill, W. G. (1986). Nonuniform recombination within the human β-globin gene cluster. *Am. J. Hum. Genet.* 38:776–8.

Winzeler, E. A., Shoemaker, D. D., Astromoff, A., Liang, H., Anderson, K., *et al.* (1999). Functional characterization of the *S. cerevisiae* genome by gene deletion and parallel analysis. *Science* 285:901–6.

Wolstenholme, D. R., and Thoday, J. M. (1963). Effects of disruptive selection. VII. A third chromosome polymorphism. *Heredity* 10:413–31.

Zeng, Z.-B. (1994). Precision mapping of quantitative trait loci. *Genetics* 136:1457–68.

Zeng, Z.-B., Kao, C.-H., and Basten, C. J. (1999). Estimating the genetic architecture of quantitative traits. *Genet. Res.* 74:279–89.

zur Lage, P., Shrimpton, A. E., Mackay, T. F. C., and Leigh Brown, A. J. (1997). Genetic and molecular analysis of *smooth*, a quantitative trait locus affecting bristle number in *Drosophila melanogaster*. *Genetics* 146:607–18.

4

Gene expression profiling in evolutionary genetics

DANIEL L. HARTL, COLIN D. MEIKLEJOHN,
CRISTIAN I. CASTILLO-DAVIS
Department of Organismic and Evolutionary Biology, Harvard University

DUCCIO CAVALIERI
Harvard Center for Genomics Research, Harvard University

JOSÉ MARIA RANZ, JEFFREY P. TOWNSEND
Department of Organismic and Evolutionary Biology, Harvard University

4.1 Introduction

Lewontin (1974) has characterized much of the history of population genetics as "the struggle to measure variation," especially genetic variation at the molecular level. His characterization portrays a time when population geneticists were severely limited in the techniques that could be applied to organisms in natural populations. Fortunately, during the past 25 years molecular biology has supported a steady stream of innovative approaches and techniques that are widely applicable to natural populations. Chief among these have been chain-termination methods of DNA sequencing (Sanger *et al.* 1977) and the polymerase chain reaction (Saiki *et al.* 1985). From these have emerged high-throughput DNA sequencing strategies resulting in the complete sequences of the genomes of innumerable organelles, viruses, prokaryotes, and agents of infectious disease, as well as the genomes of most of the key model organisms used in molecular genetics and, of course, the human genome. The availability of genomic sequences has already resulted in the new field of comparative genomics (Koonin *et al.* 2000).

By contrast, in population genetics the struggle to measure variation was largely a struggle to detect differences between genotypes of organisms within a single species. Variation within populations is important because it is essential to Darwinism to understand how genetic differences within species become transformed into differences between species over evolutionary time. Here, too, molecular techniques have eased the struggle. Although complete sequencing of multiple genomes from a single species has so far been restricted to a few organelles, viruses, and bacteria, the great interest in human single-nucleotide polymorphisms (Sachidanandam *et al.* 2001) for

The last five authors contributed equally to this work.

The Evolution of Population Biology, ed. R. S. Singh and M. K. Uyenoyama. Published by Cambridge University Press.

disease-association studies has provided impetus for genome-wide studies of polymorphism in other organisms using either low-redundancy shotgun sequencing (e.g., random threefold redundancy provides about 95% genome coverage) or high-density oligonucleotide arrays (Winzeler *et al.* 1998).

The upshot of the last 25 years is that population geneticists are no longer caught up in a struggle to measure variation. Population geneticists are rather awash in variation. The new struggle is not to measure variation, but to interpret and understand genetic variation in the context of population history, geographical structure, patterns of migration, and the internal constraints and external evolutionary forces that affect the level and distribution of genetic variation within and between genomes.

It remains a key challenge for evolutionary biology to learn how variation in genotype within populations of organisms is ultimately associated with adaptive variation in phenotype. Several new molecular approaches offer promising opportunities to meet this challenge. Among these are DNA microarrays that enable expression profiling (genome-wide analysis of the relative abundances of gene transcripts). In principle, this approach can form a bridge connecting genotype with phenotype, because specific, reproducible patterns of transcription associated with particular genotypes may also be associated with particular phenotypes and affect Darwinian fitness. Application of expression profiling to natural populations is still in its earliest stages. Hence, in this chapter we focus on some of the issues that are raised in these applications, and also give some examples.

4.2 *Déjà vu* all over again?

Although it is an admirable tradition of evolutionary biologists to put techniques developed by others to their own uses, the application to evolutionary issues often challenges the techniques and interpretations beyond the range of their original intent. Protein electrophoresis provides an example. A major advance in the study of genetic variation came with the use of starch-gel electrophoresis to study protein variation in natural populations of *Drosophila* (Hubby and Lewontin 1966, Lewontin and Hubby 1966). Starch-gel electrophoresis had been invented much earlier by Smithies (1954), who employed it to detect human genetic variation. However, when electrophoresis was used for a systematic analysis of genetic variation in natural populations, it became important to know what fraction of amino acid replacements present in a population are detected as changes in electrophoretic mobility, how many amino acid replacements differ between two electrophoretically distinct proteins encoded by alleles of a single gene, and whether polymorphic amino acid replacements affect fitness (Lewontin 1991). These issues were largely irrelevant in the original use of the technique to separate heterogeneous serum proteins (Smithies 1995). Likewise, the application of DNA sequencing to natural populations poses unique problems, because differences in sequence between alleles do not necessarily reflect differences in fitness between genotypes. The challenge of inferring population structure, population history, and evolutionary forces based on DNA sequence has stimulated important advances in theoretical population genetics (Yang 1998, Fu and Li 1999, Wakeley 1999, Nielsen and Wakeley 2001, Bustamante *et al.* 2002).

The application of expression profiling to natural populations also raises special problems not encountered in typical experiments with laboratory populations. For example, one standard kind of such experiments examines global gene expression in a knockout mutant compared against an isogenic wild-type strain grown under the same conditions (Hughes *et al.* 2000). Hence there is one variable only – the gene that has been knocked out. Another standard kind of experiment examines gene expression of a single strain under two or more conditions, such as yeast cells grown in the presence or absence of ethanol (Alexandre *et al.* 2001). Here again there is a single variable, this time environmental in origin. In both examples, the genotypes are under the experimenter's control.

With natural populations, the situation is different. Most natural populations are highly heterozygous (Lewontin 1974), hence genotypes differ within populations as well as between populations. How much of the sequence variation between organisms is reflected in differences in gene expression that can be detected experimentally? In studying natural populations, a major issue to be faced in dealing with expression profiling is that, because of heterozygosity within populations, differential expression of any given gene may result from differences in DNA sequence within regulatory domains of the gene itself (cis-acting regulatory elements), or from epistatic effects of differences in one or more other genes (trans-acting regulatory elements). Correlating sequence variation with expression-profile variation therefore requires much more than a spreadsheet of genes, sequences, and expression levels.

In addition to issues arising from epistasis, another comes from potential nonlinearity of the transcriptional response, because a relatively small change in the expression of a trans-acting regulatory gene may produce a large change in the transcriptional level of the genes that it affects. While these and other problems should not be minimized, there is nevertheless great potential for expression profiling to reveal which differences in gene expression are associated with differences in fitness, and to learn how these differences relate to environmental variables in the habitat. Understanding the contributions of sequence variation and epistasis to differences in gene expression is vital to an incorporation of expression profiling into evolutionary genetics.

4.3 DNA microarrays and their applications

Two types of DNA microarrays ("DNA chips") are commonly available. The first consists of 200 000–400 000 synthetic oligonucleotides (Chee *et al.* 1996,

Lipshutz *et al.* 1999), typically 25-mers, which are complementary to regions of genomic sequence and can be used for genotyping (Gingeras *et al.* 1998, Winzeler *et al.* 1998) or for assaying relative levels of gene expression (Lockhart *et al.* 1996, Lockhart and Winzeler 2000). The second consists of up to 20 000 DNA fragments, typically cDNA clones amplified by the polymerase chain reaction, which are used primarily for competitive RNA hybridization to assess relative levels of gene expression (DeRisi *et al.* 1997, Eisen and Brown 1999). Hybridization of mRNA samples is equivalent to a massively parallel set of Northern blots, and hybridization of a genomic DNA sample is equivalent to a whole-genome Southern blot. Hence the fundamental principle of nucleic acid hybridization with labeled probes is nothing new: microarrays are merely the application of microscale high-throughput technology to leverage these techniques.

Whether used for clinical diagnosis or studies in evolutionary genetics, DNA microarrays generate a blizzard of data concerning the relative abundance (in two or more samples of mRNA, cDNA, or genomic DNA) of molecules that have high sequence similarity to the DNA at a particular position in a microarray. The ability to distinguish between signal and background noise is determined by the hybridization methods, the number of replicates of a given experiment, and the nature of the statistical analysis. In principle, if there is adequate replication, even a small difference in abundance of an mRNA sequence between two samples can be determined with a high degree of statistical confidence. However, once a difference in relative abundance has been found, microarrays alone are limited in their ability to identify the source of the difference. This is because a significant difference in level of an mRNA species between two samples could be due to a difference at any point in a hierarchy of regulatory processes that affect transcription, RNA processing, or mRNA stability. These include differences in both cis-regulatory and trans-regulatory elements, and in some cases differential mRNA abundance reflects changes in gene copy number (Lucito *et al.* 2000). One potential pitfall in evolutionary applications comes from the ability of individual members of conserved multigene families to cross-hybridize, which confounds gene expression with gene copy number. This caveat is particularly relevant to transposable elements, whose copy number can differ dramatically between individuals and between species. Consequently, for any particular coding sequence, tracking down the source of an observed difference requires more detailed study of the gene and its regulatory elements.

To date, most applications of microarray technology have dealt with clinical diagnosis (DeRisi *et al.* 1996, Kononen *et al.* 1998, Scherf *et al.* 2000), identifying interactions among metabolic pathways that control metabolic flux (DeRisi *et al.* 1997, Wodicka *et al.* 1997), investigating control mechanisms of the cell cycle (Cho *et al.* 1998, Chu *et al.* 1998), comparing patterns of gene expression of organisms with the same genotype grown under different conditions (Hardwick *et al.* 1999, Jelinsky and Samson 1999, Hughes *et al.* 2000,

Roberts *et al.* 2000), or describing the patterns in which genes are progressively deployed during development (White *et al.* 1999, Hill *et al.* 2000). Among the first applications of DNA microarrays to evolutionary studies were those of yeast strains that had undergone adaptive evolution in laboratory culture (Ferea *et al.* 1999).

With regard to natural populations, few studies have been published that deal with the extent of variation in either levels or patterns of global gene expression among organisms isolated from natural environments. Several key questions set the agenda for molecular evolutionary biology as it enters the post-genomics era. Is there significant variation in gene expression from one organism to the next? How many genes are differentially regulated, and to what extent? What are the molecular mechanisms behind the regulatory variation? Are there particular sets of related genes whose expression tends to vary together as a result of pleiotropic effects?

4.4 Statistical analysis

Considering the large number of pairwise and higher-order interactions that are possible, the statistical analysis of genome-wide expression data is a multivariate problem of extremely high dimensionality. Several types of clustering algorithms, such as hierarchical clustering or self-organizing maps, have been suggested as exploratory tools to identify genes with correlated expression profiles in order to discover networks of coordinately expressed genes and to assign open reading frames of unknown function to regulatory clusters (Bittner *et al.* 1999, Tavazoie *et al.* 1999, Alter *et al.* 2000, Holter *et al.* 2000, Jensen and Knudson 2000, Kim *et al.* 2000). More recent approaches have exploited analysis of variance to decompose the variance in gene expression into components due to genotype, sex, environment, and other factors, as well as due to their interactions (Kerr *et al.* 2000, Gibson 2002).

For application to natural populations, the first issue for expression profiling is level of resolution. What magnitude of difference in relative intensity of signal (for example, signal from mRNA isolated from two strains of *Drosophila*) can be regarded as significantly different from background noise? Much of the earlier literature uses the rule of thumb that a difference in relative intensity is considered significantly above background if the magnitude of the larger intensity is at least twofold greater than that of the smaller intensity. One problem with this rule is that it fails to take replications into account, because a difference that is smaller than twofold but consistent across multiple independent experiments is likely to be biologically meaningful. Another problem with the twofold rule is that false positives will often occur when the weaker signal is near the background level, because the signal intensity is usually corrected by subtracting the background, which in the twofold rule results in division by a number close to zero. This is, of course, a problem with any method of analysis that uses ratios.

Another approach to assessing the significance of a difference would be a conventional Student's *t*-test in which the difference between means for the same gene across replicates is compared with the average standard deviation across replicates. This approach has the drawback that, unless the number of replicates is relatively large, the variance across replicates tends to be underestimated (Baldi and Long 2001). To overcome this limitation, Baldi and Long (2001) have developed a hierarchical Bayesian approach to the *t*-test that regards the logarithm of the ratio of expression levels for each gene as a normal distribution whose mean, given the variance, is itself assumed to be normal and whose variance is assumed to be inverse gamma. From this analysis they implement a *t*-test by adjusting the empirical variance of each log-expression value according to a local background variance associated with neighboring genes in the microarray. A type of nonparametric *t*-test has also been suggested, in which the significance level for a given gene is assessed by comparing the test statistic with those obtained from random permutations of the data (Dudoit *et al.* 2000).

For evolutionary genetics, it is important to be able to compare levels of gene expression across experiments with different genotypes, because typically genotype *A* will be compared in one experiment with genotype *B*, genotype *B* in another experiment with genotype *C*, genotype *C* in still another experiment with genotype *D*, and so forth. This design is complicated further by the fact that each experiment may be replicated a different number of times, and there may also be some ad hoc comparisons, for example, genotype *A* with genotype *C*. Any real experiment is likely to be unbalanced in its comparisons and asymmetrical in its number of replicates, if for no other reason than that not all microarray hybridization experiments yield satisfactory data, and also because repetition of failed experiments is avoided if possible because of the time and expense.

One example of such a set of comparisons is shown in Figure 4.1, taken from unpublished work of J. Townsend and D. Cavalieri. The organisms in question are a parental strain of homothallic diploid yeast (M28) and four diploid progeny obtained from germination of the spores in a single tetrad (S1, S2, F1, and F2). Each arrow indicates a single microarray hybridization, and the arrowhead points to the strain whose mRNA was labeled with cyanine 3 fluorochrome ("green") vs. cyanine 5 fluorochrome ("red"). The diagram

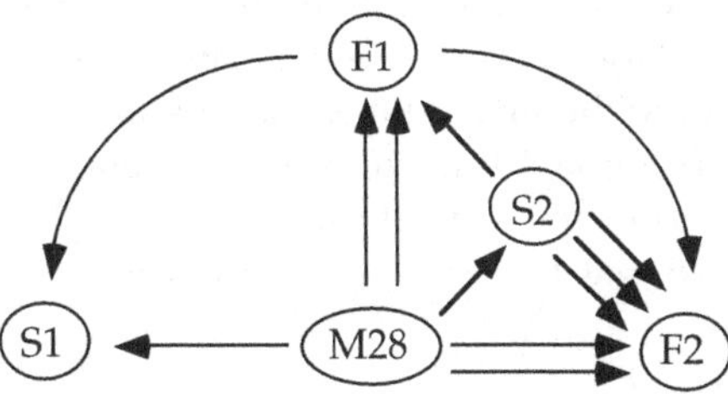

Figure 4.1. A typical unbalanced experimental design of gene-expression comparisons. M28 is a homothallic diploid natural isolate of *S. cerevisiae*, and S1, S2, F1, and F2 are four diploid progeny obtained from a single ascus.

is not included to suggest this as an ideal experimental design, but rather to show the unpredictable result of simultaneous exploratory research and technology development. The initial exploratory research indicated that, in addition to the parental strain, the progeny S2 and F2 were of greatest interest, hence these strains warranted the greatest number of additional experiments (Cavalieri *et al.* 2000). However, an optimal method of analysis would allow the multiple strains, unbalanced design, and unequal numbers of replicates to be taken into account to obtain, for each gene, an estimate of the relative expression level in each strain and a confidence interval around this estimate.

To enable the analysis of such complex experimental designs, Townsend and Hartl (2003) has implemented a Bayesian analysis known as BAGEL (the acronym stands for Bayesian Analysis of Gene Expression Levels). In this approach, the level of each signal (green or red) in each experiment is regarded as coming from a normal distribution with some unknown mean and variance. Each gene is treated as independent. This assumption is not literally true for genes that are coordinately regulated, but it serves as a useful first approximation for evaluating gene-expression levels on a gene-by-gene basis. The algorithm implemented in BAGEL is based on the hierarchical assumption that the mean and variance of the normal distribution of signal intensity at any position in the microarray are random variables with uniform distributions on the positive real numbers. These distributions are vague, and they are also "improper" in the sense that they do not integrate to 1. Such a choice gives primacy to the observations, rather than the prior distributions, in estimating the parameters of interest.

The conditional distributions of the parameters in this model, given the data, are based on computer simulations defining a Markov chain that converges to the correct stationary distribution. The mean, mode, credible interval, and other characteristics of the parameter distributions are then approximated by sampling from a long trajectory of this Markov chain. In Bayesian analysis, the 95% credible interval is the analog of the 95% confidence interval in conventional frequentist statistics, although the credible interval has the straightforward interpretation that 95% of the simulated realizations of the sample mean fall within the credible interval.

In the analysis of expression profiles of yeast isolates from natural populations discussed below, the mean levels of gene expression have been estimated using BAGEL. The mean expression levels are relative values, not absolutes. This is because the mean expression level for any gene in any treatment (e.g., genotype) is expressed as a ratio relative to the smallest mean across all treatments. Hence the expression levels are ranked with the smallest value set arbitrarily to 1. As a conservative criterion for a significant difference between any pair of estimates, we require that the 95% credible intervals be nonoverlapping.

4.5 DNA microarrays in evolutionary genetics: some specific examples

Microarray technology has only recently been used for evolutionary studies. *Saccharomyces cerevisiae* was the first eukaryotic organism for which DNA microarrays became available, and partly for this reason budding yeast is being used increasingly as a model system to study evolution in laboratory populations (Ferea *et al.* 1999, Zeyl 2000). Among other approaches, Hartwell *et al.* (1999) have advocated examining evolutionary constraints on functional modules in order to dissect the modularity of living systems, and Murray (2000) has stressed the issue of the evolvability of organisms.

Use of expression profiling to study the population genetics and molecular evolution of natural populations is only just beginning, but already it is clear that the possibilities are virtually unlimited. In this section we give some examples. Necessarily they have emerged from genomic sequencing and other research resources in model organisms (*Caenorhabditis elegans, Saccharomyces cerevisiae,* and *Drosophila melanogaster*), because these are the organisms in which DNA microarrays first became available. However, as the technology has improved and disseminated, comparable approaches are rapidly being developed in many other organisms of evolutionary interest.

4.5.1 Molecular evolution of genes expressed early and late in nematode development

Microarray technology offers a substantial bridge to connect organismic with molecular evolution. For example, global gene-expression studies on the developmental timing and tissue-specific expression of genes (White *et al.* 1999, Hill *et al.* 2000) open a large new field for evolutionary investigations. Among the issues that can be addressed are the role of genomic and developmental complexity in influencing evolutionary change, the relationship between variation in gene sequences vs. variation in gene expression, and the molecular mechanisms of intraspecific and interspecific morphological divergence.

Using gene-expression data to inform molecular evolutionary studies has already begun to yield interesting results concerning the relationships between organismic and molecular evolution. For example, it has been shown that rates of protein evolution in mammals are markedly affected by tissue-specific patterns of expression; in particular, rates of nonsynonymous substitution are negatively correlated with breadth of tissue expression, which has been interpreted as resulting from selection against mutations that would result in strongly pleiotropic effects (Duret and Mouchiroud 2001). In a comparison of genes between *C. elegans* and *C. briggsae,* Castillo-Davis and Hartl (2002) found that genes expressed during embryogenesis have significantly fewer paralogs than do genes expressed after embryogenesis is completed. Among early-expressed genes, 5–15% have paralogous copies, whereas among late-expressed genes 35–40% have paralogous copies in both genomes. It is not yet known whether the greater number of paralogs results from positive selection for tissue-specific expression or function (e.g., olfactory receptors) or from a shorter average persistence of duplicated copies of genes that are expressed early in embryogenesis due to deleterious dosage effect.

Whole-genome expression data from *C. elegans* along with comparative genomic sequence from *C. briggsae* have also afforded a test of the hypothesis that genes expressed early in embryogenesis are more conserved in sequence through evolutionary time than genes expressed later in life. For the number of nonsynomymous substitutions per nonsynomymous site, the 95% confidence intervals are 0.034–0.058 for genes expressed early and 0.040–0.067 for genes expressed late. There is clearly no significant difference, although this analysis deals with the average protein-wide rates of evolution rather than with the rates for particular functional sequence motifs. In contrast to the results for nonsynomymous substitutions, the rates of synonymous substitutions are very different between genes expressed early and late. For the number of synonymous substitutions per synonymous site, the 95% confidence intervals are 1.09–1.63 for the early genes and 0.76–0.98 for the late genes (Castillo-Davis and Hartl 2002). Genome-wide there is also a highly significant positive correlation between expression level and codon usage bias (Castillo-Davis and Hartl 2002). Because late-expressed genes have, on the average, higher expression levels than early-expressed genes (Castillo-Davis and Hartl 2002, from data in Hill *et al.* 2000), and hence more codon usage bias, a smaller rate of synonymous substitution is expected.

4.5.2 Expression profiling of natural isolates of vineyard yeast

For many years it was a mystery where yeast could be found in a natural environment, because yeast could not easily be isolated from grapes in vineyards (Mortimer 1999). Yet the organism must exist in vineyards, or in the wineries themselves, because crushed grapes begin to ferment owing to the presence of naturally occurring yeast. The mystery was resolved by the discovery that, whereas wine yeast is rarely found on grapes with an unbroken skin, viable cells are found in about one-third of damaged berries, inside of which they establish a little fermentation chamber (Mortimer 1999). Furthermore, most vineyard isolates are diploid and about 70% are homothallic (Mortimer 2000), which means that, after sporulation and the germination of a haploid spore, the mother cell changes mating type and mates with the daughter cell, forming a diploid that is completely homozygous except for the mating type locus. The diploid phase must persist long enough without sporulation to allow a significant number of mutations to occur, because there is known to be a great deal of functional heterozygosity among vineyard isolates. For example, when vineyard isolates are sporulated and their progeny tested for growth on the sugars sucrose, maltose, and galactose, approximately 67% of the isolates segregate for the inability to utilize at least one of these sugars (Cavalieri *et al.* 1998). When a homothallic diploid undergoes sporulation, each haploid spore undergoes germination, mating-type switching, and mating to form a homozygous diploid once again. This process of rendering the genome completely homozygous has been called "genome renewal" (Mortimer *et al.* 1994).

Among natural isolates of vineyard yeast isolated from around the Tuscan wine capital of Montalcino, any randomly chosen pair of strains shows a significant difference in expression level for 1–2% of their genes (Townsend *et al.* 2003). Although the percentage of differentially expressed genes is relatively small, the absolute number is large (60–120 genes), and the magnitude of the differential expression is rather large (e.g., twofold or more).

One interesting difference is found in the gene *SSU1*, which encodes a sulfite exporter. Mean expression levels in four Tuscan isolates are shown in Figure 4.2. The comparisons were carried out with a "circular" design in which each isolate is compared with the next in line, completing the circle by comparing the last with the first; each comparison was replicated twice by interchanging the fluors, and the mean relative expression levels and their 95% credible intervals were estimated using BAGEL. Each of the Tuscan isolates has an *SSU1* expression level that is different from the others, and the

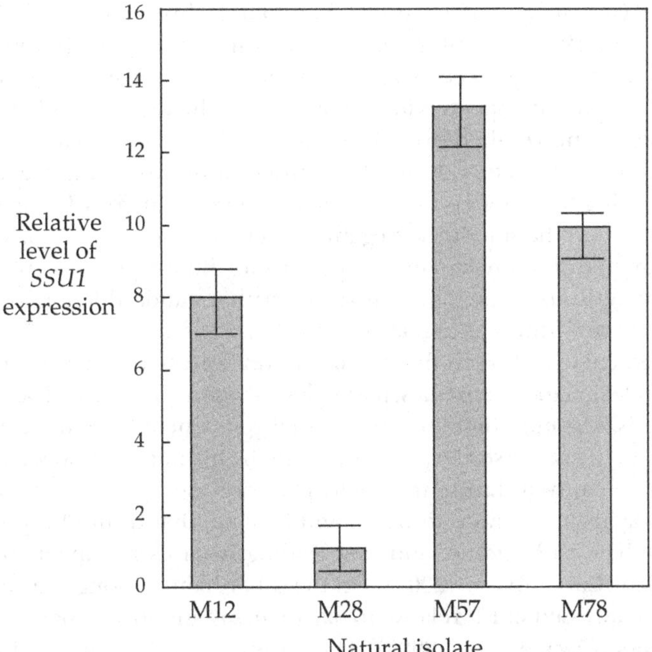

Figure 4.2. Mean expression levels of the sulfite exporter *SSU1* in four natural isolates of *S. cerevisiae* and their 95% credible intervals.

differences are all statistically significant. M28 has low expression of this sulfite exporter, and its progeny show segregation of a gene strongly affecting the closely related sulfur assimilation and methionine biosynthesis pathways (see below). It is interesting to speculate that the differences in *SSU1* expression may result from natural selection for resistance to sulfites in the environment. For as long as 200 years, Tuscan vintners have been treating vineyards with copper sulfate to inhibit the growth of molds on the grapes, and sodium sulfite, potassium metabisulfite, and sulfur dioxide are widely used during and after fermentation to stabilize the wine and kill bacteria. In this case the expression profiling may have revealed an evolutionary consequence of changes in the chemical ecology of vineyards.

The isolate M28 in Figures 4.1 and 4.2 proved to be particularly interesting. The original isolate attracted attention because it exhibited a slightly rough, delicately filigreed colony morphology. The progeny colonies showed 2:2 segregation for a smooth morphology (spores S1 and S2) or a more extremely filigreed morphology (spores F1 and F2). These are the organisms depicted in Figure 4.1.

For the full BAGEL analysis of the 12 arrays depicted in Figure 4.1, some 145 genes (~2% of the genome) had estimated levels of expression in F1 and S2 whose 95% credible intervals were nonoverlapping; 101 of these showed greater expression in F1, and 44 showed greater expression in S2. In contrast, the full BAGEL analysis of F1 and F2 showed only three genes (<0.1% of the genome) whose 95% credible intervals were nonoverlapping. The discrepancy of 145:3 is far too large to be explained by random segregation in two pairs of ascospores, hence it appeared likely that most of the differences between the F and S segregants resulted from the epistatic effects of one factor, or perhaps a small number of factors. From the standpoint of evolutionary genetics, it is worth noting that there is no detectable difference in growth rate between the filigreed and the smooth segregants under laboratory conditions, in spite of the dramatic difference in expression profile. Both types of segregants are extremely vigorous; in fact, in comparison with a standard laboratory strain, they both have a fitness advantage of 33–50%.

Detailed analysis showed that the most overexpressed genes in F1 relative to S2 encode metabolic enzymes, particularly those associated with amino acid biosynthesis. The most highly overexpressed genes include 12 in the methionine pathway, two in the serine pathway, two in the histidine pathway, and one in the arginine pathway. Entire metabolic pathways are upregulated, including the pathway from pyruvate to valine and leucine, that from phosphoribosyl pyrophosphate to histidine, and that leading from extracellular surface to methionine (Cavalieri *et al.* 2000). On the other hand, among the genes that are underexpressed in F1, relative to S2, are many amino acid permeases and transporters (Cavalieri *et al.* 2000). These results may be compared to those of Wodicka *et al.* (1997), who examined the expression profiles of a laboratory strain grown in minimal medium versus rich medium. Among 51 genes expressed more abundantly in minimal medium, 8 encoded proteins involved in arginine and methionine biosynthesis, and among 84 genes expressed much more abundantly in rich medium, 7 encoded amino acid permeases.

The gene *PHD1* is a good example of the additional power gained from taking all experiments and replicates into account, even with a design as complex as that in Figure 4.1. This gene is a principal transcriptional regulator of filamentous growth in the morphogenetic pathway induced by ammonia starvation (Gimeno and Fink 1994). Since the *MEP2* ammonium permease regulates pseudohyphal differentiation in *S. cerevisiae* (Lorenz and Heitman 1998), and *MEP2* is overexpressed sixfold in the filamentous segregants, we were somewhat surprised to find that the upregulation of *PHD1* was less than twofold (Cavalieri *et al.* 2000). However, BAGEL analysis of the results of the experiments in Figure 4.1 yield the mean levels of *PHD1* expression and the 95% credible intervals shown in Figure 4.3. The mean expression of each strain is shown immediately above the bar. Both of the filigreed segregants have a level of *PHD1* expression that is significantly greater than either of the smooth segregants, even though the difference between F2 and S1 is a factor

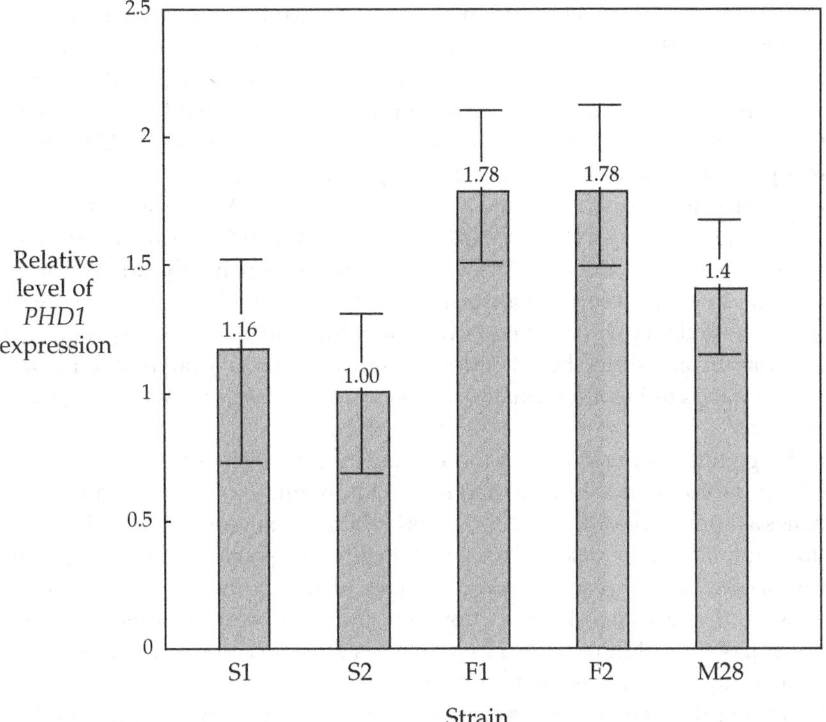

Figure 4.3. Mean expression levels of *PHD1*, a transcriptional regulator of filamentous growth, and their 95% credible intervals, estimated from a Bayesian analysis of the comparisons in Figure 4.1.

of 1.53. Hence, with sufficient replication, even relatively small differences can become significant. Although *PHD1* expression in M28 is not significantly different from either type of segregant, it is interesting that its level is intermediate, which is consistent with its less pronounced filigreed colony phenotype.

4.5.3 Variation in gene expression in *Drosophila*

Observations of Meiklejohn *et al.* (2003) and Ranz *et al.* (2003) also indicate that variation in gene-expression profiles exists between wild-type populations of *Drosophila melanogaster.* Using the standard twofold cutoff for a significant difference in replicated hybridizations, they found that 0.5–2% of approximately 4500 genes are differentially expressed between different strains, either strains collected from different parts of the world or different wild-type laboratory strains. This is in roughly the same range as found in yeast, and it will be interesting to learn whether the similarity holds as more extensive surveys of natural populations are carried out.

Thus far, the observed pattern of genome-wide expression variation follows the degree of genetic differentiation of the strains inferred from the sequences of a relatively small number of individual genes. In particular, strains from Africa (Zimbabwe) appear to harbor more sequence diversity than strains from elsewhere in the world (Begun and Aquadro 1993), and some African strains may be partially reproductively isolated from the rest of the species (Alipaz *et al.* 2001). In regard to gene-expression profiles, there is a greater number of differentially expressed genes (twofold cutoff) between African strains (32/4489 = 0.71%) and between African and laboratory strains (44/3793 = 1.16% and 69/4250 = 1.62%) than between any two laboratory strains (20/4427 = 0.45%). Eight genes that were highly variable (more than twofold in three comparisons) were analyzed by BAGEL. These eight genes showed varying patterns of variation, with some differences specific to a given strain, and some shared between Zimbabwe and laboratory strains. One gene appears to be differentially expressed at four levels between the strains analyzed.

Large differences in expression profile are associated with sex. Up to 6% of genes examined are more than fivefold upregulated or downregulated in males as compared with females. This is far larger than the differential expression found between males or between females, irrespective of their geographical origin. However, such a large number of differentially regulated genes between the sexes may be more apparent than real, because some unknown fraction of these differences undoubtedly results from allometric relationships of anatomy and cell types between the sexes.

Among the male-specific genes, a few are also differentially expressed in males from different populations. These sex-specific and geographically divergent genes are candidates for sexual selection or fast-evolving genes involved in reproduction (Civetta and Singh 1995, Wu *et al.* 1996, Swanson *et al.* 2001).

Interestingly, one gene that is highly female specific (upregulated an average of 55-fold in females) is downregulated in Zimbabwe males relative to laboratory strains. Why a female-specific gene would be downregulated in Zimbabwe males (or upregulated in laboratory males) is unclear. One possibility is that high expression of the gene has deleterious pleiotropic effects in males, but another is that the divergence is selectively neutral and results either from random genetic drift or from the manifestation in males of differential regulation in females.

Hybridization of microarrays with genomic DNA can also be used to identify differences in copy number between genotypes. For example, two genes appear to be either greatly reduced in copy number or entirely missing in Canton-S relative to a Zimbabwe strain, and one gene appears to be missing in the Zimbabwe strain relative to Canton-S. Two of these sequences are transposable elements, so differences in copy number are to be expected, but the third is unique-sequence DNA missing in Canton-S.

4.5.4 Interspecific comparisons in *Drosophila*

DNA microarrays also allow whole-genome comparisons between species that otherwise must be done on a gene-by-gene basis. Using traditional Southern blots, Schmid and Tautz (1997) found that about one-third of the open reading frames had diverged enough between *D. melanogaster* and *D. virilis* to fail in cross-hybridization. A similar figure was inferred from the success rate of *in situ* hybridizations of probes from *D. melanogaster* onto polytene chromosomes of *D. repleta* (Ranz *et al.* 2001). These species pairs both have estimated divergence times of 40 million years (Russo *et al.* 1995).

An example of this approach using microarrays comes from unpublished work of Ranz *et al.* (2003). They carried out a comparison between *D. melanogaster* and *D. simulans* in order to assess to what extent the expression profile of adults of the species has been modified during the approximately 3 million years since their divergence from a common ancestor. For this purpose, cDNA microarrays from *D. melanogaster* were used for competitive hybridization with mRNA extracted from adults of each sex of each species. As shown in the top part of Table 4.1, 4.4% of approximately 4800 spots on the microarray showed differential expression between the species using a twofold cutoff. About 75% of these genes are apparently overexpressed in *D. melanogaster*, relative to *D. simulans*. A total of 170 genes are differentially expressed in the other species in one sex only, and among these genes more than 80% are differentially expressed in males. How these findings may relate to the rapid sequence evolution of male-specific genes (Civetta and Singh 1995, 1998, Wu *et al.* 1996) has yet to be determined.

A somewhat different picture emerges when genes differentially expressed in the sexes are compared between the species. About 15% of the spots on the microarray fall into this category, and the comparative data are shown

Table 4.1. *Comparison of expression profiles of* D. melanogaster *and* D. simulans

Genes differentially expressed by species and overexpressed twofold or more in:	
Both sexes of *D. melanogaster*	34
Both sexes of *D. simulans*	11
Males only of *D. melanogaster*	114
Males only of *D. simulans*	28
Females only of *D. melanogaster*	18
Females only of *D. simulans*	10
TOTAL	215 (4.4%)
Genes differentially expressed by sex and overexpressed twofold or more in:	
Males of both *D. melanogaster* and *D. simulans*	211
Males of *D. melanogaster* only	90
Males of *D. simulans* only	63
Females of both *D. melanogaster* and *D. simulans*	251
Females of *D. melanogaster* only	182
Females of *D. simulans* only	102
TOTAL	899 (19%)

in the bottom part of Table 4.1. Among these genes, the majority (60%) are overexpressed in females. However, among those that are overexpressed in only one of the species, the majority (66%) are apparently overexpressed in males only.

We stress the use of the term "apparently overexpressed" when comparing *D. melanogaster* with *D. simulans* because the efficiency of hybridization of a probe depends not only on the relative expression level but also on the degree of sequence divergence. To distinguish between these possibilities it is also necessary to carry out competitive hybridizations with genomic DNA from the species. The results of Ranz *et al.* (2003) indicate that a relatively small fraction of the differences in Table 4.1 can be attributed to sequence divergence. Hence, each of the genes that shows differential expression between the species is a candidate for having undergone interesting and informative regulatory changes in its recent evolution.

4.6 Concluding remarks

In population genetics, much of the last third of the twentieth century was devoted to answering the question: How much genetic variation is present in natural populations and how is it maintained? The struggle to measure variation was eased first by the application of protein electrophoresis (Lewontin 1974) and then virtually eliminated by the development of automated DNA sequencing and the polymerase chain reaction. The "how much" part of the question has been answered (or at least is answerable) in any organism to which these techniques can be applied. Much more difficult are issues related to the adaptive significance of genetic variation within species and of genetic differences between species. The neutral theory (Kimura 1983) was debated fervently when data were scant and analytical methods only beginning to be developed. Much has happened both technologically and analytically, making it possible to close the book on the neutral theory and to turn to other questions. For example, how does polymorphism and divergence of regulatory elements compare with polymorphism and divergence of protein-coding sequences? Do genes with high levels of amino acid polymorphism also have high levels of expression polymorphism? Do genes that show evidence of positive selection at the amino acid level also show evidence of positive selection of their regulatory sequences? Do genes that show evidence of relaxed selection at the amino acid level also have relaxed selection on regulatory variation? More generally, considering the recent advances in functional genomics and proteomics, including expression profiling, the opportunities are auspicious for evolutionary biologists to be able to identify the genetic, cellular, developmental, and organismic changes that drive adaptive evolution and speciation.

4.7 Acknowledgments

This work was supported by NIH grants GM60035, GM58423, and HG01250 to DLH, and by a fellowship from the National Research Council of Spain to JMR. We thank John Parsch, Yun Tao, and Justin Blumenstiel of the Hartl lab for their help and advice with *Drosophila* microarrays; the Harvard Center for Genomics Research, especially Rachel Erlich, Claire Bailey, and Andrew Murray for their support and encouragement; and all the other members of the "*Drosophila* chip" consortium including Neal Silverman, Inez Alvarez-Garcia, Eric Bernstein, Yong Dai, Rich Dearborne, Stephanie Mohr, Sam Kunes, Bill Gelbart, and Tom Maniatis.

REFERENCES

Alexandre, H., Ansanay-Galeote, V., Dequin, S., and Blondin, B. (2001). Global gene expression during short-term ethanol stress in *Saccharomyces cerevisiae. FEBS Letters* 498:98–103.

Alipaz, J. A., Wu, C. I., and Karr, T. L. (2001). Gametic incompatibilities between races of *Drosophila melanogaster. Proc. R. Soc. Lond. Ser. B* 268:789–95.

Alter, O., Brown, P. O., and Botstein, D. (2000). Singular value decomposition for genome-wide expression data processing and modeling. *Proc. Natl. Acad. Sci. USA* 97:10101–6.

Baldi, P., and Long, A. D. (2001). A Bayesian framework for the analysis of microarray expression data: regularized *t*-test and statistical inferences of gene changes. *Bioinformatics* 17:509–19.

Begun, D. J., and Aquadro, C. F. (1993). African and North American populations of *Drosophila melanogaster* are very different. *Nature* 365:548–50.

Bittner, M., Meltzer, P., and Trent, J. (1999). Data analysis and integration: of steps and arrows. *Nature Genet.* 22:213–5.

Bustamante, C., Nielsen, R., Sawyer, S. A., Olsen, K. M., Purugganan, M. D., and Hartl, D. L. (2002). The cost of inbreeding in *Arabidopsis*. *Nature* 416:531–4.

Castillo-Davis, C. I., and Hartl, D. L. (2002). Genome evolution and developmental constraint in *Caenorhabditis elegans*. *Mol. Biol. Evol.* 19:728–35.

Cavalieri, D., Barberio, C., Casalone, E., Pinzauti, F., Sebastiani, F., Mortimer, R. K., and Polsinelli, M. (1998). Genetic and molecular diversity in *S. cerevisiae* natural populations. *Food Technol. Biotechnol.* 36:45–50.

Cavalieri, D., Townsend, J. P., and Hartl, D. L. (2000). Manifold anomalies in gene expression in a vineyard isolate of *Saccharomyces cerevisiae* revealed by DNA microarray analysis. *Proc. Natl. Acad. Sci. USA* 97:12369–74.

Chee, M., Yang, R., Hubbell, E., Berno, A., Huang, X. C., Stern, D., Winkler, J., Lockhart, D. J., Morris, M. S., and Fodor, S. P. A. (1996). Accessing genetic information with high-density DNA arrays. *Science* 274:610–14.

Cho, R. J., Campbell, M. J., Winzeler, E. A., Steinmetz, L., Conway, A., Wodicka, L., Wolfsberg, T. G., Gabrielian, A. E., Landsman, D., Lockhart, D. J., and Davis, R. W. (1998). A genome-wide transcriptional analysis of the mitotic cell cycle. *Mol. Cell* 2:65–73.

Chu, S., DeRisi, J., Eisen, M., Mulholland, J., Botstein, D., Brown, P. O., and Herskowitz, I. (1998). The transcriptional program of sporulation in budding yeast. *Science* 282:699–705.

Civetta, A., and Singh, R. S. (1995). High divergence of reproductive tract proteins and their association with postzygotic reproductive isolation in *Drosophila melanogaster* and *Drosophila virilis* group species. *J. Mol. Evol.* 41:1085–95.

Civetta, A., and Singh, R. S. (1998). Sex-related genes, directional sexual selection, and speciation. *Mol. Biol. Evol.* 15:901–9.

DeRisi, J., Penland, L., Brown, P. O., Bittner, M. L., Meltzer, P. S., Ray, M., Chen, Y. D., Su, Y. A., and Trent, J. M. (1996). Use of a cDNA microarray to analyse gene expression patterns in human cancer. *Nat. Genet.* 14:457–60.

DeRisi, J. L., Iyer, V. R., and Brown, P. O. (1997). Exploring the metabolic and genetic control of gene expression on a genomic scale. *Science* 278:680–6.

Dudoit, S., Yang, Y. H., Callow, M. J., and Speed, T. P. (2000). Statistical methods for identifying differentially expressed genes in replicated cDNA microarray experiments. *Dept. Statistics Technical Report No. 578*, Univ. Calif., Berkeley.

Duret, L., and Mouchiroud, D. (2001). Determinants of substitution rates in mammalian genes: expression pattern affects selection intensity but not mutation rate. *Mol. Biol. Evol.* 17:68–74.

Eisen, M. B., and Brown, P. O. (1999). DNA arrays for analysis of gene expression. *Methods Enzymol.* 303:179–205.

Ferea, T. L., Botstein, D., Brown, P. O., and Rosenzweig, R. F. (1999). Systematic changes in gene expression patterns following adaptive evolution in yeast. *Proc. Natl. Acad. Sci. USA* 96:9721–6.

Fu, Y. X., and Li, W. H. (1999). Coalescing into the 21st century: an overview and prospects of coalescent theory. *Theor. Popul. Biol.* 56:1–10.

Gibson, G. (2002). Microarrays in ecology and evolution: a preview. *Mol. Ecol.* 11:17–24.

Gimeno, C. J., and Fink, G. R. (1994). Induction of pseudohyphal growth by overexpression of *PHD1*, a *Saccharomyces cerevisiae* gene related to transcriptional regulators of fungal development. *Mol. Cell. Biol.* 14:2100–12.

Gingeras, T. R., Ghandour, G., Wang, E. G., Berno, A., Small, P. M., Drobniewski, F., Alland, D., Desmond, E., Holodniy, M., and Drenkow, J. (1998). Simultaneous genotyping and species identification using hybridization pattern recognition analysis of generic Mycobacterium DNA arrays. *PCR Meth. Appl.* 8:435–48.

Hardwick, J. S., Kuruvilla, F. G., Tong, J. K., Shamji, A. F., and Schreiber, S. L. (1999). Rapamycin-modulated transcription defines the subset of nutrient-sensitive signaling pathways directly controlled by the Tor proteins. *Proc. Natl. Acad. Sci. USA* 96:14866–70.

Hartwell, L. H., Hopfield, J. J., Leibler, S. S., and Murray, A. W. (1999). From molecular to modular biology. *Nature* 402:C47–52.

Hill, A. A., Hunter, C. P., Tsung, B. T., Tucker-Kellogg, G., and Brown, E. L. (2000). Genomic analysis of gene expression in *C. elegans*. *Science* 290:809–12.

Holter, N. S., Mitra, M., Maritan, A., Cieplak, M., Banavar, J. R., and Fedoroff, N. V. (2000). Fundamental patterns underlying gene expression profiles: simplicity from complexity. *Proc. Natl. Acad. Sci. USA* 97:8409–14.

Hubby, J. L., and Lewontin, R. C. (1966). A molecular approach to the study of genic heterozygosity in natural populations I. The number of alleles at different loci in *Drosophila pseudoobscura*. *Genetics* 54:577–94.

Hughes, T. R., Marton, M. J., Jones, A. R., Roberts, C. J., Stoughton, R., Armouor, C. D., Bennett, H. A., Coffey, E., Dai, H., He, Y. D., Kidd, M. J., King, A. M., Meyer, M. R., Slade, D., Lum, P. Y., Stepaniants, S. B., Shoemaker, D. D., Gachotte, D., Chakraburtty, K., Simon, J., Bard, M., and Friend, S. H. (2000). Functional discovery via a compendium of expression profiles. *Cell* 102:109–26.

Jelinsky, S. A., and Samson, L. D. (1999). Global response of *Saccharomyces cerevisiae* to an alkylating agent. *Proc. Natl. Acad. Sci. USA* 96:1486–91.

Jensen, L. J., and Knudson, S. (2000). Automatic discovery of regulatory patterns in promoter regions based on whole cell expression data and functional annotation. *Bioinformatics* 16:326–33.

Kerr, M. K., Martin, M., and Churchill, G. A. (2000). Analysis of variance for gene expression microarray data. *J. Comput. Biol.* 7:819–37.

Kim, S., Dougherty, E. R., Chen, Y., Sivakumar, K., Meltzer, P., Trent, J. M., and Bittner, M. (2000). Multivariate measurement of gene expression relationships. *Genomics* 67:201–9.

Kimura, M. (1983). *The Neutral Theory of Molecular Evolution*. Cambridge: Cambridge University Press.

Kononen, J., Bubendorf, L., Kallioniemi, A., Barlund, M., Schraml, P., Leighton, S., Torhorst, J., Mihatsch, M. J., Sauter, G., and Kallioniemi,O. P. (1998). Tissue microarrays for high-throughput molecular profiling of tumor specimens. *Nat. Med.* 4:844–7.

Koonin, E. V., Aravind, L., and Kondrashov, A. S. (2000). The impact of comparative genomics on our understanding of evolution. *Cell* 101:573–6.

Lewontin, R. C. (1974). *The Genetic Basis of Evolutionary Change*. New York: Columbia University Press.

Lewontin, R. C. (1991). Electrophoresis in the development of evolutionary genetics: milestone or millstone? *Genetics* 128:657–62.

Lewontin, R. C., and Hubby, J. L. (1966). A molecular approach to the study of genic heterozygosity in natural populations II. Amount of variation and degree of heterozygosity in natural populations of *Drosophila pseudoobscura*. *Genetics* 54:595–609.

Lipshutz, R. J., Fodor, S. P. A., Gingeras, T. R., and Lockhart, D. J. (1999). High density synthetic oligonucleotide arrays. *Nat. Genet.* 21(Suppl S): 20–4.

Lockhart, D. J., and Winzeler, E. W. (2000). Genomics, gene expression and DNA arrays. *Nature* 405:827–36.

Lockhart, D. J., Dong, H. L., Byrne, M. C., Follettie, M. T., Gallo, M. V., Chee, M. S., Mittmann, M., Wang, C. W., Kobayashi, M., Horton, H., and Brown, E. L. (1996). Expression monitoring by hybridization to high-density oligonucleotide arrays. *Nat. Biotechnol.* 14:1675–80.

Lorenz, M. C., and Heitman, J. (1998). The MEP2 ammonium permease regulates pseudohyphal differentiation in *Saccharomyces cerevisiae*. *EMBO J.* 17:1236–47.

Lucito, R., West, J., Reiner, A., Alexander, J., Esposito, D., Mishra, B., Powers, S., Norton, L., and Wigler, M. (2000). Detecting gene copy number fluctuations in tumor cells by microarray analysis of genomic representations. *Genome Res.* 10:1726–36.

Meiklejohn, C. D., Parsch, J., Ranz, J. M., and Hartl, D. L. (2003). Rapid evolution of male-biased gene expression in *Drosophila*. *Proc. Natl. Acad. Sci. USA* (in press).

Mortimer, R. K. (1999). On the origins of wine yeast. *Res. Microbiol.* 150:199–204.

Mortimer, R. K. (2000). Evolution and variation of the yeast (*Saccharomyces*) genome. *Genome Res.* 10:891–9.

Mortimer, R. K., Romano, P., Suzzi, G., and Polsinelli, M. (1994). Genome renewal: a new phenomenon revealed from a genetic study of 43 strains of *Saccharomyces cerevisiae* derived from natural fermentation of grape musts. *Yeast* 10:1543–52.

Murray, A. W. (2000). Whither genomics? *Genome Biol.* 1:1–6.

Nielsen, R., and Wakeley, J. (2001). Distinguishing migration from isolation: a Markov chain Monte Carlo approach. *Genetics* 158:885–96.

Ranz, J. M., Casals, F., and Ruiz, A. (2001). How malleable is the eukaryotic genome? Extreme rate of chromosomal rearrangement in the genus *Drosophila*. *Genome Res.* 11:230–9.

Ranz, J. M., Castillo-Davis, C. I., Meiklejohn, C. D., and Hartl, D. L. (2003). Sex-dependent gene expression and evolution of the *Drosophila* transcriptome. *Science* (in press).

Roberts, C. J., Nelson, B., Marton, M. J., Stoughton, R., Meyer, M. R., Bennett, H. A., He, Y. D., Dai, H., Walker, W. L., Hughes, T. R., Tyers, M., Boone, C., and Friend, S. H. (2000). Signaling and circuitry of multiple MAPK pathways revealed by a matrix of global gene expression profiles. *Science* 287:873–80.

Russo, C. A. M., Takezaki, N., and Nei, M. (1995). Molecular phylogeny and divergence times of Drosophilid species. *Mol. Biol. Evol.* 12:391–404.

Sachidanandam, R., Weissman, D., Schmidt, S. C., Kakol, J. M., Stein, L. D., Mullikin, J. C., Mortimore, B. J., Willey, D. L., Hunt, S. E., Cole, C. G., Coggill, P. C., Rice, C. M., Ning, Z. M., Rogers, J., Bentley, D. R., Kwok, P. Y., Mardis, E. R., Yeh, R. T., Schultz, B., Cook, L., Davenport, R., Dante, M., Fulton, L., Hillier, L., Waterston,

R. H., Altshuler, D. *et al.* (2001). A map of human genome sequence variation containing 1.42 million single nucleotide polymorphisms. *Nature* 409:928–33.

Saiki, R. K., Scharf, S. J., Faloona, F., Mullis, K. B., Horn, G. T., Erlich, H. A., and Arnheim, N. (1985). Enzymatic amplification of β-globin genomic sequences and restriction site analysis for diagnosis of sickle cell anemia. *Science* 230:1350–4.

Sanger, F., Nicklen, S., and Coulson, A. R. (1977). DNA sequencing with chain-termination inhibitors. *Proc. Natl. Acad. Sci. USA* 74:5463–7.

Scherf, U., Ross, D. T., Waltham, M., Smith, L. H., Lee, J. K., Tanabe, L., Kohn, K. W., Reinhold, W. C., Myers, T. G., Andrews, D. T., Scudiero, D. A., Eisen, M. B., Sausville, E. A., Pommier, Y., Botstein, D., Brown, P. O., and Weinstein, J. N. (2000). A gene expression database for the molecular pharmacology of cancer. *Nat. Genet.* 24:236–44.

Schmid, K. J., and Tautz, D. (1997). A screen for fast evolving genes from *Drosophila*. *Proc. Natl. Acad. Sci. USA* 94:9746–50.

Smithies, O. (1954). Zone electrophoresis in starch gels: group variation in the serum proteins of normal human adults. *Biochem. J.* 61:629–41.

Smithies, O. (1995). Early days of gel electrophoresis. *Genetics* 139:1–3.

Swanson, W. J., Clark, A. G., Waldrip-Dail, H. M., Wolfner, M. F., and Aquadro, C. F. (2001). Evolutionary EST analysis identifies rapidly evolving male reproductive proteins in *Drosophila*. *Proc. Natl. Acad. Sci. USA.* 98:7375–9.

Tavazoie, S., Hughes, J. D., Campbell, M. J., Cho, R. J., and Church, G. M. (1999). Systematic determination of genetic network architecture. *Nat. Genet.* 22:281–5.

Townsend, J. P., and Hartl, D. L. (2002). Bayesian analysis of gene expression levels: statistical quantification of relative mRNA level across multiple strains or treatments. *Genome Biol.* 3: research 0071.1–0071.16.

Townsend, J. P., Cavalieri, D., and Hartl, D. L. (2003). Population genetic variation in global gene expression. *Mol. Biol. Evol.* 20:955–63.

Wakeley, J. (1999). Nonequilibrium migration in human history. *Genetics* 153:1863–71.

White, K. P., Rifkin, S. A., Hurban, P., and Hogness, D. S. (1999). Microarray analysis of *Drosophila* development during metamorphosis. *Science* 286:2179–84.

Winzeler, E. A., Richards, D. R., Conway, A. R., Goldstein, A. L., Kalman, S., Mccullough, M. J., Mccusker, J. H., Stevens, D. A., Wodicka, L., Lockhart, D. J., and Davis, R. W. (1998). Direct allelic variation scanning of the yeast genome. *Science* 281:1194–7.

Wodicka, L., Dong, H. L., Mittmann, M., Ho, M. H., and Lockhart, D. J. (1997). Genome-wide expression monitoring in *Saccharomyces cerevisiae*. *Nat. Biotechnol.* 15:1359–67.

Wu, C.-I., Johnson, N. A., and Palopoli, M. F. (1996). Haldane's rule and its legacy: why are there so many sterile males? *Trends Ecol. Evol.* 11:281–4.

Yang, Z. (1998). Likelihood ratio tests for detecting positive selection and application to primate lysozyme evolution. *Mol. Biol. Evol.* 15:568–73.

Zeyl, C. (2000). Budding yeast as a model organism for population genetics. *Yeast* 16:773–84.

5

Population biology and bioinformatics

G. BRIAN GOLDING
Department of Biology, McMaster University, Hamilton

5.1 Introduction

There have been two major technological breakthroughs in the last 20 years. One has been in the power, speed and cost of computers and the second has been in the basic knowledge, sequencing abilities and technologies of molecular biology. Both areas are rapidly leading to major new innovations and are changing our daily lives. Many of the changes are unforeseen and many affect our lives in ways that were not anticipated and may not even be recognized as having anything to do with these advances.

To give but one such example of events that are shaping our lives I need only consider what happens every time that I go to the grocery store. At the check-out line, the clerk will ask me for my AIRMILES card and will "swipe" it through a scanner after ringing in my purchases. When I go to the department store, I will be asked if I subscribe to AIRMILES and if so, to supply the card so that they may record my purchases. Anytime that I go to the local hardware store, to the drugstore, to the shoe store, to the garden supply store, ... it seems, when I go to most any store they ask me for my AIRMILES card.

For those of you who may be unfamiliar with these pervasive cards (certainly this could not include anyone from Canada), let me explain a little further. The AIRMILES company supplies the AIRMILES cards free of charge to anyone that would like one. These are cards very much like any standard credit card. The AIRMILES company will put magnetic card readers in stores that will scan the magnetic strip on the back of the cards. The holder of the card then is able to collect "points" based on the amount of goods purchased. The points can be redeemed for prizes which include "air miles" on various airlines but also include many other types of gifts (from electronics to toys) all in reward for accumulating the sufficient number of points.

So what do my daily shopping habits have to do with technological revolutions? Well, consider that the AIRMILES company is paying companies to have these scanners installed and then, having done so, will pay out huge amounts

The Evolution of Population Biology, ed. R. S. Singh and M. K. Uyenoyama. Published by Cambridge University Press.

of cash for all manner of gifts to the card holders. They are not doing this out of the generosity of their hearts! Rather they are a business, and a very successful one.

Alliance Data Systems is based in Dallas, Texas, with over 7000 employees in seven cities and it operates the AIRMILES program. Alliance Data Systems has anticipated revenues of 770 million dollars for 2001 (a computer-based stock analyst recommends this company as a "BUY"). This company makes its money by selling knowledge. It sells "the spending habits" of literally millions of Canadians (and British, and Dutch, and Spanish – all of which have AIRMILES programs). This is a company that makes its living by carrying out a process known in academic circles as data mining. It collects and stores huge amounts of information. Last year it collected over 2.5 billion transactions by 72 million consumers. It then synthesizes all of that information down to relevant pieces of knowledge that other companies want to know. The concept of collecting and storing billions of bits of information and then being able to make sense of it all would not have been possible without the recent technological innovations.

Biology and particularly environmental biology is ready to go through the same revolution that molecular biology and computers have gone through and it too will have many unforeseen implications on daily lives and on how environmental biology is carried out. It, of course, depends on these two other revolutions and it too will depend strongly on data mining. Computers and their associated technologies allow vast amounts of automated information to be collected on plants and/or animals, their movements and their habitats. The molecular biology revolution allows vast amounts of information to be collected on their physiology, their genes, and their taxonomic identification. These masses of information will be collected (most likely in an automated fashion) and will be the basis of many parts of science, including population studies, in the next few years.

5.2 Background

Bioinformatics is a new science that specializes in the analysis of large quantities of biological data, particularly sequence data. Bioinformatics is the application of the information sciences to biology, particularly in the areas of genomics and molecular biology. As a discipline, it is very young, dynamic, and fast growing. It constitutes a new interdisciplinary field at the intersection of biology, mathematics, statistics, and computer sciences. It combines all four fields to gather, sort, and analyze large masses of biological information in a meaningful way. While bioinformatics provides essential tools for the Human Genome Project, it is also a key ingredient in many other areas of biotechnology such as knowledge-based drug design, genetic toxicology, forensic biology, and agricultural biotechnology. Because of its importance, its novelty and its far-reaching implications, it also encompasses important social and legal issues regarding personal privacy and equality. Bioinformatics lies at the interface of several traditionally disparate areas of research and this interdisciplinary nature is at the very essence of bioinformatics, but it has also led to a deficiency of skilled people.

The development of protein sequencing techniques in the 1950s allowed researchers to compare the amino acid sequences of proteins from related species (reviewed by Wilson *et al.* 1977). The proteins used in these studies (e.g., globin, cytochrome c, and rubisco) were often chosen because they were easily purified and relatively well conserved. The high degree of similarity that these proteins shared between related species made it easy to align them in order to perform evolutionary analyses. The advent of modern DNA sequencing techniques in the early 1980s had a major impact on how molecular sequence analyses are performed, and also on the types of questions that could be answered using these techniques. DNA sequencing revealed that eukaryotic genes were "in pieces," i.e., the coding sequences (exons) were interrupted by noncoding sequences (introns), and that a large fraction of the genetic variability found in living organisms was found in the noncoding fraction of the genomes and in silent sites (i.e., sites that can change without changing the amino acid encoded by a codon). These discoveries led to the development of computational methods which could be used to infer the location of coding regions, measure the number of substitutions in different parts of genes, infer evolutionary relationships from DNA sequences, etc. The large body of data generated by DNA sequencing techniques also necessitated the creation of national databanks so that organized records of these sequences could be kept and distributed. The diversity of the sequences obtained by DNA sequencing also required the development of new tools to search these sequence databanks and to produce multiple sequence alignments.

At the core of "traditional" bioinformatics is molecular sequence analysis. It constitutes the focus of many research projects aimed at new gene discovery which, in turn, is a prerequisite for much of modern biopharmaceutical research. Large-scale sequence searching and analysis allow for the rapid identification of novel protein-coding sequences and provide potential leads to novel protein gene-products. Data mining is defined as "exploration and analysis by automatic and semi-automatic means, of large quantities of data in order to discover meaningful patterns and rules." Such data can, for example, be used to pinpoint functional domains in biologically important molecules.

Sometimes seen as tangential, computational biology is not really distinct from bioinformatics but is generally meant to emphasize the development of methods (mathematical and statistical) to analyze data as opposed to the operation of software which carries out this analysis. Except for the fact that sufficient computational power has only recently become generally available and has been necessary for many new studies, computational biology and the methodologies of bioinformatics have a long history and deep roots within population biology. The theoretical aspects of population biology were developed by such notable scientists as Francis Galton, Karl Pearson and Ronald A. Fisher. Thus the scientists that currently study population biology are well equipped to apply the techniques of computational biology and bioinformatics to population biology studies. While population ecology and population genetics have traditionally applied mathematical modeling in their fields, the application of bioinformatics and genomics to environmental population biology will be much broader and will encompass many more fields simultaneously.

In the following few sections is an outline of just a few aspects of population biology that will feel the effects of automated methods generating large amounts of biological data.

5.3 Population-level bioinformatics

Any objective of a population-level project that involves the characterization of the genetics of a group of individuals will require bioinformatics at one or more levels. But even if the goal is not to understand the genetics of a group, but rather to study population dynamics, migration, dispersal, parentage, mating strategies, biogeography, or even conservation biology, the best path to the answers for these questions is through genetics and through molecular markers. Mitochondrial DNA variation has been used for 20 years to successfully answer questions related to evolution, population history, biogeography, and gene flow. Hence, many scientists are familiar with the power of this molecule to answer such questions but there are still many students and nonexperts who do not appreciate the need for investigators to determine the genetics of the organisms under study and hence fail to learn or to fund these techniques.

Furthermore, as we ask more difficult questions, we find that the mitochondria provide only an incomplete view since the molecule is maternally transmitted. Recently Y-chromosome variation, especially in human populations, has been used to get paternal information in order to infer male migration, etc. Similar studies have been initiated on the evolution of the Y-chromosome in dogs, cats, bovine species, and mice. In most species, migration patterns, mating patterns, and other life-history parameters differ between the sexes.

Many of these studies have taken advantage of the various genome projects that are presently under way to identify the DNA markers. While for some species there are numerous DNA markers available from genome projects, for other species they must be found with hard work. But genome projects are becoming more common for many species and, more importantly, the techniques and equipment to identify markers are becoming more common. This will mean that markers identifying each chromosome rather than just the mitochondria and Y-chromosome will soon be routinely surveyed. This will provide a complete picture of the species if we can also relate this information back to the information of the whole organism. We will need not only to know the state of the molecular markers, but also that all of the "X" markers came from habitat "A" and all of the "Y" markers come from habitat "B", etc. Because the generation of the molecular markers can be largely automated, this will put greater pressure on the population biologists to collect these precious samples and on the bioinformaticians to help synthesize these data. Adding more molecular biology to population studies should enliven the field with more data and with a precision to the data that can sometimes surprise (e.g., extra-pair mating in presumed monogamous groups).

5.4 Biodiversity

Genomic differences will accumulate as mutations occur and as polymorphisms develop within species. These differences ultimately culminate in the biological diversity that surrounds us on this planet. Humans are only beginning to realize that much of this biological diversity can be exploited in a very applied sense. The biological systems of the world form many distinct and complex ecological systems. It is these ecosystems that provide products to human society including food, fuel, fiber, biological products such as membranes, surfaces, and released byproducts such as minerals and complex organic chemicals. In addition, biodiversity is a rich source of new medicines, drugs, and genes. Many of the new drugs that have been discovered have been isolated from comparatively rare species. The declining rainforests of the world are thought to be rich areas for the discovery of more of these drugs. Despite their importance, how much diversity there is, what its nature is, how it may change over time, and how it may be best protected and utilized are all questions virtually unanswered.

The collapse of fisheries in Canada, the disruption of ecosystems by exotic invaders, the looming effects of climate change, the loss of biological diversity on a planetary scale due to the actions of humans, each of these potential impacts is reason enough to justify the energization of a commitment to increase our understanding of biodiversity. Although the need has never been greater, the number of researchers with expertise in systematics and taxonomy has collapsed. As a result of its popularity, molecular biology has depleted the ranks of those who study life at an organismic or ecosystem level by absorbing people and money but it offers its own solution to taxonomic problems.

Why does this situation exist? Biodiversity has been seen as the total complexity of all life, including not only the great variety of organisms but also their varying behavior and interactions. From this viewpoint, an effective measure of biodiversity is difficult to obtain. Indeed it is difficult to even define a measure of biodiversity that will suit all purposes (e.g., Santini and Angulo 2001). It is difficult to track individual organisms without knowledge of what type of organism they are. But the ability to identify organisms generally requires an expert in the taxonomy of each particular species and this can generally be done only with appropriate samples (such as from the male of the species rather than the female, or from an adult but not a juvenile). The rapid advances that have occurred in genomics and the powerful applications of laboratory automation and computerized databases can be harnessed to change this via molecular identification.

Pollock *et al.* (2000) have suggested that it should be possible to combine the DNA samples from multiple taxa and then to shotgun sequence the entire ensemble in a single step. They suggest that greater efficiency, automation, and more extensive sampling can occur in this fashion. They assessed the feasibility of this evolutionary genomics strategy using a simulation of 10 vertebrate mitochondrial genomes. Their experiments suggest that the method would indeed be feasible and would produce the desired sequence in the most cost-effective manner. However, this is probably a far too ambitious plan at the moment for general biodiversity studies. In part, this is because the study of biodiversity is so important and so pervasive that surveys of biodiversity will have to be done multiple times temporally as well as in multiple locations.

Less encompassing, but more practical, proposals are to use molecular information to sequence selected genes and then to use these to identify the taxonomic species involved (Weider *et al.* 1999, Soltesova *et al.* 2000, van der Kuyl *et al.* 2000, P. D. Hebert pers. comm.). Using these sequences it is possible to place any unidentified taxon (OTU) into a specific class, family, genus, species, or even into a population depending on the gene(s) chosen and the level of variation in the particular OTU. A proposal to completely automate this process by Hebert and his coworkers, if implemented, could revolutionize how species are identified and how biodiversity is measured. DNA-based identification systems offer two major advantages to conventional taxonomy: they enable the reliable identification of all life stages and they can be applied objectively. By contrast there are many situations in which skilled taxonomists are currently unable to identify life. There are, for example, often difficulties in assigning the immature stages of invertebrates or even fish to a particular species because morphological keys are reliant on adult characters. The same problem is encountered in plants. Efforts to survey rainforest regeneration are constrained by the inability of researchers to discriminate seedlings of even very divergent trees. Even where keys exist, the discrimination of closely allied species is often reliant upon the examination of morphological nuances. The use of such keys requires so much experience that misidentifications are common. The development of DNA-based systems allows the capture of taxonomic information in a form where it can be reliably applied by individuals with no prior experience of the taxonomic group. Furthermore, if all of this data is centrally stored and freely accessible it would become an indispensable biodiversity resource.

In addition to this form of taxonomic database there are currently many projects to store, present, and analyze biodiversity data. Among these are the Global Biodiversity Information Facility (GBIF; Edwards *et al.* 2000) and the Species Information Service (SIS; Smith *et al.* 2000). This growing field is now termed biodiversity informatics by Bisby (2000), but whatever it is called, it will require new talents of its practitioners and skill sets that fall within the bioinformatics realm.

5.5 Comparative biology and population bioinformatics

There are two approaches to studying biology (or biomedicine or any natural science): the experimental and the comparative. The first approach aims to test hypotheses through the manipulation of appropriate experimental systems. We might, for instance, alter nucleotide sequences upstream of selected genes and measure their expression in order to identify promoters. The second approach searches for common patterns in a variety of species, individual organisms or molecules, thus relying on "experiments of nature." The comparative approach is biology's oldest method of investigation, and it is not merely observational. The theory of evolution is based on it, as is much of what we know about basic cell biology, anatomy, physiology, behavior, and ecology.

Comparative genomics has developed out of previous work in molecular evolution. Whereas earlier studies have compared individual gene sequences, more recent work has focused on entire genomes. Interest in this area is very high because it allows us to exploit the information obtained from sequencing more than one genome. The complete genomes of over 60 organisms are now available. Differences in the life styles, the pathenogenicity and the metabolisms of these organisms can be traced back to the differences in their genomes. Because of their comparative nature, these studies cut across many organismic and disciplinary boundaries and are also a great computational challenge, since they involve the comparison of enormous and complex data sets.

The study of comparative genomics will form the basis for a fundamental understanding of how genomes change over time and how they adapt to new circumstances and to new environmental challenges. Just one recent discovery from comparative studies has been the presence of extensive lateral transfer among bacterial genomes and this will strongly alter our view of bacterial evolution. Human activities have placed new opportunities and new problems on biological communities of the earth. This is particularly true for the microbial communities. For example, the application of antibiotics has suddenly eliminated millions of infections, and to survive, bacteria have evolved antibiotic resistances. This resistance has spread by lateral transfer. Novel pollutants (e.g., synthetic nylon byproducts; Okada *et al.* 1983) have also provided new opportunities for growth. Bacteria have responded by evolving new genes that permit growth on these sources. Again this ability has spread horizontally. If humans wish to prevent these changes or to harness them, we must have an understanding of microbial genomic change. Major microbial adaptations can happen on a rapid time scale and, due to horizontal transfer, spread widely.

Not only have antibiotic resistance gene cassettes evolved, so too have novel catabolic gene cassettes (Tan 1999).

5.6 Environmental microbial bioinformatics

The use of microarrays is described elsewhere in this volume and I will, therefore, only note that these too will become major tools of the population biologist and environmentalist. It is a major challenge to detect and measure the presence (and more importantly the effects) of carcinogens and other environmental hazards. Microarrays have the potential sensitivity to detect the biological effects of such hazards. They can provide data on the expression of thousands of genes in an afternoon's experiment and will be prime candidates for data mining. The use of microarrays to detect differential gene expression due to the presence of pollutants in the environment (e.g., Afshari *et al.* 1999) has begun and will become more common in the very near future. As examples, Lu *et al.* (2001) report on 60 genes whose expression is altered in humans that have been chronically exposed to arsenic. Yoneda *et al.* (2001) report on the temporal changes in gene expression of cells exposed to smoke and hydrogen peroxide. Both these results demonstrate complex gene-expression changes in response to the insult of environmental pollutants. Many of these genes are those involved in DNA damage responses and in apoptosis. This has led to the creation of a field of study termed toxicogenomics (Nuwaysir *et al.* 1999). When these methods are applied to organisms other than humans, particularly to the comparatively sessile bacteria and plants that are more directly exposed to the pollutants that man generates, it could provide an early warning system for humans.

5.7 The shift toward environmental genomics and bioinformatics

I hope it is apparent from the few paragraphs above that the application of bioinformatics and genomics will extend well beyond the realms of medical and pharmaceutical sciences. It will impinge greatly upon how population biology is done in the next few decades.

It will also change the nature of the work produced and it will require current scientists to alter how we evaluate academic contributions. How does the casual reader evaluate each individual's contribution when most genomics papers have a large number of authors? I am, myself, a coauthor on a genomics paper with 55 other authors. In particular, how do you do this when you are trying to hire new staff or faculty? It is possible that the usual criteria of numbers and quality of primary publications will become less and less relevant. The ability to work within a large team and to provide the skills for, perhaps, only a small part of a much larger project may become more important. Paradoxically, the individuals that stand out in a crowd may not be the best choices for a team player.

These are all challenges that must be faced when a large inroad of genomics, automated technologies and bioinformatics is made into traditional population biology. Similar predictions, that computational advances would alter science, were made by Levin *et al.* (1997) a half decade ago. Has the "revolution" begun? I believe that it has – there are now more simulations, and many more databases relevant to whatever your favorite problem is than just a few years ago. But there is a large inertia in university systems, and other realities have stood between more extensive use. Genomics is expensive and ecological/environmental grants are generally smaller than medically oriented grants. Automated technologies are expensive. As medical and other high-priority fields make greater use of these technologies, their cost will slowly decrease and still useful instruments will be discarded by the medical profession in favor of the bleeding edge. In addition, unused capacity will become more readily available. And yet still population biology is somewhat stalled in this direction because we do not have people trained to take advantage of these technologies. Nor is there the tradition to leave behind individualized studies and work toward larger projects. But skilled young people are coming and the lure of larger projects will be forced by the excitement of the data that they can generate.

So is all of this in the future for population biology studies? Not really, all that I have discussed is using current technology. This technology has simply not yet been applied broadly. What the future holds will be far more exciting and far more unpredictable. Perhaps, we too will soon be handing out prizes for the accumulation of data.

REFERENCES

Afshari, C. A., Nuwaysir, E. F., and Barrett, J. C. (1999). Application of complementary DNA microarray technology to carcinogen identification, toxicology, and drug safety evaluation. *Cancer Res.* 59:4759–60.

Bisby, F. A. (2000). The quiet revolution: biodiversity informatics and the Internet. *Science* 289:2309–12.

Edwards, J. L., Lane, M. A., and Nielsen, E. S. (2000). Interoperability of biodiversity databases: biodiversity information on every desktop. *Science* 289:2312–14.

Levin, S. A., Grenfell, B., Hastings, A., and Perelson, A. S. (1997). Mathematical and computational challenges in population biology and ecosystems science. *Science* 275:334–43.

Lu, T., Liu, J., LeCluyse, E. L., Zhou, Y. S., Cheng, M. L., and Waalkes, M. P. (2001). Application of cDNA microarray to the study of arsenic-induced liver diseases in the population of Guizhou, China. *Toxicol. Sci.* 59:185–92.

Nuwaysir, E. F., Bittner, M., Trent, J., Barrett, J. C., and Afshari, C. A. (1999). Microarrays and toxicology: the advent of toxicogenomics. *Mol. Carcinog.* 24:153–9.

Okada, H., Negoro, S., Kimura, H., and Nakamura, S. (1983). Evolutionary adaptation of plasmid-encoded enzymes for degrading nylon oligomers. *Nature* 306: 203–6.

Pollock, D. D., Eisen, J. A., Doggett, N. A., and Cummings, M. P. (2000). A case for evolutionary genomics and the comprehensive examination of sequence biodiversity. *Mol. Biol. Evol.* 17:1776–88.

Santini, F., and Angulo, A. (2001). Assessing conservation biology priorities through the development of biodiversity indicators. *Riv. Biol.* 94:259–75.

Smith, A. T., Boitani, L., Bibby, C., Brackett, D., Corsi, F., da Fonseca, G. A., Gascon, C., Dixon, M. G., Hilton-Taylor, C., Mace, G., Mittermeier, R. A., Rabinovich, J., Richardson, B. J., Rylands, A., Stein, B., Stuart, S., Thomsen, J., and Wilson, C. (2000). Databases tailored for biodiversity conservation. *Science* 290:2073–4.

Soltesova, A., Spirek, M., Horvath, A., and Sulo, P. (2000). Mitochondria – tool for taxonomic identification of yeasts from *Saccharomyces sensu stricto* complex. *Folia Microbiol. (Praha)* 45:99–106.

Tan, H. M. (1999). Bacterial catabolic transposons. *Appl. Microbiol. Biotechnol.* 51:1–12.

van der Kuyl, A. C., van Gennep, D. R., Dekker, J. T., and Goudsmit, J. (2000). Routine DNA analysis based on 12S rRNA gene sequencing as a tool in the management of captive primates. *J. Med. Primatol.* 29:309–17.

Weider, L. J., Hobaek, A., Dufresne, F., Colbourne, J. K., Crease T. J., and Hebert, P. D. N. (1999). Circumarctic phylogeography of an asexual species complex I. mtDNA variation in arctic *Daphnia*. *Evolution* 53:777–92.

Wilson, A. C., Carlson, S. S., and White, T. J. (1977). Biochemical evolution. *Annu. Rev. Biochem.* 46:573–639.

Yoneda, K., Peck, K., Chang, M. M., Chmiel, K., Sher, Y. P., Chen, J., Yang, P. C., Chen, Y., and Wu, R. (2001). Development of high-density DNA microarray membrane for profiling smoke- and hydrogen peroxide-induced genes in a human bronchial epithelial cell line. *Am. J. Respir. Crit. Care Med.* 164:S85–9.

6

Beyond beanbag genetics: Wright's adaptive landscape, gene interaction networks, and the evolution of new genetic systems

RAMA S. SINGH, RICHARD A. MORTON
Department of Biology, McMaster University, Hamilton

6.1 Introduction

Diversity and evolutionary innovation are the hallmarks of life. Life appears to defy the laws of physics and chemistry and it does so with the help of energy. Life succeeds by finding ways to overcome natural and physical hindrances. Life spread over the face of the globe and filled almost every crevice and habitat. Of course there are limits to what evolution can do. Everything is not possible. For example, there must be a limit to how tall California redwoods can grow just as there must have been a limit to the size of dinosaurs. Although life's success appears almost boundless and has had continued success for more than four billion years, there are no general laws describing the limits of evolution. This is in contrast to physical and chemical laws that define constraining limits. The grandest experiment of nature, the evolution of life, has no theory equivalent to $E = mc^2$.

The basic forces of evolution, both intrinsic (mutation, migration, sexuality, selection, and random genetic drift) and extrinsic (historical contingency and major catastrophe), generate not a single outcome but a range of possibilities. We cannot predict what evolution will produce in any given situation, but only describe the factors that may have produced what has happened. Population genetics is theoretically the most advanced field in evolutionary biology. Theories of evolution are constrained by population biology, but there is no quantitative representation of Darwin's theory of evolution by natural selection. Fisher's fundamental theorem is the closest attempt to define a theory of constraint but it does not always hold. Kimura's neutral theory links the rate of evolution to rate of mutation but it applies only to a subclass of mutations, neutral mutations that do not matter to the organism.

Wright's shifting balance theory (SBT) (Wright 1931, 1932) is one theory that has loomed large in population genetics because, unlike others, it incorporates all the major factors of evolution and describes their roles in a precise sequence of events. In addition, every population geneticist and population

The Evolution of Population Biology, ed. R. S. Singh and M. K. Uyenoyama. Published by Cambridge University Press.

biologist can relate to Wright's emphasis on population structure. Thus as a theory, SBT seems to be a complete description of how evolution occurs in natural populations.

In his later years, however, Wright no longer saw his shifting balance theory as an alternative to Fisher's mass selection, but as one of four important approaches to evolution (Crow 1991). In his last paper, published in his ninety-ninth year, Wright stated: "Kimura's 'neutral' theory dealt with the exceedingly slow accumulation of neutral biochemical changes from accidents of sampling in the species as a whole. Fisher's 'fundamental theory of natural selection' was concerned with the total combined effects of alleles at multiple loci under the assumption of panmixia. Haldane gave the most exhaustive treatment of the case in which the effects of a pair of alleles are independent of the rest of the genome. I attempted to account for occasional exceedingly rapid evolution on the basis of intergroup selection (differential diffusion) among small local populations that have differentiated at random (i.e., by inbreeding). All four are valid" (Wright 1988, p. 122).

The works of Fisher, Wright, Haldane, and Kimura do not really represent four alternative views of evolution but rather deal with four major problems that were important at the time. Fisher's fundamental theorem (Fisher 1930, 1958) was a much-needed foundation linking genetics to evolution. Haldane (1932) dealt with rates of change caused by major mutations since rates of change and the sufficiency of the geological time scale were major issues in evolutionary biology at that time. Kimura's neutral theory was designed to deal with a specific problem: how to account for large amounts of molecular polymorphisms without incurring a large reduction in population fitness (Kimura 1983). Wright's theory can be seen as a mechanism accounting for occasional rapid evolution.

Our purpose here is neither to criticize nor to defend Wright's shifting balance theory of evolution. This has been done vigorously and thoroughly in a number of publications, the most comprehensive and notable being the recent exchange between Wade and Goodnight (Wade and Goodnight 1991, 1998; Goodnight and Wade 2000) arguing in support of SBT, and Coyne *et al.* (1997, 2000) against it. We would like to bypass this debate with a fourfold aim. First we argue that the problem that the shifting balance process of evolution was supposed to solve, i.e., the evolution of co-adapted genetic systems, is an important one regardless of whether or not Wright's theory provides a satisfactory answer to it. Second, we argue that epistatic gene interaction, and not random genetic drift, is the most important assumption in Wright's theory on which the theory would stand or fall. Thirdly, we review the literature of molecular genetics and genomics in model organisms such as *Drosophila*, nematode and mouse, and provide evidence for the pervasiveness of gene interaction, epistasis and pleiotropy. Finally, we suggest that diverse ideas such as those of Waddington, McClintock, Goldschmidt, Wilson, Lundquist, and others about co-adapted gene complexes will eventually provide a realistic alternative to Wright's random genetic drift as a mechanism of generating new gene combinations.

6.2 Wright's shifting balance theory

Wright's shifting balance theory was a response to two observations from selection and breeding experiments. First, that the response to selection was often limited by low fecundity and, second, that traits were often affected by gene interaction and pleiotropism. He suggested a solution to the perceived difficulty of natural selection to produce new combinations of alleles, combinations that were only adaptive when acting together. Selection was to operate at two levels, *genic* selection of the most favorable allele in an average genetic background, and *organismic* selection among different co-adapted subpopulations. SBT viewed evolutionary change in a population as movement between locally optimum peaks of fitness. Population fitness was viewed as an adaptive landscape in a multidimensional, gene frequency space. A population occupied a locally optimum peak but, in order to move to another peak of higher fitness, must move across a valley or saddle of maladaptive gene combinations. Since classical theories of selection on individuals in a very large population imply that transition between adaptive peaks is either very slow or impossible, Wright devised a mechanism consisting of several elements or stages. Stage 1: Genetic drift allows small subdivisions of the larger population to occupy a saddle of maladaptive genotypes and ultimately reach a neighboring fitness peak. Stage 2: Individual selection allows the subpopulation to climb the new peak achieving a new co-adapted genotype. Stage 3: A subpopulation that has reached a higher peak sends migrants to other subpopulations, allowing them to shift as well. Also, some have added a Stage 0, the wait for new mutations to occur that, in combination with other alleles, will produce the co-adapted genotype.

Wright's theory can be evaluated as an exercise in population genetics. That is, under what conditions of epistasis, recombination, linkage, migration, etc., can such a process of transition between fitness peaks occur? We can also ask if these conditions actually exist for populations in the real world and if such a process is actually important in evolution.

6.2.1 Shifting balance theory: critical assumptions

The three main assumptions of the SBT are: genetic polymorphisms, epistatic interactions between genes, and subdivided population structure. Wright believed that genetic drift would be ineffective in producing a favorable combination of more than two genes and seemed to suggest that SBT would apply to combinations of two alleles at a time. Even saddles can only be crossed when the equilibrium frequency of both loci was high. Thus, only if there were a large number of pairs of two-allele, polymorphic loci with epistatic interactions would SBT work. Most evolutionary biologists would vouch for the adequacy of genetic polymorphisms in natural populations, Kimura's neutral theory notwithstanding. Much like Darwin, who used the nature of genetic variation in plants and animals to support his theory of natural selection, Wright, based on his experience with the role of inbreeding and occasional crossing in animal improvement, argued that a rapidly evolving genetic system should combine the advantages of both uniparental (inbreeding) and biparental reproduction. He saw local differentiation as a mechanism of amplifying the field of variability, and selective diffusion as a mechanism of amplifying natural selection. Most population biologists would relate favorably to Wright's assumption of population structure (see Wade and Goodnight 1998), although the critical values of migration between demes remain a problem (Barton and Rouhani 1987). Of course Wright's view of population structure may not apply to all species, as, for SBT to be effective, species must be divided into many demes, each going through drift. Drift in small populations is a double-edged sword: small size makes populations likely to experience drift but it also makes them lose genetic variation and likely to go extinct.

The most critical assumption underlying SBT is that gene interaction and epistasis create alternate adaptive peaks. This is because in the three-phased, shifting balance evolution (i.e., drift of gene frequencies, appearance and increase of favorable gene combination, and diffusion to other demes) the second stage critically depends on gene interaction and epistasis. In Wright's scheme drift is simply a mechanism for generating favorable gene combinations, and migration leads to their spread in the species as a whole. The existence of a rugged fitness landscape requires epistatic gene interactions. It is the assumption of multiple fitness peaks based on different co-adapted combinations of *existing* alleles that is under question – even more so in the light of gene redundancy (discussed below). Based on his extensive experience with the genetics of mouse coat color, Wright assumed that gene interaction and epistasis were universal.

We will review molecular mechanisms and the strength of epistasis in model organisms and show that Wright's idea of the need for the evolution of new genetic systems is worthy of consideration but that his mechanism, i.e., peak shift by genetic drift, may not be right. There may well be no other fitness peaks with existing alleles. The real question is what is being explained: the macroevolution of new phenotypes, the evolution of new species, or the adaptation of a species to environmental change by change of allele frequencies.

Wright, and indeed the whole of population genetics, has been constrained by thinking of evolution in terms of gene frequency change. Changes in the frequencies of existing allelic forms of genes may not, however, adequately describe what happens when populations evolve new co-adaptive genetic systems. Loss and gain of genes and gene functions through duplication, deletion, transposition, horizontal gene transfer, and gene silencing are now seen as significant mechanisms of evolution. We propose that to adequately include the role of epistasis in evolution, Wright's gene frequency surfaces should be replaced with epistatic genetic networks. These may be more analogous to connected spiders' webs than geographic surfaces of hills and valleys. Webs suggest genetic networks without unsurpassable fitness peaks and valleys. Creation of new networks rather than movement along gene frequency surfaces may provide the basis for the evolution of new genetic systems. Evolution by changes in gene regulation and gene interaction are new and rapidly growing areas of evolutionary biology that will eventually provide realistic answers to many adaptive peak shifts.

6.2.2 Shifting balance theory: critics and defenders

The most severe criticism of shifting balance theory has been based on the role of drift. While most of his contemporaries saw drift as a nondirected, nonadaptive force of evolution, Wright saw drift as a mechanism for searching the topography of fitness surface and generating new gene combinations.

Wright (1980, 1982a,b) held that his theory was misunderstood by many of his contemporaries, including Huxley (1942), and Fisher and Ford (1947), who took the fixation of characters by drift as the essence of his theory – what came to be known as the "Sewall Wright effect" (Huxley 1942). Mayr (1954), by contrast, emphasized loss of alleles as the essence of Wright's theory. Mayr enunciated the "founder principle" and proposed the significance of founder events as a "genetic revolution." Mayr (1954) referred to "well-integrated coadapted gene complexes" which Wright thought were essentially the same as what he called peak shift. Mayr (1959) called Wright's theory, along with those of Haldane and Fisher, all versions of the "bean bag theory." Wright believed that he always emphasized the creative aspect of drift and thought that his peak shift was much like Mayr's genetic revolution except that, unlike Mayr, he believed the unit of evolution to be a subdivided population and not simply an isolated founder population.

Wright's theory has been vigorously debated (for recent exchange see Coyne *et al.* 1997, 2000, and Wade and Goodnight 1991, 1998). Relevant issues have been raised in depth by these authors and therefore we will only touch on those points which matter for the main issue of our proposal, which is that importance of epistasis and not drift or population structure is the central issue differentiating Fisher's mass selection and Wright's shifting balance.

In terms of our own criticism, we would like to raise two points. First, Wright considered drift to produce qualitatively new results – new gene combinations – and therefore the increase in the additive genetic variance by drift of nonadditive genetic variance is not the most important point. In the comparison of the two theories, mass selection vs. SBT, the amount of genetic variance is not the real issue. Second, in Wright's theory linkage does not play an important part and hence joint drift of high-frequency alleles at polymorphic loci required to produce favorable gene combination becomes exponentially impossible beyond a few loci. Therefore drift is a weak force, at best, to generate novelty in a sexually breeding population. There may be other mechanisms such as relaxation of selection (Carson 1997), environmental- or stress-induced genetic transpositions (McClintock 1957), genetic capacitation (Queitsch *et al.* 2002) and genetic imprinting (Beaudet and Jiang 2002) that can produce genetic novelty without drift or inbreeding. We therefore further propose that gene interaction and epistasis, and not population structure, should be seen as the core of Wright's theory of evolution by shifting balance and that to make it viable, the predominant emphasis on genetic drift should be replaced by other genetic mechanisms generating co-adaptive gene complexes.

6.2.3 Shifting balance theory: an alternative to mass selection?

It has become customary to contrast Wright's shifting balance theory against Fisher's mass selection and to seek evidence for the former taking the latter as the null hypothesis. Since the shifting balance process is based on three phases, drift within demes and selection within and between demes, the critics of SBT ask for evidence of all three phases in order that SBT be accepted. Such support is difficult to obtain. This is partly because most phenotypic characters do not seem to follow a fitness surface full of peaks and valleys. Generally speaking, one can think of three sorts of traits: simple metric traits showing continuous variation such as height and weight, complex phenotypic traits such as the wings of butterflies and moths, and fitness traits such as hybrid inviability and fertility. It is only the last two types that are likely to evolve by SBT.

Wright did not care for such a test of the theory as he was not against mass selection within demes, only that he questioned the adequacy of its premises: gene additivity and panmixia in all cases. Since Wright believed more in evolutionary *consistency* rather than *contingency*, his shifting balance theory, as formulated, was a theory not intended to hold everywhere but only in special cases where it provided a mechanism for rapid evolution. While for most of his career Wright did not write much about speciation and macroevolution, towards his later years he began to put more emphasis on the role of his theory in organismic selection, speciation, and macroevolution.

Wright's theory combines essential features of both Haldane's theory of major mutations and Fisher's multiple factors. Wright believed in Fisher's mass selection and put emphasis on multiple factors when describing drift within individual demes, but switched to major alleles with epistasis between loci to explain how new gene combinations arise. It is important to note that while the core of his theory required genetic drift for operation, Wright thought peak shift could occur without drift in exceptional circumstances. Thus he mentioned the occasional role of major mutations, major interactions, and change of environmental conditions and selection pressure with or without small populations.

We suggest that the critical difference between mass selection and SBT is not population size but epistasis. Epistasis depends on the nature of gene interactions. The origin of new gene combinations is the crux of shifting balance and this cannot occur if we think of epistasis as an added component on the top of additive effects (much like a surtax!). Epistatic interaction has to be the major factor, not necessarily for all genes or all alleles, but certainly for a significant fraction of the genes involved in complex adaptations. Increase in additive variance, due to conversion from epistatic effects, may be important but not as a major factor. It may create more variance for selection to operate on, but SBT involves qualitatively different combinations of alleles, which produce new co-adaptations. For this reason, we think that it is in the new molecular biology of gene networks, and not genetic drift, where one should look for mechanisms for the evolution of favorable gene combinations.

6.2.4 Shifting balance theory: linking microevolution to macroevolution

Wright's shifting balance theory can be treated at two levels: as a general theory of all adaptation, or as a specific, occasional process producing major evolutionary novelty. The critics of SBT have viewed Fisher's mass selection and Wright's shifting balance process as if they were alternative theories of evolution. This is not true. In his writings Wright appears to have presented his theory in two forms, as a general theory of evolution (an alternative to mass selection) and as a specific theory capable of explaining rapid change. Wright started with SBT as a general theory but in his later years he changed his views towards the specific sense. Thus in his seminal 1932 paper the treatment of SBT is in a general sense. He writes:

'The most general conclusion is that evolution depends on a certain balance among its factors. There must be gene mutation, but an excessive rate gives an array of freaks, not evolution; there must be selection, but too severe a process destroys the field of variability, and thus the basis for further advance; prevalence of local inbreeding within a species has extremely important evolutionary consequences, but too close inbreeding leads merely to extinction. A certain amount of crossbreeding is favorable but not too much. In this dependence on balance the species is like a living organism. At all levels of organization life depends on the maintenance of a certain balance among its factors.... The course of evolution through the general field is not controlled by direction of mutation and not directly by selection, except as conditions change, but by a trial and error mechanism consisting of a largely nonadaptive differentiation of local races (due to inbreeding balanced by occasional crossbreeding) and a determination of long time trend by intergroup selection.'

(Wright 1932, p. 170)

In this general sense, then, it is not surprising that Wright saw practically every observation of evolutionary biology, no matter how trivial, as supporting his theory.

However, there was no doubt that during his later years Wright considered his theory as a special mechanism for rapid evolution. Wright wrote four reviews (Wright 1980, 1982a,b, 1988) in which he modified his views regarding the role and the significance of his theory. In 1988 (p. 123) he wrote: "I attempted to account for occasional exceedingly rapid evolution on the basis of intergroup selection (differential diffusion) among small local populations that have differentiated at random."

Wright also saw his theory to be applicable to different hierarchical levels. He used the shifting balance process more generally to defend mainstream population genetics and to challenge non-Darwinian macroevolutionary mechanisms of evolution proposed by others. He criticized Goldschmidt's (1940) proposal for the role of macromutations in the evolution of higher taxa, and argued in favor of interdeme selection counteracting the prevailing view which favored genic selection (as opposed to group selection) but ignored interdeme selection (Maynard Smith and Williams 1976, Williams 1966, Dawkins 1976).

Because of two special features of shifting balance theory, population structure and intergroup selection, Wright saw his theory as applicable to both speciation and macroevolution. Wright treated species as the result of reproductive isolation and argued that the same situation which promotes shifting balance evolution would also lead to speciation. He saw evolution of polyploids and the fixation of translocations in small populations leading to reproductive isolation as examples of his theory.

The shifting balance process was the only population genetic theory that offered a mechanism for punctuated evolution – rapid evolution interspersed with stasis – as observed in the paleontological records. Wright held that the shifting balance process provided a satisfactory explanation for the varying patterns of evolution as brought out in Simpson's *Tempo and Mode of Evolution* (Simpson 1944) and as argued by Gould and Eldredge (1977) in their theory of "punctuated equilibrium." Wright (1980) believed in Gould's hierarchical evolution (evolution within populations, speciation, and macroevolution) but argued against the need for Gould's revival of the role of "Goldschmidtian" major mutations (Gould 1977). Wright thought that his theory was capable of explaining "explosive" evolution (Wright 1982, p. 440). Rapid evolution by moving into new or vacated niches, by acquiring new ways of life, or possibly even by major mutations (if the deleterious pleiotropic effects could be suppressed) were some of the ways in which Wright saw higher taxa arising.

Mass selection is an idealized description of the process of evolution. Mass selection is based on multifactorial genetic systems (Crow 2002). As the investigations of modern genetics unravel more and more variety of mutational alterations, genomic plasticity, regulatory changes and gene interactions, so we will look more and more for a theory, something not unlike the shifting balance theory, that can provide a mechanistic opportunity for populations to respond rapidly to changing environments. Shifting balance process is not an alternative to mass selection, but rather a special case which includes mass selection and a host of other factors that make rapid evolution possible. This is obvious from Wright's remarks to Fisher about the fundamental theorem. In his letter to Fisher, Wright points out the comparison by Fisher between natural selection and the law of increase of entropy, as the latter is only meant to apply to a closed system. This was not the case with Fisher's theorem, which ignored all other factors of evolution except natural selection. Wright would have been happy if the fundamental theorem had been stated with qualifications: "The rate of increase in fitness of any population at any time is equal to its genetic variance in fitness at that time, except as affected by mutation, migration, change of environment and the effects of random sampling" (quoted in Provine 1986, p. 272). As Provine (1986, p. 266) remarks, "both Wright and Fisher knew that their most quantitative results agreed in all significant particulars. What divided them was decidedly not the differences in quantitative analysis but their qualitative views about the process of evolution." Like mass selection, it is beyond question that a shifting balance process is needed. Its scope of operation remains to be determined.

6.3 The problem of epistasis: two views of developmental complexity

Every student of elementary genetics knows that when different mutations interact they can produce unexpected results. Variations of Mendel's ratios illustrate this fact. The question is: how common are such interactions? How important is epistasis in explaining the developmental and phenotypic diversity that we see around us? There are two views on the role of gene interaction and epistasis. One view is that the enormous diversity of living systems, and the intricate and complex adaptations of organisms, can all be explained by a population genetics theory of evolution based on mass selection, multiple, additive gene systems, and panmictic populations. According to this view, the machinery of population genetics is capable, when challenged by the environment, of grinding out the evolutionary diversity that we see around us. Beside mass selection the only other factors to be considered are the contribution, in varying degrees, of variations in breeding systems, geographic separation, reproductive isolation, and periodic extinctions.

Fisher (1930, 1958) saw the majority of mutations as additive and incremental in their effects, with gene interactions and pleiotropy as playing a minor role. This is not to say that in this view of evolution gene interactions are not important. They are, except that additive allelic effects are thought capable of modifying, molding, and fine tuning the outcome of selection such that there is no major, unattainable phenotypic domain. Organismic complexity is the result of incremental accumulation of noncomplex subsystems. In this view of evolution, if the tape of the history of life were to be replayed, as Gould (1989) put it, we would have a different outcome because of completely different *historical contingency* and not because an entirely new set of gene interactions were selected. We can say that the mass selection process, based on additively acting multiple gene systems, dictates the minimal evolutionary view of life.

In the second view, historical contingency also plays a major role but gene interactions and epistasis reign supreme. In effect, "interactions and context" are both important determining factors in evolution (Lewontin 1974). This view would hold that gene interactions and epistasis put constraints on evolution so that everything is not possible. According to this view, neither all possible organisms are possible in the grand developmental scheme of the unfolding of life, nor all possible allelic combinations are possible within a local population. Using Wright's terminology, thc fitness surface has holes, domains of genetic combinations that are developmentally not available to organisms. In this view of evolution, if the tape of the history of life were to be replayed, we will have a completely different outcome because of both *historical contingency* and the uniqueness of *genetic interactions*. We can call this interactive evolutionary view of life "Gould's view of life." Wright's shifting balance theory is in essence a much milder form of this view of evolution as it allows organisms a smooth fitness surface in which there are no holes, only peaks and valleys.

All through the post-modern synthesis era, we have been attracted to "Gould's view of life." Science does not attempt to prove the null hypothesis, but to disprove it. You want to discover new things. There has been an incessant desire to explain the complexity of life. Explanations have ranged from the role of Goldschmidtian "monstrous" mutations on one hand to more balanced views such as those of Mayr, Carson, and Wright on the other. Founder effects, genetic revolutions, shifting balance, all have one thing in common. All look for a mechanism of rapid evolution caused by changes in the breeding system, genetic system, and/or the environment. We must of course demand quantification, rigor, and proof but to say that as a parsimonious explanation mass selection would suffice is to stop doing science. A theory of evolutionary biology incorporating gene interaction and epistasis is needed for the next stage of revolution in evolutionary investigations, especially now that the genome sequencing has opened the field of molecular evolution and developmental biology to experimental manipulation in a wide variety of organisms.

6.4 Epistasis in fitness, phenotypes and molecular traits

Provine (1986) discussed in detail the two genetic representations that Wright used for fitness surfaces. The concept of population fitness is appealing as it provides a connection to evolutionary parameters. The individual (genotype) fitness surface is also appealing because it can be connected to individual adaptation. In addition, Simpson (1944), in his *Tempo and Mode in Evolution*, introduced the concept of fitness surface in terms of phenotypes. Provine states, "all three versions have enjoyed a lively existence, and all have gone under the name of 'Wright's fitness surface' or 'Wright's adaptive landscape'." Provine (1986, p. 315) further remarks, "Wright's shifting balance theory of evolution in no material way depends upon the usefulness of his fitness surfaces as heuristic devices. The shifting balance theory gains its place among the few really robust theories of evolutionary change as a result of its close relation to known effective methods of animal breeding and increasing of population structure in natural populations."

Fitness function can have several representations. It can be a discrete function, the fitness of individual genotypes. The population fitness represents the genotype fitness times the frequency of the genotype. This can approach a continuous function if the mean fitness of the population is represented against gene frequency. However, the genotype fitness does not have unique meaning if it depends on the frequency of other genotypes in the population, nor does the population fitness if it depends on the frequency of other populations in the ecosystem. One of the biggest problems in measuring epistasis in real populations and relating it to Wright's shifting balance theory is that Wright's theory related to *all possible* gene combinations among polymorphic loci (representing axes with $p = 0–1$) while real populations have *only one set* of allele frequencies. Therefore the population fitness on Wright's fitness surface is an average value, based on the fitness of individual genotypes within a population. In this representation the epistatic effect of gene interaction on fitness has already been taken into account. How then can epistasis become a significant mechanism for creating new gene combinations? In Wright's scenario this cannot happen in a large population as a few high-fitness genotypes cannot make a significant difference. In a small population, however, such high-fitness gene combinations are less likely, but if they do occur drift may take them to fixation. This is why Wright depended on drift as the main mechanism for creating new genetic combinations.

But real populations do not experience all possible genotypes even with a single set of allele frequencies. Mutations occur *sequentially* and not all at once, and so historical sequence and genomic context of mutations are important considerations for evaluating the role of interaction and epistasis in evolution (Brodie 2000, Phillips *et al.* 2000). Systematic searches for the role of epistasis in evolution involve both direct and indirect measurements. A detailed description of various, nonmolecular, approaches for measuring epistasis is given by Phillips *et al.* (2000). We will focus here on only those approaches which either have been most commonly used in the literature or are likely to provide molecular insights.

6.4.1 The nature of newly arising mutations

Most newly arising mutations are deleterious (Simmons and Crow 1977) and those that are not probably act in a predominantly additive fashion. Most surviving mutations have small effects (positive or negative) verging on neutrality

(Kimura 1968, Ohta 1976, Gillespie 1991). Of course mutations with small and additive effects individually can have significantly large epistatic effects in combinations (Moreno 1994) but such mutations are also expected to be a minority. Even deleterious mutations can interact in compensatory epistatic fashion and can evolve compensatory modifiers (Bouma and Lenski 1988). We need to know if the variance in epistatic effects is small or large (Phillips *et al.* 2000) and whether epistasis merely increases genetic variance or produces qualitatively different gene combinations (Whitlock *et al.* 1995). Recent efforts to saturate the genome of model organisms with mutations have focused on those that disrupt gene function. Perhaps surprisingly, many gene knockouts of what were thought to be critical developmental genes had only minor phenotypic effects. Double knockouts often produced synthetic lethals that completely blocked essential developmental pathways. There are now a number of examples of double gene knockouts that produce much stronger effects than either mutation has by itself (Wilkins 1997). The gene knockout approach has much to offer.

6.4.2 Evidence from linkage disequilibrium and complex polymorphisms

Supergenes or complex polymorphisms provide the best examples of epistasis within species (Ford 1975, Hedrick *et al.* 1978, Lyttle 1991, Dobzhansky 1970, Morita *et al.* 1993, Ardie 1998, Klein 1986). Such complex polymorphisms tend to occur in discrete traits whose variation is coupled with variation in some aspect of the biotic or abiotic environment.

In the pre-genomics era, the role of gene interaction and epistasis was primarily considered with regard to the maintenance of genetic variation in natural populations. Lewontin and Kojima (1960) introduced the concept of two-locus linkage disequilibrium as a mechanism of maintaining genetic variation and opened a door for experimental investigation of complex polymorphisms in natural populations (Lewontin 1974). Linkage disequilibrium provided a solution to the problem of genetic load, arising from genetic polymorphisms, as well as a mechanism for maintaining complex polymorphisms such as chromosome translocations in *Moraba scura* (Lewontin and White 1960), inversion polymorphisms in *Drosophila pseudoobscura* (Dobzhansky 1970), and the diversity of the histocompatibility complex and immunoglobulins gene in humans (Klein 1986).

However, the expectations of linkage disequilibrium in natural populations were not realized. Surveys of electrophoretic variations provided little evidence for linkage disequilibrium (Charlesworth and Charlesworth 1973; Langley *et al.* 1974). These negative results could be due to lack of selection or lack of epistatic interactions in maintaining allozyme variation. However, it is also true that the allozyme loci employed in most of these tests were very loosely linked and did not provide the best conditions for the detection of epistasis. Tests of linkage disequilibrium using fine-resolution DNA sequence variation from shorter segments of the chromosomes are more likely to produce significant results. Alcohol dehydrogenase polymorphism in *Drosophila melanogaster* (Berry and Kreitman 1993), meiotic drive systems in many species of *Drosophila* (Lyttle 1991), and the histocompatibility complex in humans (Klein 1986) are some of the best examples. Epistatic interactions are also more likely to occur in DNA sequence variants between the coding and the noncoding regulatory or intron regions (Stam and Laurie 1996).

The lack of a significant number of cases of linkage disequilibrium in natural populations need not mean that epistasis does not have a significant role in evolution. While the nature of evolutionarily significant genetic variation is assumed to be the same within and between populations, that existing within populations reflects an instant of what happens during the lifetime of a species. Epistatic interactions can be rare within populations and still play a significant role in evolution between populations and species over time.

Epistasis is a double-edged sword and may become a constraint to evolution. Complex balanced polymorphisms not only maintain genetic variation but they lock this variation within species such that there is reduced evolutionary flexibility across loci between species. Wright realized this. He believed that linkage will be broken down and allelic combinations will be homogenized in sexually breeding populations. Wright believed in average allelic effects, as implied by Fisher's fundamental theorem, and invoked general epistasis in fitness across loci unconstrained by linkage. The problem of linkage and epistasis is well researched and the diversity of opinions on how to maintain epistasis vary from those of Franklin and Lewontin (1970) who proposed the chromosome, driven by linkage and epistasis, as being the unit of evolution, to Wright's (1978) idea of functional epistasis unconstrained by linkage. The search for epistatic gene interactions, constrained or unconstrained by linkage, will be an active area of research in molecular evolutionary biology.

6.4.3 Evidence from species hybrids

Johnson (2000) has reviewed the literature of gene interactions and its role in the origin of species. Species hybrids and backcross methods (Dobzhansky 1937, Coyne *et al.* 1998, Wu and Palopoli 1994, Sawamura *et al.* 2000) have become a common approach for discovering "speciation genes" as well as gene interactions. The exact nature of gene interactions in the F1 hybrids remains to be elucidated, and we do not know if much of what we have learned about "speciation genes" from backcross studies is relevant to the hybrid sterility in F1. Results of transgenic experiments would be interesting in order to sort out the effect of so-called speciation and helper genes. However, backcross and chromosome substitution experiments have revealed between-species gene interactions among the various chromosomes. Based on the amount of genetic material and general divergence of genes on all chromosomes, we would expect to find the following relationship of pairwise chromosome interactions: auto–auto > auto–x > auto–y > x–y. Since in most between-species chromosome substitution experiments autosomes remain in heterozygous condition, their full effect on hybrid male sterility is not revealed. The auto–sex interaction appears to be the largest component affecting hybrid sterility and there is relatively little evidence for nuclear–nuclear or nuclear–cytoplasmic interactions. Epistasis is also significant in the evolution of sexually antagonistic genes (Rice 1992).

6.4.4 Epistasis and sexual traits

Our views about whether epistasis is common or rare in nature would determine where we should look for it. Assuming that organisms represent complex adapted genetic systems, we expect gene interactions to be present at different levels. These can be revealed by genetic crosses between individuals from within and between populations. We can distinguish two kinds of epistasis: *quantitative* and *qualitative*. Under quantitative epistasis we include all gene interactions that are of quantitative and incremental nature and which affect the amount of genetic variance within and between populations but which are less likely to produce qualitatively unique or novel gene combinations. Under qualitative epistasis, by contrast, we include all interactions that are novel and have large effects. If we are looking for the latter forms of epistasis, we stand a better chance of discovering them in studies involving species hybrids. Species and species hybrids, rather than individuals within populations, provide the best means of testing the adequacy of epistasis as far as Wright's theory is concerned. This is because of the evolutionary time scale involved for significant epistasis to evolve and because of the appearance of novel phenotypes in species hybrids.

6.5 Is shifting balance theory more relevant to speciation and macroevolution?

Although Wright may have treated, in his early years, his shifting balance theory as an alternative to Fisher's mass selection, he always held that his theory is more relevant at higher levels of evolution, beyond mass selection within populations, to differentiation between populations leading to speciation and macroevolution (Wright 1940, 1980, 1982, 1988). This position would appear to be supported by what we know about the nature and extent of epistasis within populations, the likelihood (or availability) of alternate fitness peaks within species, and the adequacy of population structure for peak shift except in cases of geographic isolation. Epistasis contributes to additive variance within populations and to genetic divergence between populations (Wade 2000). Population bottlenecks add to the genetic variance, but estimates of epistasis within populations are frequently quite small (Whitlock *et al.* 1995; Goodnight 2000).

Empirically it is impossible to know if alternate fitness peaks exist within populations (Whitlock *et al.* 1995), or to demonstrate that populations have reached their present fitness peaks by traversing fitness valleys (Bradshaw *et al.* 2000). Different adaptive peaks may be connected by adaptive ridges (Gaverilets 1997), or may be the result of environmental change (Whitlock 1997).

Although under question with respect to their role within populations, all three factors of SBT operate for speciation and macroevolution. These are: epistatic gene interactions that produce novel traits, environmental change and new niches, and population structure. Therefore the importance of shifting balance theory is more in providing a link between microevolution and macroevolution, through its role in population divergence and speciation, than serving as an alternative to evolution by mass selection. Wright's retrospection of his theory (1980, 1982, 1988) supports this conclusion. As Provine (1986, p.286) remarks, "Over fifty years later [since 1932] Wright would apply his distinction between periods of evolution, dominated by Fisherian modes of change, and shorter periods of intense evolutionary change, dominated by his shifting balance process, to the theory of punctuated equilibrium put forward by Eldredge and Gould (Eldredge and Gould 1972; Wright 1982)." This conclusion is also supported by the fact that sex genes and sexual traits directly connected to fitness increasingly appear to be more important in the origin of species (Singh 2000, Singh and Kulathinal 2000). Not only do sex genes show rapid rates of evolution between species (Coulthart and Singh 1988, Civetta and Singh 1995, 1998a,b, Palumbi 1999, Singh and Kulathinal 2000, Swanson and Vacquier 2002), breakdowns in species hybrids are limited to sexual/fitness traits and do not extend to nonfitness traits (Civetta and Singh 1995, 1998a,b).

6.6 Alternative views of epistasis and stability

The stability of genetic systems and their resistance to change is central to Wright's metaphor of an adaptive landscape. The nature of this fitness landscape determines how a population moves between adaptive peaks. Understanding the trajectory along the fitness surface depends on whether or not it can be represented as a constant surface dependent only on allele frequencies. This can only be true on a time scale short in comparison to the evolution of new species and ecosystem change. The problem then becomes one of changes in the relative frequencies of existing genes. This is the problem favored by beanbag genetics.

Waddington's metaphor of a development as a ball rolling in an adaptive valley is as powerful as Wright's adaptive landscape. They both visualize the same property of genetic systems, stability to perturbation. One uses a developmental landscape, the other an evolutionary landscape. Waddington (1957, 1959) was one of the first to consider the problem of conversion from one adaptive gene complex to another from the standpoint of development. He emphasized the importance of assimilating new developmental modules in evolutionary change. To do this, genetic variation that was hidden because of evolved mechanisms to resist perturbation must be revealed by environmental perturbations.

6.6.1 Functional redundancy

Genome sequencing projects have revealed an unexpected degree of functional redundancy in genomes. The analysis of gene frequency changes largely ignores the importance of functional redundancy in the relationship between genotype and phenotype. The attempt to explain genetic phenomena derived from functional redundancy by fitness surfaces requires the introduction of radical, epistatic interactions and the evolution of the fitness surface itself.

The idea of parallel or duplicate genetic pathways is an old one in genetics. Duplicate gene action was discovered in 1914 by A. F. Shull by observing a 15:1 dihybrid ratio for capsule shape in crosses of shepherd's-purse (Schull 1948). He suggested that gene duplication was involved because of chromosome doubling in the species he used. Duplicate gene action requires that redundant or partly redundant biochemical or developmental pathways alternatively lead to a similar phenotype. Generalization of duplicate gene action led to the multiple-factor hypothesis as a basis for explaining quantitative variation (Nilsson-Ehle 1909). The importance of gene duplication in evolution has not been ignored. A key problem that has been examined is how genetic redundancy can be maintained since it is unlikely that duplicate functions should be retained over evolutionary time (Thomas 1993). The essence of genetic redundancy is that two null mutations, which individually have little effect, produce a detrimental effect when combined. This is a special case of more general types of genetic interactions in which the combination of two mutations causes a synergistic effect, a phenotype which is different from that produced by either mutation alone. The effect may be suppression, in which one mutation tends to counteract the other so that the phenotype becomes more like wild type, or enhancement, in which the severity of the phenotype diverges even further from wild type. Synthetic lethals are an indication of genetic redundancy or partial redundancy in essential genetic pathways (Guarente 1993).

The extent of gene duplication creating functional similarity has been emphasized by genome sequencing, which has revealed extensive arrays of gene families (Tatusov *et al.* 1997, Remm and Sonnhammer 2000): groups of genes with nucleotide sequence similarity, often limited to common functional domains, but sometimes extending over the entire gene sequence. Another indication of unexpected amounts of functional redundancy comes from results of systematically inactivating genes. For example, gene knockout experiments in yeast have indicated that a large fraction of the genome may be individually inactivated with only minor phenotypic consequence (Winzeler *et al.* 1999). Why are there so many genes in the yeast genome with apparently small phenotypic effect when inactivated? Several explanations are possible (Wagner 2000, Featherstone and Broadie 2002).

1. Although such genes have a marginal effect on fitness, the effect is large enough so that the gene is conserved on an evolutionary time scale.
2. Such genes are functionally duplicated so that either copy can assume sufficient function.
3. The genome consists of a co-adapted network of interacting genes that is buffered against change.

The first of these ideas is most compatible with traditional views of "mass selection." Each gene is associated with an independent phenotype caused by mutation and selection can fine-tune fitness. Single and double mutants can have quite different effects so that epistatic interactions are possible. Where there are functional differences between "duplicated" genes, it is likely that mutations may have pleiotrophic effects. An example is the *Dichaete* and *SoxN* genes of *Drosophila* (Overton *et al.* 2002). These genes are members of the SoxB subgroup, part of the Sox family, a diverse group of transcriptional regulators found throughout the animal kingdom (Bowles *et al.* 2000). Both genes are expressed in overlapping compartments early in central nervous system development. Single mutations in either gene have little effect in the early medial neuroectoderm where expression overlaps. However, double mutants are severely affected, suggesting a partial functional redundancy in the region (Overton *et al.* 2002).

Possibilities for functional overlap are illustrated by structurally similar or nearly duplicate genes ("paralogs") that have similar activities (Wilkins 1997). Are paralogous genes retained because of functional differences or regulatory independence or do they represent a form of functional redundancy facilitating genetic stability and canalization? An example may clarify the problem. The *E. coli* genome has two very similar copies (*tufA* and *tufB*) of genes coding for the translation factor EF-Tu (Abdulkarim and Hughes 1996). A similar gene pair is found at homologous linkage positions in *Salmonella*, implying that duplicate functions have been retained for over 100 million years. No differences in activity or regulation have been found that could account for the conservation of these duplicated genes.

6.6.2 Interacting genetic networks

A different explanation for the marginal fitness effect of gene inactivation is that the genome is buffered against change by the complex nature of its interconnections. This does not require that a gene of marginal fitness have a similar gene elsewhere in the genome. Epistatic interactions between gene pairs are a special example of consequences arising from highly connected and interdependent genetic networks. Genomic studies of complete genomes have revealed the complexity of gene interactions, and application of network theory has shown that existing genetic networks are not random, but have evolved properties of stability and can be considered as co-adaptive systems (Wagner and Fell 2001, Jeong *et al.* 2000). The broad area of functional genomics has focused on protein expressions and metabolic networks. We will review some of them as examples of how future developments in this area may provide insight into adaptive peak shifts.

Transcriptional networks are closest to indicating those gene interactions that may be important in epistasis. They also have the possibility to reveal differences in genetic interactions between related species (Enard *et al.* 2002). The idea is to perturb the expression of a gene (perhaps by mutation) or system of genes (perhaps by changing the environment) and detect consequent changes in the levels of expression of as many genes in the genome as possible. Functional connections between genes can be revealed by the results of mutation (e.g., deletion). A large effort has gone into detecting the expression profile resulting from mutating (usually by deletion) genes in the yeast genome (Hughes *et al.* 2000). These, as well as expression profiles resulting from environmental perturbations, have revealed large networks of coregulated genes.

Protein networks have been constructed by a number of techniques such as the yeast two-hybrid system (Legrain *et al.* 2001). Observed interactions are summarized by network connections allowing the identification of functional families. Protein–protein interaction maps have been developed for the proteins of a number of genomes including bacteria such as *H. pylori* (Rain *et al.* 2001), yeast (Uetz *et al.* 2000), *C. elegans* (Walhout 2000).

Genomic studies have begun to bridge the gap between isolated, limited biochemical pathways and the metabolism of intact cells. We may never find a complete molecular description of a system as complex as a cell, but the analysis of metabolic networks represents a first step towards an integrated theory of cellular processes. Such a network would include connections between all reactions occurring under all possible conditions. Genome sequences allow interconnected biochemical pathways to be predicted and models of cell metabolism to be developed (Overbeek *et al.* 2000). Metabolic models have been made by combining genome sequences with biochemical data for several organisms such as *H. pylori* (Schilling *et al.* 2002) and *E. coli* (Edwards and Palsson 2000).

6.6.3 Network stability

Metabolic control analysis has replaced the simplistic idea of rate-limiting reactions with a more complex and interdependent view of biochemical networks based on flux control coefficients (Fell 1997). Flux control coefficients represent fractional changes in flux through a pathway per fractional change in enzyme concentration. Flux control coefficients are properties of the biochemical system as a whole. They represent the interconnectivity as well as the stability of biochemical networks to activity perturbations, for example by genetic mutation. If an enzyme has a small flux control coefficient, loss-of-function mutations in its gene are likely to be recessive since decreases in activity have little effect on pathway flux. Kacser and Burns (1981) used this idea in their explanation of dominance in terms of the systems behavior of biochemical pathways. Important questions yet to be answered are the extent to which enzymes evolve so that flux control is partitioned over the whole, and how new pathways are added to an existing system.

Although highly interconnected, gene networks do not have properties characteristic of randomly connected networks. Many connections are partly redundant and there are only a few, central connections. Specifically, there are a few genes (hubs) that interact with many other genes (Wagner and Fell 2001, Jeong *et al.* 2000, Featherstone and Broadie 2002). This property makes genome networks scale independent. That is, connectivity properties are largely independent of the scale at which the network is observed. Such networks are also resistant to perturbation, for example by mutation. This is a consequence of the fact that removing one node at random has a statistically small probability of disturbing the network's connectivity.

While such networks resist random change, they are not stable to change directed from the outside. Altering the central hubs of such networks can dramatically alter their properties. This could be relevant to the adaptation of gene complexes to new environments. Mutation may be random, but selected change may be directed towards critical genes that play key roles in parts of the network that involve adaptation. Change may act in a threshold fashion and accumulate faster once key network components change.

6.7 Evolution of new genetic systems

6.7.1 Gene transfer

Genome sequencing projects have revealed the large impact that horizontal gene transfer (HGT) has had in microbial evolution (Jain *et al.* 2002, Levin and Bergstrom 2000). HGT is detected by phylogenetic comparisons of individual genes as well as by comparing the sequence properties of groups of genes. For example, comparisons of *E. coli* and *S. typhimurium* indicate that substantially more than 10% of *E. coli MG1655* genes have entered the genome in the last ~100 million years (Lawrence and Ochman 1998, Koski *et al.* 2001). There is also considerable variation among strains of *E. coli*, which may have genomes differing by as much as a million base pairs of DNA (Hurtado and Rodríguez-Valera 1999, Ochman and Jones 2000). Much of the conserved core among strains consists of housekeeping-type genes, and the parts of the genome in flux seem to be responsible for many specific adaptations of microbial populations (Ochman *et al.* 2000, Wren 2000, Ochman and Moran 2001). Pathogenesis is an example of a trait frequently acquired by HGT. Virulence of *E. coli* strains has been independently acquired a number of times by HGT of genes, and *Salmonella* species have acquired several large "pathogenicity islands" (Bäumler 1997). Symbiotic bacteria have also used HGT to adapt to their hosts. A ~500 kb "symbiotic island" found in *M. loti* strain R7A can transfer to nonsymbiotic rhizobacteria in the environment, converting them to *Lotus* symbionts (Sullivan *et al.* 2002).

Why do bacteria so frequently use HGT as a means to adapt to new environments? A reason may be that, as clonally reproducing organisms with relatively little capacity for recombination, they need it to produce new gene complexes. Adaptive gene complexes in clonal organisms evolve by sequential substitution (Levin and Bergstrom 2000). A favorable combination of genes must be reached by stepping through a series of intermediate stages of single substitutions. If any of these are disadvantageous, the final combination can be produced only very slowly, if at all. The ability to acquire new genes or favorable combinations of genes from unrelated, but ecologically similar, bacteria will provide an enormous advantage. There is another way that bacteria have used to acquire new genetic information, by gene duplication (Wren 2000).

6.7.2 Gene loss

Gene duplication has long been thought to be the basis for the evolution of new gene function. Ohno (1970) emphasizes the role of natural selection as a "policeman," but points out that evolution is also characterized by loss of parts and specialization of systems and that gene elimination will also play an important role in the modification of existing traits. The evolution of olfactory receptor (OR) genes is an example of how the problem of moving from one adaptive peak to another might be solved. Organisms in general, and mammals in particular, vary greatly in their ability to detect odor. Reduction in odor detection such as that observed for primates in comparison with other mammals is not simply due to loss of OR genes. Differences in anatomical structures such as the nasal morphology, olfactory epithelium, the olfactory bulb, and integrative areas of the brain may also be involved (Moulton 1967). The perception of odors affects many traits, such as predator–prey interactions and reproductive behavior. Olfaction could well be viewed as a species-specific biological system. Evolution of new abilities and even the loss of old abilities require changes in many different patterns of gene expression and interaction. Gene elimination coupled with positive selection following gene duplication appears to be a major molecular mechanism in the complex process of odor evolution. More than 70% of primate OR genes are nonfunctional pseudogenes in humans. In contrast, over 95% are functional in monkeys (Rouquier *et al.* 1998, 2000). However, novelty cannot be entirely built from null mutations. In the OR gene cluster, weak positive selection may explain patterns of polymorphism and divergence between humans and chimpanzees (Gilad *et al.* 2000). Large clusters of OR genes are scattered in many locations about the human genome, locations that are not always conserved between species (Rouquier *et al.* 1998). Polymorphism is also present within human populations. For example, individuals can have 7–11 copies of an OR-containing DNA segment that are part of larger, subtelomeric duplications. This OR-containing segment is present at common locations in all humans sampled, indicating that its chromosomal dispersion pre-dated the expansion of human populations. But it is not found at these locations in other primates, indicating that it has changed both copy number and position during primate evolution (Trask *et al.* 1998).

Lynch and Conery (2000) suggested a model of speciation that involves loss of different copies of duplicated genes. Hybridization of populations in which different copies of a duplicated gene had become a pseudogene produces a heterozygous F1, but subsequent segregation produces individuals having two pseudogenes and no functional copy. They suggest that such a process of "divergent resolution" could be the basis of reproductive isolation. Taylor *et al.* (2001) extended this idea to genome duplications, suggesting possible examples in fish and plant speciation.

6.7.3 Gene addition

We fully appreciate the key role of gene duplication in evolution. From genome duplication to gene duplication, it has been acknowledged as the source of new genetic functions. But it is little appreciated as a source of adaptation. Wright was fully aware of gene duplication, but did not see it as relevant to the problem of adaptive peak shift. Even though he saw SBT as relevant to speciation and macroevolution, he did not see the process as a problem in evolving new gene functions, but rather as a problem in allele frequency change, perhaps because adaptations involved in speciation did not seem to require new genetic functions.

Gene duplication has taken place on a number of different scales from the duplication of complete or major parts of genomes such as yeast (Seoighe and Wolfe 1998), *Arabidopsis* (Simillion *et al.* 2002) and fish (Van de Peer 2002) to the tandem duplication of single genes. As an example of the extent of gene duplication, of the order of 5% of human genomic DNA are segmental duplications that have occurred within the last ~40 million years (Bailey *et al.* 2002). Bailey *et al.* (2002) suggest that much of this duplicated DNA may contain genes involved in primate-specific adaptations. Recent comparisons of the mosquito (*A. gambiae*) and *Drosophila* (*D. melanogaster*) genomes revealed considerable differential expansion of many gene families (Zdobnov *et al.* 2002). Less than 50% of the genes in either one of these species could be identified as orthologous to a gene present in the other, indicating a vast degree of gene innovation since they diverged ~250 million years ago (Zdobnov *et al.* 2002).

Acquisition of new gene function following gene duplication is a recognized fact of evolution. A key problem in the evolution of new gene function is that in the initial stages a duplicated gene with completely overlapping and identical function is unlikely to be retained (Lynch and Conery 2000). An exception would be if there were advantages to increased activity provided by gene duplication (Brown *et al.* 1998). Pseudogenes are stark reminders of the fate of most duplicate genes. Only if a new function is acquired can the duplicate gene be maintained. A number of molecular mechanisms may be imagined for the acquisition of a new function following gene duplication (Cooke *et al.* 1997). A duplicated gene may acquire a function in a different pathway. Thus, the second function may become essential or at least advantageous in some other context, but yet be able to compensate for the first if it is disrupted (Tautz 1992). At the level of gene regulation, a duplicated gene may add a new promoter element that allows it to be expressed at a different time or in a different tissue. If the new function is adaptive, the duplicate gene becomes conserved, although redundant from the standpoint of the original function, or adaptive mutations may allow the function of a duplicated gene to diverge. Gene conversion may aid in evolution of new functions. If, after gene duplication, one copy accumulates deleterious mutations, gene conversion presents these to selection, thereby avoiding negative epistatic interactions present if mutated one at a time (Hansen *et al.* 2000). Similar considerations apply to the evolution of new modules within existing genetic networks. Modular addition of new connections seems the most reasonable mechanism for evolving new co-adapted genetic networks.

Major shifts in evolution require new gene complexes that must be formed by the addition of adaptive modules to existing genetic networks. But how likely is it that such processes are a major factor in the adaptation of populations to new environments and in the evolution of new species? Does peak shift by acquiring new genes outweigh peak shift by the replacement of one group of interacting alleles by another group of interacting alleles? Of course at some stage the process requires that individuals in a population shift from having predominantly one group of alleles to another group of alleles. But the shift from individuals lacking a gene to ones having a new gene is entirely different from changing interacting alleles of existing genes. Recent evidence indicates that new gene functions have been created in that model of speciation, the *Drosophila melanogaster* complex. *Drosophila melanogaster* has been found to have at least one gene that its closely related sibling, *Drosophila simulans*, does not have. *Sdic* is a sperm-specific dynein made from two different genes by a complicated process of duplication and rearrangement (Nurminsky *et al.* 1998). In fact, the X-chromosome of *D. melanogaster* contains an additional 70 kb of DNA in the *Sdic* region (Nurminsky *et al.* 1998). Another novel gene, *jingwei*, is found only in the sibling species *D. teissieri* and *D. yakuba* (Long and Langley 1993). It was derived from *Adh* mRNA by chimeric fusion with parts of an unrelated gene. The extent and importance of such novel genes in the adaptive differences between closely related species remains to be determined. Genome sequencing should help clarify this question.

6.8 Conclusions

Selective neutrality, additive gene action, gradual evolution, and allopatric speciation can all be considered as null hypotheses. Fisher's "fundamental theorem of natural selection" with multiple loci, in large panmictic populations, is also a null hypothesis, in that it approximates the operation of evolution under minimal, ideal conditions. Wright's shifting balance theory of evolution provides an alternative to Fisher's mass selection and an avenue to study evolution in its essential details. All genes are not additive nor are they all of small incremental effect in their action. Gene interactions and epistasis are essential characteristics of co-adapted gene complexes. Mutations may take on different fitness values in a new, changing environment. Populations may be small and/or subdivided in nature with restricted gene flow between them.

Wright proposed random genetic drift as a mechanism for natural populations to explore the fitness landscape and to produce new, adaptive gene combinations. But random genetic drift may not be an effective means of producing new gene combinations. Decreasing genetic variation may not be the best way to produce new genotypes. Alternatively, creation of new gene functions may hold the key to producing new gene combinations. Individual genotypes are the basis of selection and so it is a gene network rather than allele frequency vectors that provides the basis for connecting genes to fitness surfaces. A fitness surface connected with interacting sets of gene networks provides a bridge across fitness valleys. Buildup of new gene interactions as a result of new major mutations, rearrangement of existing alleles as in the case of repeated cycles of population expansion and constriction (*sensu* Carson 1997), or environmentally induced changes in gene regulation (Queitsch *et al.* 2002), would allow natural populations to explore their genetic and fitness surroundings without losing genetic variation due to drift. An occasional new, beneficial network module can provide the basis for genetic assimilation (*sensu* Waddington 1957).

A modified form of Wright's theory based on epistatic gene networks will provide a basis for a comprehensive theory of evolution by gene regulation (*sensu* King and Wilson 1975). Such a modified theory can also provide room for Goldschmidt's view of evolution, which in retrospect looks less "monstrous" than before. Such a theory of evolution, interspersing gradual evolution with occasional rapid selective sweeps, would also provide a basis for Gould's punctuated equilibrium (Gould 1980). Much like the history of life, with epochs of gradual evolution interspersed with adaptive radiation and major extinctions, the history of genome evolution can be seen as consisting of gradual evolution by selection of multiple genes with additive gene effects, interspersed with occasional rapid evolution involving major mutations, gene duplication, new gene combinations, and new gene modules. In other words, mass selection for metric traits, shifting balance for complex traits. The evolution of new genetic systems is a major problem in evolutionary biology and a renewal of its evolutionary importance seems appropriate in the present climate of rapid progress in the fields of genomics, proteomics, and gene networks.

The metaphor of the adaptive landscape must be changed. The new adaptive peak does not exist as a possibility in the old landscape. New dimensions must be added to the old landscape along which the population evolves to reach a novel co-adapted peak. The idea of genetic redundancy is central to this process, for without it, additions to the old network are likely too improbable or maladaptive. Problems in the evolution of new developmental pathways and morphological innovation now have counterparts in the molecular domain. How can something new be cobbled from something old? The answer appears to lie in functional innovation from functional redundancy. How to refashion a spider's web to better capture a fly?

A new metaphor, more appropriate for complex evolving genetic networks, would be a collection of spider webs floating on a fitness sea. Each web represents a different individual in the population. Individuals are different, not only because they contain different alleles at each genetic locus, but also because they contain different loci. Each web representing a series of genetic interactions is different. Each is in a process of building and rebuilding. Some differences between webs are nonallelic because a new gene leads to new interactions and connections among previous members. Most often these new connections are detrimental holes in the web that causes it to sink in the sea. Rather than thinking of evolution in terms of replacing different nodes in a single web with altered genes, we need to visualize the process of building new webs from the old. Individuals may differ in the number and nature of functional redundancies. Allelic substitutions among partly redundant copies could allow a web to slide along the surface to a new fitness peak. Rather than a static surface, it modulates and varies, in time as well as by virtue of the number of webs that try to occupy its surface.

As Wright believed in consistency and not contingency, and as his commitment was to shifting balance and not drift, the replacement of drift by gene interaction network as the major mechanism of producing new gene combinations would remove criticism leveled against drift as a mechanism of evolutionary innovation. It would also bolster shifting balance theory as a mechanism of evolution by providing a link to new developments in molecular and developmental evolutionary biology of the phenotype on one hand and to evolution in meta-populations on the other.

REFERENCES

Abdulkarim, F., Hughes, D. (1996). Homologous recombination between the *tuf* genes of *Salmonella typhimurium*. *J. Mol. Biol.* 260:506–22.

Ardie, K. G. (1998). Putting the brake on drive: meiotic drive of *t* haplotypes in natural populations of mice. *Trends Genet.* 14:189–93.

Bailey, J. A., Gu, Z., Clark, R. A., Reinert, K., Samonte, R. V., Schwartz, S., Adams, M. D., Myers, E. W., Li, P. W., and Eichler, E. E. (2002). Recent segmental duplications in the human genome. *Science* 297:1003–7.

Barton, N. H., and Rouhani, S. (1987). The frequency of shifts between alternative equilibria. *J. Theoret. Biol.* 125:397–418.

Bäumler, A. (1997). The record of horizontal gene transfer in *Salmonella*. *Trends Microbiol.* 5:318–22.

Beaudet, A. L., and Jiang, Y.-H. (2002). Perspective: a rheostat model for a rapid and reversible form of imprinting-dependent evolution. *Am. J. Hum. Genet.* 70:1389–97.

Berry, A., and Kreitman, M. (1993). Molecular analysis of an allozyme cline: alcohol dehydrogenase in *Drosophila melanogaster* on the east coast of North America. *Genetics* 143:869–93.

Bouma, J. E., and Lenski, R. E. (1988). Evolution of a bacteria/plasmid association. *Nature* 335:351–2.

Bowles, J., Schepers, G., and Koopman, P. (2000). Phylogeny of the SOX family of developmental transcription factors based on sequence and structural indicators. *Dev. Biol.* 227:239–55.

Bradshaw, W. E., Fujiyama, S., and Holzapfel, C. M. (2000). Adaptation to the thermal climate of North America by the pitcher-plant mosquito, *Wyeomyia smithii*. *Ecology* 81:1262–72.

Brodie, E. D. III. (2000). Why evolutionary genetics does not always add up. In J. B. Wolf, E. D. Brodie, III and M. J. Wade (eds) *Epistasis and the Evolutionary Process*, pp. 3–19. Oxford: Oxford University Press.

Brown, C. J., Todd, K. M., and Rosenzweig, R. F. (1998). Multiple duplications of yeast hexose transport genes in response to selection in a glucose-limited environment. *Mol. Biol. Evol.* 15:931–42.

Carson, H. L. (1997). The Wilhelmine E. Key 1996 Invitational Lecture. Sexual selection: a driver of genetic change in Hawaiian Drosophila. *J. Hered.* 88(5): 343–52.

Charlesworth, B., and Charlesworth, D. (1973). A study of linkage disequilibrium in populations of *Drosophila melanogaster*. *Genetics* 73:351–9.

Civetta, A., and Singh, R. S. (1995). High divergence of reproductive tract proteins and their association with postzygotic reproductive isolation in *Drosophila melanogaster* and *Drosophila virilis* group species. *J. Mol. Evol.* 41:1085–95.

Civetta, A., and Singh, R. S. (1998a). Sex-related genes, directional sexual selection and speciation. *Mol. Biol. Evol.* 15:901–9.

Civetta, A., and Singh, R. S. (1998b). Sex and speciation: genetic architecture and evolutionary potential of sexual vs. nonsexual traits in the sibling species of the *Drosophila melanogaster* complex. *Evolution* 52:1080–92.

Cooke, J., Nowak, M. A., Boerlijst, M., and Maynard Smith, J. (1997). Evolutionary origins and maintenance of redundant gene expression during metazoan development. *Trends Genet.* 13:360–4.

Coulthart, M. B., and Singh, R. S. (1988). High level of divergence of male-reproductive-tract proteins between *Drosophila melanogaster* and its sibling species, *D. simulans*. *Mol. Biol. Evol.* 5:82–191.

Coyne, J. A., Barton, N. H., and Turelli, M. (1997). Perspective: a critique of Sewall Wright's shifting balance theory of evolution. *Evolution* 51:643–71.

Coyne, J. A., Barton, N. H., and Turelli, M. (2000). Is Wright's shifting balance process important in evolution? *Evolution* 54:306–17.

Coyne, J. A., Simeonidis, S., and Rooney, P. (1998). Relative paucity of genes causing inviability in hybrids between *Drosophila melanogaster* and *D. simulans*. *Genetics* 150(3):1091–103.

Crow, J. F. (1991). Was Wright right? *Science* 253:973.

Crow, J. F. (2002). Here's to Fisher, additive genetic variance, and the Fundamental Theorem of Natural Selection. *Evolution* 56:1313–16.

Crow, J. F., Engels, W. R., and Denniston, C. (1990). Phase three of Wright's shifting balance theory. *Evolution* 44:233–47.

Dawkins, R. (1976). *The Selfish Gene.* Oxford: Oxford University Press.

Dobzhansky, Th. (1937). *Genetics and the Origin of Species.* New York: Columbia University Press.

Dobzhansky, Th. (1970). *Genetics and the Evolutionary Process.* New York: Columbia University Press.

Edwards, J. S., and Palsson, B. O. (2000). The *Escherichia coli* MG1655 *in silico* metabolic genotype: its definition, characteristics, and capabilities. *PNAS, USA* 97:5528–33.

Eldredge, N. H., and Gould, S. J. (1972). Punctuated equilibria: an alternative to phyletic gradualism. In T. J. M. Schopf (ed.) *Models in Paleontology.* San Francisco: W. H. Freeman.

Enard, W., Khaitovich, P., Klose, J., Zöllner, S., Heissig, F., Giavalisco, P., Nieselt-Struwe, K., Muchmore, E., Varki, A., Ravid, R., Doxiadis, G. M., Bontrop, R. E., and Pääbo, S. (2002). Intra- and interspecific variation in primate gene expression patterns. *Science* 296:340–3.

Featherstone, D. E., and Broadie, K. (2002). Wrestling with pleiotrophy: genomic and topological analysis of the yeast gene expression network. *BioEssays* 24: 267–74.

Fell, D. (1997). *Understanding the Control of Cell Metabolism.* London: Portland Press.

Fisher, R. A. (1930, 1958). *The Genetical Theory of Natural Selection.* New York: Dover.

Fisher, R. A., and Ford, E. B. (1947). The spread of a gene in natural conditions in a colony of the moth *Panaxia dominula*. *Heredity* 1:143–74.

Ford, E. B. (1975). *Ecological Genetics.* London: Chapman and Hall.

Franklin, I., and Lewontin, R. C. (1970). Is the gene the unit of selection? *Genetics* 65:707–34.

Gavrilets, S. (1997). Hybrid zones with Dobzhansky-type epistatic selection. *Evolution* 51:1027–35.

Gilad, Y., Segre, D., Skorecki, K., Nachman, M. W., Lancet, D., and Sharon, D. (2000). Dichotomy of single nucleotide polymorphism haplotypes in olfactory receptor genes and pseudogenes. *Nat. Genet.* 26:221–4.

Gillespie, J. H. (1991). *The Causes of Molecular Evolution.* Oxford: Oxford University Press.

Goldschmidt, R. B. (1940). *The Material Basis of Evolution.* New Haven, CT: Yale University Press.

Goodnight, C. J. (2000). Modeling gene interaction in structured populations. In J. B. Wolf, E. D. Brodie III and M. J. Wade (ed.) *Epistasis and the Evolutionary Process,* pp. 129–45. Oxford: Oxford University Press.

Goodnight, C. J., and Wade, M. J. (2000). The ongoing synthesis: a reply to Coyne, Barton and Turelli. *Evolution* 54(1):317–24.

Gould, S. J. (1977). The return of hopeful monsters. *Nat. Hist. Magazine,* June–July 1977:22–30.

Gould, S. J. (1980). Is a new and general theory of evolution emerging? *Paleobiology,* 6(1):119–30.

Gould, S. J. (1989). *Wonderful Life: the Burgess Shale and the Nature of History.* New York: Norton.

Gould, S. J., and Eldredge, E. (1977). Punctuated equilibria: the tempo and mode of evolution reconsidered. *Paleontology* 3:115–51.

Guarente, L. (1993). Synthetic enhancement in gene interaction: a genetic tool come of age. *Trends Genet.* 9:362–6.

Haldane, J. B. S. (1932). *The Causes of Evolution.* London: Longmans, Green.

Hansen, T. F., Carter, A. J., and Chiu, C. H. (2000). Gene conversion may aid adaptive peak shifts. *J. Theor. Biol.* 207:495–511.

Hedrick, P., Jain, S. K., and Holden, L. (1978). Multilocus systems in evolution. *Evol. Biol.* 11:101–84.

Hughes, T. R., Marton, M. J., Jones, A. R., Roberts, C. J., Stoughton, R., Armour, C. D., Bennett, H. A., Coffey, E., Dai, H., He, Y. D., Kidd, M. J., King, A. M., Meyer, M. R., Slade, D., Lum, P. Y., Stepaniants, S. B., Shoemaker, D. D., Gachotte, D., Chakraburtty, K., Simon, J., Bard, M., and Friend, S. H. (2000). Functional discovery via a compendium of expression profiles. *Cell* 102:109–26.

Hurtado, A., and Rodríguez-Valera, F. (1999). Accessory DNA in the genomes of representatives of the *Escherichia coli* reference collection. *J. Bact.* 181: 2548–54.

Huxley, J. S. (1942). *Evolution: the Modern Synthesis.* New York: Harper.

Jain, T., Rivera, M. C., Moore, J. E., and Lake, J. (2002). Horizontal gene transfer in microbial genomes evolution. *Theor. Pop. Biol.* 61:489–95.

Jeong, H., Tombor, B., Albert, R., Oltval, Z. N., and Barabási, A.-L. (2000). The large scale organization of metabolic networks. *Nature* 407:651–4.

Johnson, N. A. (2000). Gene interactions and the origin of species. In J. B. Wolf, E. D. Brodie III and M. J. Wade (eds) *Epistasis and the Evolutionary Process,* pp. 197–212. Oxford: Oxford University Press.

Kacser, H., and Burns, J. A. (1981). The molecular basis of dominance. *Genetics* 97:639–66.

Kimura, M. (1968). Evolutionary rate at the molecular level. *Nature* 217:624–26.

Kimura, M. (1983). *The Neutral Theory of Molecular Evolution.* Cambridge: Cambridge University Press.

King, M. C., and Wilson, A. C. (1975). Evolution at two levels in humans and chimpanzees. *Science* 188:107–16.

Klein, J. (1986). *Natural History of the Major Histocompatibility Complex.* New York: John Wiley.

Koski, L. B., Morton, R. A., and Golding, G. B. (2001). Codon bias and base composition are poor indicators of horizontally transferred genes. *Mol. Biol. Evol.* 18: 404–12.

Langley, C. H., Tobari, Y. N., and Kojima, K.-I. (1974). Linkage disequilibrium in natural populations of *Drosophila melanogaster. Genetics* 78:921–36.

Lawrence, J. G., and Ochman, H. (1998). Molecular archaeology of the *Escherichia coli* genome. *PNAS, USA* 95:9413–17.

Legrain, P., Wojcik, J., and Gauthier, J.-M. (2001). Protein–protein interaction maps: a lead towards cellular functions. *Trends Genet.* 17:346–52.

Levin, B. R., and Bergstrom, C. T. (2000). Bacteria are different: observations, interpretations, speculations, and opinions about the mechanisms of adaptive evolution in prokaryotes. *PNAS, USA* 97:6981–5.

Lewontin, R. C. (1974). *The Genetic Basis of Evolutionary Change.* New York: Columbia University Press.

Lewontin, R. C., and Kojima, K.-I. (1960). The evolutionary dynamics of complex polymorphisms. *Evolution* 14:458–72.

Lewontin, R. C., and White, M. J. D. (1960). Interaction between inversion polymorphisms of two chromosome pairs in the grasshopper, *Moraba scurra. Evolution* 14:116–29.

Long, M., and Langley, C. H. (1993). Natural selection and the origin of *jingwei*, a chimeric processed functional gene in *Drosophila. Science* 260:91–4.

Lynch, M., and Conery, J. S. (2000). The evolutionary fate and consequences of duplicate genes. *Science* 290:1151–5.

Lyttle, T. W. (1991). Segregation distorters. *Annu. Rev. Genet.* 25:511–57.

Maynard Smith, J., and Williams, G. C. (1976). Reply to Barash. *Am. Nat.* 10:897.

Mayr, E. (1954). Change of genetic environment and evolution. In J. Huxley, A. C. Hardy and E.B. Ford (eds) *Evolution As a Process,* pp. 157–80. New York: Macmillan.

Mayr, E. (1959). Where are we? *Cold Spring Harbor. Symp. Quant. Biol.* 24:1–14.

McClintock, B. (1957). Chromosome organization and gene expression. *Cold Spring Harbor Symp. Quant. Biol.* 16:13–47.

Moreno, G. (1994). Genetic architecture, genetic behavior, and character evolution. *Annu. Rev. Ecol. Syst.* 25:31–44.

Morita, T., Kubota, H., Satta, Y., and Matsushiro, A. (1993). Evolution of the mouse t haplotype. In N. Takahata and A. G. Clark (eds) *Mechanisms of Molecular Evolution,* pp. 151–8. Sunderland, MA: Sinauer.

Moulton, D. G. (1967). Olfaction in mammals. *Am. Zool.* 7:421–9.

Nilsson-Ehle, H. (1909). Kreuzungsuntersuchungen an Hafer und Weizen. *Lund Univ. Aarskr. N. F.* 5(2):1–22.

Nurminsky, D. I., Nurminskaya, M. V., DeAguiar, D., and Hartl, D. L. (1998). Selective sweep of a newly evolved sperm-specific gene in *Drosophila. Nature* 396:572–5.

Ochman, H., and Jones, I. B. (2000). Evolutionary dynamics of full genome content in *Escherichia coli. EMBO J.* 19:6637–43.

Ochman, H., and Moran, N. A. (2001). Genes lost and genes found: evolution of bacterial pathogenesis and symbiosis. *Science* 292:1096–8.

Ochman, H., Lawrence, J. G., and Grolsman, E. A. (2000). Lateral gene transfer and the nature of bacterial evolution. *Nature* 405:299–304.

Ohno, S. (1970). *Evolution by Gene Duplication.* Berlin: Springer-Verlag.

Ohta, T. (1976). Role of very slightly deleterious mutations in molecular evolution and polymorphism. *Theor. Pop. Biol.* 10:254–75.

Overbeek, R., Larsen, N., Pusch, G. D., D'Souza, M., Selkov, E. Jr., Kyrpides, N., Fonstein, M., Maltsev, N., and Selkov, E. (2000). WIT: integrated system for high throughput genome sequence analysis and metabolic reconstruction. *Nuc. Acids Res.* 28:123–5.

Overton, P. M., Meadows, L. A., Urban, J., and Russell, S. (2002). Evidence for differential and redundant function of the Sox genes *Dichaete* and *SoxN* during CNS development in *Drosophila. Development* 129:4219–28.

Palumbi, S. R. (1999). All males are not created equal: fertility differences depend on gamete recognition polymorphisms in sea urchins. *Proc. Natl. Acad. Sci. USA* 96(22):12632–7.

Phillips, P. C., Otto, S. P., and Whitlock, M. C. (2000). Beyond the average: the evolutionary importance of gene interactions and variability of epistatic effects. In J. B. Wolf, E. D. Brodie, III and M. J. Wade (eds) *Epistasis and the Evolutionary Process*, pp. 20–38. New York: Oxford University Press.

Provine, B. (1986). *Sewall Wright and Evolutionary Biology*. Chicago: University of Chicago Press.

Queitsch, C., Sangster, T. A., and Lindquist, S. (2002). Hsp90 as a capacitor of phenotypic variation. *Nature* 417:618–24.

Rain, J.-C., Selig, L., De Reuse, H., Battaglia, V., Reverdy, C., Simon, S., Lenzen, G., Petel, F., Wojcik, J., Schachter, V., Chemama, Y., Labigne, A., and Legrain, P. (2001). The protein–protein interaction map of *Helicobacter pylori. Nature* 409:211–15.

Remm, M., and Sonnhammer, E. (2000). Classification of transmembrane protein families in the *Caenorhabditis elegans* genome and identification of human orthologs. *Genome Res.* 10:1679–89.

Rice, W. R. (1992). Sexually antagonistic genes: experimental evidence. *Science* 256:1436–9.

Rouquier, S., Blancher, A., and Giorgi, D. (2000). The olfactory receptor gene repertoire in primates and mouse: evidence for reduction of the functional fraction in primates. *PNAS, USA* 97:2870–4.

Rouquier, S., Taviaux, S., Trask, B. J., Brand-Arpon, V., van den Engh, G., Demaille, J., and Giorgi, D. (1998). Distribution of olfactory receptor genes in the human genome. *Nat. Genet.* 18:243–50.

Sawamura, K., Davis, A. W., and Wu, C. I. (2000). Genetic analysis of speciation by means of introgression into *Drosophila melanogaster. PNAS, USA* 97:2652–5.

Schilling, C., Schilling, C. H., Covert, M. W., Famili, I., Church, G. M., Edwards, J. S., and Palsson, B. O. (2002). Genome-scale metabolic model of *Helicobacter pylori* 26695. *J. Bact.* 184:4582–93.

Schull, A. F. (1948). *Heredity*. New York: McGraw-Hill.

Seoighe, C., and Wolfe, K. H. (1998). Extent of genome rearrangement after genome duplication in yeast. *PNAS, USA* 95:4447–52.

Simillion, C., Vandepoele, K., Van Montagu, M. C., Zabeau, M., and Van De Peer, Y. (2002). The hidden duplication past of *Arabidopsis thaliana. PNAS, USA* 99:13627–32.

Simmons, M. J., and Crow, J. F. (1977). Mutations affecting fitness in *Drosophila* populations. *Annu. Rev. Genet.* 11:49–78.

Simpson, G. G. (1944). *Tempo and Mode in Evolution.* New York: Columbia University Press.

Singh, R. S. (2000). Toward a unified theory of speciation. In R. S. Singh and C. B. Krimbas (eds) *Evolutionary Genetics: From Molecules to Morphology,* pp. 570–608. Cambridge: Cambridge University Press.

Singh, R. S., and Kulathinal, R. J. (2000). Sex gene pool evolution and speciation: a new paradigm. *Genes Genet. Syst.* 75:119–30.

Stam, L. F., and Laurie, C. C. (1996). Molecular dissection of a major gene effect on a quantitative trait: the level of alcohol dehydrogenase expression in *Drosophila melanogaster. Genetics* 144:1559–64.

Sullivan, J. T., Trzebiatowski, J. R., Cruickshank, R. W., Gouzy, J., Brown, S. D., Elliot, R. M., Fleetwood, D. J., McCallum, N. G., Rossbach, U., Stuart, G. S., Weaver, J. E., Webby, R. J., De Bruijn, F. J., and Ronson, C. W. (2002). Comparative sequence analysis of the symbiosis island of *Mesorhizobium loti* strain R7A. *J. Bact.* 184: 3086–95.

Swanson, W. J., and Vacquier, V. D. (2002). The rapid evolution of reproductive proteins. *Nat. Rev. Genet.* 3(2):137–44.

Tatusov, R. L., Koonin, E. V., and Lipman. D. J. (1997). A genomic perspective on protein families. *Science* 278:631–7.

Tautz, D. (1992). Redundancies, development and the flow of information. *BioEssays* 14:263–6.

Taylor, J. S., Van de Peer, Y., and Meyer, A. (2001). Genome duplication, divergent resolution and speciation. *Trends Genet.* 17:299–301.

Thomas, J. H. (1993). Thinking about genetic redundancy. *Trends Genet.* 11: 395–9.

Trask, B. J., Friedman, C., Martin-Gallardo, A., Rowen, L., Akinbami, C., Blankenship, J., Collins, C., Giorgi, D., Iadonato, S., Johnson, F., Kuo, W. L., Massa, H., Morrish, T., Naylor, S., Nguyen, O. T., Rouquier, S., Smith, T., Wong, D. J., Youngblom, J., and van den Engh, G. (1998). Members of the olfactory receptor gene family are combined in large blocks of DNA duplicated polymorphically near the ends of human chromosomes. *Hum. Mol. Genet.* 7:13–26.

Uetz, P., Giot, L., Cagney, G., Mansfield, T. A., Judson, R. S., Knight, J. R., Lockshon, D., Narayan, V., Srinivasan, M., Pochart, P., Qureshi-Emili, A., Li, Y., Godwin, B., Conover, D., Kalbfleisch, T., Vijayadamodar, G., Yang, M., Johnston, M., Fields, S., and Rothberg, J. M. (2000). A comprehensive analysis of protein–protein interactions in *Saccharomyces cerevisiae. Nature* 403:623–7.

Van de Peer, Y., Taylor, J. S., Joseph, J., and Meyer, A. (2002). Wanda: a database of duplicated fish genes. *Nuc. Acids Res.* 30:109–12.

Waddington, C. H. (1957). *The Strategy of the Genes.* New York: Macmillan.

Waddington, C. H. (1959). Canalization of development and genetic assimilation of acquired characters. *Nature* 183:1654–5.

Wade, M. J. (2000). Epistasis as a genetic constraint within populations and an accelerant of adaptive divergence among them. In J. B. Wolf, E. D. Brodie III, and M. J. Wade (eds) *Epistasis and the Evolutionary Process,* pp. 213–31. Oxford: Oxford University Press.

Wade, M. J., and Goodnight, C. J. (1991). Wright's shifting balance theory: an experimental study. *Science* 253:1015–18.

Wade, M. J., and Goodnight, C. J. (1998). Perspective: the theories of Fisher and Wright in the context of metapopulations: when nature does many small experiments. *Evolution* 52(6):1537–53.

Wagner, A. (2000). Robustness against mutations in genetic networks of yeast. *Nat. Genet.* 24:355–61.

Wagner, A., and Fell, D. A. (2001). The small world inside large metabolic networks. *Proc. R. Soc. Lond. B.* 268:1803–10.

Walhout, A. J. M., Sordella, R., Lu, X., Hartley, J. L., Temple, G. F., Brasch, M. A., Thierry-Mieg, N., and Vidal, M. (2000). Protein interaction mapping in *C. elegans* using proteins involved in vulval development. *Science* 287:116–22.

Whitlock, M. C. (1997). Founder effect and peak shifts without genetic drift. *Evolution* 51:1044–8.

Whitlock, M. C., Phillips, P. C., Moore, F. B.-G., and Tonsor, S. J. (1995). Multiple fitness peaks and epistasis. *Annu. Rev. Ecol. Syst.* 26:601–29.

Wilkins, A. S. (1997). Canalization: a molecular genetic perspective. *BioEssays* 19:257–62.

Williams, G. C. (1966). *Adaptation and Natural Selection.* Princeton, NJ: Princeton University Press.

Winzeler, E. A., Shoemaker, D. D., Astromoff, A., Liang, H., Anderson, K., Andre, B., Bangham, R., Benito, R., Boeke, J. D., Bussey, H., Chu, A. M., Connelly, C., Davis, K., Dietrich, F., Dow, S. W., Bakkoury, M. E., Foury, F., Friend, S. H., Gentalen, E., Giaever, G., Hegemann, J. H., Jones, T., Laub, M., Liao, H., Liebundguth, N., Lockhart, D. J., Lucau-Danila, A., Lussier, M., M'Rabet, N., Menard, P., Mittmann, M., Pai, C., Rebischung, C., Revuelta, J. L., Riles, L., Roberts, C. J., Ross-MacDonald, P., Scherens, B., Snyder, M., Sookhai-Mahadeo, S., Storms, R. K., Véronneau, S., Voet, M., Volckaert, G., Ward, T. R., Wysocki, R., Yen, G. S., Yu, K., Zimmermann, K., Philippsen, P., Johnston, M., and Davis, R. W. (1999). Functional characterization of the *S. cerevisiae* genome by gene deletion and parallele analysis. *Science* 285:901–6.

Wren, B. W. (2000). Microbial genome analysis: insights into virulence, host adaptation and evolution. *Nat. Rev. Genet.* 1:30–9.

Wright, S. (1931). Evolution in Mendelian populations. *Genetics* 16:97–159.

Wright, S. (1932). The roles of mutation, inbreeding, crossbreeding, and selection in evolution. *Proc.* 6th *Int. Cong. Genet.* 1:356–66.

Wright, S. (1940). Breeding structure of populations in relation to speciation. *Am. Nat.* 74:232–48.

Wright, S. (1978). *Evolution and the Genetics of Populations*: Vol. 4: *Variability Within and Among Natural Populations.* Chicago: University of Chicago Press.

Wright, S. (1980). Genic and organismic selection. *Evolution* 34(5):825–43.

Wright, S. (1982a). Character change, speciation and the higher taxa. *Evolution* 36(3):427–43.

Wright, S. (1982b). The shifting balance theory and macroevolution. *Annu. Rev. Genet.* 16:1–19.

Wright, S. (1988). Surfaces of selective values revisited. *Am. Nat.* 131:115–23.

Wu, C.-I., and Palopoli, M. F. (1994). Genetics of postmating reproductive isolation in animals. *Annu. Rev. Genet.* 28:283–308.

Zdobnov, E. M., von Mering, C., Letunic, I., Torrents, D., Suyama, M., Copley, R. R., Christophides, G. K., Thomasova, D., Holt, R. A., Subramanian, G. M., Mueller, H. M., Dimopoulos, G., Law, J. H., Wells, M. A., Birney, E., Charlab, R., Halpern, A. L., Kokoza, E., Kraft, C. L., Lai, Z., Lewis, S., Louis, C., Barillas-Mury, C., Nusskern, D., Rubin, G. M., Salzberg, S. L., Sutton, G. G., Topalis, P., Wides, R., Wincker, P., Yandell, M., Collins, F. H., Ribeiro, J., Gelbart, W. M., Kafatos, F. C., and Bork, P. (2002). Comparative genome and proteome analysis of *Anopheles gambiae* and *Drosophila melanogaster*. *Science* 298:149–59.

PART III

PHENOTYPES TO FITNESS: GENETICS AND ECOLOGY OF POPULATIONS

7

Density-dependent selection

FREDDY BUGGE CHRISTIANSEN
Department of Genetics and Ecology, University of Aarhus

7.1 Introduction

The classical studies of the interplay between natural selection on a character and Mendelian inheritance of variant traits employed simple models, which in essence are those used in most introductory texts on population genetics. Those models therefore form the common reference in discussions of the genetical effects of selection. In their classical models Fisher (1922) and Haldane (1924, 1932) formulated the effect of selection in terms of constant individual fitnesses in a population that reproduces by random mating (Ewens 1979), a situation that mirrors the action of zygotic selection in a random mating population with nonoverlapping generations (Christiansen and Prout 2000). These models form the root of a proliferation of theories incorporating a variety of genetic phenomena, theories that give to the population a genetic description of the action of natural selection (Crow and Kimura 1970, Ewens 1979, Christiansen 1999, Bürger 2000).

Changes in the physical environment can be studied by simply changing the fitness values in time or space, and the methods from the classical analyses are readily adapted to accommodate this extension (Haldane 1948, Haldane and Jayakar 1963). The dependence of individual fitness on population composition, usually referred to as frequency-dependent selection, requires more profound alterations in the models. Some effects of population composition, most notably sexual selection and other effects related to mating, are inherently frequency dependent, but others should be viewed as effects of the biotic environment on the individual. Such effects hence depend on the density and composition of the population, and the ensuing phenomenon is density-dependent selection.

Students of *Drosophila* control densities in their experiments to make them repeatable, an effect quantified by Lewontin (1955) in a study of the influence of larval density on their viabilities. The accompanying phenomenon of composition dependence was also demonstrated, and laboratory populations of

The Evolution of Population Biology, ed. R. S. Singh and M. K. Uyenoyama. Published by Cambridge University Press.

Drosophila have been widely used as models for studying density- and composition-dependent selection (Prout 1980, Mueller 1997, Mueller and Joshi 2000).

Population density effects on fitness were originally discussed by Fisher (1930). He used the Malthusian growth rate as a descriptor of individual fitness, but argued that the increase in fitness due to natural selection would merely be accompanied by an episode of increased population size, and that the average fitness would in the long run be very close to zero. This argument was put into a more formal framework by MacArthur (1962), a work usually considered the initiation of theoretical investigations of density-dependent selection. Although Robert MacArthur in this and subsequent publications established the use of Mendelian genetics in evolutionary deliberations in ecology, the approach was preceded by a quarter century by Kostitzin's (1936, 1938a,b,c,d) work. Historical reasons aside, both of these pioneering works are worth discussing because of their very different approach to the study of natural selection when selection is described as variation in ecological parameters.

7.2 The origin

MacArthur's (1962) argument may be formulated in terms of the logistic model for population growth (Box 7.1), although his model and argument were more general. He showed that in a population which may be assumed to reach its equilibrium population size fairly rapidly, the final outcome of selection is determined exclusively by the carrying capacity associated with the fitness characteristics of the genotypes. MacArthur and Wilson (1967) reiterated and strengthened the conclusion and termed the effect K selection in reference to the carrying capacity parameter that describes the equilibrium population size (Box 7.1). In a population that cannot be assumed to be at equilibrium, for instance in a species living in ephemeral habitats, variation in the intrinsic growth rate becomes important and r selection results, a point also raised by Lewontin (1965). These results have been interpreted as evidence that density-dependent selection, viewed as K selection, gives qualitatively different results from the classical population genetic models, viewed as r selection.

Kostitzin (1936) also investigated density-dependent selection based on the logistic model for population growth (a simplified version is given in Box 7.2), or rather on Lotka's (1925) and Volterra's (1926) population models of interacting species. In Kostitzin's formulation the equation for change in gene frequency is of the same form as the equation for density-independent fitnesses (see the last equation in Box 7.2). His results are a more explicit statement of Fisher's (1930) argument, that the change in average fitness causes the equilibrium population size to increase, while by definition the Malthusian growth rate, the average fitness, remains close to zero in a population close to equilibrium.

Box 7.1 MacArthur's model

Suppose two alleles, A and a, at an autosomal locus segregating in a population reproducing by random mating. The genotypic frequencies among zygotes are thus in Hardy–Weinberg proportions:

$$\begin{array}{ccc} AA & Aa & aa \\ p^2 & 2pq & q^2 \end{array}$$

where p is the frequency of A and q is that of a $(p+q=1)$. The population size is N and the numbers of the three genotypes are N_{AA}, N_{Aa}, N_{aa}, respectively; $p = (N_{AA} + \frac{1}{2}N_{Aa})/N$. The fitnesses of the three genotypes are given in terms of the intrinsic growth rate r and the carrying capacity K, and assuming small differences between the three genotypes (weak selection), the change in their numbers may be approximated by

$$\frac{dN_{AA}}{dt} \approx p^2 N r_{AA} \frac{K_{AA} - N}{K_{AA}},$$

$$\frac{dN_{Aa}}{dt} \approx 2pq N r_{Aa} \frac{K_{Aa} - N}{K_{Aa}},$$

$$\frac{dN_{aa}}{dt} \approx q^2 N r_{aa} \frac{K_{aa} - N}{K_{aa}}.$$

The assumption of weak selection further provides the approximations

$$\frac{dp}{dt} \approx \bar{N}r \frac{\bar{K} - N}{\bar{K}}$$

and

$$\frac{dN}{dt} \approx pq[p(r_{AA} - r_{Aa}) + q(r_{Aa} - r_{aa})] \frac{\bar{K} - N}{\bar{K}} + pq[p(K_{AA} - K_{Aa}) + q(K_{Aa} - K_{aa})] \frac{\bar{r}N}{\bar{K}^2},$$

where the bar over a parameter signifies its average value in the population. By the assumption of weak selection the rate of gene frequency change is much slower than that of the population size. Thus N will rapidly approach $\bar{K}$, and in that state the influence of the genetic variation in the intrinsic growth rate r will vanish. The change in gene frequency will therefore be determined by the variation in the carrying capacity K.

Box 7.2 Kostitzin's model

Using the same model as in Box 7.1 let the fitnesses of the three genotypes be given in terms of the birth rate b and the density-dependent death rate $d + \delta N$, where d is the density-independent death rate and δ is a coefficient describing the dependence of the death rate on density. The changes in their numbers are then

$$\frac{dN_{AA}}{dt} = p^2 bN - (d_{AA} + \delta_{AA}N)\, N_{AA},$$

$$\frac{dN_{Aa}}{dt} = 2pqbN - (d_{Aa} + \delta_{Aa}N)\, N_{Aa},$$

$$\frac{dN_{aa}}{dt} = q^2 bN - (d_{aa} + \delta_{aa}N)\, N_{aa}$$

(variation in birth rate makes the model considerably more complicated; see e.g., Christiansen and Prout 2000). Again, assumption of weak selection provides the approximations

$$\frac{dN}{dt} \approx bN - (\bar{d} + \bar{\delta}N)\,N, \text{ and}$$

$$\frac{dp}{dt} \approx pq[p(d_{AA} - d_{Aa}) + q(d_{Aa} - d_{aa})]$$
$$+ pq[p(\delta_{AA} - \delta_{Aa}) + q(\delta_{Aa} - \delta_{aa})]N.$$

The population size will rapidly approach $N = (b - \bar{d})/\bar{\delta}$, but whether or not the population size is close to this value, the genotypic fitnesses depend both on the density-independent death rate d and on the density-dependent death rate coefficient δ. Thus, we may as well write the gene frequency equation as

$$\frac{dp}{dt} \approx pq\{p[(d_{AA} + \delta_{AA}N) - (d_{Aa} + \delta_{Aa}N)]$$
$$+ q[(d_{Aa} + \delta_{Aa}N) - (d_{aa} + \delta_{aa}N)]\}.$$

The formal results reached by Fisher, Kostitzin, and MacArthur are, of course, the same. Their mathematical models are similar. Their biological models are different, however, in that they disagree on the ecological description of density-dependent growth. Fisher and Kostitzin formulate density-dependent growth in the Malthusian tradition used by Verhulst (1838) and Pearl and Reed (1920); that is, population growth is modeled in terms of birth and death rates. MacArthur, by contrast, uses the heuristic parameterization used by Gause (1934) in his analysis of experiments on Protozoa. That parameterization is very attractive because it describes observable population characteristics, but it is not evident that it is useful in evolutionary discussions.

The correspondence between the parameterizations in the simple models given in Boxes 7.1 and 7.2 is that the average intrinsic growth rate of the population is $\bar{r} = b - \bar{d}$ and the average carrying capacity is approximately $\bar{K} = \bar{r}/\bar{\delta}$, where b and d are the birth and death rates of individuals in the population, and δ is the density-dependent death rate coefficient, adding a death rate δN at the population density N. Thus, the reason for the variation in the interpretation of the results is that in Kostitzin's model carrying capacity is proportional to intrinsic growth rate. This may seem a strong influence of density-independent birth and death rates on the equilibrium population size, but it may nevertheless be more acceptable than considering the carrying capacity an independent descriptor of conditions in a crowded population (Christiansen and Fenchel 1977). The covariation of r and K has even been observed in selection experiments in *Escherichia coli* and *Paramecium primaurelia*, where selection for growth rate gave a correlated response in the equilibrium population size (Luckingbill 1978, 1979). The description of density-dependent growth in terms of birth and death rates emphasizes that the carrying capacity is merely the equilibrium population size; that is, the population size at which birth and death are in balance. Natural selection is defined in terms of birth and death rates, and the effect of density dependence is simply to modify these rates smoothly as population density changes.

7.3 Nonoverlapping generations

MacArthur's contribution triggered work on density-dependent selection in both population genetics and evolutionary ecology. Anderson (1971), Roughgarden (1971), and Clarke (1972) formulated discrete time versions of MacArthur's model as extensions of Haldane's classical models (Box 7.3), and Roughgarden in particular illustrated nicely the action of r and K selection as described by MacArthur and Wilson (1967).

The simplicity of these discrete generation models has made them standard reference models for investigations in evolutionary ecology, especially where interspecific interactions complement the intraspecific effects described by density-dependent selection. However, the simplification achieved in the classical population genetic models is due to the discreteness of breeding, and in this sense the density-dependent model is overly simplified, because the assumption of nonoverlapping generations need not have repercussions for the assumption of how density dependence works. Having realized this, Poulsen (1979) formulated a discrete generation model where the mortality of individuals throughout their development is density dependent (Box 7.4). In this model the growth in a monomorphic population censused at a particular age follows a logistic growth curve, but as soon as

Box 7.3 Anderson's and Roughgarden's models

Again assume two alleles segregating in a random mating population, now with nonoverlapping generations. The change in gene frequency is

$$p_{t+1} - p_t = p_t q_t \frac{p_t(w_{AA} - w_{Aa}) + q_t(w_{Aa} - w_{aa})}{p_t^2 w_{AA} + 2p_t q_t w_{Aa} + q_t^2 w_{aa}},$$

where the ws describe the survival to maturity of the genotypes. The growth of a monomorphic population is described by the difference equation

$$N_{t+1} - N_t = N_t R \frac{K - N_t}{K},$$

where $1 + R$ is the intrinsic growth factor and K the carrying capacity. This model may be viewed as an approximation to the logistic model whenever R is small ($R \ll 1$). The fitnesses of the three genotypes in the Anderson–Roughgarden model are given in terms of these parameters as

$$w_{AA} = 1 + R_{AA} \frac{K_{AA} - N_t}{K_{AA}}$$

with similar expressions for w_{Aa} and w_{aa}.

genotypic variation in the density-dependent death rate is allowed, selection becomes both density and frequency dependent (Poulsen 1979, Christiansen and Fenchel 1977).

For a large intrinsic growth factor ($1 + R$) the dynamics of Anderson's and Roughgarden's models deviate considerably from those of the logistic model for growth, whether in the logistic form (Boxes 7.1 and 7.2), in the discrete form (Box 7.4) or in the form used by Clarke (1972). For large Rs it shows a nonmonotonic growth curve, cycles or chaotic behavior (and for very large Rs it crashes), a behavior which is common in many discrete generation models (May 1976). Such behavior is typical for population dynamical models where the influence of density on growth rate is not immediate but occurs after a certain time lag (May 1973). In Poulsen's model the density effects are instantaneous and the dynamics smooth, whereas in the Anderson–Roughgarden model, and in many other models formulated as difference equation models (Box 7.5), the delay is maximal, in that the crowding effects in a given generation are determined by the density of the parental population.

The analysis of evolutionary changes due to density-dependent selection is difficult in a population where the size oscillates as regular cycles or chaos.

Box 7.4 Poulsen's model

Using the model in Box 7.3 let the density-dependent death rates of the three genotypes during their development from newly formed zygotes at age 0 to sexually mature adults at age T be given by $d + \delta n$, where $n(\tau)$ is the population size at age τ, $0 \leq \tau \leq T$. The changes in their numbers are then

$$\frac{dn_{AA}(\tau)}{d\tau} = -[d_{AA} + \delta_{AA} n(\tau)]\, n_{AA}(\tau),$$

$$\frac{dn_{Aa}(\tau)}{d\tau} = -[d_{Aa} + \delta_{Aa} n(\tau)]\, n_{Aa}(\tau),$$

$$\frac{dn_{aa}(\tau)}{d\tau} = -[d_{aa} + \delta_{aa} n(\tau)]\, n_{aa}(\tau).$$

Generation $t + 1$ starts with

$$n_{AA}(0) = p_t^2 BN_t, \quad n_{Aa}(0) = 2p_t q_t BN_t, \quad \text{and} \quad n_{aa}(0) = q_t^2 BN_t,$$

and the numbers of mature adults of the three genotypes become $n_{AA}(T)$, $n_{Aa}(T)$, and $n_{aa}(T)$, which in turn determine p_{t+1} and N_{t+1}. The intrinsic growth factor describing growth in population size at low densities is given by

$$1 + R = B[p^2 \exp(-d_{AA}T) + 2pq \exp(-d_{AA}T) + q^2 \exp(-d_{AA}T)].$$

An explicit expression for the equilibrium population size is not known in general, but in a monomorphic population it is $\hat{N} = K$ where

$$K = \frac{d(BD - 1)}{\delta B(1 - D)} \quad \text{with} \quad D = \exp(-dT).$$

This is determined using the recurrence equation

$$N_{t+1} = \frac{BDN_t}{1 + (\delta/d)\, B(1 - D)\, N_t},$$

and the equilibrium $\hat{N} = K$ is globally stable when $BD > 1$.

With more stable dynamics, slow evolutionary processes may be studied by assuming that the population size is close to equilibrium (Boxes 7.1 and 7.2). Such slow processes may be studied in oscillating populations by viewing the density oscillations as a varying environment (Christiansen 1984) and using the classical methods (Haldane and Jayakar 1963, Karlin and Liberman 1974). These methods require evaluation of the stationary distribution of the

Box 7.5 Models with time lag

The mortality description in Poulsen's model may be extended to include delayed effects of density. In a monomorphic population survival is thus described by

$$\frac{dn(\tau)}{d\tau} = -\left[d + \delta \int_0^\tau n(\xi)\psi_\tau(\xi)\, d\xi\right] n(\tau),$$

where $\psi_\tau(\xi) \geq 0$, $0 \leq \xi \leq \tau$, and $\int_0^\tau \psi_\tau(\xi)\, d\xi = 1$.

Poulsen's model is recovered when the entire mass of the distribution ψ is concentrated in τ. Maximal time lag is obtained if the mass of ψ is concentrated in 0, and the result is in essence the Ricker model with the recurrence equation

$$N_{t+1} = BDN_t \exp(-\delta BN_t T),$$

where $D = \exp(-dT)$.

The equilibrium population size is $\hat{N} = K$ where

$$K = \frac{b-d}{\delta B} \quad \text{with} \quad b = \frac{\log B}{T},$$

and this equilibrium is globally stable when $1 < BD < 2$; for $BD > 2$ sustained oscillations occur (see May 1976).

environment, and they may become more manageable by instead considering Lyapunov exponents (Ferriere and Gatto 1995).

7.4 Exploitative competition

The consideration of density-dependent selection is but the first stage in the description of individual fitness as a function of the biotic environment. Barring commensalism and amensalism, the species that influence the fitness of a particular species are themselves affected by that species. This interesting fraction of the biotic environment coevolves with the species of interest, whereas commensals and amensals may be considered part of the physical environment. Roughgarden (1977) analyzed simple models for coevolution in such a network of species and showed that the parallel to MacArthur's (1962) one-species result is that evolution at one locus maximizes or minimizes the population size of a given species, depending on the properties of the interactions in the network. In these models conditions for the existence of polymorphic equilibria and their stability closely resemble those of classical population genetics (Levin and Udovic 1977). More complete models of species interactions allow for interactions between genotypes of pairs of species, and I will briefly outline competition between species because of its intimate relation to classical density-dependent selection.

In Poulsen's model density-dependent mortality leads to frequency-dependent selection, a phenomenon expected to occur in any discrete generation model where survival is described as the outcome of a density-dependent process. In addition, both Kostitzin and Poulsen allowed the density of the various genotypes to exert different influences on the mortality of a given genotype, and their models were therefore formulated as intraspecific competition models that parallel the Lotka–Volterra models of interspecific competition. Selection in these generalized models is therefore inherently frequency dependent, and the main difference from the simple density-dependent models is, as in any frequency-dependent selection model, that a generalization of Fisher's (1930) fundamental theorem of natural selection does not exist.

The study of selection due to intraspecific competition took off when Roughgarden (1972) tied competition for food to variation in a phenotypic character (jaw size in *Anolis* lizards) using the exploitative competition model formulated by MacArthur and Levins (1967). The evolutionary dynamics of these models are dominated by the frequency dependence of the fitnesses. For instance, a one-locus two-allele polymorphic equilibrium may be stable even though at equilibrium the fitness of the heterozygote is inferior to that of both homozygotes (Christiansen and Loeschcke 1980). Matessi and Jayakar (1981) were nevertheless able to formulate a maximization principle for the evolutionary change that facilitates the analysis of these models.

MacArthur and Levin's model describes the dynamics of a continuum of resources exploited by predators, and by assuming a fast dynamics of the resource species, the dynamics of the predator species may be described as in Box 7.6. Fenchel and Christiansen (1977), inspired by competition among species of *Hydrobia* snails, considered this model for two species. We studied the initial increase of mutants having small effects in the two species. Given enough possibilities for mutation this analysis indicates the evolutionary path taken by the mean resource utilization of each of the two species, a technique that was later formalized in adaptive dynamics (Hofbauer and Sigmund 1998). Assuming a Gaussian spectrum of resource qualities and Gaussian utilization functions, we showed that evolutionarily stable coexistence is possible when the resource spectrum is wider than the utilization functions ($\sigma > \omega$ in the model of Box 7.6). The two species will then eventually place themselves symmetrically in characteristic positions around the resource optimum. In monomorphic species with this configuration any mutant with a small effect can invade the population, and we therefore predicted that the species would expand its niche by becoming polymorphic. The initial selection for these polymorphs, however, is an order of magnitude weaker than the selection for a rare allele that moves the mean utilization of resources in the first stages

Box 7.6 MacArthur and Levin's model

Suppose a continuum of resources ρ with abundance $\mathcal{S}(\rho)$, $-\infty < \rho < \infty$. Individuals of type i seek resources according to the utilization function $\mathcal{U}_i(\rho)$. Both spectra are assumed to be Gaussian, $\mathcal{S}$ with mean 0 and variance σ^2, and $\mathcal{U}_i$ with mean m_i and variance ω^2:

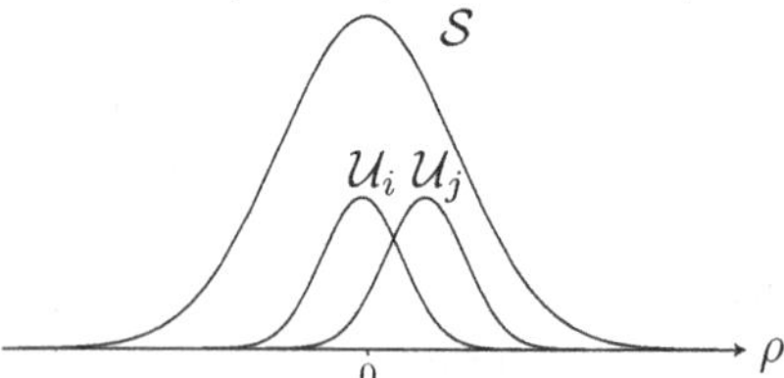

Using a discrete generation model closely related to the Anderson–Roughgarden model, the fitness (or the growth rate) of individuals of type i in a population where individuals of type j are present is

$$w_i = 1 + V[K_i - \gamma_{ii} N_i - \gamma_{ij} N_j] \, ,$$

where V is a constant, in that the proportionality between the intrinsic growth factor and the carrying capacity is natural in MacArthur and Levin's model (Christiansen and Fenchel 1977, Christiansen and Loeschcke 1980). The parameters of the fitnesses are

$$K_i = \exp\left[-\frac{m_i^2}{2\sigma^2 + 2\omega^2}\right] \quad \text{and} \quad \gamma_{ij} = \exp\left[-\frac{(m_i - m_j)^2}{4\omega^2}\right].$$

The types are genotypes of various species, and the fitness expression is readily extended to include a more reasonable number of types.

of the coevolutionary process. At the final stage, each species thus shows a polymorphic evolutionarily attainable stable trait (Christiansen 1991).

7.5 Density-independent selection

Density-dependent survival may occur in some life stages, whereas other stages may exhibit selection independently of population size (Poulsen 1979, Prout 1980). In Poulsen's model (Box 7.4) the death rate parameters may depend on the age of the individual, and its development may begin with a stage where the density-dependent death rate coefficient is independent of genotype. This early density-independent selection could be caused by differential mortality of eggs or seeds, and similar variations in survival might occur towards the end of the developmental period before the time of breeding. In insects, the larval stage could be subject to density-dependent survival,

Box 7.7 Classical selection

In Poulsen's model (Box 7.4) assume that the density-dependent death rate $d+\delta n$ is independent of genotype. Then $n_{AA}(T) = \Delta n_{AA}(0)$, $n_{Aa}(T) = \Delta n_{Aa}(0)$, and $n_{aa}(T) = \Delta n_{aa}(0)$ where

$$\Delta = \frac{BD}{1+(\delta/d)\,B(1-D)\,n(0)} \quad \text{with} \quad n = n_{AA} + n_{Aa} + n_{aa}.$$

Assume an early density-independent mortality, such that development in generation $t+1$ starts with

$$n_{AA}(0) = u_{AA}\, p_t^2 BN_t,\; n_{Aa}(0) = u_{Aa}\, 2p_t q_t BN_t, \text{ and } n_{aa}(0) = u_{aa}\, q_t^2 BN_t.$$

Further assume a late density-independent mortality giving the size

$$N_{t+1} = v_{AA} n_{AA}(T) + v_{Aa} n_{Aa}(T) + v_{aa} n_{aa}(T)$$

and the gene frequency

$$p_{t+1} = \frac{v_{AA} n_{AA}(T) + \frac{1}{2} v_{Aa} n_{Aa}(T)}{N_{t+1}}$$

in the breeding population.

and the egg and pupa could show density-independent mortality. In addition, certain aspects of sexual selection could well be density independent. In such a model with no genetic variation in density-dependent mortality (Box 7.7), the genetic changes are determined by the fitnesses $w_{AA} = u_{AA}v_{AA}$, $w_{Aa} = u_{Aa}v_{Aa}$, and $w_{aa} = u_{aa}v_{aa}$, where u and v are the early and late probabilities of survival, respectively. The change in population size in response to the evolutionary change in the gene frequencies will nevertheless depend on whether the breeding population, the population of mature individuals, or the population just starting development are censused (Prout 1980). The breeding population is greatly reduced by late mortality, whereas early mortality has a more indirect effect on the sizes of mature and breeding populations.

7.6 Structured habitats

Classical population genetic models assuming a panmictic population rarely need to refer to the interaction between natural selection and the regulation of population density. Even when considering a Y-chromosome driver the qualitative result is well described by the conclusion that the population will eventually consist of males only and will therefore go extinct. In reality it will probably dwindle away. The only situations where variations in absolute population size enter population genetic reflections on selection are when the effect of random genetic drift is in focus and when populations are compared. The first situation is indeed relevant, but let us focus on the second situation which is of much wider relevance in discussions of selection. Whenever populations in different habitats exchange migrants, the sizes of local populations are compared.

Dempster (1955) addressed this problem in his discussion of Levene's (1953) multiple-niche model. Levene assumed a panmictic population that occupies a number of different habitats with independent regulation of the population size, in that the number of breeding individuals that originate from a given habitat is independent of the genetic composition of the population. In such a population polymorphism may be maintained at a two-allelic locus even though the average survival does not exhibit overdominance. Dempster argued that this result was due to the assumption of a characteristic number of breeders: if the number of breeders from a habitat is proportional to the number of survivors, the model becomes a classical model where the genotypic fitnesses are given by the average survival in the population. The comparison of the two models, however, has an extra twist because Dempster's argument refers to the absolute fitnesses in the habitats, whereas Levene's model is well posed in terms of relative fitnesses (Walsh 1984). Levene's model of population structure was termed soft selection and Dempster's hard selection (Christiansen 1975), inspired by Wallace's (1968) discussion of similar phenomena.

Prout's (1980) analysis of the effect of density-independent selection shows that Dempster's model could describe the action of late viability selection. Levene's model is difficult to produce within the model sketched in Box 7.7 (Christiansen 1985), and it seems to provide an accurate description of only selection that does not influence population dynamics. One example might be male sexual selection in a polygynous species where all females are mated. The model nevertheless provides a simple and convenient approximation of a situation where the considered variation has little impact on the relative sizes of the subpopulations. For very high fecundities the number of mature adults in a subpopulation becomes virtually independent of the seeding size (Box 7.4), a situation expected to occur in many annual plants (Holsinger and Pacala 1990). In addition, the seemingly qualitative difference between the hard and soft selection models of population structure is only present in a panmictic population. When mating is local and the exchange of individuals among subpopulations limited, the difference between the models is of a more quantitative nature (Christiansen 1975), and evidently the difference vanishes as subpopulations approach complete isolation.

An account of the origin of individuals in the local breeding population provides a sufficient description of the genetic effects of migration among subpopulations – really the fractions of individuals with the various origins suffice, the so-called backward migration rates (Malécot 1948, 1969, Bodmer and Cavalli-Sforza 1968). Even the sizes of the subpopulations do not enter explicitly into the description. The sizes of the local populations in Levene's model merely describe their relative contribution to the breeding population. Description of the local population dynamics, by contrast, requires specification of both the number of individuals that emigrated and the number of immigrants. Migration rates in classical population genetics are therefore derived quantities and much simpler than the migration rates used in demography. For instance, the only way to model sources and sinks is by having sources contribute large immigrant fractions and sinks small fractions, but this in general does not differ from assuming large and small population sizes or migration with a preferred direction in a geographically structured population.

A model of density-independent selection in a subdivided population based on the model in Box 7.7 provides an immediate opportunity to include subpopulations in habitats that cannot sustain a population on their own (Christiansen 1985). Such sinks, with their supposedly special environment, contribute to the gene pool of the species and thereby its evolution. Evolution at the boundaries of the range of a species is a classic example where such explicit dynamical consideration of local populations is needed. Kirkpatrick and Barton (1997) address this question using a model resembling that in Box 7.7, in that they model density-independent selection and assume that the local equilibrium population size is an increasing function of the local average intrinsic rate of increase. Their approach allows the study in marginal populations of the balance between selection for local adaptation and immigration from more central populations, and it shows that this balance is important for the evolutionary definition of a species' range.

7.7 Population dynamics and selection

The explicit modeling initiated by Kostitzin and MacArthur of the population dynamical aspects of natural selection has, in addition to providing key contributions to evolutionary theory, changed our view of the ecological processes in natural populations. The synthesis of genetic and ecological dynamics created a foundation on which to consider variation among individuals as an important aspect of population dynamics and to see the ensuing evolutionary changes as a part of the interaction between organisms and their environment. The current increase in attention to parasites as important determinants for the population dynamics of a species has reemphasized the importance of genetic and evolutionary considerations. Malaria has changed the genetic constitution of human populations, and agriculture appears to engender an endless genetic battle between cereals and their fungal pathogens. These aspects of ecology will undoubtedly see major developments in the coming years, enhanced by the increasing ease with which the presence of parasites and diseases can be diagnosed by molecular genetic techniques.

The biological interpretations of the theoretical results reached by Kostitzin and MacArthur represent the widest range of possibilities. Introducing genetic variation into the logistic model of population dynamics, they deemed the effect of density dependence in selection as respectively trivial and of crucial importance. This emphasizes that the choice of parameterization of fitness effects is essential in evolutionary models. Just as in a computer simulation and in life: "... what you get out of it depends on what you put into it" (Lehrer 1959). Another example of the important connection between model assumptions and the biological interpretation of the results is found in simple models of selection on two loci. It is extremely unlikely that stable polymorphism at two loci will result in linkage equilibrium, that is, independence between the alleles at the two loci in the gametes of the population. Nevertheless, simple selection models commonly exhibit such equilibria. For weak selection and loose linkage, however, a stable polymorphic equilibrium will show a small deviation from linkage equilibrium. Thus, a simple model like the symmetric viability model (Lewontin and Kojima 1960), which may possess a linkage-equilibrium polymorphism, mimics a biologically plausible situation in a theoretically convenient framework, but the existence of linkage equilibrium at the steady state can never be interesting. That property was built into the model. In a similar way, it is often convenient to use genotypic carrying capacity parameters to summarize fitness variation when studying a population close to population size equilibrium, but then it cannot be surprising that selection is well described by the variation in those parameters.

Within population genetics the consideration of the population dynamical aspects of natural selection has had the strongest impact on theoretical investigations of geographically structured populations. Although observations on patterns of genetic variation may give an impression of the amount of migration among populations, they can only provide limited information on the dynamical aspects of migration. The genetic parameters are the backward migration frequencies that provide a very limited description of the dynamical interactions among local populations. This description, however, is an evolutionarily sufficient description of these interactions.

Density-dependent selection, a key concept in population biology, has become a key concept in evolutionary population genetics. The richness of individual interactions whose evolutionary effects can be modeled using the tools of theoretical ecology has contributed significantly to our understanding of natural selection as an evolutionary mechanism.

7.8 Acknowledgments

Comments on the manuscript from Dave Parker and Else Løvdal are gratefully acknowledged.

REFERENCES

Anderson, W. W. (1971). Genetic equilibrium and population growth under density-regulated selection. *Am. Nat.* 105:489–98.

Bodmer, W. F., and Cavalli-Sforza, L. L. (1968). A migration matrix model for the study of random genetic drift. *Genetics* 59:565–92.

Bürger, R. (2000). *The Mathematical Theory of Selection, Recombination, and Mutation.* Chichester: Wiley.

Christiansen, F. B. (1975). Hard and soft selection in a subdivided population. *Am. Nat.* 109:11–16.

Christiansen, F. B. (1984). Evolution in a temporally varying environment: density and composition dependent genotypic fitnesse. In K. Wohrmann and V. Loeschcke (eds) *Population Biology and Evolution,* pp. 115–24. Berlin: Springer-Verlag.

Christiansen, F. B. (1985). Selection and population regulation with habitat variation. *Am. Nat.* 126:418–29.

Christiansen, F. B. (1991). On conditions for evolutionary stability for a continuously varying character. *Am. Nat.* 138:37–50.

Christiansen, F. B. (1999). *Population Genetics of Multiple Loci.* Chichester: Wiley.

Christiansen, F. B., and Fenchel, T. M. (1977). *Theories of Populations in Biological Communities.* Berlin: Springer-Verlag.

Christiansen, F. B., and Loeschcke, V. (1980). Evolution and intraspecific exploitative competition I. One-locus theory for small additive gene effects. *Theor. Pop. Biol.* 19:378–419.

Christiansen, F. B., and Prout, T. (2000). Aspects of fitness. In R. S. Singh and C. B. Krimbas (eds) *Evolutionary Genetics: From Molecules to Morphology,* pp. 146–56. Cambridge: Cambridge University Press.

Clarke, B. (1972). Density-dependent selection. *Am. Nat.* 106:1–13.

Crow, J. F., and Kimura, M. (1970). *An Introduction to Population Genetics Theory.* New York: Harper & Row.

Dempster, E. R. (1955). Maintenance of genetic heterogeneity. *Cold Spring Harbor Symp. Quant. Biol.* 20:140–3.

Ewens, W. J. (1979). *Mathematical Population Genetics.* Berlin: Springer-Verlag.

Fenchel, T. M., and Christiansen, F. B. (1977). Selection and interspecific competition. In F. B. Christiansen and T. M. Fenchel (eds) *Measuring Selection in Natural Populations.* Vol. 19 of *Lecture Notes in Biomathematics,* pp. 477–98. Berlin: Springer-Verlag.

Ferriere, R., and Gatto, M. (1995). Lyapunov exponents and the mathematics of invasion in oscillatory or chaotic populations. *Theor. Pop. Biol.* 48:126–71.

Fisher, R. A. (1922). On the dominance ratio. *Proc. R. Soc. Edinburgh* 42:321–431.

Fisher, R. A. (1930). *The Genetical Theory of Natural Selection.* Oxford: Clarendon Press.

Gause, G. F. (1934). *The Struggle for Existence.* New York: Hafner.

Haldane, J. B. S. (1924). A mathematical theory of natural and artificial selection. *Trans. Camb. Phil. Soc.* 23:19–41.

Haldane, J. B. S. (1932). A mathematical theory of natural and artificial selection. Part IX. Rapid selection. *Proc. Camb. Phil. Soc.* 28:244–8.

Haldane, J. B. S. (1948). The theory of a cline. *J. Genet.* 48:277–84.

Haldane, J. B. S., and Jayakar, S. D. (1963). Polymorphism due to selection of varying direction. *J. Genet.* 58:237–42.

Hofbauer, J., and Sigmund, K. (1998). *Evolutionary Games and Population Dynamics.* Cambridge: Cambridge University Press.

Holsinger, K. E., and Pacala, S. W. (1990). Multiple-niche polymorphisms in plant populations. *Am. Nat.* 135:301–9.

Karlin, S., and Liberman, U. (1974). Random temporal variation in selection intensities: case of large population size. *Theor. Pop. Biol.* 6:355–82.

Kirkpatrick, M., and Barton, N. H. (1997). Evolution of species' range. *Am. Nat.* 150:1–23.

Kostitzin, V. A. (1936). Sur les équations différentielles du problème de la sélection mendelienne. *C. R. Acad. Sci. Paris* 203:156–7.

Kostitzin, V. A. (1938a). Équations différentielles générales du problème de sélection naturelle. *C. R. Acad. Sci. Paris* 206:570–2. English translation in Scudo and Ziegler (1978).

Kostitzin, V. A. (1938b). Sélection naturelle et transformation des espèces du point de vue analytique, statistique et biologique. *C. R. Acad. Sci. Paris* 206:1442–4. English translation in Scudo and Ziegler (1978).

Kostitzin, V. A. (1938c). Sur les coefficients mendeliens d'hérédité. *C. R. Acad. Sci. Paris* 206:883–5. English translation in Scudo and Ziegler (1978).

Kostitzin, V. A. (1938d). Sur les points singuliers des équations différentielles du problème de la sélection naturelle. *C. R. Acad. Sci. Paris* 206:976–8. English translation in Scudo and Ziegler (1978).

Lehrer, T. (1959). *An Evening Wasted with Tom Lehrer.* London: Decca Records.

Levene, H. (1953). Genetic equilibrium when more than one niche is available. *Am. Nat.* 87:331–3.

Levin, S. A., and Udovic, J. D. (1977). A mathematical model of coevolving populations. *Am. Nat.* 111:657–75.

Lewontin, R. C. (1955). The effects of population density and composition on viability in *Drosophila melanogaster. Evolution* 9:27–41.

Lewontin, R. C. (1965). Selection for colonizing ability. In H. G. Baker and G. L. Stebbins (eds) *The Genetics of Colonizing Species*, pp. 77–91. New York: Academic Press.

Lewontin, R. C., and Kojima, K. (1960). The evolutionary dynamics of complex polymorphism. *Evolution* 14:458–72.

Lotka, A. J. (1925). *Elements of Physical Biology.* Baltimore, MD: Williams and Wilkins. Reprinted as Lotka (1956).

Lotka, A. J. (1956). *Elements of Mathematical Biology.* New York: Dover.

Luckingbill, L. S. (1978). *r* and *K* selection in experimental populations of *Escherichia coli. Science* 202:1201–3.

Luckingbill, L. S. (1979). Selection and the r/K continuum in experimental populations of protozoa. *Am. Nat.* 113:427–37.

MacArthur, R. H. (1962). Some generalized theorems of natural selection. *Proc. Natl. Acad. Sci. USA* 48:1893–7.

MacArthur, R. H., and Levins, R. (1967). The limiting similarity convergence and divergence of coexisting species. *Am. Nat.* 101:377–85.

MacArthur, R. H., and Wilson, E. O. (1967). *The Theory of Island Biogeography.* Princeton, NJ: Princeton University Press.

Malécot, G. (1948). *Les Mathématiques de l'Hérédité.* Paris: Masson et Cie. Translated as Malécot (1969).

Malécot, G. (1969). *The Mathematics of Heredity.* San Francisco, CA: W. H. Freeman.

Matessi, C., and Jayakar, S. D. (1981). Coevolution of species in competition: a theoretical study. *Proc. Natl. Acad. Sci. USA* 78:1081–4.

May, R. M. (1973). *Stability and Complexity in Model Ecosystems.* Princeton, NJ: Princeton University Press.

May, R. M. (1976). Simple mathematical models with very complicated dynamics. *Nature (London)* 261:459–67.

Mueller, L. D. (1997). Theoretical and empirical examination of density-dependent selection. *Annu. Rev. Ecol. Syst.* 28:269–88.

Mueller, L. D., and Joshi, A. (2000). *Stability in Model Populations.* Princeton, NJ: Princeton University Press.

Pearl, R., and Reed, L. S. (1920). On the rate of growth of the population of the United States since 1790 and its mathematical representation. *Proc. Natl. Acad. Sci. USA* 6:275–88.

Poulsen, E. T. (1979). A model for population regulation with density- and frequency-dependent selection. *J. Math. Biol.* 8:325–43.

Prout, T. (1980). Some relationships between density independent selection and density dependent growth. *Evol. Biol.* 13:1–68.

Roughgarden, J. (1971). Density-dependent natural selection. *Evolution* 52:453–68.

Roughgarden, J. (1972). Evolution of niche width. *Theor. Pop. Biol.* 5:297–332.

Roughgarden, J. (1977). Selection and interspecific competition. In F. B. Christiansen and T. M. Fenchel (eds) *Measuring Selection in Natural Populations.* Vol. 19 of *Lecture Notes in Biomathematics,* pp. 449–517. Berlin: Springer-Verlag.

Scudo, F. M., and Ziegler, J. R. (1978). *The Golden Age of Theoretical Ecology: 1923–1940.* Vol. 22 of *Lecture Notes in Biomathematics.* Berlin: Springer-Verlag.

Verhulst, J. H. (1838). Notice sur la loi que population suit dans son accroissement. *Corr. Math. Phys.* 10:113–21.

Volterra, V. (1926). Variazione e fluttuazioni del numero d'individui in specie animali conviventi. *Mem. Accad. Nazionale Lincei (Ser VI)* 2:31–113. English translation in Scudo and Ziegler (1978).

Wallace, B. (1968). Polymorphism, population size, and genetic load. In R. C. Lewontin (ed.) *Population Biology and Evolution,* pp. 87–108. Syracuse, NY: Syracuse University Press.

Walsh, J. B. (1984). Hard lessons for soft selection. *Am. Nat.* 124:518–26.

8

Nonsynonymous polymorphisms and frequency-dependent selection

BRYAN CLARKE
Institute of Genetics, University of Nottingham

8.1 Introduction

In discussions about the causes of biological diversity, viewpoints have often been grouped into 'isms' (neutralism, selectionism, gradualism, punctuationism), and debated as if they were coherent warring parties. This kind of usage seems more political than scientific. Fortunately, the current group of 'isms' is disappearing. Experiments prove hypotheses wrong, however passionately their proponents believe in them. The latest approximations to the truth do not often emerge as victories, or even as compromises, but rather as mosaics of old and new.

Vestiges of the 'isms' still remain. For example, it has been customary either to argue that evolutionary replacements and polymorphisms are both predominantly driven by selection or that both are predominantly neutral.[1] Yet, as first pointed out by Fisher (1930), most replacements could be driven by selection even if most polymorphisms were neutral (because selected replacements would contribute little to the standing variation). It is worth considering Fisher's argument, which did not allow for many polymorphisms maintained by selection, to see whether it is compatible with current data.

The debates between "neutralists" and "selectionists" were initially about amino acid replacements in evolution. Here it has transpired that the selectionist view was nearer the truth. When Motoo Kimura was asked what proportion of selectively driven amino acid replacements would make him seriously uncomfortable, he replied "Ten percent." I think he would be uncomfortable now. However, he would find solace in the realm of noncoding sequences, even though many questions about them remain unanswered. For example, although we know that synonymous substitutions can affect the efficiency and

[1] Here, the term "replacement" means an evolutionary change from one nucleotide or amino acid to another. Some have used the same term to designate mutations or morphs that change amino acids (as in "replacement polymorphisms"). For the latter case, "nonsynonymous" avoids possible confusion.

The Evolution of Population Biology, ed. R. S. Singh and M. K. Uyenoyama. Published by Cambridge University Press.

accuracy of translation, we do not know how important are their effects on splicing, on the three-dimensional structure and stability of messenger RNAs, on the synthesis of noncoding RNAs or on the binding of transcription factors.

This chapter is about nonsynonymous coding polymorphisms.[2] I shall argue that many of them, perhaps even the majority, are maintained either by balancing selection[3] or by the balance between selection and migration. The actual proportion of selectively maintained polymorphisms is still uncertain, but we can make some rough estimates based on molecular data, and can see the way to get better ones. Studying the ecology of selective agents allows us to assess the relative importance of different balancing systems.

8.2 How many nonsynonymous polymorphisms are there?

Lewontin has earned his place in the Pantheon, along with Hubby and Harris, for demonstrating the remarkably high level of protein polymorphism, and for appreciating the evolutionary problems that it poses. The early estimates, based on small samples, have turned out to be remarkably robust. Within populations of outcrossing organisms, about 25–30% of loci are polymorphic in electrophoretic mobility, although the proportion varies greatly with different breeding systems, population sizes, and historical bottlenecks. Since fewer than half the changes in amino acids are detectable by standard electrophoresis, it might be supposed that the proportion of polymorphic loci should be much greater than 30%. However, polymorphic sites are clumped. Loci with high levels of electrophoretic variants also tend to have high levels of other amino acid polymorphisms.

Data from the genome programs for humans and *Drosophila melanogaster*, combined with electrophoretic data and surveys of single nucleotide polymorphisms (e.g., Moriyama and Powell 1996, Wang *et al.* 1998, Marth *et al.* 2001), can give us a rough idea about the total number of nonsynonymous polymorphic nucleotide sites in each organism. Humans, with 3.2 billion nucleotides, more than 3 million single nucleotide polymorphisms (SNPs) and at least 30 000 genes, have about 50 000 nonsynonymous polymorphic nucleotide sites (Kruglyak and Nickerson 2001). Because the surveys of SNPs are based on small samples of chromosomes, those currently detected are biased towards polymorphisms with two or more alternatives at high frequencies. Marth *et al.* (2001) estimate that 75% of detected SNPs have their minor alleles at frequencies of 20% or more. The figure of 50 000 may be conservative. Cargill *et al.* (1999) suggest that it is as high as 100 000. Nonetheless,

[2] By convention, a locus is described as polymorphic if at least two alleles occur at frequencies higher than 1%. This value is chosen because alleles at higher frequencies are unlikely to be kept in the population by a balance between mutation and selection.

[3] I use the term "balancing selection" to include all the selective forces that keep alternative alleles in a population, even if there is no equilibrium (as in cases that produce stable limit cycles).

because SNPs are not randomly distributed, but strongly aggregated in clumps, the number of *loci* polymorphic for amino acids is probably no more than 20 000.

Drosophila melanogaster has 180 million nucleotides and more than 14 000 genes. Because detailed surveys of SNPs have not been made, we can only estimate very roughly the numbers of nonsynonymous polymorphisms. *D. melanogaster* appears to have an average of about one SNP every 250 nucleotides, suggesting a total of 750 000 SNPs. The density of genes per unit length of DNA in *melanogaster* is five times higher than it is in humans, and so the proportion of SNPs in the coding regions should be correspondingly greater, at about 20%. Aquadro *et al.* (2001) estimate that the nonsynonymous/synonymous ratio in *D. melanogaster* is about 0.05, suggesting approximately 7500 nonsynonymous polymorphic sites. Singh and Rhomberg (1987) found that the proportion of enzyme loci that are polymorphic is somewhat higher in *D. melanogaster* than it is in humans, 0.67 as against 0.47. Extrapolating this proportion to all genes gives a total of about 9000 nonsynonymous polymorphic loci. A similar calculation applied to humans, using the value of 0.47, suggests that they have about 14 000 nonsynonymous polymorphic loci, an estimate that is compatible with the figures given above.

These calculations apply to each species as a whole. The number of nonsynonymous polymorphisms within a single population is likely to be about half the specific total. Humans and *melanogaster* are both genetically less variable than some of their close relatives. In any organism, of course, the number of loci with nonsynonymous polymorphisms cannot exceed the number of genes, which is in the order of 10^4 or less.

8.3 How many of these polymorphisms are maintained by natural selection?

Before trying to answer this question, we need to consider what might limit the numbers of selected polymorphisms.[4]

8.3.1 Genetic load

At first sight, the problem of simultaneously maintaining 10^4 polymorphic loci seems insuperable. If selection acts on each locus independently, mortality will be multiplicative (Lewontin and Hubby 1966, Hubby and Lewontin 1966). No population with 10^4 polymorphisms could possibly support the numbers of gametes and zygotes necessary to compensate for the ensuing deaths or failed births. Various escapes have been suggested (linkage disequilibrium, epistasis,

[4] Here, a polymorphism is regarded as actively maintained if some of its alleles are subject to balancing forces that include selection, even if the selection happens sporadically. No polymorphism lasts forever, and balancing agents can only prolong its existence.

truncation selection and so on). All of them involve a lack of independence between the polymorphisms, so that a single death or failed birth can remove alleles at more loci. An extreme example, in which nearly unlimited numbers of polymorphic sites can be maintained without any serious problems of genetic load, is a conflict between selection and migration. If two adjacent populations live in different habitats, each favoring different alleles at many loci, a chromosome carried from one to the other will be disadvantageous, and removed by selection. The single demise of a 'foreign' chromosome will eliminate as many foreign alleles as it possesses. These alleles will be replaced by new migrants. Many polymorphisms can be maintained because, at least in first-generation crosses, there is complete gametic disequilibrium. What happens in subsequent generations will depend on how much recombination occurs in the time between the arrival of a chromosome and the elimination of its contents, which in turn depends on the degree of linkage between selected sites, the amount of selective interaction between loci, and the strength of selection.

If the rate of migration is large enough, this system grades into Levene's (1953) model of two niches within a single population, which can also maintain polymorphism. Although Levene did not say so, his model is a frequency-dependent one. The required negative feedback comes from the fact that the density-dependent regulation of numbers occurs independently within each niche. Here the numbers are subject to the most extreme possible form of density dependence. They are fixed. The frequency-dependent selection occurs because an allele favored in one niche will lose any advantage as its numbers rise. If the separation of density-dependent controls does not occur, so that the numbers are regulated at the level of the entire population, the frequency dependence disappears, and Levene's system can no longer maintain polymorphism (Dempster 1955).

Simple cases like this should be easily detectable by the geographical pattern of allelic frequencies, and by the consequential gametic disequilibrium. Situations that are more realistic are likely to be less obvious. A patchwork of habitats varying in their selective demands, with populations exchanging genes at different rates, can produce complicated patterns in which disequilibria are difficult to interpret even when natural selection is strong. Intuition suggests that as things get more complicated the problem of genetic load gets bigger, but the quantitative nature of this relationship is obscure, and likely to vary with the circumstances.

Gametic disequilibrium associated with genetic linkage can, of course, also happen without any conflict between migration and selection, and with only indirect relevance to the question of load. It may be produced by drift or epistasis. In the genomes of European people there seem to be islands of linkage disequilibrium separated by hotspots of recombination. The former may be up to 200 kilobases in length, and the latter 1–2 kilobases (see reviews by Goldstein 2001, Gabriel *et al.* 2002). Each island has few haplotypes, but may have several SNPs. If the islands were distributed at random through the genome, we would expect two or three nonsynonymous polymorphisms per island. However, both the amount of linkage disequilibrium and the amount of polymorphism vary greatly between chromosomal regions.

Other factors can help to mitigate the problem of load. One of them is inherent in the nature of the selective process. Selective agents act on phenotypes rather than genotypes. When working at the molecular level it is easy to forget that every selected gene contributes to one or more characters, be they biochemical, physiological, anatomical, or ecological. Selection may discriminate between genotypes, chromosomes, genomes, populations, or species, but in every case it acts through phenotypic differences. Rarely indeed does phenotypic variation involve only one gene. For example, if we expose fruit flies, over several generations, to a medium containing higher than normal concentrations of ethanol, the flies evolve tolerance. At the alcohol dehydrogenase locus, selection favors the Fast allozyme. Despite being favored, the allele contributes little to the additional tolerance (McKenzie and McKechnie 1978, Middleton and Kacser 1983, Bokor and Pecsenye 1998). Alleles of other genes have larger effects. Some of them may change controlling regions, others may alter amino acids, but all, including *Adh,* are quantitative trait loci (QTLs) for ethanol tolerance, and all are selected by ethanol.

If several QTLs contribute additively to a character that is selected towards an optimum, the loci are inevitably epistatic with respect to fitness. Although this can substantially reduce the genetic load per locus, it does not promote polymorphism, except to a small degree when the population gets near the optimum (Singh and Lewontin 1966). However, there are much greater opportunities when the quantitative character is *itself* under balancing frequency-dependent selection. Such selection on metrical characters is not often discussed, and workers in population genetics generally ignore it, but there is experimental evidence that it occurs. The evidence comes from studies of avian predators hunting artificial prey. The birds were offered prey with a quasi-normal distribution of color or shape. They ate disproportionately more prey individuals with values near the mean, and fewer with values at the extremes. When the distribution of shape was 'inverted' so that prey with values at the extremes were commoner, the predators reversed their behavior, and took disproportionately more of the extreme prey (Allen 1973, Shelton 1986, Mani *et al.* 1990). There is no reason in principle why parasites or intraspecific competitors should not select metrical characters in a similar manner, but the tests, as far as I am aware, have not been made. Simulations suggest that character-based balancing selection can maintain many polymorphisms among the QTLs without serious problems of genetic load (Mani *et al.* 1990). For each selected character, of course, the larger the number of QTLs, the weaker will be the selection per locus. As Lewontin (1974) pointed out, there is an ecological cost to frequency-dependent selection, in mortality or reduced fecundity, even if at equilibrium the variant phenotypes have equal selective values. In this particular case, however, many genetic polymorphisms can be maintained at a small cost, particularly if the intensity of selection is density dependent.

The ecological cost of maintaining polymorphisms is greatly diminished when selection is sporadic. Natural environments are continually in flux. For most of the time, populations are not in equilibrium, but either resting in a selective vacuum, or desperately (so to speak) trying to keep up with environmental changes. Agents of selection come and go. Predators, parasites, and competitors fluctuate in numbers and influence. Diseases sweep through and then disappear. Climates change. Many of these agents, even the last, can directly or indirectly produce frequency-dependent selection. In the intervals between bouts of selection, gene frequencies can drift. All that is needed is an occasional kick to keep the relevant alleles in the population.

All these considerations suggest that genetic loads are not necessarily a problem, not even in maintaining nonsynonymous polymorphisms at 10^4 loci.

8.3.2 Polymorphism and recombination

In *Drosophila melanogaster* and humans there are positive associations between the rate of recombination in a chromosomal region and its proportion of SNPs. This happens because selective sweeps and background selection are more effective at eliminating adjacent variation when the rate of recombination is low (Charlesworth *et al.* 1993). Kreitman and Akashi (1995) have argued that the low levels of polymorphism in regions of reduced crossing-over mean that we should reject balancing selection as a general way of maintaining genetic variation. However, even in these regions SNPs are still to be found, often arranged into a few alternative haplotypes. The fourth chromosome of *Drosophila melanogaster* is an example. Originally believed wholly to lack recombination and polymorphism, it has turned out to have both, albeit in patches and at relatively low levels (Wang *et al.* 2002).

Bearing in mind that synonymous and noncoding polymorphisms are those more likely to be eliminated by selective sweeps and background selection, and that overall they greatly outnumber nonsynonymous ones, the correlation between variety and recombination does not exclude the active maintenance of nonsynonymous polymorphisms. As Kreitman and Akashi pointed out, neither does it exclude *any* balanced polymorphisms that are transient over evolutionary time scales. The fact that many polymorphisms have not accumulated "neutral" alleles in linkage disequilibrium may argue against long-term heterozygous advantage, but does not exclude forms of frequency dependence in which a turnover of alleles is expected. Kreitman and Akashi did not remark that the local rate of recombination itself is subject to modification (Chinnici 1971). In chromosomal regions rich in selected polymorphic genes, recombination can rise locally if adjacent loci are periodically subject to opposing forces of selection, or fall if particular allelic associations are consistently favored.

8.3.3 Polymorphism and constraint

There are widespread negative correlations between the rigor of selective constraints (i.e., the extent to which suboptimal phenotypes are eliminated) and the incidence of polymorphisms. This fact is often taken as evidence that most polymorphisms are neutral. However, this is not a logical progression. In any molecule there are bound to be conflicts between different functions and requirements. Constraints imposed by one requirement can very easily limit responses to another.

Where constraints are strong, there will be few positions in the molecule where polymorphisms can occur without disrupting some other adjustment. The population will be pushed towards uniformity. When other constraints are weak, there will be more scope for the action of selective factors favoring polymorphism. The negative correlations are inevitable whether the polymorphisms are balanced or neutral.

8.3.4 The problem of testing for selection

In his most famous book, Lewontin (1974) described the approach of "ecological geneticists" in the following terms: "It is their hope that by studying polymorphisms in large, countable, nonsecretive and amenable organisms like snails, moths, butterflies and man, they will be able to establish case after case of selective polymorphism. They are frankly partisan in their belief that polymorphism is in general balanced, but such bias is necessary for the success of this research strategy, for it is a strategy of confirmation rather than exclusion."

There is truth here, but it is not the whole truth. Without a belief in the efficacy of natural selection, there is indeed little motivation to carry out the necessary observations and experiments. However, the "bias" among the ecological geneticists came from the fact that neutrality was rejected in cases like the shell polymorphism in *Cepaea*, which had been widely believed to be the product of random processes (Cain and Sheppard 1954, and other cases reviewed by Endler 1986). An ecological geneticist could argue, conversely, that there has been a bias in favor of the neutral theory because it removes the obligation to do experiments. The "confirmation" of selection is just the rejection of neutrality, followed by the demonstration that a selective agent causes differential mortality or differential reduction in the numbers of offspring.

Information about DNA sequences now lets us detect selection with less labor. Three recent studies of *Drosophila* DNA, based on ratios of synonymous to nonsynonymous changes, suggest that about half the nonsynonymous substitutions during the divergence of species have been driven by natural selection (Bustamente *et al.* 2002, Fay *et al.* 2002, Smith and Eyre-Walker 2002). Methods using DNA sequences can also detect balancing selection on individual polymorphisms (Kreitman and Akashi 1995), and the tests have been positive in a dozen cases. Several of these polymorphisms were already known to be selected, including HLA in humans, ADH in *Drosophila melanogaster*, and self-incompatibility in crucifers (reviews by Kreitman and Akashi 1995, Hughes 1999, and Uyenoyama and Takebayashi, this volume). Although tests on DNA have the advantage that they locate the selection directly to a polymorphic locus, they have the disadvantages of being insensitive (i.e., lacking power), and of telling us little about the mechanisms of balance.

How far is it generally legitimate to extrapolate from evolutionary substitutions to polymorphisms? I have already remarked, following Fisher (1930), that there is no *necessary* correspondence between the two. In practice, of course, there may be such a correspondence. Evidence can be found by studying the amino acid changes that are involved in each case. If nonsynonymous polymorphisms, unlike evolutionary substitutions, are predominantly neutral, then we expect their amino acid replacements to be more conservative, in the sense that the chemical and functional differences involved should be smaller. We already know, however, that enzyme polymorphisms often involve changes that alter the functions of the enzymes concerned, and are not disproportionately conservative. For example, in the early days of studies on allozymes, Harris (1971) tabulated 23 human polymorphisms, 16 of which showed clear differences in enzyme activity between common variants. Chasman and Adams (2001), using a computational method for detecting changes of function, concluded that about 30% of natural nonsynonymous polymorphisms involve such changes. They estimated that an average human is heterozygous for functionally different alleles at about 9000 loci. Because of biases inherent in the method, their figures are probably underestimates. Of course, the functional changes need not be important to survival or reproduction *in vivo*. Whether or not this is so, the only evidence of differences in selective constraints between nonsynonymous polymorphisms and evolutionary substitutions indicates that the polymorphisms are significantly *more* likely to involve changes of function (Sunyaev *et al.* 2000). Thus, the possibility remains that more than 50% of nonsynonymous polymorphisms are under selection.

There is a way to test this possibility. It is to take a sample of protein polymorphisms such as those in *Drosophila pseudoobscura* described by Lewontin and Hubby, and to test them one by one for evidence of selection, making allowance for the sensitivity of the tests, and identifying, where possible, the selective agents. *Pace* Lewontin, this is a strategy not of confirmation but of sampling, with the same validity as Lewontin and Hubby's method of estimating the proportion of polymorphic proteins. At the worst, it should give us a minimum estimate of the proportion of selected polymorphic loci, just as Lewontin and Hubby gave us a minimum estimate of polymorphic loci as a whole.

If we can find a putative agent of selection, it becomes possible to intensify it, ideally in natural populations. Polymorphic phenotypes responding differentially to this intensification should also be capable of responding, at a lower level and over a longer period, to the same selective agent in undisturbed conditions. The extrapolation is valid only if there are no thresholds for the selection.

8.4 Is the selection balancing?

Detectable selection acting on a polymorphism that has persisted for more than a hundred generations strongly suggests that balancing forces are, or have been, at work. Accepting that a significant proportion of nonsynonymous polymorphisms are under selection, there are three serious contenders for roles as general balancing systems: heterozygous advantage, migration-selection balance and frequency-dependent selection. Other possible mechanisms, such as selective conflicts between haploid and diploid phases, and maternal–fetal interactions, do not seem to be common enough to explain the very high incidence of nonsynonymous polymorphisms. Before discussing the relative merits of the three main contenders, there is a need to consider the agents of selection, and to ask what may be their general properties.

8.4.1 Detecting selective agents

The factors that exercise selection can be any of those that affect the survival or reproduction of organisms. They include the physical and chemical nature of the environment, the level of available resources, and the presence of predators, parasites, competitors, and symbionts.

Identifying selective agents associated with a particular polymorphism can be developed systematically. There are three ways of doing it. The most obvious is to find correlations between gene frequencies and environmental variables. Sometimes the geographical distribution of gene frequencies immediately suggests selective factors. For example, the occurrence of sickle-cell, thalassemia, Duffy blood groups and glucose-6-dehydrogenase alleles in malarious parts of the world has provoked studies of selection by the disease (Allison 1955, Hill 1996, Weatherall 1997). The association of brown- and pink-shelled *Cepaea nemoralis* with woodlands has inspired experiments on selection by visual predators (Sheppard 1951).

An alternative approach is to start with a factor that is expected to be selective, and then to find the genetic variation on which it acts. This regime has been applied with great success to heavy-metal tolerance in plants (Antonovics and Bradshaw 1970, Shaw 1990), as well as antibiotic resistance in bacteria, DDT resistance in insects, and warfarin resistance in rodents (for complementary reviews of resistance to man-made disturbances, see Bishop and Cook 1981 and Palumbi 2001). A problem is that the relevant genetic variation may reveal itself as a quantitative character, not easily broken down to the level of individual genes. This problem is eased by new methods, albeit still laborious, that find QTLs.

A third way of seeking out selective factors is through the phenotypes produced by the segregating alleles themselves. Sometimes, as in mimetic polymorphisms, the phenotypic differences are extremely obvious, and immediately suggest the agents involved. At other times, detailed biochemical or physiological investigations may be needed. This last approach is particularly appropriate to the study of enzyme polymorphisms (Clarke 1975, Lewontin 2000).

Once a putative selective agent has been identified, it is necessary to demonstrate how it acts on the allelic differences. A simple correlation is not good evidence of selection. We must observe and manipulate the selective agents, and show their mechanical connections with the polymorphism. Moreover, it is not enough to show by analyzing DNA sequences that selection has acted on a particular locus. Two biologically interesting questions remain: "What selection?" and "How does it work?" They are more difficult to answer, but they lead us into the fundamentals of the evolutionary process.

For the present task, which is to generalize about the likelihood of balancing selection, we do not have to accumulate individual studies of selected polymorphisms. We can enquire about the general properties of selective agents.

8.4.2 The ecology of selective agents

The most important factors in balancing polymorphisms are likely to be other organisms, notably predators, parasites and competitors. Abiotic factors, such as temperature, pH, humidity and so on, cannot by themselves produce balance except through heterozygous advantage. However, they may interact with other factors, such as migration, to do so. The justification for these general statements has already been given (e.g., by Clarke 1979) and will not be repeated here. It suffices to say that few ecological factors affecting the survival or reproduction of organisms act independently of their numbers. Correspondingly, few agents of selection are unaffected by the numbers and frequencies of phenotypes. The assumption of constant selective values has no justification except mathematical convenience.

Recent work has added greatly to our understanding of how the agents of selection behave, and it has strengthened the case for their frequency and density dependence. Here I shall concentrate on studies that have been published since the general reviews by Ayala and Campbell (1974), Clarke (1979) and Clarke and Partridge (1988).

8.4.2.1 Predators

The widespread tendency of predators to concentrate disproportionately on common forms of prey, and to overlook rare ones, is now well established (see reviews by Allen 1988, Sherratt and Harvey 1993). Its sources in the psychology of predators are discussed by Guildford (1992) and Marples and Kelly (2001). The book by Edmunds (1974) describes the precise and complicated mechanisms that prey have evolved as defences against predators.

When predation acts against common morphs within a species, the balancing force has been termed *apostatic selection* (Clarke 1962). When it affects the relative densities of different prey species it has been called *switching* (Murdoch 1969). If prey are very numerous the behavior may go into reverse, so that the predators disproportionately overeat rare phenotypes, and select for uniformity rather than polymorphism (Allen *et al.* 1998). The nature of the background, the degree of crypsis, and whether the prey are clumped can also affect the likelihood of balance. At low densities, the clumping of prey seems to diminish the frequency dependence (Gianino and Jones 1989), but not at high densities (Church *et al.* 1997).

A useful experimental technique has been to make pictures of prey and their backgrounds on computer screens, and to study the behavior of predators in response to them. Birds and humans have been used as predators, and both have exerted apostatic selection when the prey were cryptic but not when they were more conspicuous (Tucker and Allen 1988). Using humans as predators, Cook and Kenyon (1991) found that the likelihood of apostatic selection also depended on prey density and the nature of the background. Bond and Kamil (2002) allowed "virtual prey" to evolve in response to selection by blue jays (*Cyanocitta cristata*). The prey became more cryptic and phenotypically more variable.

Observations of natural populations indicate that crypsis is not always necessary for apostatic selection. The animal kingdom has a multitude of species with nonmimetic color and pattern polymorphisms whose morphs are distinct because of obvious visual "tricks" that make them look different from each other. An example is the extravagant shell polymorphism in Cuban land snails of the genus *Polymita*. There are individuals with bright red, bright yellow, white, or black shells. Others have dazzling combinations of these colors (Fernández Milera and Martínez Fernández 1987). Like many species with striking visible polymorphisms, *Polymita* has a specialist predator, in this case the Cuban kite, *Chondrohierax wilsonii*.

The habit of concentrating on common forms of prey is not restricted to selection on visible characters. Under experimental conditions, mice that hunt by smell will concentrate on the commoner variety (Soane and Clarke 1973), and it is reasonable to expect scent polymorphisms in their prey. There should also be predator-maintained polymorphisms in shape, sound, tactile features and, among aquatic organisms, electrical properties. Driver and Humphries (1988) have made a detailed case for the value of diverse ("protean") behavior when avoiding predators, although not all the resulting behavioral polymorphisms are likely to be genetic.

Since the time of Fisher (1930) it has been known that polymorphic Batesian mimics are subject to frequency-dependent selection. Any morph that is rare compared with its distasteful model will gain an advantage, which will diminish as the frequency of the morph rises. The number of such cases is very large, particularly among insects. Pfennig *et al.* (2001) report an experimental demonstration of the frequency dependence in Batesian mimicry (see also Mallett and Joron 2000, Holloway *et al.* 2001).

Polymorphism can be promoted when two separate enemies, predators or parasites, act differentially, each causing 'density-dependent' mortality on its preferred morph (Losey *et al.* 1997). In principle, polymorphism could also be the result of selection by prey on predators. Intelligent prey should learn to avoid common phenotypes of their enemies (Paulson 1973).

8.4.2.2 Parasites

The cause of frequency-dependent predation lies in the memory of the predators. Parasites and pathogens[5] also have memories, but they are usually carried in genes rather than neurons. Natural selection adjusts the parasites to the commonest form of host. Variant hosts may therefore gain an advantage (Haldane 1949, O'Brien and Evermann 1988, Jeffery and Bangham 2000). This form of balancing selection is particularly effective when the parasites have short generations, and can evolve within a single host (Clarke 1976, McMichael and Klenerman 2002, Moore *et al.* 2002). Some parasites acquire host proteins, and use them as disguises against immune reactions, taking advantage of the fact that organisms cannot normally react to their own molecules. If a parasite carries its disguise when it moves from one host to another, variant hosts may be favored because they can attack it.

Coevolution between parasites and hosts can produce sequential bursts of change, first in the hosts, then in the parasites, then in the hosts, and so on. Alternatively, it can generate frequency-dependent selection, with polymorphisms in either or both of the hosts and parasites. Which of these outcomes actually occurs will depend on the generation times of hosts and parasites, on the frequencies of mutations to resistance or tolerance in the hosts and to virulence or infectivity in the parasites, and on the ecological details of the associations. The widespread occurrence of polymorphisms at loci involved in the interactions between parasites and hosts testifies to the selective importance of these interactions, which are inherently frequency dependent. For example, there are striking polymorphisms in the surface antigens of malarial parasites, on which there is evidence of strong selection (Escalante *et al.* 1998, Polley and Conway 2001).

Following Damian (1969), medical researchers have become excited about "molecular mimicry," antigen-sharing between parasite and host. It is of interest because humans, while reacting to antigens in parasites, may produce antibodies against their own proteins, causing auto-immune diseases, such as insulin-dependent diabetes mellitus, various forms of arthritis, and Reiter's syndrome (see, for example, Ramsingh *et al.* 1997). It is tempting to ascribe

[5] I use the term parasite to include any organism that lives on or in its host, and that harms it. The term pathogen is usually restricted to microbes.

these cases to selection driving the parasites to mimic host antigens, but it is possible that some of the resemblances are fortuitous (Quaratino *et al.* 1995, Mason 1998). Even fortuitous resemblances, however, can generate selection on hosts and parasites. Molecular mimicry is not restricted to interactions with the immune system, but may involve host enzymes and receptors (Stebbins and Galán 2001; see also Ewald and Cochran this volume).

Plants and their parasites have evolved complementary polymorphisms in resistance and virulence (Flor 1956, Barrett 1988, Burdon and Thrall 1999, Stahl *et al.* 1999). Although in some systems there are clear "gene-for-gene" relationships (with a resistance allele in the host for each avirulence allele in the parasite), in others there are more complicated interactions. Many of them seem to involve balancing frequency dependence, which is not always strong enough by itself to maintain polymorphism (Brunet and Mundt 2000). Molecular techniques are now revealing the mechanisms of plant defences (Dangl and Jones 2001). Bacteria and phages have similar interactions, some of them also producing frequency-dependent balance. The widespread polymorphisms in restriction enzymes are examples (Levin 1988, Frank 1994).

8.4.2.3 Competitors

Intraspecific competition can, in principle, maintain balancing frequency-dependent polymorphisms (see reviews by Bell 1997 and Christiansen, this volume). The likelihood of balance in a population with several 'niches' is greater when each morph can choose its appropriate niche, or when there is some assortative mating within the niches. The latter case can be regarded either as polymorphism or as incipient speciation.

There have been many experimental observations of balancing frequency-dependent selection on competing genotypes of *Drosophila* (for example, by Lewontin and Matsuo 1963, see review by Antonovics and Kareiva 1988). The agents of selection have often been obscure, but likely candidates include spatial or temporal differences in the chemical environment of the experimental medium. The temporal changes take place within a single generation creating, in effect, two or more niches ("early" and "late"), each of which can favor a different genotype. The genotypes may then reach maturity at different times, producing a tendency to assortative mating, and thereby favoring balance (Borash *et al.* 1998). Smith and Skúlason (1996) give an account of putative resource polymorphisms in other animals (see also Schluter 2000).

Several recent studies of microbial populations have demonstrated the origin and maintenance of polymorphism by cross-feeding between genotypes (Rozen and Lenski 2000), and through spatial heterogeneity in the environment (reviewed by Rainey *et al.* 2000). Both kinds of systems show balancing frequency dependence.

8.4.2.4 Self-incompatibility

Many organisms have devices to prevent inbreeding. One mechanism is self-incompatibility, often achieved by a polymorphic locus with many alleles, arranged so that organisms with the same alleles cannot mate. The result is that the bearers of rare alleles can gain an advantage because they have greater chances of successful fertilization (Wright 1939). Self-incompatibility has been best studied in plants, and there are reviews by Charlesworth and Awadalla (1998), Charlesworth (2000) and Uyenoyama and Takebayashi (this volume). It has been suggested that the MHC polymorphism, in which there is evidence of nonrandom mating, also acts as a self-incompatibility system that promotes outcrossing (Potts *et al.* 1994). These systems, and many other strongly selected polymorphisms such as that determining shell color and banding in *Cepaea*, and those associated with Batesian mimicry, share one striking feature. They are organized into 'supergenes,' with haplotypes that differ at several or many interacting loci. The selection on each of the complexes can maintain many nonsynonymous polymorphisms.

8.5 Heterozygous advantage

Hybrids between populations are often more vigorous than their parents because the expression of deleterious recessive alleles is suppressed. This heterosis can accelerate the introgression of genes from one population to another, and can last for several or many generations before it is destroyed by recombination. For an experimental demonstration of this phenomenon, see Saccheri and Brakefield (2002). The existence of heterosis makes it difficult to detect heterozygous advantage. An apparently greater fitness of heterozygotes at a locus may be due to deleterious recessives at other loci in linkage disequilibrium with it. We can only be sure of genuine heterozygous advantage when the selective agent has been identified, and there is a clear mechanism connecting it to the advantage. Such cases are few. Notable among them are human polymorphisms associated with malaria (Allison 1955, Hill 1996, Weatherall 1997), and the polymorphism in rodents for resistance to warfarin (reviewed by Bishop 1981). A case can be made for heterozygous advantage in the polymorphisms in the visual pigments of some New World monkeys (reviewed by Yokoyama 1997, Wolf 2002). For example, the squirrel monkey, *Saimiri sciurius*, has a sex-linked polymorphism with three alleles determining pigments sensitive to different wavelengths of light. Males and homozygous females are dichromatic, but females heterozygous for two of the pigments have full color vision. It might be supposed that these individuals would have an advantage capable of maintaining the polymorphism. However, trichromacy can have its costs. Visual acuity may be reduced, and "color-blind" individuals can detect camouflaged objects missed by trichromats. It is possible that animals with a particular kind of vision gain an advantage when they are uncommon, because they exploit a visual "niche" differing from that of the majority. Whether this happens must inevitably depend on particular ecological circumstances.

Other polymorphisms that have been attributed variously to heterozygous advantage and frequency-dependent selection are those in the vertebrate major histocompatibility complex (Takahata and Nei 1990, Hughes 1999). There is no doubt that these loci are under selection. In the parts of the molecules that bind foreign antigens, the frequency of nonsynonymous substitutions is greater than that of synonymous substitutions. Moreover, some of the polymorphisms have persisted far longer than would be expected under neutrality. If appropriate assumptions are made, the data are formally compatible with both heterozygous advantage and some models of frequency-dependent selection. However, since the alleles differ in their resistance to pathogens, the likelihood of constant selective values is small.

Lewontin *et al.* (1978) pointed out that heterozygous advantage could maintain more than three alleles at a locus only under very restrictive conditions. There must be a high degree of symmetry in the selective values of the homozygotes. Takahata and Nei (1990) put forward some computer models of heterozygous advantage at the MHC loci. The subset of their models that assume asymmetrical fitnesses supports an average of 3.1 alleles, compared with the 10 or more in human and mouse populations.

Heterozygous advantage may have a limited ability to maintain polymorphism in the long term because selection destroys it. As Spofford (1969) pointed out, a duplication by unequal crossing-over can be favored because it avoids the production of disadvantageous homozygotes. Gu and Nei (1999) remark that MHC loci are subject to frequent duplication and deletion, which would be expected from Spofford's argument.

Whether or not visual pigments and MHC proteins involve heterozygous advantage, the proportion of polymorphisms maintained in this way seems to be small. Nonetheless, neutral nonsynonymous polymorphisms could be maintained for long periods by heterosis due to deleterious alleles at closely linked loci (associative overdominance; see Zouros 1993). Such polymorphisms will eventually decay through recombination, but the process may take a very long time if they lie in chromosomal inversions or other regions where crossing-over is reduced. The difficulty of demonstrating by indirect methods that there is heterozygous advantage emphasizes, once again, the need to identify selective agents.

8.6 Balance between selection and migration

A selected locus can be maintained in a state of polymorphism if one allele is favored in one place, an alternative allele is favored in another, and genes can be exchanged between the two. There are limitations to this process. If the rate of migration is too small compared with the strength of selection the immigrant alleles may never reach polymorphic frequencies. If the rate of migration is too large, the organisms in the two places will become members of the same population. The situation then corresponds to Levene's (1953) frequency-dependent model of two niches (see above).

In natural situations, balance between selection and migration is likely to generate clines, but a cline, even if it is correlated with some environmental gradient, does not necessarily indicate selection. It may have come about by drift or invasion. A cline and an environmental gradient will inevitably be correlated unless they are exactly at right angles. For good statistical evidence ($p < 0.01$) of a link, seven or more independent clines are needed, all correlated in the same way with the environmental factor. This is found, for example, in *Cepaea nemoralis,* where there are many clines in the frequencies of brown or pink shells from woods to grasslands (Cain and Sheppard 1954). In *Cepaea,* however, it is certain that the polymorphism is not due solely to a balance between migration and selection, since other balancing factors are known, and even isolated populations are polymorphic.

Thompson and Cunningham (2002) investigated geographical heterogeneity in the interactions between a moth, *Greya,* and a plant, *Lithophragma.* The relationship between them varied, by locality, from mutualism to antagonism. Thompson and Cunningham argued that the resulting differences in selection are important in maintaining diversity. Burdon and Thrall (1999) have reviewed the more general issues of patchiness in plant–pathogen interactions.

At present it is difficult to assess the general importance of selection-migration balance in promoting polymorphism. That it has an important role seems inevitable. Bell (1997) has estimated that about 50% of the genotypic variance within species of cultivated plants is attributable to variation in their relative performance in different environments. Once more, however, we are faced with the need to discover the relevant selective agents, so that we can study their distributions in space.

8.7 Conclusions

In outbred eukaryotes, there may be as many as 10^4 nonsynonymous polymorphisms whose morphs are under selection. The search for balancing forces has produced more examples of frequency-dependent selection than of other plausible mechanisms, such as heterozygous advantage, or the balance between selection and migration. The case for frequency dependence as a major factor in maintaining nonsynonymous polymorphisms seems overwhelming. This does not mean, of course, that directional and unbalancing selection are less frequent than balancing selection. They must be a great deal commoner. Nonsynonymous polymorphisms represent a small fraction of genetic variants. We observe them because they have not been eliminated.

Much remains to be learned. We need to find selective agents, and to study their behavior, so that we can make better generalizations about the causes of genetic diversity. It would be desirable to carry out experiments on natural populations, or at any rate on populations as close as possible to the natural state, even though such experiments are difficult to do. It is not enough just to detect that selection has occurred. We need to understand the mechanisms of its action. This will take us into the realm of ecology.

It would be a great advance if we could discriminate between frequency dependence and heterozygous advantage, or between different forms of frequency dependence, simply by examining DNA sequences. We do not know if this is possible, because most theoretical models assume constant selective values. Even if we could identify frequency-dependent selection in this way, we would still need to study the agents that cause it.

8.8 Envoi

Dick Lewontin is a very enjoyable person to disagree with, because he takes disagreement as a challenge rather than a threat. Thank you, Dick, for the pleasure of it, which has always been spiced with admiration.

8.9 Acknowledgments

I am very grateful to John Brookfield, Brian Charlesworth, Ann Clarke, Peter Clarke, Jim Murray, and Paul Sharp for valuable discussions, and for critically reading the manuscript. Deborah Charlesworth kindly gave advice about self-incompatibility. I am indebted to the Leverhulme Trust for an Emeritus Research Fellowship.

REFERENCES

Allen, J. A. (1973). *Apostatic selection: the response of wild passerines to artificial polymorphic prey.* Ph.D. thesis, University of Edinburgh.

Allen, J. A. (1988). Frequency-dependent selection by predators. *Phil. Trans. R. Soc. Lond. Ser. B* 319:485–503.

Allen, J. A., Raison, H. E., and Weale, M. E. (1998). The influence of density on frequency-dependent selection by wild birds feeding on artificial prey. *Proc. R. Soc. Lond. Ser. B* 265:1031–5.

Allison, A. C. (1955). Aspects of polymorphism in man. *Cold Spring Harbor Symp. Quant. Biol.* 20:239–44.

Antonovics, J., and Bradshaw, A. D. (1970). Evolution in closely adjacent plant populations. VIII. Clinal patterns at a mine boundary. *Heredity* 25:349–62.

Antonovics, J., and Kareiva, P. (1988). Frequency-dependent selection and competition. *Phil. Trans. R. Soc. Lond. Ser. B* 319:601–13.

Aquadro, C. F., DuMont, V. B., and Reed, F. A. (2001). Genome-wide variation in human and fruitfly: a comparison. *Curr. Opinions in Genet. Dev.* 11:627–34.

Ayala, F. J., and Campbell, C. A. (1974). Frequency-dependent selection. *Annu. Rev. Ecol. Syst.* 5:115–38.

Barrett, J. A. (1988). Frequency-dependent selection in plant–fungal interactions. *Phil. Trans. R. Soc. Lond. Ser. B* 319:473–83.

Bell, G. (1997). *Selection: the Mechanism of Evolution.* New York: Chapman and Hall.

Bishop, J. A. (1981). A neoDarwinian approach to resistance: examples from mammals. In J. A. Bishop and L. M. Cook (eds) *Genetic Consequences of Man Made Change.* London: Academic Press.

Bishop, J. A., and Cook, L. M. (eds) (1981). *Genetic Consequences of Man Made Change.* London: Academic Press.

Bokor, K., and Pecsenye, K. (1998). Comparative influence of *Odh* and *Adh* loci on alcohol tolerance in *Drosophila melanogaster. Genet. Select. Evol.* 30:503–16.

Bond, A. B., and Kamil, A. C. (2002). Visual predators select for crypticity and polymorphism in virtual prey. *Nature* 415:609–13.

Borash, D. J., Gibbs, A. G., Joshi, A., and Mueller, L. D. (1998). A genetic polymorphism maintained by natural selection in a temporally varying environment. *Am. Nat.* 151:148–56.

Brunet, J., and Mundt, C. C. (2000). Disease, frequency-dependent selection, and genetic polymorphisms: experiments with stripe rust and wheat. *Evolution* 54:406–15.

Burdon, J. J., and Thrall, P. H. (1999). Spatial and temporal patterns in coevolving plant and pathogen associations. *Am. Nat.* 153:S15–S33.

Bustamente, C. D., Nielsen, R., Sawyer, S. A., Olsen, K. M., Purugganan, M. D., and Hartl, D. L. (2002). The cost of inbreeding in *Arabidopsis. Nature* 416:531–4.

Cain, A. J., and Sheppard, P. M. (1954). Natural selection in *Cepaea. Genetics* 39:89–116.

Cargill, M., and 16 others (1999). Characterization of single nucleotide polymorphisms in coding regions of human genes. *Nat. Genet.* 22:231–8.

Charlesworth, B., Morgan, M. T., and Charlesworth, D. (1993). The effect of deleterious mutations on neutral molecular variation. *Genetics* 134:1289–303.

Charlesworth, D. (2000). Unlocking the secrets of self-incompatibility. *Curr. Biol.* 10:R184–6.

Charlesworth, D., and Awadalla, P. (1998). The molecular population genetics of flowering plant self-incompatibility polymorphisms. *Heredity* 81:1–9.

Chasman, D., and Adams, R. M. (2001). Predicting the functional consequences of nonsynonymous single nucleotide polymorphisms: structure-based assessment of amino acid variation. *J. Mol. Biol.* 307:683–706.

Chinnici, J. P. (1971). Modification of recombination frequency in *Drosophila.* I. Selection for increased and decreased crossing-over. *Genetics* 69:71–83.

Church, S. C., Jowers, M., and Allen, J. A. (1997). Does prey dispersion affect frequency-dependent predation by wild birds? *Oecologia* 111:292–6.

Clarke, B. (1962). Balanced polymorphism and the diversity of sympatric species. In D. Nichols (ed.) *Taxonomy and Geography.* Oxford: Systematics Association.

Clarke, B. (1975). The contribution of ecological genetics to evolutionary theory: detecting the direct effects of natural selection on particular polymorphic loci. *Genetics* 79:101–13.

Clarke, B. (1976). The ecological genetics of host–parasite relationships. In A. E. R. Taylor and R. M. Muller (eds.) *Genetic Aspects of Host–Parasite Relationships.* Oxford: Blackwell.

Clarke, B. (1979). The evolution of genetic diversity. *Proc. R. Soc. Lond. Ser. B* 205:453–74.

Clarke, B., and Partridge, L. (eds) (1988). *Frequency-dependent Selection.* London: The Royal Society (also published in the *Phil. Trans. R. Soc. Lond. Ser. B* 319:459–640).

Cook, L. M., and Kenyon, G. (1991). Frequency-dependent selection with background heterogeneity. *Heredity* 66:67–73.

Damian, R. T. (1969). Molecular mimicry: antigen sharing between parasite and host and its consequences. *Am. Nat.* 98:129–49.

Dangl, J. L., and Jones, J. D. G. (2001). Plant pathogens and integrated defence responses to selection. *Nature* 411:826–33.

Dempster, E. R. (1955). The maintenance of genetic heterogeneity. *Cold Spring Harbor Symp. Quant. Biol.* 20:25–32.

Driver, P. M., and Humphries, D. A. (1988). *Protean Behaviour: the Biology of Unpredictability.* Oxford: Clarendon Press.

Edmunds, M. (1974). *Defence in Animals: a Survey of Anti-predator Defences.* Harlow: Longman Group.

Endler, J. A. (1986). *Natural Selection in the Wild.* Princeton, NJ: Princeton University Press.

Escalante, A. A., Lal, A. A., and Ayala, F. J. (1998). Genetic polymorphism and natural selection in the malaria parasite *Plasmodium falciparum. Genetics* 149:189–202.

Fay, J. C., Wyckoff, G. J., and Wu, C.-I. (2002). Testing the neutral theory of molecular evolution with genomic data from *Drosophila. Nature* 415:1024–27.

Fernández Milera, J. M., and Martínez Fernández, J. R. (1987). *Polymita.* Havana: Editorial Cientifico-Técnica.

Fisher, R. A. (1930). *The Genetical Theory of Natural Selection.* Oxford: Clarendon Press.

Flor, H. H. (1956). The complementary genetic systems of flax and flax rust. *Adv. Genet.* 8:29–54.

Frank, S. A. (1994). Polymorphism of bacterial restriction-modification systems: the advantage of diversity. *Evolution* 48:1470–77.

Gabriel, S. B., and 17 others (2002). The structure of haplotype blocks in the human genome. *Science* 296:2225–9.

Gianino, J. S., and Jones, J. S. (1989). The effects of dispersion on frequency-dependent predation of polymorphic prey. *Heredity* 62:265–8.

Goldstein, D. B. (2001). Islands of linkage disequilibrium. *Nat. Genet.* 29:109–11.

Gu, X., and Nei, M. (1999). Locus specificity of polymorphic alleles and evolution by a birth-and-death process in mammalian MHC genes. *Mol. Biol. Evol.* 16:147–156.

Guildford, T. (1992). Predator psychology and the evolution of prey coloration. In M. C. Crawley (ed.) *Natural Enemies: the Population Biology of Predators, Parasites and Diseases.* Cambridge: Blackwell Scientific Publications.

Haldane, J. B. S. (1949). Disease and evolution. *La Ricerca Scientifica Supplement* 19:68–76.

Harris, H. (1971). Protein polymorphism in man. *Can. J. Genet. Cytol.* 13:381–96.

Hill, A. V. S. (1996). Genetics of infectious disease resistance. *Curr. Opinions in Genet. Dev.* 6:348–53.

Holloway, G., Gilbert, F., and Brandt, A. (2001). The relationship between mimetic imperfection and phenotypic variation in insect colour patterns. *Proc. R. Soc. Lond. Ser. B* 269:411–16.

Hubby, J. L., and Lewontin, R. C. (1966). A molecular approach to the study of genic heterozygosity in natural populations. I. The number of alleles at different loci in *Drosophila pseudoobscura*. *Genetics* 54:577–94.

Hughes, A. (1999). *Adaptive Evolution of Genes and Genomes.* New York: Oxford University Press.

Jeffery, K. J. M., and Bangham, C. R. M. (2000). Do infectious diseases drive MHC diversity? *Microbes and Infection* 2:1335–41.

Kreitman, M., and Akashi, H. (1995). Molecular evidence for natural selection. *Annu. Rev. Ecol. Syst.* 26:403–22.

Kruglyak, L., and Nickerson, D. A. (2001). Variation is the spice of life. *Nat. Genet.* 27:234–6.

Levene, H. (1953). Genetic equilibrium when more than one ecological niche is available. *Am. Nat.* 87:331–3.

Levin, B. R. (1988). Frequency-dependent selection in bacterial populations. *Phil. Trans. R. Soc. Lond. Ser. B* 319:459–72.

Lewontin, R. C. (1974). *The Genetic Basis of Evolutionary Change.* New York: Columbia University Press.

Lewontin, R. C. (2000). The problems of population genetics. In Singh, R. S. and Krimbas, C. B. (eds) *Evolutionary Genetics from Molecules to Morphology.* Cambridge: Cambridge University Press.

Lewontin, R. C., and Hubby, J. L. (1966). A molecular approach to the study of genic heterozygosity in natural populations. II. Amount of variation and degree of heterozygosity in natural populations of *Drosophila pseudoobscura*. *Genetics* 54:595–609.

Lewontin, R. C., and Matsuo, Y. (1963). Interaction of genotypes determining viability in *Drosophila busckii*. *Proc. Natl. Acad. Sci. USA* 49:270–8.

Lewontin, R. C., Ginzburg, L. R., and Tuljapurkar, S. D. (1978). Heterosis as an explanation for large amounts of genic polymorphism. *Genetics* 88:149–70.

Losey, J. E., Ives, A. R., Harmon, J., Ballantyne, F., and Brown, C. (1997). A polymorphism maintained by opposite patterns of parasitism and predation. *Nature* 388:269–72.

Mallett, J., and Joron, M. (2000). Evolution of diversity in warning colour and mimicry: polymorphisms, shifting balance and speciation. *Annu. Rev. Ecol. Syst.* 30:201–33.

Mani, G. S., Shelton, P. R., and Clarke, B. (1990). A model of quantitative traits under frequency-dependent balancing selection. *Proc. R. Soc. Lond. Ser. B* 240:15–28.

Marples, N. M., and Kelly, D. J. (2001). Neophobia and dietary conservatism: two distinct processes? *Evol. Ecol.* 13:641–55.

Marth, G., Yeh, R., and Minton, M. (2001). Single nucleotide polymorphisms in the public domain: how useful are they? *Nat. Genet.* 27:371–2.

Mason, D. (1998). A very high level of cross-reactivity is an essential feature of the T-cell receptor. *Immunology Today* 19:395–404.

McKenzie, J. A., and McKechnie, S. W. (1978). Ethanol tolerance and the *Adh* polymorphism of *Drosophila melanogaster*. *Nature* 272:75–6.

McMichael, A., and Klenerman, P. (2002). HLA leaves its footprints on HIV. *Science* 296:1410–11.

Middleton, R. J., and Kacser, H. (1983). Enzyme variation, metabolic flux and fitness: alcohol dehydrogenase in *Drosophila melanogaster*. *Genetics* 105:633–50.

Moore, C. B., John, M., James, I. R., Christiansen, F. T., Witt, C. S., and Mallal, S. A. (2002). Evidence of HIV-1 adaptation to HLA-restricted immune responses at a population level. *Science* 296:1439–43.

Moriyama, E. N., and Powell, J. R. (1996). Intraspecific nuclear DNA variation in *Drosophila*. *Mol. Biol. Evol.* 13:261–77.

Murdoch, W. W. (1969). Switching in general predators: experiments on predator specificity and stability of prey populations. *Ecol. Monographs* 39:335–54.

O'Brien, S. J., and Evermann, J. F. (1988). Interactive influence of infectious disease and genetic diversity in natural populations. *Trends Ecol. Evol.* 3:254–9.

Palumbi, S. R. (2001). Humans as the world's greatest evolutionary force. *Science* 293:1786–90.

Paulson, D. R. (1973). Predator polymorphism and apostatic selection. *Evolution* 27:269–77.

Pfennig, D. W., Harcombe, W. R., and Pfennig, K. S. (2001). Frequency-dependent Batesian mimicry. *Nature* 410:323.

Polley, S. D., and Conway, D. J. (2001). Strong diversifying selection on domains of the *Plasmodium falciparum* apical membrane antigen 1 gene. *Genetics* 158:1505–12.

Potts, W. K., Manning, C. J., and Wakeland, E. K. (1994). The role of infectious disease, inbreeding and mating preferences in maintaining MHC genetic diversity: an experimental test. *Phil. Trans. R. Soc. Lond. Ser. B* 346:369–78.

Quaratino, S., Thorpe, C. J., Travers, P. J., and Londei, M. (1995). Similar antigenic surfaces, rather than sequence homology, dictate T-cell epitope molecular mimicry. *Proc. Natl. Acad. Sci, USA* 92:10398–402.

Rainey, P. B., Buckling, A., Kassen, R., and Travisano, M. (2000). The emergence and maintenance of diversity: insights from experimental bacterial populations. *Trends Ecol. Evol.* 15:243–7.

Ramsingh, A. I., Chapman, N., and Tracy, S. (1997). Coxsackieviruses and diabetes. *BioEssays* 19: 793–800.

Rozen, D. E., and Lenski, R. E. (2000). Long-term experimental evolution in *Escherichia coli*. VIII. Dynamics of a balanced polymorphism. *Am. Nat.* 155:24–35.

Saccheri, I. J., and Brakefield, P. M. (2002). Rapid spread of immigrant genomes into inbred populations. *Proc. R. Soc. Lond. Ser. B* 269:1073–8.

Schluter, D. (2000). *The Ecology of Adaptive Radiation*. Oxford: Oxford University Press.

Shaw, A. J. (ed.) (1990). *Heavy Metal Tolerance in Plants: Evolutionary Aspects*. Boca Raton, FL: CRC Press.

Shelton, P. R. (1986). *Some studies of frequency-dependent selection on metrical characters*. Ph. D. thesis, University of Nottingham.

Sheppard, P. M. (1951). Fluctuations in the selective value of certain phenotypes in the polymorphic land snail *Cepaea nemoralis* L. *Heredity* 5:125–34.

Sherratt, T. N. and Harvey, I. F. (1993). Frequency-dependent food selection by arthropods: a review. *Biol. J. Linn. Soc.* 48:167–86.

Singh, M., and Lewontin, R. C. (1966). Stable equilibria under optimising selection. *Proc. Natl Acad. Sci. USA* 56:1345–8.

Singh, R. S., and Rhomberg, L. R. (1987). A comprehensive study of genic variation in natural populations *Drosophila melanogaster*. II. Estimates of heterozygosity and patterns of geographic differentiation. *Genetics* 117:255–71.

Smith, N. G. C., and Eyre-Walker, A. (2002). Adaptive protein evolution in *Drosophila*. *Nature* 415:1022–4.

Smith, T. B., and Skúlason, S. (1996). Evolutionary significance of resource polymorphisms in fishes, amphibians, and birds. *Annu. Rev. Ecol. Syst.* 27:111–33.

Soane, I. D., and Clarke, B. (1973). Evidence for apostatic selection by predators using olfactory cues. *Nature* 241:62–4.

Spofford, J. B. (1969). Heterosis and the evolution of duplications. *Am. Nat.* 103:407–32.

Stahl, E. A., Dwyer, G., Mauricio, R., Kreitman, M., and Bergelson, J. (1999). Dynamics of disease resistance polymorphism at the *Rpm1* locus of *Arabidopsis*. *Nature* 400:667–71.

Stebbins, C. E., and Galán, J. E. (2001). Structural mimicry in bacterial virulence. *Nature* 412:701–5.

Sunyaev, S., Ramensky, V., and Bork, P. (2000). Towards a structural basis of human nonsynonymous single nucleotide polymorphisms. *Trends Genet.* 16:198–200.

Takahata, N., and Nei, M. (1990). Allelic genealogy under overdominant and frequency-dependent selection and polymorphism of major histocompatibility complex loci. *Genetics* 124:967–78.

Thompson, J. N. and Cunningham, B. M. (2002). Geographic structure and dynamics of coevolutionary selection. *Nature* 417:735–8.

Tucker, G. M., and Allen, J. A. (1988). Apostatic selection by humans searching for computer-generated images on a colour monitor. *Heredity* 60:329–34.

Wang, D. G., and 26 others (1998). Large-scale identification, mapping, and genotyping of single-nucleotide polymorphisms in the human genome. *Science* 280:1077–82.

Wang, W., Thornton, K., Berry, A., and Long, M. (2002). Nucleotide variation along the *Drosophila melanogaster* fourth chromosome. *Science* 295:134–7.

Weatherall, D. J. (1997). Thalassaemia and malaria revisited. *J. Trop. Med. Parasitol.* 91:885–90.

Wolf, K. (2002). Visual ecology: coloured fruit is what the eye sees best. *Curr. Biol.* 12:R253–5.

Wright, S. (1939). The distribution of self-fertility alleles in populations. *Genetics* 24:538–52.

Yokoyama, S. (1997). Molecular genetic basis of adaptive selection: examples from colour vision in vertebrates. *Annu. Rev. Genet.* 31:315–36.

Zouros, E. (1993). Associative overdominance: evaluating the effects of inbreeding and linkage disequilibrium. *Genetica* 89:35–46.

9

Why $k = 4Nus$ is silly

JOHN H. GILLESPIE
Section of Evolution and Ecology, University of California, Davis

9.1 Introduction

The rate of adaptive evolution is a primary concern of evolutionary biologists. This rate is influenced by a number of factors; cited most often are the strength of selection, the mutation rate, the population size and the rate of change of the environment. When forced to write down a mathematical expression for the substitution rate at a particular locus, most population geneticists are likely to write

$$k = 4Nus. \tag{9.1}$$

The usual justification is that $2Nu$ new advantageous mutations, on average, enter the population each generation and a fraction, $2s$, of these new mutations are fixed. Wright (1949) appears to be the first to have used this formula while Kimura (1971) was the first to apply it to molecular evolution. Equation 9.1 has been used by Kimura and Ohta (1971) to argue that the molecular clock is incompatible with molecular evolution driven by natural selection. They find the assumption that

> ...in the course of evolution three parameters N_e, s_1 and u are adjusted in such a way that their product remains constant per year over diverse lineages...

sufficiently unpalatable that natural selection should be rejected in favor of neutrality. The equation is also central to discussions about whether evolution proceeds faster in small or large populations, with the obvious implications to theories of speciation.

When tested against simulations, $k = 4Nus$ does not perform very well. For example, Gillespie (1999, 2001) has shown that the rate of substitution (of sites in an infinite-sites, no-recombination model) is a concave rather than a linear function of the population size, N. The reason for this is well understood. Given that one substitution has occurred, the subsequent substitution must begin as a mutation of the first substituting allele. More explicitly, if

The Evolution of Population Biology, ed. R. S. Singh and M. K. Uyenoyama. Published by Cambridge University Press.

the frequency of the first substituting site is p_t, then the average number of mutations entering the population that are candidates for the subsequent substitution is $2Np_t u$, which is less than $2Nu$. As N increases, substitutions want to crowd together but are inhibited from doing so by the lowered effective mutation rate, hence the concavity. Gerrish and Lenski (1998) have called this *clonal interference* and it is but one instance of many where intuition based on two-allele results ($k = 4Nus$ comes from a two-allele model) fails when more alleles are present.

Although clonal interference is one reason for doubting the veracity of $k = 4Nus$, there is another more compelling reason, which is the subject of this chapter. In deriving $k = 4Nus$, there is an implicit assumption that s is the same for each substitution. While there may be situations where this assumption is warranted, the general paradigm of Darwinian evolution would have each substitution improving a species' fit to its environment. In a static environment, the selection coefficients of a sequence of substitutions should decrease, leading to a cessation of substitutions. As a consequence (and again in a static environment), we would expect the long-term rate of substitution to be effectively zero and, trivially, independent of u, s and N.

Of course, environments are not constant and it is their change that drives adaptive evolution. Yet, there is no term in $k = 4Nus$ that reflects the rate of change of the environment. One remedy would be to assume that the environment changes at discrete points in time with a small number of substitutions following each change. This simple modification leads to a radically different equation for the rate of substitution. To see this, let the counting process M_t record the changes in the environment. That is, $M_t = n$ means that in a time span of t generations there were n changes in the environment. Let X_i be the number of substitutions following the ith change in the environment. Assume that the X_i are independent, identically distributed random variables. The total number of substitutions in t generations may now be written as

$$S_t = X_1 + X_2 + \cdots + X_{M_t}. \tag{9.2}$$

From this we get a rate of substitution,

$$k = \lim_{t \to 0} \frac{E\{M_t\}E\{X\}}{t} = \lambda E\{X\}, \tag{9.3}$$

where λ is the rate of change of the environment and the subscript for X has been suppressed.

Equation 9.3 hides a number of details. For example, the value of $E\{X\}$ depends on N and u, although, as will be shown in the next two sections, the dependence is much weaker than the linear dependence in $k = 4Nus$. $E\{X\}$ also depends on λ for the obvious reason that $E\{X\}$ is the mean number of substitutions between two successive changes in the environment. The dependency on λ is fairly simple for slowly changing environments where we can picture widely separated bursts of substitutions. However, if λ is so large that the environment changes before a substitution is completed, then Equation 9.3 no longer holds. This, too, will be explored in this chapter.

In a head-to-head competition between the models giving rise to $4Nus$ and $\lambda E\{X\}$, I would argue that the latter comes much closer to our usual sense of adaptive evolution: changes in the environment lead to gene substitutions; faster changes lead to more substitutions. It comes very close to Fisher's (1958) view of evolution in response to the "deterioration of the environment" and Van Valen's (1974) "red queen hypothesis." However, it is quite different from Wright's models based on adaptive landscapes and is certainly very different from whatever model may have led to $k = 4Nus$.

9.2 TIM and the house-of-cards model

The house-of-cards model is perhaps the simplest model of evolution that does not unduly jar our biological intuition. Under this model, each new mutation has a selection coefficient chosen from a symmetric distribution centered on zero with standard deviation σ. This model was first used to describe molecular evolution by Ohta and Tachida (1990), who referred to it as the *fixed model.* The model has been the subject of a number of papers (Tachida 1991, 1996, Gillespie 1993).

If the selection coefficient for a particular allele never changes, then the rate of substitution for the house-of-cards model becomes essentially zero when $2N\sigma$ is larger than about four or five. (This surprising and important result will be discussed below.) In order to keep things moving, the selection coefficients must change with time. These changes are intended to model the effects of a fluctuating environment. They can be added in various ways. Here we will make the simple assumption that a selection coefficient remains constant for a geometrically distributed period of time before changing to a new value. Each new value is chosen independently from the same symmetric distribution with mean zero and standard deviation σ. The assumption that selection coefficients remain constant for a geometrically distributed time means that M_t in Equation 9.2 is a discrete-time version of the Poisson process.

The house-of-cards model with temporally changing selection coefficients is often called the TIM model after the authors of one of the first papers to investigate its properties (Takahata *et al.* 1975). The rate of substitution for a TIM model as a function of population size is given in Figure 9.1 for the case of normally distributed selection coefficients. Notice that the rate quickly becomes insensitive to the population size as N increases. For this model, $k = \lambda E\{X\}$ would be a better descriptor than $k = 4Nus$ if it could be shown that $E\{X\}$ becomes insensitive to population size as N increases.

The complete dynamics of the TIM model are complex and not well understood. However, if the mutation rate is sufficiently small, so small that there is very little standing variation, and if the times between the changes in the environment are long relative to the time required to complete a small number of

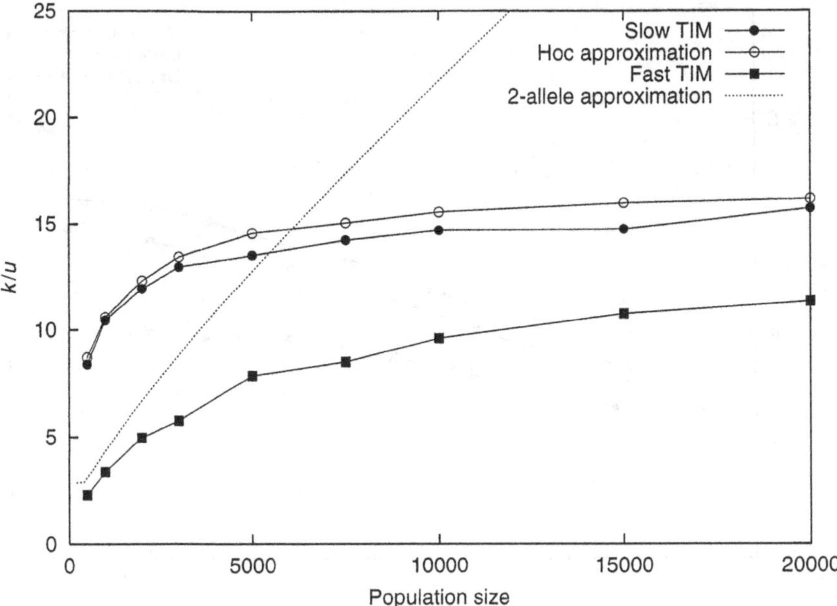

Figure 9.1. The rate of substitution divided by the mutation rate, k/u, for various haploid models. The curve labeled "Slow TIM" gives k/u for the TIM model with normally distributed selection coefficients, $u = 2.5 \times 10^{-5}$, $\sigma = 0.1$ and geometrically distributed times between changes in selection coefficients with mean 5000. "Fast TIM" is an identical simulation except that the selection coefficients change every generation. "Hoc approximation" uses $E\{X\}$ from Figure 9.2 and $\lambda = 1/5000$ to approximate k/u with Equation 9.3. "2-allele approximation" graphs Equation 9.4. Each point is based on 5000 samples. All simulations were written in the Squeak dialog of smalltalk (`www.squeak.org`).

substitutions, then its properties may be described using those of the house-of-cards model with appropriate initial conditions. The initial conditions have the population fixed for an allele with a random selection coefficient. This allele represents an allele that was fixed because of its high fitness, but a change in the environment has changed its fitness to a random quantity with the same distribution as that of newly arising mutations. Thus, we expect new mutations to arise that are more fit, one of which may even replace the initial allele. More substitutions may follow until the burst of substitutions effectively stops. The random variable X represents the number of substitutions in one such burst. The burst officially ends with the next change in the environment.

Our first job is to see if $k = \lambda E\{X\}$ is a reasonable approximation to the rate of substitution under the TIM model. We will do this only for normally distributed selection coefficients. The comparison is made in Figure 9.1, where $\lambda = 1/5000$ and the values of $E\{X\}$ are obtained from computer simulations of the house-of-cards model as reported in Figure 9.2. In Figure 9.1 k/u from a direct simulation of the TIM model is given by the curve labeled "Slow TIM" while that for the house-of-cards approximation is given by the curve labeled

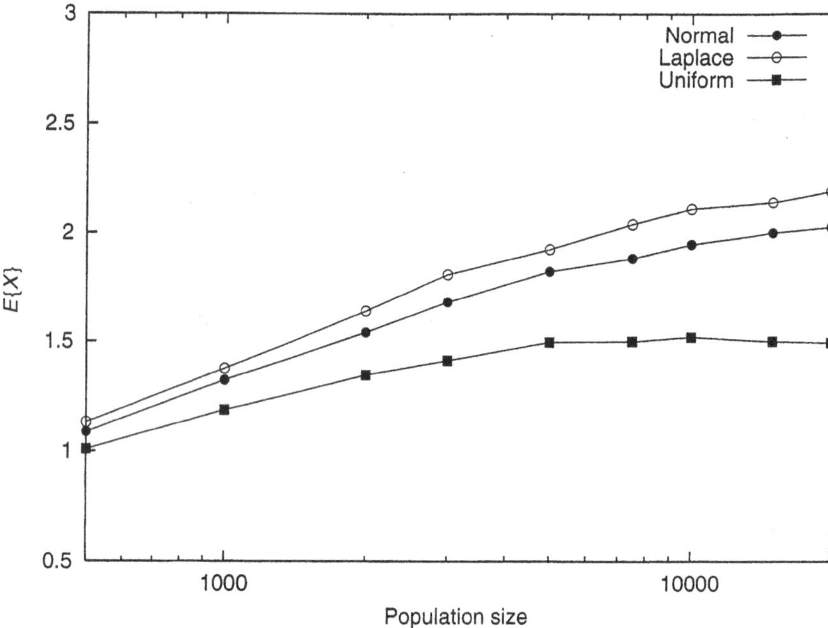

Figure 9.2. The mean number of substitutions in haploid house-of-cards models with $u = 2.5 \times 10^{-5}$ for three distributions of selection coefficients with mean zero and standard deviation 0.1. Each of the 5000 replicates was run for a geometrically distributed number of generations with mean 5000.

"Hoc approximation." Clearly, the approximation is quite good, although the approximation tends to underestimate k.

Of particular interest is the relationship between k and N, which comes entirely from the functional dependence of $E\{X\}$ on N (λ does not depend on N). The mean values of the X, as determined by computer simulation, are given in Figure 9.2 for three fitness distributions and a sequence of population sizes ranging from 500 to 20 000. $E\{X\}$ initially increases as log N, but for larger N its increase with N is much less than logarithmic. The three distributions illustrated in Figure 9.2 suggest that distributions with fatter tails, the Laplace distribution in this case, experience more substitutions than those with thinner tails. (The uniform, with the fewest substitutions, has no tail.)

For all three distributions, the number of substitutions following a change in the environment is very small. For the Laplace, normal and uniform distributions, only 2%, 1.22% and 0.08% of the bursts involve more than four substitutions, respectively, when $N = 20\,000$. This despite the fact that, on average, $5000 \times 20\,000 \times 2.5 \times 10^{-5} = 2500$ new mutations enter the population. At the other end, 3.36%, 4% and 7.76% of the bursts result in no substitutions at all. The perplexing part of this is the fact that the fitness of the initial allele is chosen at random. It has no fitness advantage accrued from being the first allele

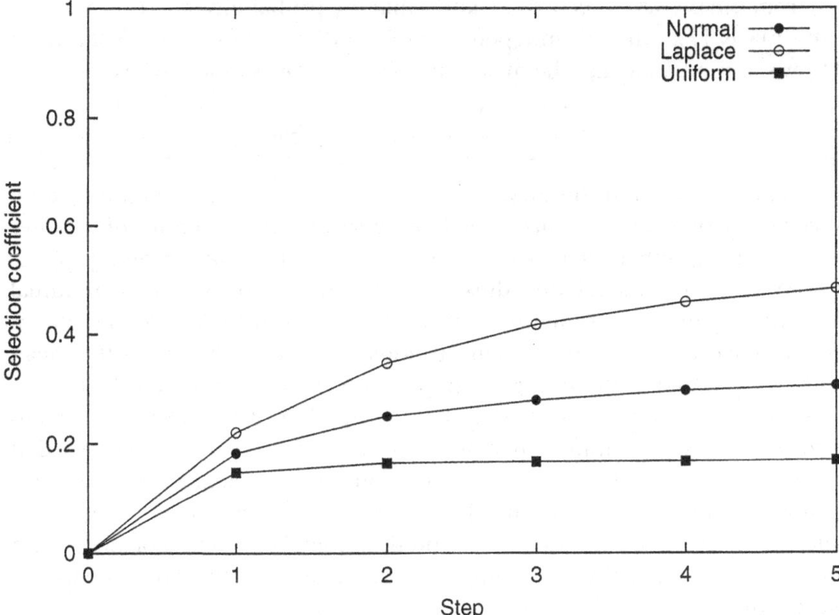

Figure 9.3. The mean value of the selection coefficients of successive substitutions for the house-of-cards model with the same parameters as those in Figure 9.2, except that the population size is fixed at $N = 20\,000$.

in the simulation. This observation alone suggests that the fitness increase from the first substitution might be, in some sense, disproportionately large.

The mean values of the selection coefficients of successive substitutions, indexed by their step number, are given in Figure 9.3. As expected, the increase in the selection coefficient decreases with each substitution as evidenced by the concavity of the curves in Figure 9.3. The uniform distribution is the most extreme in this regard, as we would expect.

The initial logarithmic increase in $E\{X\}$ with N, the small number of substitutions, and the pattern of increases in the selection coefficients may all be partially understood by appealing to the theory of records (Glick 1978, Arnold *et al.* 1998). This theory is concerned with the record properties of a sequence of IID random variables, $Y_1, Y_2, \ldots, Y_n$. A record occurs on the ith draw when

$$Y_i = \max(Y_1, Y_2, \ldots, Y_i).$$

The theory describes both the values of the records and the times at which they occur. In a sufficiently large population, an allele will fix only if its fitness is greater than that of the previously fixed allele. Thus, alleles that fix are like records. The analogy is not perfect as all records are counted whereas all alleles with higher fitnesses are not fixed. Nonetheless, it is worthwhile seeing to what extent the properties of records mimic those of fixed alleles.

From the theory of records we know that the probability of a record on the nth draw is $1/n$, which is independent of the distribution of Y_i. From this it follows immediately that the mean number of records in n draws is

$$1 + \frac{1}{2} + \frac{1}{3} + \cdots + \frac{1}{n} \sim \log n,$$

where, by convention, the first draw is counted as a record. As n increases, records become rare because it is harder to be the maximum of a larger collection of numbers than of a smaller collection. The same reasoning applies to the fixation of alleles, and thus we should not be surprised at the initial logarithmic increase of the number of substitutions with N. (Nor should we be surprised if the number of substitutions increases with $\log u$ as the mean number of mutations entering the population each generation is $2Nu$.)

Perhaps we should be surprised that the number of substitutions plateaus after the initial logarithmic increase. This is most likely due to the fact that the fitness increments of successive mutations become smaller with each successive substitution (as seen in Figure 9.3). The fixation probability of an allele decreases with this increment and thus a smaller fraction of alleles with higher fitnesses actually fix, leading to a systematic departure from the theory of records.

One oddity of records is that the mean times between records are infinite. For example, even though the probability that a record occurs on the second draw is one-half, the mean time to the second record (recalling that the first draw is always called a record) is infinite. In a population genetics context, this implies that the mean time until a mutation appears that is more fit than the initial allele is also infinite. In a very large population, the mean time until the first substitution in the house-of-cards model is "effectively" infinite; hence, the rate of substitution is effectively zero. (In finite populations some mutations will always fix due to genetic drift, which is the reason for the "effective" qualification.) This is why the house-of-cards model appears to stop evolving when $N\sigma$ is greater than four or five. These mean properties should not be confused with the distribution of the number of fixations. Even though the mean time between fixations is effectively infinite, the probability that no fixations occur is small.

These properties of records and, by analogy, substitutions under the house-of-cards model, make it clear why only a few substitutions follow a change in the environment. They also help explain why the rate equation $k = \lambda E\{X\}$, with the values of $E\{X\}$ coming from the house-of-cards model, describes so very well the rate of substitution under the TIM model.

9.3 Related models

There are other models for the burst of substitutions following a change in the environment. Two early examples are "a simple stochastic gene substitution model" (Gillespie 1982) and the mutational landscape model (Gillespie 1984). Both of these differ from the house-of-cards model in that there is recurrent mutation to a finite number of alleles that are one mutational step away from the fixed allele rather than nonrecurrent mutations to an infinity of alleles. This guarantees that there is a unique most-fit mutationally accessible allele that will eventually fix in the population (if the environment doesn't change beforehand).

In the simple model, there are n alleles with randomly assigned fitnesses that mutate to each other at the same rate. The labeling of the alleles is done such that the most fit allele is A_1, the next most fit is A_2, and so forth. Evolution is assumed to start with the mth most fit allele fixed in the population. Using SSWM (strong selection, weak mutation) limits (Gillespie 1991, Chapter 5), it is possible to show that the probability that the jth allele, $j < m$, is the first to fix is

$$\frac{s_j - s_m}{\sum_{k=1}^{m-1}(s_k - s_m)},$$

where s_i is the selection coefficient of the ith allele. (That is, its fitness is $1 + s_i$.) Thus, it is likely that the first few substitutions involve larger jumps in fitness while subsequent substitutions involve smaller jumps in fitness. The mean time to the first substitution is

$$\frac{1}{4Nu\sum_{k=1}^{m-1}(s_k - s_m)}.$$

Thus, N plays a role in the time to fixation but not in the choice of which alleles fix (as is the case for most models).

Because of the simple pattern of mutation, the mean number of substitutions following an environmental change is

$$E\{X\} = \frac{1}{2} + \frac{1}{m+1} + \frac{1}{2}\sum_{j=2}^{m}\frac{j+3}{j(j+1)}.$$

This result is remarkable in that it is independent of the population size, the mutation rate, and the distribution used to assign fitnesses. (By contrast, $E\{X\}$ under the house-of-cards model is weakly dependent on these parameters.) All of this comes about because of the role played by extreme value theory in determining the final answer. The simple model shares important properties with the house-of-cards model: each burst of substitutions involves only a small number of alleles, and the fitness jumps in the early substitutions are larger than those of subsequent substitutions.

The mutational landscape model represents a step toward a more realistic mutational structure. Under this model, each allele mutates to a finite number of other alleles. Each of these mutant alleles in turn mutates to their own finite set of unique alleles. Thus, there is an infinity of alleles in the model, but only a finite number are available at any time.

The mutational landscape model behaves much like the others: each evolution involves a small number of alleles, and the fitness jumps in the early substitutions are the largest. The model is more difficult to analyze, so most of the available results come from computer simulations, although Orr (2002) announced some exciting new insights into the mathematical structure of these models that yields, in his hands, some new analytic results.

These models may be used for the distribution of X and the rate equation $k = \lambda E\{X\}$ still applies. A more radical departure occurs if we demand that the fitnesses change continuously through time, say by making the selection coefficients stationary stochastic processes. For example, let the fitness of the ith allele be $1 + s_i(t)$, where $s_i(t)$ is a stationary Gaussian process with mean zero and variance σ^2. In this case we might expect the representation of the number of substitutions given in Equation 9.2 to fail. Curiously, it doesn't. If we jump into the evolution at a random time, there is likely to be a most fit allele in high frequency and some deleterious alleles at low frequencies. The selection coefficient of the most fit allele will slowly move toward zero because of the stationarity assumption. At some point its fitness will fall below that of some other allele; that allele will then increase in frequency to become the new most fit and most frequent allele. These interchanges often involve more than two alleles, leading to a small burst of substitutions, at the end of which some allele is comfortably at the top of the heap. In this regard, the model behaves very much like the TIM model of the previous section if we equate the changes in M_t with the times of crossings of selection coefficients that precede bursts of substitutions. There is one difference, however. The times at which selection coefficients cross form a point process that is more regular than a Poisson process, $\text{Var}\{M_t\} < E\{M_t\}$.

The reason for the regularity of M_t is obscure. The mathematical problem may be stated as follows. Consider a sequence of continuous stationary Gaussian stochastic processes $\xi_1(t), \xi_2(t), \ldots, \xi_n(t)$. Consider the point process of events that occur whenever the subscript of the process with the maximum value changes. Some smoothness conditions are required on the $\xi_i(t)$ so that the crossings of the processes are isolated. Through simulation, I have noticed that the index of dispersion of this point process is less than one and decreases as n increases. This has consequences for the index of dispersion of the substitution process, which is

$$R_t = \frac{\text{Var}\{S_t\}}{E\{S_t\}} = E\{X\}I_t + \frac{\text{Var}\{X\}}{E\{X\}},$$

where I_t is the index of dispersion of the point process recording changing in the environment. The regularity of the times of crossing implies that I_t is less than one. If the burst sizes are small, then R_t may be less than one as well. Note that in the TIM models of the previous section, M_t is a Poisson process, which implies that $I_t = 1$ and

$$R_t = E\{X_i\} + \frac{\text{Var}\{X_i\}}{E\{X_i\}} \geq 1.$$

From these two cases we are led to the rather obvious conclusion that the pattern of environmental change has a profound effect on the pattern of substitutions.

9.4 Faster fluctuations

It could be argued that the models of the previous two sections do not properly capture the intuition behind $k = 4Nus$ because the environment changes too slowly. That is, the fitness increments illustrated in Figure 9.3 decrease with each substitution because the environment doesn't change during each burst of substitutions. Were the environment to change on a time scale that is shorter than the time between substitutions, then, on average, each substitution will have the same selective advantage. This average advantage could be the s in $4Nus$ and thus this formula would apply.

While the logic of this argument appears reasonable, in fact the dynamics don't necessarily lead to $k = 4Nus$, as evidenced by the curve labeled "Fast TIM" in Figure 9.1. This curve summarizes the results of simulations of the TIM model with the environment changing every generation. Obviously, faster fluctuations do not necessarily lead to a linear relationship between k and N.

A two-allele approximation of k under the TIM model is given by

$$k = \frac{Nu\sigma^2}{\log(Nu\sigma^2)} \tag{9.4}$$

(Takahata and Kimura 1979, Gillespie 1993). The curve labeled "2-allele approximation" in Figure 9.1 shows that the two-allele approximation does not fare well over a large range of population sizes. Once again, two-allele results are misleading in multiple-allele models.

The reason for the strong concavity in k for the haploid TIM model with fast fluctuations may be inferred from the drift coefficient of its approximating diffusion model,

$$E\{dx_i\} = x_i \left[\sigma^2(\mathcal{F} - x_i) - \frac{\theta}{2}\right] dt, \tag{9.5}$$

where x_i is the frequency of the ith allele, $\theta = 2Nu$ and $\mathcal{F}$ is the homozygosity of the population. As N increases, $\mathcal{F}$ decreases, which makes it harder for new alleles to enter the population. This property is not present in the two-allele approximation. There are various other models with drift coefficients similar to that of the TIM model. Among these are the symmetric overdominance and SAS-CFF models. The rate of substitution under these models also becomes nearly independent of population size for large N (Gillespie 1999, 2001).

Figure 9.4 illustrates the dependence of the rate of substitution in the TIM model on λ. The curve labeled "Hoc approximation" uses $k = \lambda E\{X\}$, where, once again, $E\{X\}$ is obtained from house-of-cards simulations with geometrically distributed generations of mean λ^{-1}. There is a marked transition from slow to fast dynamics when $\lambda \approx 0.01$, which corresponds to a change in the

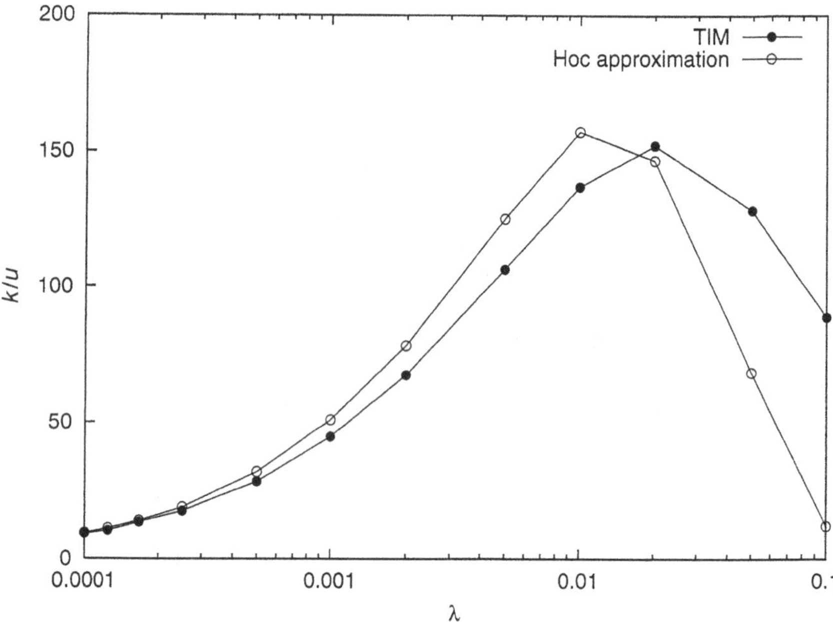

Figure 9.4. The rate of substitution as a function of λ for the TIM model and the house-of-cards approximation. The parameters are the same as for Figure 9.3, except for the values of λ.

environment every 100 generations, on average. As $\sigma = 0.1$, we can infer that when $\sigma/\lambda \approx 10$, the dynamics shift from those with isolated bursts to those where fitnesses change sufficiently fast that the burst model breaks down, taking $k = \lambda E\{X\}$ with it.

The fact that the house-of-cards approximation lies above the TIM curve is easily understood by appealing to Equation 9.5. The house-of-cards approximation is based on the assumption that $\mathcal{F} \approx 1$. In fact, as λ increases in the TIM simulations, so does $E\{\mathcal{F}\}$. Equation 9.5 shows that this will make it more difficult for new mutations to enter the population, which explains why k is lower for the TIM model than for the house-of-cards approximation.

9.5 Conclusions

A fundamental error is being made in calling the probability that a fixing mutation arises in a single generation the rate of fixation. That is, the commonplace

$$k = 2Nu \times 2s \qquad (9.6)$$

is flawed. If we knew s *a priori*, then it is certainly true that the probability of a fixing allele arising in the current generation is approximately $4Nus$. However,

Equation 9.6 implies that we know a lot more. It assumes the existence of an infinite sequence of positive selection coefficients, whose values are independent of the evolutionary history of the locus. Were $k = 4Nus$ not so ensconced in population genetics theory, such an assumption would likely be dismissed out of hand. Models of adaptive evolution that attempt to mimic a certain biological situation generally have the property that s decreases with each substitution, at least until the next change in the environment. This is true for Fisher's (1958) "geometrical model" and Orr's (1998, 1999) extensions of it. It is also true of house-of-cards models and the mutational landscape model. Equation 9.6 does not apply to these models.

For example, in the house-of-cards model, let the random variable representing the selection coefficient of the first allele be U_0 and that of the ith mutation derived from this allele be U_i. The most natural way to write Equation 9.6 in this case is

$$k = 2Nu \times 2 \times \frac{1}{2} E\{U_i - U_0 | U_i > U_0\} = 2Nu\sqrt{\frac{2\sigma^2}{\pi}}, \tag{9.7}$$

which is incorrect. The mean number of mutations entering the population in t generations is $2Nut$; the mean number of mutations whose fitness exceeds that of previous mutations grows as $\log(2Nut)$ (from the theory of records). The mean number of substitutions will be less than this, so the rate of substitution is

$$k = \lim_{t\to\infty} \frac{E\{S_t\}}{t} < \lim_{t\to\infty} \frac{\log(2Nut)}{t} = 0,$$

which is correct for sufficiently large populations. Thus, the application of the logic of Equation 9.6 is inappropriate for this model. By extension, it should be inappropriate for many other models as well.

One interpretation of Equation 9.6 is that it views evolution as *stateless*. That is, the probability that a fixing mutation arises in a particular generation may be determined by only knowing the parameters of the process; the state of the population is irrelevant. By contrast, the house-of-cards model and the simple stochastic substitution model and its descendants all use the state of the population to determine the probability of the appearance of a fixing mutation.

The amount of state information required to determine k depends on the details of the model. The episodic models in a constant or slowly changing environment with sufficiently small mutation rates only require the selection coefficient of the currently fixed allele, which enters through the calculation of $E\{X\}$ (λ is a parameter). By contrast, the TIM model with a rapidly changing environment requires knowledge of the complete state of the population. (This is not apparent in Equation 9.5, which only depends on $\mathcal{F}$, but is apparent when the diffusion coefficient is examined.) Knowledge of the state of the environment might be required as well. If the environment is autocorrelated, then the history of the population and the environment might be needed in order to describe its future.

Faced with these problems, it is tempting to accept an entirely different view: each of our models implicitly assumes that its substitution process is stationary. Thus, k is well defined as

$$k = \lim_{t\to\infty} \frac{S_t}{t}.$$

Given this, we could interpret s in $k = 4Nus$ as

$$s = \frac{k}{4Nu}.$$

But this is not particularly interesting or useful as we are, in fact, interested in the dependence of k on N and u. The dependence cannot be discovered without an analysis using, at the very least, the state of the population.

The representation

$$S_t = X_1 + X_2 + \cdots + X_{M_t}$$

captures a view of evolution that fits well with our notions of adaptive evolution. The rate equation,

$$k = \lambda E\{X\},$$

has a simple interpretation and its two components, the rate of change of the environment and the mean number of substitutions per change, are fundamental concepts in Darwinian evolution. Moreover, it leads to a more interesting spectrum of interpretations of sequence data. For example, under $k = 4Nus$, rate differences between lineages are likely to be attributed to differences in N, while under $k = \lambda E\{X\}$ they are likely to be attributed to different rates of change in the environment. The latter immediately suggests further experiments, which may well lead to a functional understanding of the evolution.

$\lambda E\{X\}$ also appears to be more compatible with the molecular clock than is $4Nus$. Many different models suggest that $E\{X\}$ may be small and insensitive to the details of models and, as a consequence, similar across species. λ is more likely to vary across species, though this variation is probably much less than the variation in population size. One could argue that, because most evolution must be in response to a species' biological rather than physical environment, and because the biological environments are themselves evolving, that λs are naturally constrained to be similar across species.

We should entertain the notion that the time scale of environmental change may be much shorter than the time scale of molecular evolution. If so, it necessitates abandonment of the representation of S_t given above. In fact, this view has a lot to recommend it. Our current state of knowledge does not allow us to speculate on the time scale of environmental change experienced by a locus. The issue is further complicated by the fact that the relevant environment for many loci is almost certainly determined by evolving epistatic interactions. Whatever the origin, the rate of substitution in a rapidly changing environment does not have properties that are compatible with $k = 4Nus$. In particular, rate of substitution is quite insensitive to both the population size and the mutation rate. This is a fertile area for future work. Many properties of the TIM model appear to become independent of N as $N \to \infty$. A clever use of boundary properties may well yield some interesting results for this model.

The analysis of this chapter suggests that we should abandon $k = 4Nus$ and begin a quest for a better descriptor of adaptive molecular evolution. While $k = \lambda E\{X\}$ has an obvious appeal, we cannot fully embrace it until we are willing to accept that the time scale of fitness change is longer than the time required to complete a burst of substitutions. Thus, a fundamental problem must be to discover the value of σ/λ. It is difficult to imagine any investigation of more importance. The fact that this question is seldom posed and that the route to its solution is so obscure is a statement about the deep chasm separating what we need to know and what we are able to discover. This is made more jarring because the role of population size, which is the focus of much of the discussion of sequence data, may be grossly overestimated. This is likely to be true for the sorts of selected mutations discussed here as well as for linked neutral mutations as described in Gillespie (2001).

9.6 Acknowledgments

This chapter has benefitted from the comments of Chuck Langley and Michael Turelli and from the editorial hand of Robin Gordon. The research reported here was funded in part by the NSF grant DEB-0089716.

REFERENCES

Arnold, B. C., Balakrishnan, N., and Nagaraja, H. N. (1998). *Records*. New York: John Wiley.

Fisher, R. A. (1958). *The Genetical Theory of Natural Selection*. New York: Dover.

Gerrish, P. J., and Lenski, R. E. (1998). The fate of competing beneficial mutations in an asexual population. *Genetica* 102/103:127–44.

Gillespie, J. H. (1982). A randomized SAS-CFF model of selection in a random environment. *Theor. Popul. Biol.* 21:219–37.

Gillespie, J. H. (1984). Molecular evolution over the mutational landscape. *Evolution* 38:1116–29.

Gillespie, J. H. (1991). *The Causes of Molecular Evolution*. New York: Oxford University Press.

Gillespie, J. H. (1993). Substitution processes in molecular evolution. I. Uniform and clustered substitutions in a haploid model. *Genetics* 134:971–81.

Gillespie, J. H. (1999). The role of population size in molecular evolution. *Theor. Popul. Biol.* 55:145–56.

Gillespie, J. H. (2001). Is the population size of a species relevant to its evolution? *Evolution* 55:2161–9.

Glick, N. (1978). Breaking records and breaking boards. *Am. Math. Monthly* 85:2–26.

Kimura, M. (1971). Theoretical foundations of population genetics at the molecular level. *Theor. Popul. Biol.* 2:174–208.

Kimura, M., and Ohta, T. (1971). Protein polymorphism as a phase of molecular evolution. *Nature* 229:467–9.

Ohta, T. and Tachida, H. (1990). Theoretical study of near neutrality. I. Heterozygosity and rate of mutant substitution. *Genetics* 126:219–29.

Orr, H. A. (1998). The population genetics of adaptation: the distribution of factors fixed during adaptive evolution. *Evolution* 52:935–49.

Orr, H. A. (1999). The evolutionary genetics of adaptation: a simulation study. *Gen. Res.* 74:207–14.

Orr, H. A. (2002). The population genetics of adaptation: the adaptation of DNA sequences. *Evolution* 56:1317–30.

Tachida, H. (1991). A study on a nearly neutral mutation model in finite populations. *Genetics* 128:183–92.

Tachida, H. (1996). Effects of the shape of distribution of mutant effect in nearly neutral mutation models. *J. Gen.* 75:33–48.

Takahata, N., Ishii, K., and Matsuda, H. (1975). Effect of temporal fluctuation of selection coefficient on gene frequency in a population. *Proc. Natl. Acad. Sci. USA* 72:4541–5.

Takahata, N., and Kimura, M. (1979). Genetic variability maintained in a finite population under mutation and autocorrelated random fluctuation of selection intensity. *Proc. Natl. Acad. Sci. USA* 76:5813–17.

Van Valen, L. (1974). Molecular evolution as predicted by natural selection. *J. Mol. Evol.* 3:98–101.

Wright, S. (1949). Adaptation and selection. In G. L. Jepson, G. G. Simpson, and E. Mayr (eds.). *Genetics, Paleontology, and Evolution*, pp. 365–89. Princeton, NJ: Princeton University Press.

10

Inferences about the structure and history of populations: coalescents and intraspecific phylogeography

JOHN WAKELEY
Department of Organismic and Evolutionary Biology, Harvard University, Cambridge

10.1 Introduction

Population geneticists and phylogeneticists view tree structures differently. To the phylogeneticist, tree structures are the objects of study and the branching patterns a tree displays are inherently significant. Phylogeneticists are interested in the relationships among species or other taxa, and these histories are tree-like structures. To the population geneticist, particularly to the student of coalescent theory, individual tree structures are usually not of interest. Instead attention is focused on the characteristics of populations or species, and intraspecific trees, or gene genealogies, are a stepping stone on the path to such knowledge. This difference in approach divides workers who study current and historical population structure into two groups: those who ascribe significance to single gene trees and those who focus on summary properties of gene trees over many loci. The purpose of this chapter is to give some perspective on this division and to suggest ways of identifying the domain of application of coalescents and intraspecific phylogeography in terms of the histories of populations or species. This is not meant to be divisive. In the not too distant future, we can hope that these complementary approaches will be unified, as models catch up with data and a science of population genomics is realized.

10.1.1 Population genetics history

Theoretical population genetics was born out of the tension between Biometricians (or Darwinians) and Mendelians in the early decades of last century. We often trace our field back to the famous paper of Fisher (1918) which settled this dispute; see Provine (1971). In short, the Biometricians, represented by W. F. R. Weldon and Karl Pearson, had for decades been measuring quantitative traits and considering such things as the correlation of traits between parents and offspring. They maintained that natural selection acted on these

The Evolution of Population Biology, ed. R. S. Singh and M. K. Uyenoyama. Published by Cambridge University Press.

continuous characters and that change in these was slow; discrete variation was unimportant to evolution. After the rediscovery of Mendel's laws in 1900, William Bateson, Hugo de Vries, and other Mendelians argued for the importance of discrete variations in evolution. Their views were directly opposed to those of the Biometricians; selection on continuous variation could not result in significant evolutionary steps, which were discontinuous. In hindsight we might say that the Biometricians' mistake was to confuse the continuity of traits with that of the underlying variation, and the Mendelians' error was to equate the mechanism of inheritance with that of evolution itself. In any case, it is clear that the two camps agreed only on one point: continuous variation and Mendelian inheritance were incompatible.

This fundamental conflict was resolved mathematically by Fisher (1918). Specifically, Fisher showed that continuous variation could be explained by the action of many Mendelian loci of small effect. In the decade or so after this remarkable start, the major results of this new branch of science, which was called theoretical population genetics, were laid down by Fisher (1930), Haldane (1932), and Wright (1931). Following the birth of theoretical population genetics, the mathematical theory was extended and the facts of genetics were reconciled with Darwin's theory of evolution. During the Modern Synthesis, these avenues of research were merged into the neo-Darwinian theory of evolution, providing a series of well-justified, more or less qualitative explanations of patterns of speciation, adaptation, and geographic variation. Two of the major architects of the Modern Synthesis were Dobzhansky (1937) and Mayr (1942). Our modern understanding of evolution is grounded in neo-Darwinism. During the next few decades, many workers contributed to the theory, although Malécot (1948) and Kimura (1955a,b) certainly stand out. By 1960 the mathematical theory of population genetics had developed a very high degree of sophistication, although for the most part, as Lewontin (1974) notes, this was in the absence of genetic data.

It wasn't until the mid 1960s that population genetics finally confronted genetic data (Harris 1966, Lewontin and Hubby 1966). Since then, we have seen a grand shift in population genetics from the forward-looking view of the classical theory of Fisher, Haldane, and Wright to the backward-looking view of the coalescent or genealogical approach; see Ewens (1990) for a review of this transformation. The modern approach focuses on inferences from samples of genetic data and, often to great advantage, recasts theoretical problems in terms of genealogies. Significant works along the path to this include Ewens (1972), which describes the distribution of the counts of alleles in a moderate-sized sample from a large population, and Watterson (1975), which describes the distribution of the number of polymorphic nucleotide sites in either a moderate or a large sample from a large population. The retrospective approach came fully to life in the early 1980s with the introduction of the coalescent process by Kingman (1982a,b,c), Hudson (1983b), and Tajima (1983). The present relative lack of concern for the structures of particular gene genealogies traces back to the constant-size, single-population, neutral coalescent model described in these works, which is discussed in detail in Section 10.2 below.

10.1.2 Phylogenetics and intraspecific phylogeography

Charles Darwin's famous book contains just one figure: a hypothetical phylogenetic tree. Long before Darwin (1859) and Wallace (1858) put forward the idea of descent with modification, biologists had employed trees to depict the relationships among species and higher taxa. Tree structures are a natural way to represent such affinities, which are groups nested within other groups. Prior to Darwin and Wallace, however, trees had been employed strictly as convenient organizational tools to represent systematic affinities. For example, the classification system put forward by Linnaeus (1735) is a branching structure which delineates relationships, yet Linnaeus rejected the idea of evolution. When the idea of descent with modification gained acceptance as the explanation for biological diversity, these tree structures gained a new significance. They were no longer an expedient, but rather represented the actual histories of groups of species. The development of phylogenetics since Darwin and Wallace has been strongly influenced by the concept of trees as history. In addition phylogenetic theory and methodology have been shaped by the evolutionary idea that descendant species which trace back to a common ancestor will inherit any unique characteristics that ancestral species had evolved.

Until the last 30 years or so, the role of theory in phylogenetics and in population genetics could not have been more different. Although there is now a lot of overlap of approach, historical differences do persist. Theoretical population genetics has always been firmly grounded in traditional applied mathematics and probability theory. In this sense population genetics has many parallels with physics. The theoretical framework is mathematical and statistical, and there is broad acceptance of this framework and its attendant models within the field of biology.

In contrast, within the field of phylogenetics there has been widespread skepticism of such approaches, particularly statistical ones. This is most evident in the cladistic approach, which practitioners credit to Hennig (1965, 1966). This approach seeks to identify the phylogenetic tree which disagrees the least with the data at hand. The criterion for it is parsimony: pick the tree that requires the fewest character state changes. The tree is then considered a potentially true statement about history. It is a phylogenetic hypothesis which predicts what further study should uncover and which thus may be shown to be false. It is not viewed as an estimate of some unknown quantity. This approach is understandable if Hennig's view is accepted: that the phylogeneticist can directly observe (the results of) history through careful study of the morphology and development of a group of organisms, by identifying shared, uniquely derived characters, or synapomorphies. Sound arguments against the cladistic approach have been made in response to seeing the blind application of the parsimony method to data which have not been subject to the careful prior study Hennig envisioned, and which are more labile than complex morphological features. Thus, with the introduction of model-based approaches, like the maximum likelihood method of Felsenstein (1981), the recent history of phylogenetics has been a progressive acceptance of the mathematical and statistical theory. However, this process of acceptance is still ongoing.

Coincident with the emergence of the backward-looking, genealogical approach to population genetics, phylogenetic methods began to be applied to intraspecific data. This was greatly facilitated by the nonrecombining nature of the first molecule examined – animal mitochondrial (mt) DNA – and the growing technical ability during the 1970s and 1980s to assay samples of mtDNA from natural populations. The result was a new and active subfield of evolutionary biology called intraspecific phylogeography, or just phylogeography (Avise *et al.* 1987, Avise 1989, 2000). A number of new methods of historical inference have resulted from this approach (Neigel *et al.* 1991, Neigel and Avise 1993, Templeton *et al.* 1995, Templeton 1998). The hallmark of phylogeography is that inferences are drawn from intraspecies or organismal gene trees which are reconstructed from data. The focus on gene trees as indicators of population structure, population history, and speciation has provided a much needed bridge between phylogenetics and population genetics (Hey 1994, Avise 2000). However, there is still a gulf between workers schooled in population genetics and those who favor traditional phylogenetics or cladistics. Bluntly put, the latter group tends to place too much emphasis on single gene genealogies whereas the former group places too little. Drawing conclusions from single genealogies can be problematic because each is only a single point in the space of all possible genealogies. Under some kinds of population histories, this will cause serious errors in inference. Conversely, focusing too much on the standard, structure-less, history-less coalescent model gives a picture of the utility of single gene trees that is too discouraging.

10.2 Gene genealogies and the coalescent

In the early 1980s, the ancestral process known as the coalescent was described. Kingman (1982a,b,c) provided a mathematical proof of the result. Hudson (1983b) and Tajima (1983) introduced this genealogical approach to population geneticists and derived many biologically relevant results. Nordborg (2001) provides a recent review; see also Hudson (1990) and Donnelly and Tavaré (1995). Kingman found a simple ancestral process to hold for samples from a wide variety of different types of populations, in the limit of large population size and providing that the genetic lineages in the population are exchangeable (Cannings 1974). Exchangeable lineages are ones whose predicted properties are unchanged if they are relabeled or permuted (Kingman

1982b, Aldous 1985). With the assumption that all variation is neutral, the familiar Wright–Fisher model (Fisher 1930, Wright 1931) of a population with nonoverlapping generations fits this criterion, as does the overlapping generation model of Moran (1958). Different populations will differ in how the actual population size is related to the effective population size that determines the rate of the coalescent process. The standard coalescent involves two very important assumptions besides exchangeability. For this model to hold, the population must be of constant effective size over time and there must be no population subdivision.

When time is measured in units of $2N_e$ generations for a population of diploid organisms, or in units of N_e generations for a population of haploid organisms, the time to a coalescent event is exponentially distributed with mean

$$E(t_k) = \frac{2}{k(k-1)} \tag{10.1}$$

where k is the number of ancestral lineages present. Under the coalescent model, each of the $\binom{k}{2}$ possible pairs of lineages coalesces with rate 1. Without recombination, which will be treated later, a sample of size n will go through exactly $n-1$ coalescent events to reach the common ancestor of the entire sample. Thus, every genealogy has $n-1$ coalescent intervals, beginning with the most recent, $k = n$, and ending with the most ancient, $k = 2$. Figure 10.1 shows an average coalescent genealogy; that is, with the lengths of the coalescent intervals drawn in proportion to Equation 10.1. The more recent coalescent intervals tend to be much shorter than the ancient ones, and on average the final coalescent interval represents more than half of the total time from the present back to the most recent common ancestor of the sample. Because the time scale of the coalescent process depends inversely on N_e, we expect genealogies to be longer when the effective size is larger.

As we trace the ancestry of the lineages back in time, because each pair that exists has the same rate of coalescence, when a common ancestor event happens each pair is equally likely to be the one that coalesces. The structure of trees under the coalescent is determined by this process of joining random pairs of lineages. The result is, if we think in forward time for the moment starting at the root of the tree, a random-bifurcating tree topology. This results from the fact that there is no structure to the coalescent process – that all lineages are exchangeable – and the resulting trees are likewise unstructured. Without intralocus recombination, all the sites at a single genetic locus will share the same genealogy. Loci that segregate independently of each other will have uncorrelated genealogies, both in terms of the coalescent times and topology. Considering topological structure, if we took a sample of three items, and labeled them A, B, and C, then each of the three possible rooted tree topologies – $((A, B), C)$, $((A, C), B)$, $((B, C), A)$ – is equally likely to occur. If

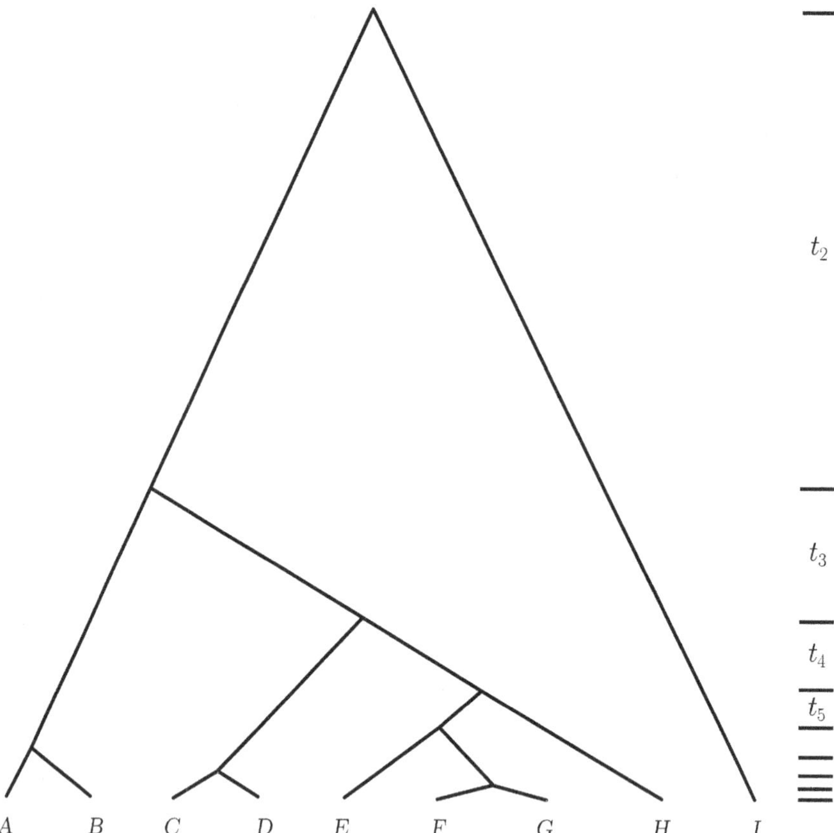

Figure 10.1. A hypothetical coalescent genealogy of a sample of size $N = 9$. The lengths of the coalescent intervals, t_n through t_2, are drawn in proportion to their expected values given by Equation 10.1.

we take a large sample of independently segregating loci, we expect to observe equal numbers of each of these three trees.

10.3 The axes of genealogical variation: tree size and branching pattern

As a starting point in talking about demographic history, we can take the standard, coalescent process as a null model. The underlying, exchangeable population genetic models, such as the Wright–Fisher model, are familiar to most biologists and their use as null models is not uncommon. This establishes predictions for what we should observe in a sample of sequences from a population. With reference to the discussion of the coalescent above, we are interested in two kinds of genealogical variation: (1) variation in the total length of the tree, and (2) variation in the branching pattern. The length of a genealogy is the sum of the lengths of all its branches. Under the standard coalescent model, this is given by the sum of $n - 1$, independent exponential times with different parameters. We expect this distribution to be realized when a large number of independent loci are sampled. The branching pattern of a genealogy specifies $2n - 3$ partitions of the n sampled sequences, tips, or leaves of the tree. That is, each branch in the genealogy divides the members of the sample into two groups, the ones on either side of the branch. The genealogy or branching pattern at each sampled locus will be a random draw from the rather large universe of all possible random-bifurcating trees.

It is very important to note that our ability to observe the length and topology of genealogies is mediated by mutation. Even without any variation, genealogies will come in different sizes and shapes; we just won't know it. We rely on mutations occurring along the branches of the tree to produce the sequence polymorphisms that provide clues about history. The rate of mutation per locus is typically very small, somewhere around 10^{-4} to 10^{-6} per generation, and mutation events in different generations are independent. Therefore, the number of mutations that occur along a genetic lineage of length t will be Poisson distributed with expectation tu, where u is the mutation rate per generation. When time is rescaled as in the coalescent, this becomes $T\theta/2$, where $T = t/(2N_e)$ and $\theta = 4N_e u$. In the standard coalescent model, the parameter θ is equal to the expected number of nucleotide differences between two randomly chosen gene copies. The randomness of the mutation process is an important factor in determining among-locus variation in the observable indicator of tree length: the number of polymorphic sites in the sample. The letter S is used to denote the number of these segregating sites in a sample. Even when the genealogies at different loci are all identical in size there will be Poisson variation around the expectation due to the randomness of the mutation process. This imposes a lower bound on the variation in S among loci, namely that the variance will be equal to the mean.

Our ability to uncover genealogical topology also depends on mutation. We become aware of particular branches in the tree when mutations occur on them. When the mutation rate at each nucleotide site at a genetic locus is small, and recombination is absent or very unlikely, the infinite-sites mutation model of Watterson (1975) is a good approximation to the mutation process. Under this model, each time a new mutation occurs, it happens at a previously unmutated site. The assumption of no recombination guarantees that all sites in a sample of DNA sequences will share the same bifurcating topology, but this is not the most important aspect of Watterson's (1975) model. If each site mutates at most once in the history of the sample, then each polymorphism is the result of a single mutation event on some branch in the tree, and the partitions of the sample made by the branch and by the polymorphism are identical. Correlation in genealogical topologies among loci will be represented in sequence data by the repetition of such site frequency patterns at many loci.

10.4 The effects of population structure and population history on genealogies

This section describes the effects on the size and shape of genealogies of deviations from the assumptions of the standard coalescent model, particularly changes in effective size over time and two kinds of population subdivision. These effects are summarized in Figure 10.2. The thin lines in the figure represent population boundaries, and thin, dashed lines indicate incomplete barriers to the movement of individuals. The genealogies of these samples of size four are drawn using thick lines. For each historical scenario, (a) through (d), hypothetical genealogies are shown for samples from two independently segregating loci. This illustrates the effects of population structure and population history on the sizes and shapes of genealogies. Note that "shape" here refers only to topological structure and not to the relative lengths of different parts of a tree. In brief, changes in population size through time change the distribution of tree sizes by making the coalescence rate time dependent, but do not affect the topology of trees. Population subdivision alters the distribution of tree lengths, but it also can have dramatic effects on the shape of trees because it makes some common ancestor events much more likely than others.

10.4.1 Population growth

If one population is twice as big as another, the former has one half the rate of coalescence as the latter. On average, trees will be twice as big in the larger population as in the smaller one. When a single population has grown in size, the rate of coalescence responds proportionately. Looking back in time, the rate of coalescence will be low until the time of growth, then it will increase. The predictions of the standard coalescent for the relative sizes of ancient and recent coalescent intervals pictured in Figure 10.1 will no longer hold. Instead, the more recent intervals will be relatively longer and the more ancient intervals will be relatively shorter. If growth is rapid and relatively recent, genealogies will tend to be star shaped, that is, to have small internal branches (Slatkin and Hudson 1991). Population growth by itself will not alter the probabilities of genealogical topologies, because when a coalescent event occurs each pair of lineages still has an equal chance of being the one that coalesces.

If population growth is rapid enough, it is well approximated by a single abrupt change in population size. In this case, the ancestral process has two additional parameters: T_C, the time of change in population size measured in units of $2N_e$ (current effective size) generations, and $Q = N_{eA}/N_e$, the ratio of the ancestral and current effective population sizes. Between the present and time T_C, each pair of lineages coalesces with rate equal to one, whereas before T_C the rate is Q per pair of lineages. Of course, this model also describes

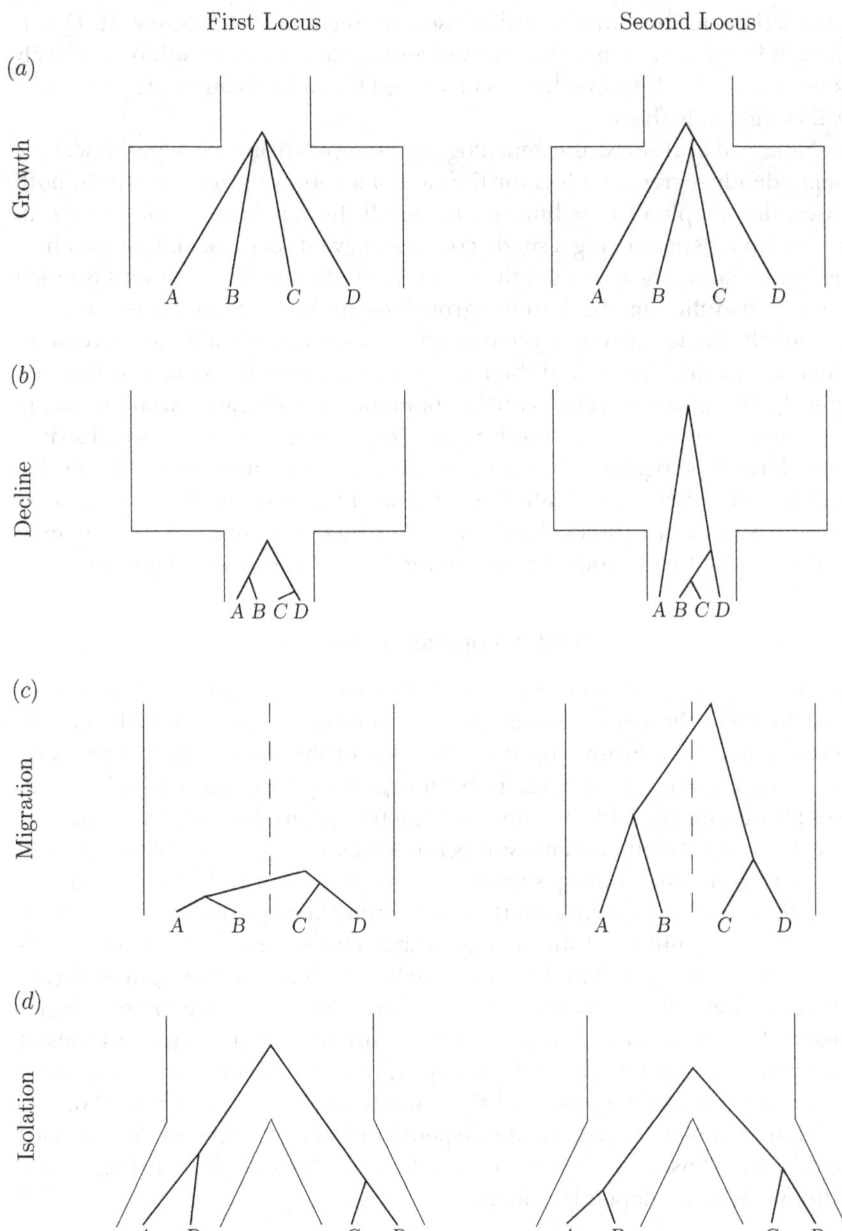

Figure 10.2. Two hypothetical genealogies for a sample of size four at two independently segregating loci under the four population models discussed in the text: (*a*) population growth, (*b*) population decline, (*c*) equilibrium migration, and (*d*) isolation without gene flow. Thin lines indicate population boundaries.

population decline, which is discussed in Section 10.4.2 below. If $Q < 1$, growth has occurred and the more recent coalescent times will be relatively long, and if $Q > 1$ decline has occurred and the most recent coalescent times will be relatively short.

Figure 10.2(a) shows the genealogies of samples from two hypothetical, independently segregating loci for the case of an abrupt growth event. In both cases, the sample of four lineages traces all the way back to the change in size without experiencing a single coalescent event. Because the recent effective size is large, the expected time back to the first coalescent event is much greater than the time back to the growth event. When the lineages arrive in the much smaller ancestral population, they experience a great increase in the rate of coalescence, and the common ancestor of the sample is reached quickly. Therefore, most trees will be about the same size, and variation among them will be much less than in Kingman's coalescent. However, the distribution of tree topologies will be the same as in the standard, constant-size model, and trees at different loci will differ in branching pattern. Thus, the two genealogies in Figure 10.2(a) have different structures. On the left, samples *A* and *B* are the first to coalesce, on the right it is *B* and *C* which are first.

10.4.2 Population decline

Turning rapid growth on its head, we have the case of rapid decline in Figure 10.2(b). Here there will be a relatively higher rate of coalescence during the recent part of the history, up until the time of the decline in effective size. As above, the event is assumed to be abrupt, simply for ease of explanation. Samples at some loci, like the one on the left in Figure 10.2(b), will trace back to a most recent common ancestor before reaching the event. These trees will be short. If multiple lineages trace their ancestry back to the decline in size, then the rate of coalescence for those remaining lineages decreases in proportion to the magnitude of the change in size. The ancient coalescent intervals will be much elongated in this case, which is depicted on the right in Figure 10.2(b). Therefore, there will be a lot of variation in the size of genealogies among loci, more than in the standard, constant-N_e coalescent. In terms of tree structure, again because the lineages are exchangeable, genealogies will be random-bifurcating trees and there will be the same very low level of correlation in branching pattern at independent loci that is seen in the standard coalescent. Thus, as in Figure 10.2(a), the genealogies in Figure 10.2(b) are different at two independent loci.

10.4.3 Equilibrium migration

Population subdivision introduces structure to genealogies, structure that may correlate with geography, and causes the tree topologies at different loci to be correlated. Subdivision will also affect variation in the sizes of genealogies among loci, but the direction of this effect depends on whether migration can occur among subpopulations or demes, as this section supposes, or not, as in Section 10.4.4 below. For simplicity, assume that a population is subdivided into D demes and conforms to the symmetric island model of Wright (1931). The demes are of equal size, N, and the fraction of each deme that is replaced by migrants each generation is the same and equal to m. This is by far the most commonly employed model of a subdivided population in both empirical and theoretical studies. The term equilibrium migration refers to the fact that this constant-rate migration is supposed to have been ongoing for long enough that the effects of any prior history are erased. In Wright's island model, migrants are equally likely to come from any deme in the population. Thus, this model does not include explicit geography. Populations that adhere to the assumptions of the island model will not display the correlation between geography and genetic variation known as isolation by distance (Wright 1943). They will show different levels of polymorphism within vs. between demes, and powerful nonparametric tests to detect subdivision have been developed (Hudson *et al.* 1992). In the case of just two populations, the island model can be considered an explicit model of geography. This simple case is considered here in order to illustrate the effects of equilibrium migration on genealogies.

The parameters that determine the pattern of genetic variation in a sample of n_1 sequences from one deme and n_2 sequences from another are θ and $M = 4Nm$. If π_w and π_b are the average number of pairwise nucleotide differences within and between populations, respectively, then for the D-deme island model we have

$$E(\pi_w) = D\theta, \tag{10.2}$$

$$E(\pi_b) = D\theta\left(1 + \frac{1}{2M}\right) \tag{10.3}$$

(Li 1976). For the two-deme model, we put $D = 2$ in Equations 10.2 and 10.3. There are two surprising aspects of these equations. First, the expected value of π_w does not depend on the rate of migration (Slatkin 1987, Strobeck 1987). This is a special property of the symmetric island model: the tendencies of within-deme pairwise coalescence times to be short if neither of the pair is a migrant and to be long if one of them is a migrant average out perfectly to give Equation 10.2. If any asymmetries are introduced into the model, this result no longer holds. Second, the effect of subdivision depends on the product of the deme size and the migration rate, which is captured in the scaled migration rate M. As M grows large, the expectation of π_b converges on that of π_w, and the population will appear panmictic. This surprising result traces back to Wright (1931), and explains why populations that are obviously not panmictic sometimes show no evidence of subdivision. That is, M can be large even when the per-generation rate of migration, m, is small. Equations for the variances of π_w and π_b both within and among loci can be found (Wakeley 1996a,b), and these both depend on the scaled migration rate. When M is large, the variances become those expected in a panmictic population, and as M decreases the variances of pairwise differences grow.

The predictions of Equations 10.2 and 10.3 can be extended to levels of polymorphism in larger samples: under equilibrium migration, levels of genetic variation will be larger on average for multi-deme samples than for single-deme samples. The effect of this will be greater when M is small. In the sample (n_1, n_2) from two demes, coalescent times among the n_1 sequences from deme one, and among the n_2 sequences from deme two, will tend to be shorter than coalescent times between sequences from different demes. This means that the topological structure of genealogies will no longer be the random-bifurcating trees predicted by the standard coalescent. There will be a tendency towards trees which have a branch that divides the sample exactly into the n_1 and n_2 sequences taken from each deme, for example trees in which the demic samples are reciprocally monophyletic. Again, this tendency will be more pronounced if the scaled migration rate between the two demes is small. Thus, the genealogies for two independent loci on the right and left of Figure 10.2(c) both show this kind of topology. In addition, variation in levels of polymorphism among loci will depend inversely on the scaled migration rate, M; for example, see Hey (1991). So, for the same average rate of polymorphisms under equilibrium migration, some loci will have very short and some very long histories. This is also displayed in Figure 10.2(c).

10.4.4 Isolation without gene flow

Equilibrium migration is just one of a multitude of possible explanations for the occurrence of subdivision. In fact, it is probably uncommon for a population to remain stably subdivided, both in the sizes of demes and in the rates and patterns of migration, for long enough to reach equilibrium. One of the earliest tenets to emerge from phylogeographic studies is that most species appear to have experienced dramatic shifts in demography over time and space (Avise 1989). Confining ourselves for the moment to models with discrete demes, the polar opposite of equilibrium migration is isolation and divergence without genetic exchange. This isolation model posits an ancestral population that splits into two descendant populations at some time, T_D, in the past and after that time the two populations do not exchange migrants. The isolation model can be compared with the migration model in Section 10.4.3 to illustrate the striking differences between equilibrium and nonequilibrium population subdivision.

In general, each population in the isolation model might be of a different size, and we would have $\theta_1 = 4N_1u$, $\theta_2 = 4N_2u$, and $\theta_A = 4N_Au$ as parameters (Wakeley and Hey 1997). However, for purposes of comparison with the equilibrium migration model of Section 10.4.3, we assume that $\theta_1 = \theta_2 = \theta_A$. In this case, the average numbers of pairwise differences within and between demes have expected values

$$E(\pi_w) = \theta \tag{10.4}$$

$$E(\pi_b) = \theta(1 + T_D) \tag{10.5}$$

(Li 1977). Aside from a constant scaling factor (D), equilibrium migration and isolation without gene flow make identical predictions about average levels of genetic variation within and between demes where $T_D = 1/(2M)$. In other words, if π_w and π_b are measured from data, then both models could be fit and their parameters estimated, but π_w and π_b would not serve to distinguish between migration and isolation. The most obvious difference between the two models is in the interpretation of the pattern of polymorphism. Under the isolation model, genetic variation between demes in a sample is a snapshot for a particular T_D. If the population were sampled again at a later date, $T_D + T$, the level of divergence would be greater. Equation 10.3, in contrast, holds for all time, and represents a dynamic balance achieved between ongoing genetic drift and migration.

In addition to this difference in interpretation, variation in levels of genetic variation among loci will be different under migration and isolation even when the average levels are the same (Li 1976, 1977, Takahata and Nei 1985, Wakeley 1996a). The variances are larger under migration than under isolation, and the difference grows with $T_D = 1/(2M)$. This results from the fact that under migration, coalescent events between samples from different demes can occur at any time, mediated by migration, whereas under isolation there can be no interdeme coalescent events until the lineages trace back into the ancestral population. In the extreme of a very long divergence time in the isolation model ($T_D \gg 1$), difference between $E(\pi_b)$ and θT_D will be negligible. In this case the distribution of the number of segregating sites among loci will approach a Poisson distribution, with mean and variance equal to θT_D. In contrast, in the extreme of a very low migration rate in the migration model, the variance of the number of segregating sites among loci will be much greater than the mean (Wakeley 1996a). Thus, the trees for two independent loci under isolation in Figure 10.2(d) are more similar in size than those shown in Figure 10.2(c) for migration. Equilibrium migration and isolation without gene flow share the prediction that genealogical trees will tend towards reciprocal monophyly, and this is also displayed in Figure 10.2(d).

10.5 Domains of application: coalescents and phylogeography

The above discussion illustrates some general principles about the effect of population structure and population history on the sizes and shapes of genealogies. To summarize:

1. population growth/decline tends to decrease/increase variation in tree size among loci but does not affect variation in tree shape relative to the standard coalescent model,
2. both equilibrium and nonequilibrium population subdivision (migration vs. isolation above) alter the structure of genealogies such that genealogies at independently segregating loci will tend to share topological features, and
3. migration increases variation in tree size among loci whereas isolation decreases it.

This section investigates how the strengths of these trends depend on the parameters of a population. The goal is to identify population histories for which the analysis of single gene genealogies is likely to be fruitful and those for which it will be less useful to refer to any specific genealogy. Simulations are used to determine the distribution of tree size and shape among loci. The parameters are those discussed above in Section 10.4 and the quantities used to measure variation in the size and shape of genealogies are described below.

10.5.1 Measures of variation in tree size

The most straightforward measure of the size of a genealogy is the number of segregating sites, S. A sample from any population will have some expected value of S and some variance. For example, in the case of a sample of n sequences under the standard, constant size, unstructured coalescent with infinite-sites mutation,

$$E(S) = \theta \sum_{i=1}^{n-1} \frac{1}{i} \tag{10.6}$$

$$V(S) = \theta \sum_{i=1}^{n-1} \frac{1}{i} + \theta^2 \sum_{i=1}^{n-1} \frac{1}{i^2} \tag{10.7}$$

(Watterson 1975). When we sample a large number of loci, we should find that the mean and variance among them would conform to Equations 10.6 and 10.7. This, of course, assumes that the sample size, n, and the mutation parameter, θ, are the same at every locus. However, this assumption is made only as a matter of convenience in comparing different population structures and histories below; it would be straightforward to allow for differences in θ and n among loci.

There are many ways in which we could compare levels of variation in S, our measure of tree size, among loci. The standardized measure,

$$\Omega = \frac{\widehat{V(S)} - \bar{S}}{\widehat{V(S)}} \tag{10.8}$$

will be used here, in which $\bar{S}$ is the average number of segregating sites and $\widehat{V(S)}$ is the observed variance of S among loci. Given a multilocus data set, Ω

is easy to compute. The expectation of Ω is given approximately by

$$E(\Omega) \approx \frac{V(S) - E(S)}{V(S)} \tag{10.9}$$

The number of segregating sites, S, is a compound random variable (see Section 10.3). Thus we can intuitively partition $V(S)$ into contributions due (1) to variation in tree size and (2) to variation in the mutation process. If there is no variation in the size of genealogies among loci, then all of the variation in S will be due to the Poisson mutation process and the expected value of Ω will be zero. Instead, if the variation in tree size among loci is much greater than the mean, then $V(S)$ will be large and Ω will be close to its upper bound of one. Thus, Ω is a normalized measure which can be compared under different assumptions about the population. Our null model, the standard coalescent, predicts a fairly high value of Ω, depending of course on θ and n. If $\theta = 10$ and $n = 20$, which are the values used in simulations below, Equation 10.9 gives $E(\Omega) = 0.82$.

10.5.2 Measures of correlation in branching pattern

There is also a multitude of ways we could compare genealogical topologies among loci. If we knew the true trees or if we were very confident about our trees reconstructed from data, then we could use a tree comparison metric like that of Robinson and Foulds (1981). Alternatively, if we are not confident about our reconstructed trees or do not wish to make explicit reference to them, we could use some measure of the correlation in haplotype patterns among loci such as coefficient of linkage disequilibrium (Lewontin and Kojima 1960). This measures gametic associations between alleles at two loci, but multilocus statistics are also possible (Smouse 1974). Here, because of the focus on simple two-deme models of subdivision, we will instead consider the co-occurrence of identical data partitions among loci, that is the observation of identical patterns of polymorphism among members of the sample at several loci. This presupposes that the same individuals were assayed at all genetic loci.

Assuming that the infinite-sites mutation model holds, each polymorphic site in a sample divides the members of the sample into two groups, ones which retain the ancestral base at the site and ones which have inherited the mutant base. As noted in Section 10.3, the one-to-one correspondence between mutation events and polymorphic sites in the sample, and the observation of a pattern in the data guarantee the existence of a branch in the genealogy of the sample, one that divides the sample exactly as the polymorphism does. For example, a mutation event on the shortest internal branch in the genealogy in Figure 10.1, the one which exists only during t_5, would make a polymorphic site at which samples E, F, and G would show the mutant base and samples A, B, C, D, H, and I would show the ancestral base.

In the standard coalescent model, we would not expect to see this pattern repeated at another, independent locus sequenced in the same individuals because the fraction of random-bifurcating trees that contain such a branch is very small. However, all genealogies contain n external branches, on which singleton polymorphisms can arise, so we would expect to see these partitions, i.e., all n kinds of singletons, repeated at many loci. Thus, there is a negative correlation between the allele frequency at a polymorphic site and the chance that the same pattern will be found at other loci.

In a sample from a subdivided population, we expect sites which divide the sequences along deme-sample lines to tend to be repeated at multiple loci. There might be a fairly low overall concordance of whole tree topologies among loci, because of the variability of within-deme patterns of common ancestry, but some branches would tend to be repeated. For the simple two-deme models considered here, these repeated branches will be the ones that divide the sample into the n_1 and n_2 sequences sampled from demes one and two. A statistic that will be sensitive to the co-occurrence of single partitions across loci is $max(p_i)$, in which p_i is the fraction of loci that show at least one polymorphic site with partition i. Singleton partitions are excluded in the calculation of $max(p_i)$ because all loci are expected to show these regardless of population structure and history. This measure will be sensitive to the effects of subdivision as it is modeled here. As the level of subdivision increases, the partition most frequently observed across loci will be the one that corresponds exactly to the two demes' samples, and $max(p_i)$ will approach one. We take the null distribution of $max(p_i)$ to be that found under the standard coalescent. This will depend on the sample size and on θ. For $\theta = 10$ and $n = 20$, used in the simulations below, the standard coalescent gives $max(p_i) \approx 0.04$.

10.5.3 Simulations of population structure and population history

The usual coalescent simulations were performed (Hudson 1990), adding a change in size, cf. Hudson (1990), or migration/isolation, cf. Wakeley (1996b), as indicated. The statistics Ω and $max(p_i)$ were computed for each simulation replicate. In addition to simulations under the standard coalescent model, a small set of parameter values was chosen to illustrate the effects of population structure and population history on the joint distribution of Ω and $max(p_i)$. The sample size was $n = 20$ when there was no structure, and $n_1 = n_2 = 10$ under migration and isolation. Only one case each of growth and decline is presented: ($\theta = 100.0$, $Q = 0.01$, $T_C = 0.1$) and ($\theta = 0.25$, $Q = 100.0$, $T_C = 0.1$). These were selected to represent extreme growth and extreme decline respectively, and the values of θ were chosen so that the average number of polymorphic sites per locus would be the same under both models. Several levels of subdivision were investigated for equilibrium migration and isolation without gene flow. Under migration these were $M = 0.5, 0.25, 0.01$ with $\theta = 5.0$, and under isolation they were

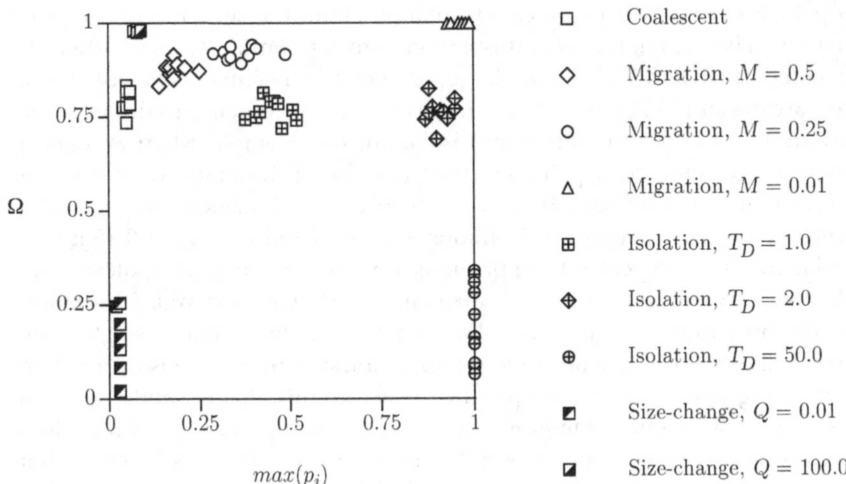

Figure 10.3. The results of the simulations described in the text. Each point in the scatterplot is the pair of $(max(p_i), \Omega)$ values for a single simulation replicate.

$T_D = 1.0, 2.0, 50.0$ with $\theta = 10.0$. These parameter sets were chosen in consideration of Equations 10.2 through 10.5, so that the expected numbers of pairwise differences within and between the two demes would be equivalent in the two models for three different levels of differentiation. One hundred independent loci were surveyed in the sampled individuals.

The results are shown in Figure 10.3. Only ten simulation replicates were performed for each set of parameters, as this was enough to distinguish the cases, and the results of all replicates are plotted in Figure 10.3. Simulations under the standard coalescent model cluster around the values $\Omega = 0.82$ and $max(p_i) \approx 0.04$ mentioned above. Under population growth and decline, the value of $max(p_i)$ is nearly unchanged from the constant-size case, but the value of Ω changes drastically. This accords well with the discussion in Section 10.4 above. The minor differences in $max(p_i)$ between these and the standard coalescent result from the fact that singleton polymorphisms are ignored in computing $max(p_i)$, and there are a lot more singletons under population growth than under population decline. This is essentially the same as the mutation rate effect on Ω that can be seen for the standard coalescent from Equations 10.6 and 10.7; as θ grows, so does the expected value of Ω. In sum, under this model of dramatic growth we expect the size of even a single genealogy to accurately represent the history of the population but, because there is no structure to the population, the topology of the tree contains little or no information about historical demography. Under decline, neither the size nor the shape of a single genealogy will be informative about history.

Subdivided populations vary both in Ω and in $max(p_i)$. Under both equilibrium and nonequilibrium subdivision, the repetition of genealogical topologies across loci provides information about the structure of the population. That is, migration and isolation converge on $max(p_i) = 1$ when M becomes small and T_D becomes large, respectively. Two interesting aspects of this are evident in Figure 10.3. First, the rates of convergence to this extreme are different under migration and isolation. For example, when we expect the average number of pairwise differences between demes to be twice as big as that within demes ($M = 0.5$ or $T_D = 1.0$; see Equations 10.2 to 10.5), simulations give $max(p_i) \approx 0.18$ under migration and $max(p_i) \approx 0.45$ under isolation. This is expected from previous work on genealogical topologies under the two models (Tajima 1983, Takahata and Slatkin 1990, Wakeley 1996b). In the present context it means that, other things being equal, single gene trees will be more informative about population structure under isolation than under migration. The second point is related to this; that is, subdivision has to be quite strong under migration for $max(p_i)$ to approach one. Even when the average number of pairwise differences between demes is 50 times that within demes, about four out of 100 loci will not show the (n_1, n_2) partition that defines the samples. That equilibrium migration is a highly variable process can also be seen in values for Ω, which approach one as M decreases. In contrast, as T_D increases between two isolated demes, Ω decreases, but a very long divergence time is required for Ω to be close to zero.

The measures Ω and $max(p_i)$ appear to distinguish well among the models. In addition, they serve to illustrate how single gene trees might or might not be representative of population structure and population history in terms of the parameters of the models. The broad empty area of Figure 10.3, for lower values of Ω and intermediate values of $max(p_i)$, is an artifact of the simplicity of the models considered here. Populations that follow the isolation model but have a small value of θ_A relative to θ_1 and θ_2 can produce values in this range.

10.6 Conclusions

While reconstructing a genealogy is not a necessary step in population genetic inference, it can be quite informative under some circumstances. There is a difference of approach in this regard between workers who use coalescent techniques and those who practise intraspecific phylogeography. While this dichotomy is far from complete, it is real enough. Coalescent technicians do not usually make reference to particular gene trees. This is part of the culture of coalescents: that gene trees are unobservable random quantities which certainly shape genetic variation but whose branching patterns do not contain much information about population history. This view is most reasonable when populations conform to the standard coalescent model. When trees are referred to explicitly, it is typical to "integrate" over them in making inferences (Kuhner *et al.* 1995, Grifiths and Tavaré 1996). In contrast, the first step in a phylogeographic analysis is to reconstruct a gene tree from data, and inferences are based upon this inferred tree. This sensibility about the significance of inferred trees was received and adapted from the field of phylogenetics. At the intraspecific level, roughly speaking, the circumstances favorable to using inferred gene trees are those in which random genetic drift is relatively unimportant compared with nonequilibrium factors like the splitting of populations.

Only the simplest nonequilibrium model was considered here: a single population that split into two isolated demes at some time in the past. This kind of history has the qualities necessary for the single-tree approach to be most fruitful; that is, small Ω and large $max(p_i)$. However, most of the branches in the genealogies under this model, those for the intrademe patterns of common ancestry, will be discordant among loci. A more ideal scenario for the single-gene-tree approach is the stepping-stone model of range expansion considered by Slatkin (1993), which is a history of multiple isolation events. If a single sample was taken from each subpopulation, then we might expect the population tree to be reproduced at many loci. Of course, this too will depend upon the population splits being separated enough in time for the effect of drift to be negligible. Otherwise, even without migration, a gene tree may be different from the population tree (Neigel and Avise 1986, Pamilo and Nei 1988). This will be an issue as well for continuously distributed populations that have undergone range expansions; the movement of individuals will have to be restricted for historical structure to be evident in gene tree topologies.

This treatment has assumed no recombination within loci and free recombination between loci. Intralocus recombination will decouple sites' histories. Multiple genealogies will be realized in the history of a single locus and these will be correlated along the sequence (Hudson 1983a, Kaplan and Hudson 1985). Restricted interlocus recombination will make genealogies across sampled loci correlated. Both of these processes should tend to increase $max(p_i)$. Intralocus recombination increases the number of chances a locus has to realize a given partition, and restricted recombination between loci will cause branches to be shared across loci. They should have opposite effects on Ω, though. Intralocus recombination will lower the variation in tree sizes because there will be more independence among sites. The increased correlation among loci caused by restricted interlocus recombination, conversely, will increase the variance of tree size. Intralocus recombination is quite problematic for inferred gene-tree approaches since the genealogy is no longer a bifurcating tree (Hein 1993). It also represents a significant computational hurdle to coalescent inference methods which make explicit use of linkage patterns (Grifiths and Marjoram 1996).

The entire field of population genetics will benefit from increased exchange between coalescents and phylogeography. There is growing overlap already. On the one hand, the importance of coalescent approaches is evident in Avise's (2000) book about phylogeography. On the other, one of the currently most used coalescent inference programs, GENETREE (Bahlo and Grifiths 2000), produces an inferred genealogy. The future availability of multilocus genetic data will serve as a further bridge between these two approaches.

10.7 Acknowledgments

It has been my pleasure of the past few years to be a colleague to Dick Lewontin. I am thankful to him for the inspiration to do good work, and to Rama Singh for the invitation to contribute to this volume in Dick's honor. I also thank Monty Slatkin for comments on the manuscript. This work was supported by grant DEB-9815367 from the National Science Foundation.

REFERENCES

Aldous, D. J. (1985). Exchangeability and related topics. In A. Dold and B. Eckmann (eds) *École d'Été de Probabilités de Saint-Flour XII – 1983*, pp. 1–198. Vol. 1117 of *Lecture Notes in Mathematics*. Berlin: Springer-Verlag.

Avise, J. C. (1989). Gene trees and organismal histories: a phylogenetic approach to population biology. *Evolution* 43:1192–208.

Avise, J. C. (2000). *Phylogeography: the History and Formation of Species*. Cambridge, MA: Harvard University Press.

Avise, J. C., Arnold, J., Ball, R. M., Bermingham, E., Lamb, T., Neigel, J. E., Reeb, C. A., and Saunders, N. C. (1987). Intraspecific phylogeography: the mitochondrial DNA bridge between population genetics and systematics. *Annu. Rev. Ecol. Syst.* 18:489–522.

Bahlo, M., and Grifiths, R. C. (2000). Inference from gene trees in a subdivided population. *Theor. Popul. Biol.* 57:79–95.

Cannings, C. (1974). The latent roots of certain Markov chains arising in genetics: a new approach. I. Haploid models. *Adv. Appl. Prob.* 6:260–90.

Darwin, C. (1859). *On the Origin of Species*. London: Murray.

Dobzhansky, T. (1937). *Genetics and the Origin of Species*. New York: Columbia University Press.

Donnelly, P., and Tavaré, S. (1995). Coalescents and genealogical structure under neutrality. *Annu. Rev. Genet.* 29:401–21.

Ewens, W. J. (1972). The sampling theory of selectively neutral alleles. *Theor. Popul. Biol.* 3:87–112.

Ewens, W. J. (1990). Population genetics theory – the past and the future. In S. Lessard (ed.) *Mathematical and Statistical Developments of Evolutionary Theory*, pp. 177–227. Amsterdam: Kluwer Academic Publishers.

Felsenstein, J. (1981). Evolutionary trees from DNA sequences: a maximum likelihood approach. *J. Mol. Evol.* 17:368–76.

Fisher, R. A. (1918). The correlation between relatives on the supposition of Mendelian inheritance. *Trans. R. Soc. Edin.* 52:399–433.

Fisher, R. A. (1930). *The Genetical Theory of Natural Selection*. Oxford: Clarendon.

Grifiths, R. C., and Marjoram, P. (1996). Ancestral inference from samples of DNA sequences with recombination. *J. Comp. Biol.* 3:479–502.

Grifiths, R. C., and Tavaré, S. (1996). Monte Carlo inference methods in population genetics. *Math. Comput. Modelling* 23:141–58.

Haldane, J. B. S. (1932). *The Causes of Natural Selection.* London: Longmans Green.

Harris, H. (1966). Enzyme polymorphism in man. *Proc. R. Soc. Lond. Ser. B* 164:298–310.

Hein, J. (1993). A heuristic method to reconstruct the history of sequences subject to recombination. *J. Mol. Evol.* 36:396–405.

Hennig, W. (1965). Phylogenetic systematics. *Annu. Rev. Entomol.* 10:97–116.

Hennig, W. (1966). *Phylogenetic Systematics.* Urbana: University of Illinois Press.

Hey, J. (1991). A multi-dimensional coalescent process applied to multi-allelic selection models and migration models. *Theor. Popul. Biol.* 39:30–48.

Hey, J. (1994). Bridging phylogenetics and population genetics with gene tree models. In B. Schierwater, G. P. Wagner, and R. DeSalle (eds) *Molecular Ecology and Evolution: Approaches and Applications,* pp. 435–49. Basel, Switzerland: Birkhäuser Verlag.

Hudson, R. R. (1983a). Properties of a neutral allele model with intragenic recombination. *Theor. Popul. Biol.* 23:183–201.

Hudson, R. R. (1983b). Testing the constant-rate neutral allele model with protein sequence data. *Evolution* 37:203–17.

Hudson, R. R. (1990). Gene genealogies and the coalescent process. In D. J. Futuyma and J. Antonovics (eds) *Oxford Surveys in Evolutionary Biology,* pp. 1–44. Vol. 7. Oxford: Oxford University Press.

Hudson, R. R., Boos, D. D., and Kaplan, N. L. (1992). A statistical test for detecting geographic subdivision. *Mol. Biol. Evol.* 9:138–51.

Kaplan, N. L., and Hudson, R. R. (1985). The use of sample genealogies for studying a selectively neutral *m*-loci model with recombination. *Theor. Popul. Biol.* 28:382–96.

Kimura, M. (1955a). Solution of a process of random genetic drift with a continuous model. *Proc. Natl. Acad. Sci. USA* 41:144–50.

Kimura, M. (1955b). Stochastic processes and the distribution of gene frequencies under natural selection. *Cold Spring Harbor Symp. Quant. Biol.* 20:33–53.

Kingman, J. F. C. (1982a). The coalescent. *Stochastic Process. Appl.* 13:235–48.

Kingman, J. F. C. (1982b). Exchangeability and the evolution of large populations. In G. Koch and F. Spizzichino (eds) *Exchangeability in Probability and Statistics,* pp. 97–112. Amsterdam: North-Holland.

Kingman, J. F. C. (1982c). On the genealogy of large populations. *J. Appl. Prob.* 19A:27–43.

Kuhner, M. K., Yamato, J., and Felsenstein, J. (1995). Estimating effective population size and mutation rate from sequence data using Metropolis-Hastings sampling. *Genetics* 140:1421–30.

Lewontin, R. C. (1974). *The Genetic Basis of Evolutionary Change.* New York: Columbia University Press.

Lewontin, R. C., and Hubby, J. L. (1966). A molecular aproach to the study of genic diversity in natural populations II. Amount of variation and degree of heterozygosity in natural populations of *Drosophila pseudoobscura. Genetics* 54:595–609.

Lewontin, R. C., and Kojima, K. (1960). The evolutionary dynamics of complex polymorphisms. *Evolution* 14:450–72.

Li, W.-H. (1976). Distribution of nucleotide difference between two randomly chosen cistrons in a subdivided population: the finite island model. *Theor. Popul. Biol.* 10:303–8.

Li, W.-H. (1977). Distribution of nucleotide difference between two randomly chosen cistrons in a finite population. *Genetics* 85:331–7.

Linnaeus, K. (1735). *Systema Naturae.*

Malécot, G. (1948). *Les Mathématiques de l'Hérédité.* Paris: Masson. Extended translation: *The Mathematics of Heredity.* San Francisco: W. H. Freeman (1969).

Mayr, E. (1942). *Systematics and the Origin of Species.* New York: Columbia University Press.

Moran, P. A. P. (1958). Random processes in genetics. *Proc. Camb. Phil. Soc.* 54:60–71.

Neigel, J. E., and Avise, J. C. (1986). Phylogenetic relationships of mitochondrial DNA under various demographic models of speciation, In E. Nevo and S. Karlin (eds) *Evolutionary Processes and Theory*, pp. 515–34. New York: Academic Press.

Neigel, J. E., and Avise, J. C. (1993). Application of a random walk model to geographic distribution of animal mitochodrial DNA variation. *Genetics* 135:1209–20.

Neigel, J. E., Ball, M., and Avise, J. C. (1991). Estimation of single generation migration distances from geographic variation in animal mitochodrial DNA. *Evolution* 45:423– 32.

Nordborg, M. (2001). Coalescent theory. In D. J. Balding, M. J. Bishop, and C. Cannings (eds) *Handbook of Statistical Genetics.* Chichester: John Wiley.

Pamilo, P., and Nei, M. (1988). The relationships between gene trees and species trees. *Mol. Biol. Evol.* 5:568–83.

Provine, W. B. (1971). *The Origins of Theoretical Population Genetics.* Chicago: University of Chicago Press.

Robinson, D. F., and Foulds, L. R. (1981). Comparison of phylogenetic trees. *Math. Biosci.* 53:131–47.

Slatkin, M. (1987). The average number of sites separating DNA sequences drawn from a subdivided population. *Theor. Popul. Biol.* 32:42–49.

Slatkin, M. (1993). Isolation by distance in equilibrium and non-equilibrium populations. *Evolution* 47:264–79.

Slatkin, M., and Hudson, R. R. (1991). Pairwise comparisons of mitochondrial DNA sequences in stable and exponentially growing populations. *Genetics* 129:555–62.

Smouse, P. E. (1974). Likelihood analysis of recombination disequilibrium in multiple-locus gametic frequencies. *Genetics* 76:557–65.

Strobeck, C. (1987). Average number of nucleotide differences in a sample from a single subpopulation: a test for population subdivision. *Genetics* 117:149–53.

Tajima, F. (1983). Evolutionary relationship of DNA sequences in finite populations. *Genetics* 105:437–60.

Takahata, N., and Nei, M. (1985). Gene genealogy and variance of interpopulational nucleotide differences. *Genetics* 110:325–44.

Takahata, N., and Slatkin, M. (1990). Genealogy of neutral genes in two partially isolated populations. *Theor. Popul. Biol.* 38:331–50.

Templeton, A. R. (1998). Nested clade analysis of phylogeographic data: testing hypotheses about gene flow and population history. *Mol. Ecol.* 7:381–97.

Templeton, A. R., Routman, E., and Phillips, C. (1995). Separating population structure from population history: a cladistic analysis of the geographical distribution of mitochondrial DNA haplotypes in the tiger salamander, *Ambystoma tigrinum. Genetics* 140:767–82.

Wakeley, J. (1996a). The variance of pairwise nucleotide differences in two populations with migration. *Theor. Popul. Biol.* 49:39–57.

Wakeley, J. (1996b). Distinguishing migration from isolation using the variance of pairwise differences. *Theor. Popul. Biol.* 49:369–86.

Wakeley, J., and Hey, J. (1997). Estimating ancestral population parameters. *Genetics* 145:847–55.

Wallace, A. R. (1858). On the tendency of varieties to depart indefinitely from the original type. *Proc. Linn. Soc. Lond.* 3:53–62.

Watterson, G. A. (1975). On the number of segregating sites in genetical models without recombination. *Theor. Popul. Biol.* 7:256–76.

Wright, S. (1931). Evolution in Mendelian populations. *Genetics* 16:97–159.

Wright, S. (1943). Isolation by distance. *Genetics* 28:114–38.

11

The population genetics of life-history evolution

BRIAN CHARLESWORTH

Institute of Cell, Animal and Population Biology, University of Edinburgh

11.1 Introduction

The subject of this chapter is how natural selection acts on age-structured populations, in which age-specific patterns of survival probabilities and reproductive rates are the fundamental parameters describing the life history of a genotype. This is a large subject, and the chapter does not offer a comprehensive review. Instead, I will concentrate on some aspects on which I myself have worked. My interest in the subject was stimulated as a beginning postdoc in Dick Lewontin's laboratory, over 30 years ago, when he and Tim Prout expressed some well-justified skepticism about the logical basis of the use of Fisher's "Malthusian parameters" (Fisher 1930, Chapter 2) for modeling selection in populations with overlapping generations (Charlesworth 1970). I have contributed, somewhat fitfully, to this subject during much of my subsequent career.

The fundamental problem is how to describe selection in the context of a population divided into different age groups. This provides the basis for asking the more exciting question of how to understand the evolution of life histories, which will be briefly discussed at the end of this chapter. Dick Lewontin was, of course, one of the pioneers of theoretical work on life-history evolution (Lewontin 1965); his paper, together with Hamilton's seminal paper on the evolution of senescence (Hamilton 1966), motivated me to pursue further work on selection with age structure. My aim was to examine to what extent the widespread assumption that the growth rate of a population can be used as a fitness measure could be justified in terms of population genetics. A more complete review of work on selection with age structure, and its relevance to life-history evolution, is given by Charlesworth (1994).

The Evolution of Population Biology, ed. R. S. Singh and M. K. Uyenoyama. Published by Cambridge University Press.

11.2 Constructing a model of selection

11.2.1 Demographic parameters

Standard demographic concepts (e.g., Stearns 1992, Chapter 2; Charlesworth 1994, Chapter 1) underpin the theory of selection in diploid, sexually reproducing age-structured populations. The population is assumed to be sufficiently large that its behavior can be treated deterministically. In a purely demographic context, a life history can be described in terms of discrete time intervals, such that individuals of a given age x have an age-dependent probability, $P(x)$, of survival to the next time interval, when they will be aged $x + 1$. For simplicity, sex differences will be ignored initially and only the female component of the population considered. The justification for this is that female fertility is rarely limited by the number of males in the population, so that population growth can be realistically modeled by considering only the female component of the population. Female fecundity at age x, $m(x)$, is defined as the expected number of daughters produced by a female aged x. Their chance of survival to the next time interval, when they would be of age 1 and their surviving parents $x + 1$, is defined as $P(0)$. The net probability of surviving from conception to age x is $l(x) = P(0)\,P(1)\ldots P(x{-}1)$. The environment is assumed to be constant, so that these life-history parameters are independent of time. The life history can be summed up by the compound function $k(x) = l(x)\,m(x)$, which represents the expected number of daughters produced at age x by a newly formed female individual.

In this case, it is well known that (under rather light conditions on the life-history parameters; Charlesworth 1994, p. 23) the population rapidly approaches a constant exponential rate of growth in size, corresponding to a constant age structure (i.e., constant proportions of individuals in the different age classes). This growth rate, r, is the *intrinsic rate of increase* or *Malthusian parameter*, and is the (unique) real number that satisfies the *Euler–Lotka equation*:

$$\sum_{x=1}^{\infty} \mathrm{e}^{-rx}\, k(x) = 1. \tag{11.1}$$

This is exact for species that reproduce at discrete time intervals, such as annually breeding temperate zone birds or mammals. It provides an approximation for continuously reproducing species such as human beings or *Drosophila*, which can be made arbitrarily good by increasing the number of age classes into which the life history is divided. Integration over a continuous range of ages then replaces summation, but nothing fundamental is changed.

11.2.2 Genetic parameters

If genetic variation at a single locus with alleles $A_1, A_2 \ldots A_n$ is allowed, a given female genotype $A_i A_j$ has its own set of life-history characteristics, i.e., a set of values of $k_{ij}(x) = l_{ij}(x) m_{ij}(x)$. Application of the Euler–Lotka equation to a hypothetical population with the demographic parameters of genotype $A_i A_j$ enables an intrinsic rate of increase r_{ij} to be defined for this genotype. It is easy to see that, if the population reproduces asexually, so that females produce only daughters, each genotype constitutes a separate subpopulation that reproduces independently of the other genotypes. The genotype with the highest value of r_{ij} will thus eventually outcompete all the others; this result has often been used by ecologists to justify the use of the intrinsic rate of increase as a measure of fitness.

It is, however, by no means obvious that this will work in a sexually reproducing population with separate sexes. Males and females contribute equally to the gene pool, at least for autosomal loci, and so male demographic parameters need to be considered. But the absolute reproductive successes of males must in general depend on the age structure and genotypic composition of the female population, and it is unclear at first sight how to model their input into the population.

The problem can be solved in the following way, using as a starting point the single-locus model introduced above. Let $M_{fij}(x)$ be the total number of offspring produced at a given time by an $A_i A_j$ female aged x; this function is assumed to be independent of time. $M_{mij}(x, t)$ is the corresponding total number of offspring produced at time t by an $A_i A_j$ male aged x; as just mentioned, M_{mij} will in general depend on t. Let $N_{fij}(x, t)$ and $N_{mij}(x, t)$ be the numbers of $A_i A_j$ females and males aged x at time t. The total number of zygotes produced by the population at time t, $B(t)$, is then given by

$$B(t) = \sum_{ij} \sum_{x=1}^{\infty} N_{fij}(x, t)\, M_{fij}(x) = \sum_{ij} \sum_{x=1}^{\infty} N_{mij}(x, t)\, M_{mij}(x, t). \qquad (11.2)$$

This indicates that the values of the $M_{mij}(x, t)$ are constrained by the state of the population. The nature of the constraint is very simple if we assume that matings of fertile individuals occur randomly with respect to age and genotype, as happens if males shed sperm or pollen into a pool from which ova are fertilized at random, or if females choose mates at random from a group formed by all the fertile males. In both cases, we can define a quantity $\theta_{ij}(x, t)$, which characterizes either the probability that a male contributes to the pool of gametes or that he obtains a mate (Charlesworth 1994, pp. 108–9). The relative values of the $\theta_{ij}(x, t)$ for different genotypes can in principle be independent of the composition of the population, and hence of t, unless there is frequency-dependent selection. The values for each genotype can then be defined relative to some arbitrary standard genotype, and treated as time-independent functions, $\theta_{ij}(x)$. From Equation 11.2 , this implies that we can write

$$M_{mij}(x, t) = \theta_{ij}(x)\, B(t) / B^{*}(t) \qquad (11.3)$$

where

$$B^*(t) = \sum_{kl} \sum_{x=1}^{\infty} N_{mkl}(x, t)\theta_{kl}(x).$$

11.3 Determining the effects of selection

11.3.1 The selection equations

For further progress, we make the further assumption that there are no differences in age-specific survival probabilities between males and females, and that the $\theta_{ij}(x)$ functions for each male genotype are proportional to the corresponding female fecundity functions, $M_{fij}(x)$. It is then easily seen that the allele frequencies will be the same for male and female gametes, so that the frequency of allele A_i among all gametes produced at time t can be written as $p_i(t)$. If the sex ratio among zygotes is independent of parental age and genotype, the $M_{fij}(x)$ and hence $M_{mij}(x)$ functions for each genotype are proportional to the corresponding $m_{ij}(x)$ functions. If mating is at random with respect to both age and genotype, the frequency of the ordered genotype $A_i A_j$ among zygotes produced at time t is $p_i(t)\, p_j(t)$.

Putting all these assumptions together, we obtain the very simple recursion relation:

$$B(t)\, p_i(t) = \sum_{x=1}^{\infty} B(t-x)\, p_i(t-x) \sum_j p_j(t-x) k_{ij}(x). \tag{11.4}$$

This bears an obvious resemblance to the standard recursion equation for selection at a single locus in discrete-generation population genetics, which is simply the special case of this for a single age class (Crow and Kimura 1970, pp. 179–80). However, although we have got rid of the problem of the dependence of the male fertility function on time, this result does not help us to find a fitness measure analogous to that used in discrete-generation models, since we are confronted with a nonlinear set of high-dimensional equations that are intractable analytically, as far as an exact solution is concerned.

11.3.2 Equilibrium conditions

Several approaches to dealing with this problem have been used in the literature. The simplest, and in some ways the most general, is to consider equilibrium conditions. If the allele frequencies have converged to equilibrium, such that $p_i(t) = p_i$ (with p_i independent of time), the population can be treated as genetically homogeneous, and hence will approach an asymptotic growth rate, r, given by the equivalent of Equation 11.1:

$$\sum_{x=1}^{\infty} e^{-rx} \sum_{ij} p_i p_j k_{ij}(x) = 1. \tag{11.5}$$

This implies that $B(t-x)/B(t)$ tends to exp $(-rx)$, and so Equation 11.4 becomes

$$\sum_{j} p_j \sum_{x=1}^{\infty} e^{-rx} k_{ij}(x) = 1. \tag{11.6}$$

This is identical in form to the standard discrete-generation equation for equilibrium in a multiallelic system under selection:

$$\sum_{j} p_j \, w_{ij} = \overline{w} \tag{11.7}$$

where w_{ij} is the fitness of genotype $A_i A_i$ and $\overline{w}$ is the population mean fitness (Crow and Kimura 1970, p. 272).

This suggests that the fitness of a genotype in an equilibrium population can be defined as

$$w_{ij} = \sum_{x=1}^{\infty} e^{-rx} k_{ij}(x). \tag{11.8}$$

From Equation 11.5 it can be seen that this definition implies that the population mean fitness is equal to one.

It is easy to generalize this result to more complex situations, such as mutation-selection balance or multiple loci with selection and recombination (Charlesworth 1994, Chapter 3). A problem with its application is, however, that r is determined by Equation 11.5, which involves the allele frequencies which are themselves to be determined. One solution to this problem is to regard r as a quantity that can be determined empirically, e.g., when applying this equation to data on natural or laboratory populations. The other is to solve the set of simultaneous equations (Equations 11.5 and 11.7), or their equivalents for more complex genetic models. In general, this can only be done numerically, but analytical results can be obtained for the case of mutation-selection balance (Charlesworth 1994, p. 126), or a single-locus system with two alleles (Charlesworth 1994, p. 122). In the latter case, assuming a polymorphic equilibrium with alleles A_1 and A_2, r satisfies the equation

$$(1 - w_{11})(1 - w_{22}) = (1 - w_{12})^2. \tag{11.9}$$

This has only a single real root, and the existence of a polymorphic equilibrium requires either heterozygote superiority or inferiority for the w_{ij} with this value of r, as in the discrete-generation case. These conditions can be shown to correspond to heterozygote superiority or inferiority in the r_{ij} (Charlesworth 1994, pp. 122–3). As in the discrete-generation case, the equilibrium frequency of A_1 is given by

$$p_1 = \frac{(w_{12} - w_{22})}{(2w_{12} - w_{11} - w_{22})}. \tag{11.10}$$

A reasonable assumption for natural populations is that r is approximately zero, due to density-dependent mortality or fertility, in which case Equation 11.8 reduces to the life-time reproductive success of a genotype, which is commonly employed by empirical evolutionary ecologists as a fitness measure (Clutton-Brock 1988). However, if there is density-dependent selection, so that there are genotypic differences in the sensitivities of age-specific mortality or reproductive rates to population density, it is necessary to include their functional dependence on the numbers of individuals in the relevant age groups in the equations. This means that we are confronted with a similar problem with respect to the equilibrium population density to that for determining r (Charlesworth 1994, pp. 123–5).

11.3.3 Stability analyses

The second approach is to consider stability conditions, either global or local. Only limited progress has been made with global stability analyses, and this has been confined to the case of a single locus with two alleles in a density-independent population (Norton 1928, Pollak and Kempthorne 1971, Charlesworth 1980, pp. 186–8). The most complete treatment is that of Norton (1928) for the continuous-time case. By a difficult analysis, he showed that the asymptotic outcome of selection is determined entirely by the relations between the r_{ij}. Allele A_2 tends to fixation if r_{22} is the largest of the set (r_{22}, r_{12}, r_{11}); the allele frequencies tend to the neighborhood of the equilibrium given by Equations 11.9 and 11.10 if there is heterozygote superiority in the r_{ij}. If there is heterozygote inferiority in the r_{ij}, the allele whose initial frequency lies between the equilibrium and one tends to fixation. A local stability analysis of the polymorphic equilibrium can also be carried out for the discrete-time model (see Charlesworth 1994, pp. 162–6), and gives corresponding results. It is interesting to note that the approach to equilibrium in the case of heterozygote advantage is through a series of damped oscillations in allele frequencies, in contrast to the monotonic approach found in the discrete-generation case (Nagylaki 1992, Chapter 4).

The local stability method can also be applied to the case of the invasion of a population that is fixed for a given allele (A_1) by a rare, nonrecessive mutant (A_2). Equation 11.4 then yields the relation

$$p_2(t) = \sum_{x=1}^{\infty} p_2(t-x)e^{-r_{11}x}\, k_{12}(x) + O(p_2^2). \tag{11.11}$$

Neglecting the second-order terms, this leads to an Euler–Lotka equation defining the asymptotic rate of change of the logarithm of p_2 as the real root of

$$\sum_{x=1}^{\infty} e^{-(r+r_{11})x} k_{12}(x) = 1 \tag{11.12}$$

that is, $r = r_{12} - r_{11}$.

This provides a measure of the selective advantage of A_2 when rare, and shows that A_2 will invade a population fixed for A_1 when $r_{12} > r_{11}$. By symmetry, A_1 will invade a population fixed for A_2 when $r_{12} > r_{22}$, consistent with the results on heterozygote superiority described above. The case of invasion by a recessive mutant allele is harder to analyze, since second-order terms in mutant allele frequency can no longer be neglected, but it can be shown that A_2 will invade an A_1 population if $r_{22} > r_{11}$. Its selective advantage is, however, not equal to $r_{22} - r_{11}$ (Charlesworth 1994, p. 152). This analysis can easily be extended to the case of an initial population held at a stationary population size by density dependence; in this case, the demographic parameters for the invading mutant heterozygote are evaluated at the population size for the stationary initial population (Charlesworth 1994, p. 153). The predominant homozygote necessarily has an intrinsic rate of zero, so that the mutant will invade only if the heterozygote has a positive growth rate under the prevailing intensity of density-dependent factors.

11.3.4 Weak selection

The last method is to assume weak selection, and to seek approximations for the rates of change of allele frequencies that can be used to study the global dynamics of the system. This method was first applied by Haldane (1927, 1962), and was generalized by Charlesworth (1974, 1994, Chapter 4). The starting point is to select an arbitrary genotype as a standard, with life history $k_s(x)$. Weak selection implies that any other genotype $A_i A_j$ has a life history $k_{ij}(x)$ that differs from $k_s(x)$ by an amount $\varepsilon_{ij}(x)$ such that $|\varepsilon_{ij}(x)| < \varepsilon$, where the measure of selection intensity, ε, is sufficiently small that second-order terms in ε can be neglected. If this is the case, in a density-independent population the rate of change per time interval in the frequency of allele A_i at time t, $\Delta p_i(t)$, is asymptotically of order ε. In turn, this implies that the rate of change in the growth rate of the total number of zygotes, $B(t)$, is asymptotically second order in ε (Charlesworth 1994, pp. 137–9). These results can be used to show that Δp_i rapidly approaches the value

$$\Delta p_i = p_i(r_i - \bar{r}) + O(\varepsilon^2) \tag{11.13}$$

where

$$r_i = \sum_j p_j \, r_{ij}$$

and

$$\bar{r} = \sum_{ij} p_i p_j r_{ij}.$$

The leading term in Equation 11.13 is identical to the expression introduced by Fisher (1930 Chapter 2), where r_{ij} corresponds to the Malthusian parameter of genotype $A_i A_j$. This formulation was later extensively used by Kimura in his studies of selection theory (Kimura 1958, Crow and Kimura 1970). It is also similar in form to the selection equation for the discrete-generation case with weak selection. However, neither Fisher nor Kimura made any mention of a restriction to the case of weak selection.

It can also be shown that the rate of change of the population size asymptotically obeys the equation

$$\Delta^2 \ln B = 2 \sum_i p_i (r_i - \bar{r})^2 + O(\varepsilon^3), \qquad (11.14)$$

that is, the leading term in the rate of change of the rate of growth of log population size becomes equal to the additive genetic variance in intrinsic rate of increase (the fundamental theorem of natural selection: Fisher 1930, Chapter 2; Kimura 1958; Nagylaki 1992, Chapter 4).

This approach can be extended, using some specializing assumptions, to the case of density-dependent populations, and yields results similar to those which have been derived for the discrete-generation case (Charlesworth 1994, pp. 147–9). As in that case, it turns out that the direction of change in allele frequencies is determined by differences among the carrying capacities of the different genotypes. This result depends, however, on the assumption that mortality and/or reproductive rates are decreasing functions of the number of individuals in a specific age class or set of age classes (the *critical age group*). The carrying capacity of a genotype is then defined as the equilibrium number of individuals in the critical age group, for a population with the life history of that genotype.

11.3.5 Reconciling the fitness measures

The results of the exact analysis of equilibrium populations, and the analysis of conditions for local and global stability and of weak selection, give somewhat different interpretations of what is meant by fitness in an age-structured population. The dynamical analyses suggest that we can use differences in intrinsic rates of increase among genotypes to predict rates of change of allele frequencies, as a first-order approximation when selection is weak or when a rare allele is invading a population. The equilibrium fitness expression, Equation 11.8, is not explicitly related to the intrinsic rate of increase. This means that, if selection is strong, there is no such thing as a constant fitness measure for a genotype in an age-structured population. For example, with heterozygote advantage in a biallelic system, a rare allele will initially increase in frequency at a rate governed by the difference between the intrinsic rate of the heterozygote and the prevailing homozygote, but neither its dynamics near equilibrium nor its equilibrium frequency are simply determined by the genotypic intrinsic rates. In turn, this implies that attempts to fit population trajectories to the standard constant fitness models will reveal discrepancies, if selection is strong. But with sufficiently weak selection, such discrepancies will not be apparent (Charlesworth 1994, pp. 178–9), as can be seen as follows.

First, note that the weak selection equations 11.13 and 11.14 give rates of change in time units of one time unit, not generations as with the standard discrete-generation model. To make them comparable with the discrete-generation case, we need to convert to a time scale of generations. This can be done as follows. The absolute value of the derivative with respect to r of the left-hand side of the Euler–Lotka equation (11.1) provides a measure of generation time:

$$T = \sum_{x=1}^{\infty} x \, \mathrm{e}^{-rx} \, k(x).$$

In fact, T is the mean age of mothers in a population that has reached its asymptotic growth rate (Charlesworth 1994, p. 30).

When there are genotypic differences in life-history parameters, we can again select a standard genotype, and use its generation time T_s as a measure for the population as a whole, with an error of order ε. If Equations 11.13 and 11.14 are multiplied by T_s, we then obtain expressions for rates of change of allele frequencies and population growth rate that are of the same order of accuracy as the original equations, but which now give rates per generation. Furthermore, a Taylor expansion of Equation 11.8 shows that the equilibrium fitness, w_{ij}, of genotype $A_i A_j$ can be written as

$$\sum_{x=1}^{\infty} \mathrm{e}^{-rx} \, k_{ij}(x) = 1 + (r_{ij} - r) \sum_{x=1}^{\infty} x \, \mathrm{e}^{-r_{ij}x} k_{ij}(x) + O(\varepsilon^2)$$

which (after some further manipulation) yields

$$w_{ij} - 1 = T_s(r_{ij} - r_s) + O(\varepsilon^2).$$

With weak selection, the per-generation rate of change in allele frequencies, given by Equation 11.13, can therefore be approximated using differences in the fitness function defined by Equation 11.8. Equivalently, genotypic differences in this fitness function are proportional to the differences in the intrinsic rates of increase, with a proportionality constant equal to the generation time for a standard genotype, to terms of first order in the strength of selection.

11.3.6 Sex differences and nonrandom mating

These results have all been derived under some specific assumptions, notably random mating with respect to age and genotype, and proportionality of the male and female $k(x)$ functions for different genotypes. Nonrandom mating with respect to age is very likely to occur; it can be shown to produce deviations from Hardy–Weinberg frequencies among the zygotes, which are second order with respect to the selection intensity ε (Charlesworth 1994, pp. 114–16). A similar result holds for sex differences in the life-history parameters. For equilibrium populations, Equation 11.8 can be still applied as a first-order approximation, with respect to the magnitude of sex differences in life-history parameters, by replacing $k_{ij}(x)$ by the arithmetic mean of the male and female functions (Charlesworth and Charlesworth 1973, Charlesworth 1994, pp. 120–1).

For nonequilibrium populations, both sex differences and nonrandom mating with respect to age lead to time-dependent terms of order ε appearing in the male $k_{ij}(x)$ functions. It turns out that a first-order approximation for the change in allele frequencies can be derived by choosing values for the male $k_{ij}(x)$ functions that are determined in some standard population with constant genotypic composition, and which therefore do not change with time. Corresponding genotypic intrinsic rates of increase can be assigned to males with these $k_{ij}(x)$ functions; to terms in first order in ε, the arithmetic means of the male and female intrinsic rates of increase for each genotype can then be used in the equivalent of Equation 11.13 (Charlesworth 1994, pp. 143–5). This is very similar to the result for discrete-generation populations (Nagylaki 1979). Results for invasion by rare alleles with arbitrary selection intensities can also be derived (Charlesworth 1994, p. 150).

Little work has been done on populations that mate nonrandomly with respect to genotype (but see Pollak and Kempthorne 1970). If a constant Wrightian inbreeding coefficient, F (Crow and Kimura 1970, p. 88), is used to describe departures from Hardy–Weinberg frequencies among zygotes, the methods used here will yield similar predictions to those obtained with the standard discrete-generation models that use the assumption of constant F (Nagylaki 1992, p. 63). With weak selection, this would be expected to provide an adequate first-order approximation for studying selection in an inbred population (Charlesworth *et al.* 1997, p. 171).

11.3.7 Multiple loci and quantitative traits

The models described up to now have involved only a single-locus system. Exact treatments of multilocus systems are even more intractable than those for the single-locus case, and have never been attempted, to my knowledge. When selection is weak in relation to recombination frequency, however, it is possible to show that a two-locus system converges to a state in which the coefficient of linkage disequilibrium becomes of order ε, and changes at a rate which is second order in ε (Charlesworth 1994, pp. 146–7). This is equivalent to the "quasi-linkage equilibrium" situation for discrete-generation models, which has proved to be a powerful tool for the analysis of multilocus systems (Barton and Turelli 1991, Nagylaki 1993). While this has not been formally extended to multilocus systems with age structure, similar results to those for discrete-generation equations should apply. Based on this assumption, it is possible to use the single-locus results described above to derive an equation for selection on a multivariate set of additively inherited quantitative traits (Lande 1982, Charlesworth 1993), which are similar to the standard discrete-generation expression

$$\boldsymbol{\Delta}\bar{\boldsymbol{z}} = \boldsymbol{G}\nabla r_{\bar{z}} + O(\varepsilon^2).$$

Here, $\bar{\boldsymbol{z}}$ is the vector of mean values for each of the traits, $\nabla r_{\bar{z}}$ is the gradient vector of the derivatives of the intrinsic rate of increase of the population with respect to the components of $\bar{\boldsymbol{z}}$, $\boldsymbol{G}$ is the additive genetic variance–covariance matrix, and ε is now a measure of the overall strength of selection on the traits. Conveniently, this term vanishes at equilibrium (Charlesworth 1993). As in the single-locus case, this can be converted into a per-generation rate of change by multiplying both sides by the generation time of a standard genotype, T_s.

11.3.8 Variable environments

It is hard to provide analytical treatments of selection in variable environments for anything other than invasion by rare alleles. As pointed out by Charlesworth (1980, p. 175), the type of approach that yields Equations 11.11 and 11.12 can be generalized to temporally varying environments. This leads to the conclusion that a rare, nonrecessive mutant allele will invade if heterozygotes are associated with a higher mean population growth rate (on a log scale) than that for the predominant homozygote. This is similar to the classic result for discrete generations (Haldane and Jayakar 1963). The problem is how to calculate this mean growth rate when there is an arbitrary pattern of temporal variation in life-history parameters. It has often been assumed in the ecological literature that the relevant measure is the mean of the intrinsic rates of increase over each environment that the population encounters (e.g., Schaffer 1974). However, this is true only if the intervals between environmental changes are so long that the population has time to approach its asymptotic growth rate in each environment; in general, more complex approximations are needed (Tuljapurkar 1990). A recent application of the invasion criteria to the case of a temporally varying environment is described by Turelli *et al.* (2001).

The case of spatially variable environments is easier to deal with, and it is straightforward to derive conditions that are similar in form to the classical invasion conditions for the discrete-generation case (Crow and Kimura 1970, pp. 281–4), for both hard and soft selection (Charlesworth 1994, pp. 158–9).

11.4 Conclusions

One important conclusion from the study of selection in age-structured populations is that most of the basic results from models of discrete-generation populations (or continuous-time models that ignore age structure) still apply, especially if selection is weak. This also applies to other aspects of population genetics; results such as the equality of the neutral mutation rate and the rate of substitution of neutral mutations, and concepts such as effective population size, can be extended to age-structured populations (Charlesworth 1994, 2001a). For many purposes, evolutionary geneticists can therefore continue to ignore the complications introduced by age structure, as they have mostly done for the last 70 years.

11.4.1 Measuring fitness

There are, however, some important qualifications to this conclusion, which mean that the effort devoted to obtaining the results described above has not been completely wasted. In the first place, it is clear that there is no unique measure of fitness when selection is relatively strong. While differences in genotypic intrinsic rates of increase can be used to predict the initial increase in frequency of a nonrecessive rare allele, this does not hold true near polymorphic equilibria. The frequencies of genotypes in such equilibrium populations are, in fact, determined by the fitness measure of Equation 11.8, as was first noted for the case of a biallelic single locus by Norton (1928). For experimentalists dealing with examples of strong selection, it is important to be clear about this, as otherwise erroneous inferences of frequency-dependent selection might be made (Charlesworth 1994, p. 180). The results on the initial increase of rare alleles can be used to justify the use of intrinsic rate of increase for determining optimal life histories, using the same logic as in ESS theory (Charlesworth 1994, pp. 186–8).

Furthermore, Equation 11.8, and its extension to the case of sex differences in life-history parameters (Charlesworth 1994, p. 120), provide a rigorous basis for estimating the relative fitness of genotypes, which replaces the ad hoc methods which were previously in use (e.g., Bodmer 1968). This is useful for human geneticists (Charlesworth and Charlesworth 1973), especially if they wish to estimate mutation rates for dominant or sex-linked recessive deleterious alleles from trait frequencies and the relative fitnesses of normal and affected individuals (Vogel and Motulsky 1997). A recent application of the formula to the characterization of the fitness effects of mutations in *Caenorhabditis elegans* is described by Peters and Keightley (2000).

11.4.2 Dependence of fitness on demography

Equation 11.8 also has the important implication that the relative fitnesses of genotypes may depend on the demographic state of the population, and hence may be sensitive to environmental changes that alter the population's growth rate and age structure, even if these changes do not affect the relative values of the age-specific mortality and reproductive rates of different genotypes. This follows from the fact that the analysis leading to Equation 11.8 shows that genetic equilibrium is formally impossible unless the population has attained a state of constant growth and age structure, except when the $k_{ij}(x)$ functions for different genotypes have the same relative values for each value of x (Charlesworth 1994, pp. 117–9).

Changes in the environment which lead to different values of the equilibrium population growth rate, r, will in general lead to different relative fitness values as determined by Equation 11.8, such that genotypes with relatively high $k_{ij}(x)$ functions early in life are favored if r is positive, and genotypes with relatively low $k_{ij}(x)$ functions early in life are favored if r is negative (Charlesworth 1994, pp. 129–31). Fluctuations in population density, such as occur in microtine rodents, can therefore drive changes in genotype frequencies (Charlesworth 1994, pp. 126–9). Longer-term shifts in genotype frequencies can be induced in populations that are tightly regulated by density-dependent factors, and so are always near stationarity in population size, if changes in mortality patterns change the relative proportions of young and old individuals in the population. These results exemplify the general dependence of relative fitnesses in age-structured populations on the overall demographic structure of the population, emphasized by workers such as Williams (1957), Lewontin (1965) and Hamilton (1966).

Some examples of changes in the fitnesses of carriers of human genetic diseases in response to recent changes in demography are described by Charlesworth and Charlesworth (1973) and Charlesworth (1994, pp. 132–4). The recent declines in population growth and mortality rates in human populations have led to selection placing a greater emphasis on late-life survival and reproduction, so that diseases with a late age of onset, such as Huntington's chorea, are now more disadvantageous in terms of their net fitness effects than they were a century ago, at least in the affluent section of the world. If this demographic situation persists, a gradual decline in the frequency of such diseases is expected, although the process will be very slow. Faster changes may occur in genetic diseases affected by polymorphic alleles with age-dependent expression, such as Alzheimer's disease, which is in part controlled by the polymorphic apo-lipoprotein E gene (Charlesworth 1996).

11.4.3 Life-history evolution

This effect of demography on the relative fitnesses of genotypes which differ in their age-specific patterns of mortality and reproductive rates leads on to the last, and probably most important, aspect of this work. The population genetics results described here provide a rational and quantitative basis for understanding how selection acts on life-history traits in sexually reproducing populations, a topic that is now the subject of a large theoretical and experimental literature (Roff 1992, Stearns 1992, Charlesworth 1994, Chapter 5). Use of the initial invasion criterion of Equation 11.12, for example, provides a justification for using the maximization of intrinsic rate of increase or carrying capacity (when there is density dependence) as a criterion for determining the evolutionarily stable life-history strategy under an assumed set of tradeoffs between different life-history parameters (the ESS: Maynard Smith 1982). This can in turn be related to the results of quantitative genetics models of life-history evolution (Lande 1982, Charlesworth 1990, 1993, Abrams *et al.* 1993), which show that the ESS for a life history corresponds approximately to the mean attained at equilibrium under selection acting on a multivariate set of life-history traits.

The evolution of ageing is one aspect of life-history evolution for which the population/quantitative genetics approach has been especially fruitful, and where it has helped to evaluate the plausibility of different processes that may bring about increased mortality rates and decreased reproduction rates as age advances (Rose 1991, Charlesworth 2000). As first discussed in detail by Medawar (1952), and placed on a more secure quantitative basis by Wiliams (1957) and Hamilton (1966), the evolution of senescence reflects the fact that net fitness is more sensitive to changes in mortality or reproduction that occur early in life than to changes late in life. Various different realizations of this result have been proposed. The population genetics analyses cast doubt on Medawar's idea that senescence may evolve because of selection to postpone the age of onset of genetic diseases controlled by rare mutations. This is because the selection coefficients on the modifiers involved are of the order of the mutation rate to the disease alleles (Charlesworth 1994, p. 200). This suggests that senescence has evolved either because of selection favoring increases in survival or reproduction early in life at the expense of pleiotropic reductions later in life, or else because of the accumulation at higher frequencies of late-acting deleterious mutations compared with early-acting ones, or by both mechanisms (Rose 1991, Charlesworth 1994, pp. 198–200). The theory also provides testable predictions concerning age-specific patterns of components of genetic variance in mortality and reproduction rates under these models (Charlesworth and Hughes 1996, Charlesworth 2001b); in principle, these can be used to evaluate the importance of these processes in causing the evolution of aging. While there is evidence for the action of both mechanisms (Zwaan 1999), more empirical research is needed before any final verdict can be reached.

11.5 Acknowledgments

None of the work described here was supported by funding from competitive research grants; I am thankful for the opportunity of avoiding the necessity of justifying it to my peers. I thank the Royal Society for my current support.

I am deeply grateful to Dick Lewontin for his inspiring guidance early in my career, and to Deborah Charlesworth for her comments on the manuscript and for putting up with me for so long.

REFERENCES

Abrams, P. A., Harada, Y., and Matsuda, H. (1993). On the relationship between quantitative genetic and ESS models. *Evolution* 47:982–5.

Barton, N. H., and Turelli, M. (1991). Natural and sexual selection on many loci. *Genetics* 127:229–55.

Bodmer, W. F. (1968). Demographic approaches to the measurement of differential selection in human populations. *Proc. Natl. Acad. Sci. USA* 59:690–9.

Charlesworth, B. (1970). Selection in populations with overlapping generations. I. The use of Malthusian parameters in population genetics. *Theor. Pop. Biol.* 1: 352–70.

Charlesworth, B. (1974). Selection in populations with overlapping generations. VI. Rates of change of gene frequency and population growth rate. *Theor. Pop. Biol.* 6:108–32.

Charlesworth, B. (1980). *Evolution in Age-structured Populations.* 1st ed. Cambridge: Cambridge University Press.

Charlesworth, B. (1990). Optimization models, quantitative genetics, and mutation. *Evolution* 44:520–38.

Charlesworth, B. (1993). Natural selection on multivariate traits in age-structured populations. *Proc. R. Soc. Lond. B* 251:47–52.

Charlesworth, B. (1994). *Evolution in Age-structured Populations.* 2nd ed. Cambridge: Cambridge University Press.

Charlesworth, B. (1996). Alzheimer's disease and evolution *Curr. Biol.* 6:20–2.

Charlesworth, B. (2000). Biological and evolutionary perspectives on aging. In J. E. Morley, H. J. Armbrecht, R. M. Coe and B. Vellas (eds) *The Science of Geriatrics,* pp. 31–48. Paris: Serdi.

Charlesworth, B. (2001a). The effect of life-history and mode of inheritance on neutral genetic variability. *Genet. Res.* 77:153–66.

Charlesworth, B. (2001b). Patterns of age-specific means and genetic variances of mortality rates predicted by the mutation-accumulation theory of ageing. *J. Theor. Biol.* 210:47–66.

Charlesworth, B., and Charlesworth, D. (1973). The measurement of fitness and mutation rate in human populations. *Ann. Hum. Genet.* 37:175–87.

Charlesworth, B., and Hughes, K. A. (1996). Age-specific inbreeding depression and components of genetic variance in relation to the evolution of senescence. *Proc. Natl. Acad. Sci. USA* 93:6140–5.

Charlesworth, B., Nordborg, M., and Charlesworth, D. (1997). The effects of local selection, balanced polymorphism and background selection on equilibrium patterns of genetic diversity in subdivided populations. *Genet. Res.* 70: 155–74.

Clutton-Brock, T. H. (1988). *Reproductive Success.* Chicago, IL: University of Chicago Press.

Crow, J. F., and Kimura, M. (1970). *An Introduction to Population Genetics Theory*. New York: Harper and Row.

Fisher, R. A. (1930). *The Genetical Theory of Natural Selection*. Oxford: Oxford University Press.

Haldane, J. B. S. (1927). A mathematical theory of natural and artificial selection. Part IV. *Proc. Camb. Phil. Soc.* 23:607–15.

Haldane, J. B. S. (1962). Natural selection in a population with annual breeding but overlapping generations. *J. Genet.* 58:122–4.

Haldane, J. B. S., and Jayakar, S. D. (1963). Polymorphism due to selection of varying direction. *J. Genet.* 58:237–42.

Hamilton, W. D. (1966). The moulding of senescence by natural selection. *J. Theor. Biol.* 12:12–45.

Kimura, M. (1958). On the change of population mean fitness by natural selection. *Heredity* 12:145–67.

Lande, R. (1982). A quantitative genetic theory of life history evolution. *Ecology* 63:607–15.

Lewontin, R. C. (1965). Selection for colonizing ability. In H. G. Baker and G. L. Stebbins (eds) *The Genetics of Colonizing Species*, pp. 77–94. New York: Academic Press.

Maynard Smith, J. (1982). *Evolution and the Theory of Games*. Cambridge: Cambridge University Press.

Medawar, P. B. (1952). *An Unsolved Problem of Biology*. London: H. K. Lewis.

Nagylaki, T. (1979). Selection in dioecious populations. *Ann. Hum. Genet.* 14:143–50.

Nagylaki, T. (1992). *Introduction to Theoretical Population Genetics*. Berlin: Springer-Verlag.

Nagylaki, T. (1993). The evolution of multilocus systems under weak selection. *Genetics* 134:627–47.

Norton, H. T. J. (1928). Natural selection and Mendelian variation. *Proc. Lond. Math. Soc.* 28:1–45.

Peters, A. D., and Keightley, P. D. (2000). A test for epistasis among induced mutations in *Caenorhabditis elegans*. *Genetics* 156:1635–47.

Pollak, E., and Kempthorne, O. (1970). Malthusian parameters in genetic populations. Part I. Haploid and selfing models. *Theor. Pop. Biol.* 1:315–45.

Pollak, E., and Kempthorne, O. (1971). Malthusian parameters in genetic populations. Part II. Random mating populations in infinite habitats. *Theor. Pop. Biol.* 2:351–90.

Roff, D. A. (1992). *The Evolution of Life Histories: Data and Analysis*. London: Chapman and Hall.

Rose, M. R. (1991). *The Evolutionary Biology of Aging*. Oxford: Oxford University Press.

Schaffer, W. M. (1974). Optimal reproductive effort in fluctuating environments. *Am. Nat.* 121:418–431.

Stearns, S. C. (1992). *The Evolution of Life Histories*. Oxford: Oxford University Press.

Tuljapurkar, S. (1990). *Population Dynamics in Variable Environments*. Berlin: Springer-Verlag.

Turelli, M., Schemske, D. W., and Bierzychudek, P. (2001). Stable two-allele polymorphisms maintained by fluctuating fitnesses and seed banks: protecting the blues in *Linanthus parryae*. *Evolution* 55:128–1298.

Vogel, F., and Motulsky, A. G. (1997). *Human Genetics: Problems and Approaches.* Berlin: Springer-Verlag.
Williams, G. C. (1957). Pleiotropy, natural selection and the evolution of senescence. *Evolution* 11:398–411.
Zwaan, B. J. (1999). The evolutionary genetics of ageing and longevity. *Heredity* 82:589–97.

12

Gene–environment complexities: what is interesting to measure and to model?

PETER TAYLOR
Program on Critical and Creative Thinking, Graduate College of Education, University of Massachusetts, Boston

12.1 Preamble – 1974

The year 1974 saw the publication of two influential works by Richard Lewontin. In different ways both addressed the measurement and characterization of genetic variation and asked whether the resulting knowledge is interesting – what can we explain or do with it?

The Genetic Basis of Evolutionary Change (1974a) was firmly positioned within the population genetic tradition of viewing evolution as a change of gene frequencies in a population over time. In this light it was obviously important to characterize the amount of genetic variation and account for its maintenance. Lewontin's book masterfully synthesized research on genetic diversity in laboratory and natural populations in relation to models of selection or its absence. At the same time he drew attention to some troublesome themes for evolutionary biology. It was not variation as such that should count, but variation that resulted in differential fitness among the variants. Yet measurements of the components of fitness – survival and reproduction – were possible only when the phenotypic effect of a single allelic substitution was large, not when the effects of gene substitutions made only small differences. This led Lewontin to remark that: "What we can measure is by definition uninteresting and what we are interested in is by definition unmeasurable" (Lewontin 1974a, p. 23). The problem of relating models of selection to observations becomes astronomically worse when there are multiple, linked loci (Lewontin 1974a, p. 317). He concluded that population genetics should shift its attention to the fitness effects of long segments of chromosomes. Such effects were interesting evolutionarily and could be measured.

The idea that many genes may contribute small effects to a trait derives from a different research tradition, quantitative genetics, which is the subject of the other publication, "The analysis of variance and the analysis of causes" (Lewontin 1974b). Quantitative genetics concerns itself not with any specific genes having discrete ("qualitative") effects but with the statistical analysis of

The Evolution of Population Biology, ed. R. S. Singh and M. K. Uyenoyama. Published by Cambridge University Press.

continuous ("quantitative") traits varying within populations (Falconer 1989, p. 1). Traditionally, the "populations" that quantitative genetics deals with have consisted of the varieties manipulated by plant and animal breeders, who use statistics to estimate the rate of improvement in desired traits that could be expected from possible matings or crosses. Human populations have also been the subject of quantitative genetic analysis, but Lewontin (1974b) argues that this is definitely not interesting for human genetics; such analysis provides no basis for effective environmental or clinical interventions. A proper understanding of the statistical technique at the heart of quantitative genetics, the analysis of variance ("ANOVA"), shows why quantitative genetics can provide little insight into underlying causes. In any analysis of variance the ranking of varieties for the trait in question can change as the environment they develop in changes. Moreover, the degree of such reordering, the "genotype by environment interaction," is dependent on the range of environments and sample of varieties in the data under consideration. It would be more interesting, Lewontin concluded, to characterize and understand the "norms of reaction," the different responses of genetically defined varieties across the full range of environments they could experience.

Although Lewontin considers norms of reaction to be an important concept for analyzing evolution in changing environments, the analysis of variance and causes paper arose not from his research on genetic variation as much as from his critique of Arthur Jensen's quantitative genetic work on IQ (Jensen 1969, 1970, Lewontin 1970a,b). Whereas Jensen interpreted the gap in average IQ among racially defined human populations as based in genetic differences, Lewontin argued that American society had by no means explored the full range of possible environments that might boost intelligence. Lewontin's critical social commentary on science sets the scene for this essay. I do not review progress with respect to the recommendations of his 1974 publications, namely, that measurements and models of the fitness effects (or social significance) of genetic variation should integrate whole chromosome segments and responses across environments. Instead, with the intention of stimulating readers' thinking about gene–environment complexities, I introduce three lines of research in human biology and sociology. At the end I circle back to evolutionary biology to suggest a lesson that might be drawn from noticing ways that interesting things can be measured and modeled about humans: in particular, contingent situations.

12.2 "Environment" in the age of DNA

Everyone "knows" that genes and environment, nature and nurture, interact. But, in this "age of DNA," genetics is often seen as the way to expose the important or root causes of behavior and disease and as the best route to effective therapeutic technologies. Widespread public attention is given to claims that social policies and actions built on broader bases are scientifically or economically ineffective. *The Bell Curve* (Herrnstein and Murray 1994) provides the most notable recent episode.

At a different level, the dominance of genetics is also reflected within social studies of science and technology. Critical light has been shed on the history, semantic complexity, politics, ethics and other dimensions of genetics; very little interpretive scholarship concerns the sciences of, for example, educational interventions or psychological development. In general, the "environment" is underexamined and construed in simple terms. For example, many commentators on science in its social context – myself included – have invoked phenylketonuria (PKU) to demonstrate that "genetic" does not mean unchangeable. However, Diane Paul's history of screening for PKU shows that the certainty of severe mental retardation has been replaced by a chronic disease with a new set of problems. Moreover, even in this case where the condition has a clear-cut link to a single changed gene, complexities of the social environment have shaped the ways that society makes use of knowledge about that condition (Paul 1997).

The "mystique" of genetics (Nelkin and Lindee 1995) is by no means fading, but several scientific currents are bringing the environment, in different variants, back into the picture. In evolutionary biology, a great deal of attention is now given to the plasticity of phenotypes across a range of environments (Sultan 1992). Developmental biology, filling the gap between genes and the characters they shape, is experiencing a renaissance. Although the field still focuses mainly on embryological or early development, the influence of the environment is now acknowledged even for those stages (van der Weele 1995, Gilbert 1997). Behavioral genetics, once firmly directed towards establishing the "heritability" of traits, now highlights the effects of "nonshared" environmental influences, i.e., those not experienced equally by members of the same family. Among such nonshared influences, Sulloway (1996) has argued that birth order may be a key factor in explaining conformity to or rebellion against authority in intellectual and other spheres of social life.

Many questions arise once one tries to make sense of the ways scientists conceptualize the "environment." What meanings are given to the term, and how have these changed over time and in response to criticism? What is measured and what is explained? What methodologies are employed for collecting data and making inferences? What is the status of the different natural and social sciences involved? How are these colored by past and present associations with political currents?

With such questions in mind, I have begun to examine the development and reception of three areas of epidemiology. (I use this term broadly to denote research that correlates traits in general, not only disease incidence, to antecedent factors in defined populations and attempts to determine the causal processes by which the traits develop over time.) Each approach complicates the persistent, albeit often qualified, contrasts: inborn and unchangeable vs. environmental and changeable; and biological vs. social. The areas are:

1. research on gestational programming, which has identified associations between nutrition during critical periods *in utero* and diseases of late life, including heart disease, diabetes, and death by suicide (Barker 1994);
2. life events and difficulties research, which has exposed relationships between severe events and difficulties over a person's life course and the onset of mental or physical illness (Harris 2000); and
3. "reciprocal causation" models of IQ development in which there is a matching of traits and the changing environments in which traits develop so as to allow both high heritability and large gains from one generation to the next (Dickens and Flynn 2001).

In this essay I do not delve deeply into any of these approaches, but provide an introduction and overview sufficient, I hope, to bring more attention to the complexities of the "environment" and to the ways scientists account for the development of behavioral and medical conditions over any individual's lifetime. As part of exploring the significance of the three approaches, I identify various ways that they challenge each other as well as challenging more traditional accounts of gene–environment interactions from behavioral geneticists and from critics of biological determinism, such as Lewontin.

12.3 Five approaches to gene–environment complexities

12.3.1 Gestational programming

Several research groups, most notably Barker's group at the University of Southampton, have located data on body size and body shape at birth for cohorts of individuals and related these data to diseases arising in these individuals later in life (Barker 1994, 1995a, Scrimshaw 1997). Associations have been found between nutrition during critical periods *in utero* and diseases of late life, including heart disease, diabetes, and death by suicide (Barker *et al.* 1995). The associations stand out even after allowing for confounding associations between socioeconomic status, low birth weight, and adult diseases. It appears that, through "gestational programming" of biochemical patterns and cell distribution within organs, disease susceptibility can be inborn, yet with origins that are environmental, not genetic (Figure 12.1).

Within epidemiology, gestational programming has been subject both to critical commentary (Paneth 1994, Paneth and Susser 1995, Kramer and Joseph 1996) and to confirmation by former sceptics (Frankel *et al.* 1996). A major objection has been that gestational programming does not explain temporal and international trends in coronary heart disease. For example, heart disease rose in countries like Scotland, Finland or Norway, where birth weights have been among the world's highest. Work on Finnish data suggests

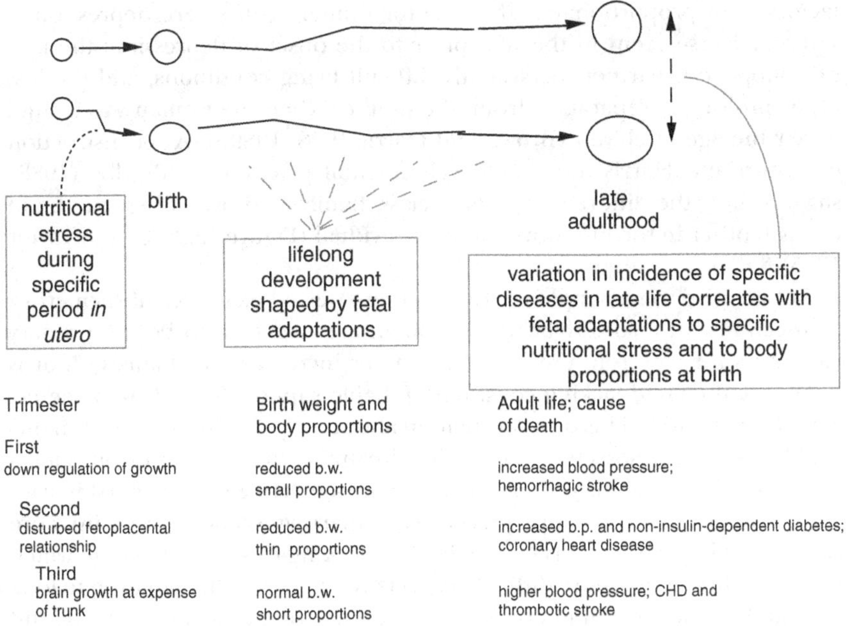

Figure 12.1. A model and observations in which nutrition during critical periods *in utero* correlates with diseases of late life. (Adapted from Barker 1994, fig. 9.4.)

ways to resolve such apparent inconsistencies. Tracing the ways that Westernization plays out over a number of generations, it turns out that women who were born small, but who, with increasing affluence, became overweight for their size, tended to have thin offspring who, although well nourished, had higher risk of heart disease (Forsén *et al.* 1997). Barker's group is now examining such contingencies and combining their findings with mechanisms of low growth rate during different periods of pregnancy (Barker 1995a) and with factors related to body weight and growth in childhood and adult life (Barker, pers. comm; see also Frankel *et al.* 1996).

12.3.2 Life events and difficulties

Another line of research from England, initiated by the sociologists Brown and Harris in the late 1960s, investigates how severe events and difficulties during people's life course influence the onset of mental and physical illnesses (Brown and Harris 1989a, Harris 2000). The most sustained research in this tradition involves explaining depression in working-class women. For a district of London in the early 1970s, Brown and Harris used interviews, ratings of transcripts (done blind, that is, without knowledge of whether the woman became depressed), and statistical analyses. They identified four factors as disproportionately the case for women with severe depression: a severe, adverse event in the year prior to the onset of depression; the lack of a supportive partner; persistently difficult living conditions; and the loss of, or prolonged separation from, the mother when the woman was a child under the age of eleven (Brown and Harris 1978, 1989b). A reconstruction of Brown and Harris' work by developmental psychologist Bowlby (1988) suggests how the different aspects of class, family, and psychology can build on each other in the life course of the individual (Figure 12.2; see also Taylor 1995).

Let me give some simplified and overgeneralized examples of such cross-connections: In a society in which women are expected to be the primary caregivers for children, the loss of a mother increases the chances of, or is linked to, the child lacking consistent, reliable support for at least some period. (Bowlby added his own speculation about early childhood "attachment" problems.) An adolescent girl in such a disrupted family or sent from such a family to a custodial institution is likely to see a marriage or partnership with a man as a positive alternative, yet such early marriages tend to break up more easily. Working-class origins tend to lead to working-class adulthood, in which living conditions are more difficult, especially if a woman has children to look after and provide for on her own. And, in these circumstances, accidents and other severe events are more likely. The consequence of a severe event is often, unless there is a supportive partner, the onset of depression (see also Brown and Moran 1997).

The life events and difficulties methodology attempts to integrate "the quantitative analyses of epidemiology and the [in] depth understanding of the case history approach" (Brown and Harris 1989a, p. x). The case history interviews allow contextual rating of events. For example, the death of a relative after a long illness has a different meaning from the sudden death of a relative on whom one depended for financial or emotional support. This methodology has been applied for many illnesses, including heart attacks, disorders of menstruation, and multiple sclerosis, but most commonly for psychological disorders. Different contributing factors are identified in different illnesses, which is to be expected. Further inquiry has been motivated by the results being less clear cut than for the 1970s London depression study and by considerable unexplained variation remaining. Another concern for life events and difficulties researchers has been achieving recognition and adoption of their approach in the United States, where conventional check-list surveys still dominate the study of associations between life events and illnesses (Brown and Harris 1989a, pp. x–xi, Brown 1989).

Before introducing the third area of epidemiology – "reciprocal causation" models of IQ development – it will help if I provide overviews of more traditional accounts of gene–environment interactions from behavioral geneticists and scientific critics of biological determinism.

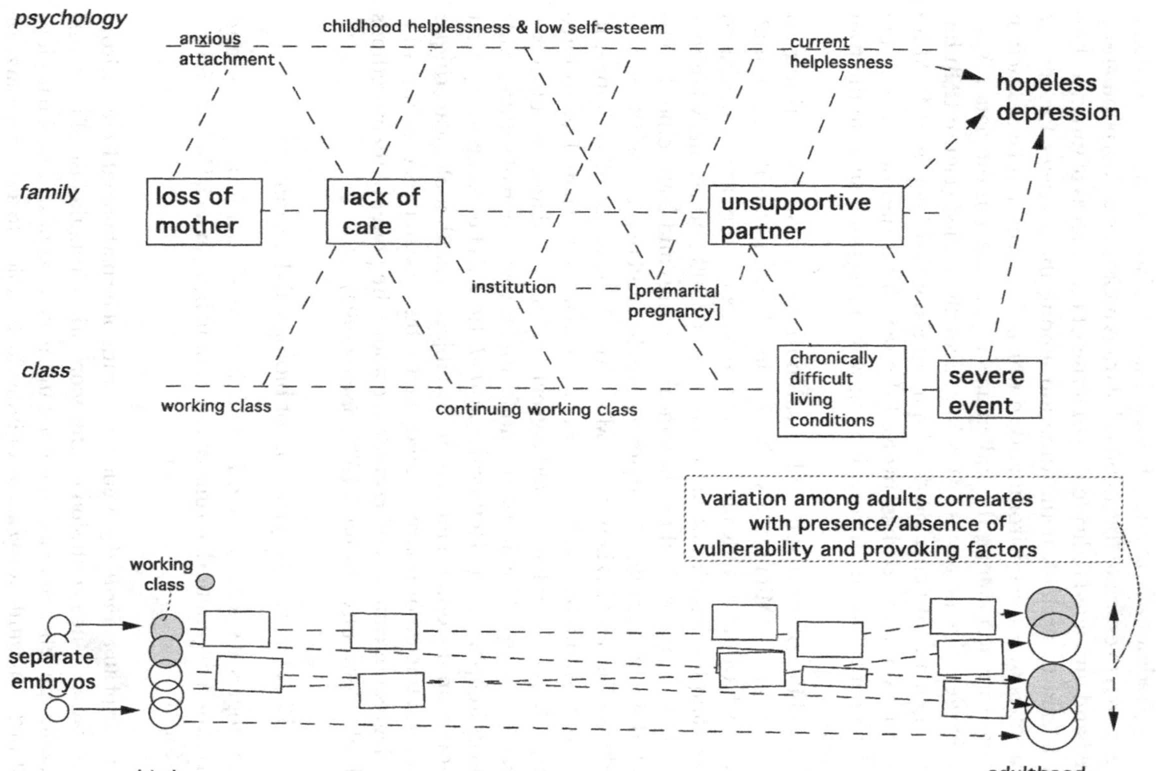

Figure 12.2. Pathways to severe depression in a study of working-class women. The dashed lines in the top diagram (based on Bowlby 1988) indicate that each strand tends to build on what has happened earlier in the different strands.

12.3.3 Behavioral genetics

The field of behavioral genetics attempts to identify the contribution of inherited factors on specific behaviors and on general psychological measures, most notably IQ. Traditionally, the field has used statistical tools derived from quantitative genetics to estimate "heritability" of traits measured across populations of related and unrelated individuals. (In terms of the analysis of variance, heritability is said to be high if the variation among the averages for genetic types/varieties over the environments in which they grow is a large fraction of the total variation among the individuals represented in the data.) As in quantitative genetics, no genes or DNA are actually studied (Figure 12.3) but a model of genetic influence on development is implied (Figure 12.4). More recently, however, research has involved the search for sites on the genome that make a contribution, in combination with many other sites, to the trait in question.

The credibility of the first line of research rose in the late 1980s, riding on the results from the Minnesota study of twins. Compared with earlier studies, recent behavioral genetic methodology has been more careful and based on larger samples. Significant heritability (up to 50%; occasionally higher) has been found for standard psychological measures and many other behaviors, including divorce rates, male homosexuality, and depression (Bouchard *et al.* 1990, McGue and Lykken 1992). Moreover, the residual effects appear to relate more to within-family differences rather than to the shared family environment, a finding that has elicited a great deal of speculation about causes (Bouchard *et al.* 1990, Plomin 1990) and further investigation of within-family differences in upbringing (Hetherington *et al.* 1994). The search for sites on the genome, on the other hand, has been subject to methodological critique and several retracted or nonreplicated claims, but researchers continue with varying degrees of caution and confidence about the power of their methods to yield reliable results (Aldhous 1992, *Science* 1994).

12.3.4 Scientific criticism of biological determinism

The main angles of scientific opposition to the field of behavioral genetics and its contribution to genetic determinist views about human social behavior have been as follows.

1. In the case of an *individual*, genetic causes cannot be partitioned from environmental causes. Much confusion on this score arises from the use of the terms heritability, genetic, and environmental in the context of statistical partitioning of variation within a *population* of individuals subject to a specific range of environments (Lewontin 1974b).
2. Heritability is not logically or empirically related to differences among the average values of groups nor to difficulty of changing the trait in question (Lewontin 1982, Block 1995).

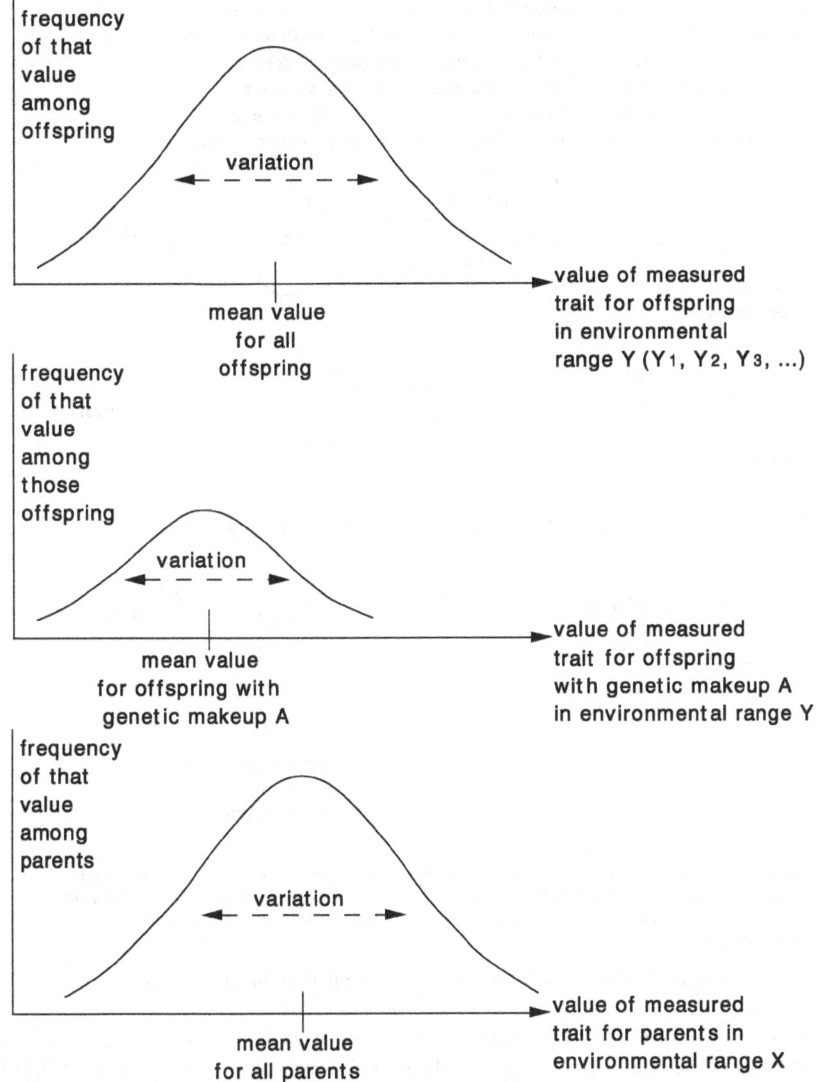

Notes: Parental values or means do not enter the calculation of heritability. The ranking of parents and their offspring need not be the same, especially when environmental range Y differs from environmental range X.

Figure 12.3. The basic statistical quality of heritability as defined by quantitative genetics.

3. Behavioral geneticists, although aware of the preceding two points, rarely incorporate them into their interpretations and ongoing research (Schiff and Lewontin 1986, pp. 220–2). They quickly discount (Plomin *et al.* 1990, p. 350), or do not even mention (Bouchard and Propping 1993), results showing that IQ of adopted children, although correlated, is on average significantly higher than that of their birth mothers.

Humans cannot be bred to produce multiple individuals of the same genetic type all grown in the same range of environments Y. Heritability can be estimated indirectly by comparing the average variation between many pairs of identical (monozygotic) twins raised in the same family (in presumably the same environment*) with the average variation between many pairs of identical twins raised apart (in environments presumably sampled from the full range of environments).

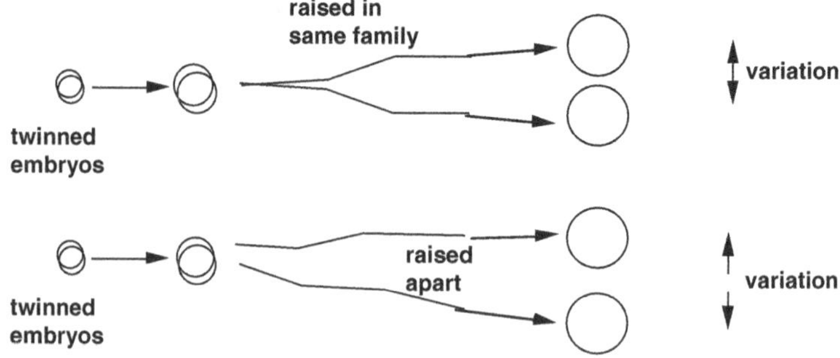

By implication – not experiment – the general case of two unrelated individuals:

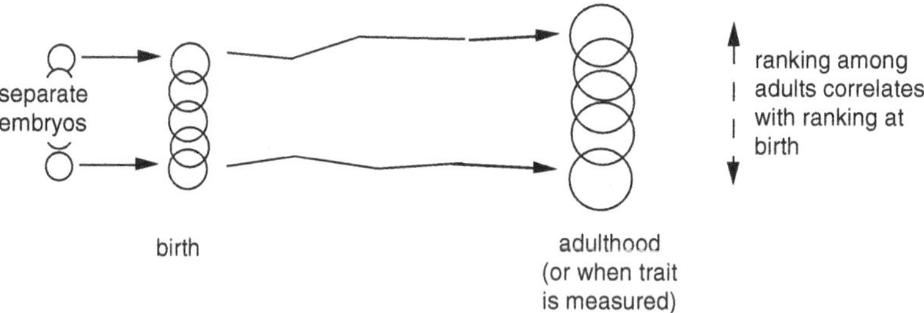

*A refined version of this analysis does not assume that the environment within a family is the same, but compares variation between pairs of monozygotic twins, nonmonozygotic twins, and sibs raised in same family so as to estimate the relative effect of being raised in different families with within-family environmental differences.

Figure 12.4. The model of development implicit in quantitative genetics.

4. Environmental or social factors can influence psychological traits, IQ, and other measurable behaviors greatly, as indicated, for example, by the effect on IQ of adoption up the socioeconomic scale (Schiff and Lewontin 1986) and by the Flynn effect – the steady improvement over time of IQ scores in most countries (Flynn 1987).
5. Behavioral genetic analyses have been based on flawed methodology and unreliable data (Kamin 1974, Devlin *et al.* 1997).

Despite their belief in the validity of these points, some critics of biological determinism have expressed a sense of vulnerability. Stewart (1979), for example, asked what would happen to their critique if a methodologically tight study demonstrated a clear DNA–behavior connection in the etiology, say, of schizophrenia in some sufferers (see Gottesman 1991). The tighter methodology and results in recent behavioral genetics can only add to these concerns. Furthermore, significant caveats are now attached to the examples often quoted to demonstrate that change is quite possible. In addition to the account of Paul (1997) on PKU screening, Woodhead (1988) has examined the contextual factors that contribute to sustained effects after early educational interventions such as Headstart programs in the USA.

12.3.5 Models of reciprocal causation of any individual's developing traits and environment

Recent "reciprocal causation" models of IQ development attempt to reconcile two observations that might appear incompatible, namely, high heritability reported for IQ (Neisser *et al.* 1996) and large gains from one generation to the next (Flynn 1987). These models allow for both observations through a matching of traits and the changing environments in which traits develop (Dickens and Flynn 2001). Such matching would occur when the higher IQ child seeks out or is provided with "cognitively enriching experiences" beyond those in a standard school setting. Matching means that small differences at birth can be amplified, especially if every individual's environment follows society-wide trends that result from many other individuals' changes (Figure 12.5). Furthermore, the environment can play a significant role without lowering heritability or resting on a single "factor-X" to account for the gains from one generation to the next.

Plausible parameters inserted into reciprocal causation models not only yield high heritability and generational gains, but allow for decay of IQ gains after Headstart and other short-term enrichment programs (i.e., nonmatching environments) end. Particular sets of data have yet, however, to be fitted to the models or used to discriminate among alternative forms of reciprocal causation (Loehlin 2002).

12.4 Some ways that the approaches challenge each other

To draw out the significance of research into gestational programming (GP), life events and difficulties (LED), and reciprocal causation (RC) models of IQ development, let me map various ways that these three approaches challenge each other. I will also map ways that they challenge more traditional accounts of gene–environment interactions emerging from behavioral genetics (BG) and from criticism of biological determinism (CBD). Notice that, because CBD challenges BG, the description of CBD above has given a preview of this mutual interrogation of approaches.

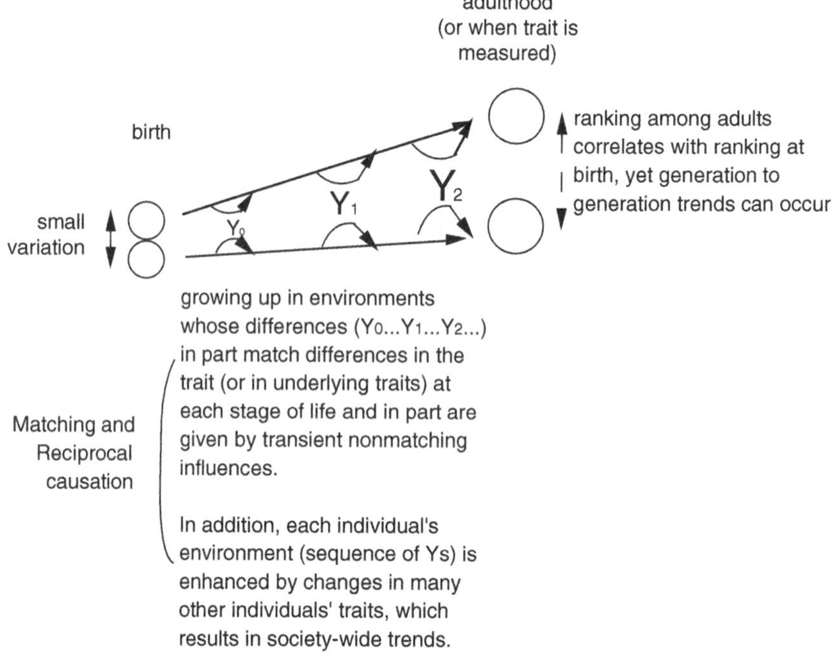

Figure 12.5. The reciprocal causation model of Dickens and Flynn (2001), in which there is a matching of traits and the changing environments in which traits develop.

12.4.1 Gestational programming

GP challenges LED. To the extent that LED research seeks to account for more of the variation among individuals who experience comparable life events, it may be productive for it to incorporate the effects of gestational nutrition. GP and LED, as well as BG, have examined heart disease (Forsén *et al.* 1997, Neilson *et al.* 1989, Ewart 1994, Barker 1995b), so this would be an obvious locus of comparison among the approaches.

GP challenges BG. The research design and analyses of BG rely heavily on the independence of environments for twins separated in infancy (Bouchard and Propping 1993). The findings of gestational programming indicate that separated twins have significant shared environments, namely, the gestational environment. Behavioral geneticists may, therefore, need to reconsider their reports of high heritabilities for behavioral differences and the method of twin studies may have to be reevaluated. With respect to the question of environmental influences experienced by biological or adopted siblings, the shared or nonshared gestational environments also need to be taken into consideration.

GP challenges CBD. Critics of genetic determinist views now face in GP a biological determinism that has clear environmental causes. That is, once the individual has experienced the adverse nutrition regime as a fetus, it is predisposed towards the corresponding diseases of older age. Should the critics view these findings of GP favorably, or develop a critique equivalent to that of behavioral genetics outlined above?

GP challenges RC. In the same way that GP has begun to resolve some inconsistencies among findings in different countries, RC could address the historical contingencies of social change in different countries. (In this vein Woodhead 1988 summarizes studies explaining how the IQ increases produced by Headstart programs tend to be transient yet in the long term the children end up with significantly higher high-school graduation rates, employment, and many other socially valued measures.)

12.4.2 Life events and difficulties research

LED challenges GP. As indicated earlier in the case of Finland, GP is beginning to resolve apparent inconsistencies by incorporating the contingencies of changing gestational conditions from one generation to the next. In this work, as well as in attempts to incorporate factors from childhood and adult life, it could be productive for GP researchers to address both the findings and case history approach of LED research.

LED challenges BG and CBD. The environmental analyses offered by critics of biological determinism and by behavioral geneticists are based on the possibility of finding correlations between the trait studied and distinct environmental factors. The analyses in BG and CBD differ mostly around whether the factors are shared by all in a family, e.g., socioeconomic status, or not. (Although Sulloway, 1996, works outside behavioral genetics, his examination of a within-family difference for scientists, namely, their birth order, also relies on analysis of correlations.) LED methodology goes beyond correlational analysis by teasing apart the *sequence* of events in *different individuals'* lifetimes that render the individual vulnerable or protected from proximal causes that provoke onset (or recurrence) of the condition in question. In this way, LED research addresses malleability or immalleability of behavioral outcomes without, in principle, requiring that genetic contributions are either ruled out or privileged as explanatory factors (Taylor 1995).

LED challenges RC. Although RC models include a series of steps in an individual's development over time, each step is still formulated in terms around correlations of distinct, pre-identified factors. In contrast, factors in LED emerge after events are rated according to their expected significance to the individual in their specific context. An equivalent context-sensitive method may help develop RC models to a point at which they can be fitted to particular sets of data and these data can be used to discriminate among alternative forms of the models.

12.4.3 Reciprocal causation models

RC challenges GP. For certain physical conditions, such as heart disease, the effect of gestational environment can be separated from later life experience, but the reciprocal causation during an individual's life course may need to be incorporated to account for the unexplained variation among individuals, especially in the cases of IQ or behavioral conditions.

RC challenges LED. When LED researchers focus on explaining the onset (or recurrence) of the condition in question, considerable unexplained variation remains, leaving room for further inquiry into differential vulnerability. The relationship between personality and subsequent severe life events invites attention to cycles of reciprocal causation, which may, as RC models assume, amplify small initial differences.

RC challenges BG. Behavioral geneticists are confident that high (within-group) heritabilities and large IQ differences between the averages for (socially defined) racial groups are connected because the groups differ genetically and genes are the primary determinant of IQ. They see no plausible environmental factor that differs among groups and could account for the average difference. In the case, however, of large gains from one generation to the next (Flynn 1987), environmental conditions must be invoked; the gains cannot be related to genetic changes in the populations over the time span of only one generation. By logical extension, RC models challenge BG's focus on genetic explanations of IQ development and variation.

RC challenges CBD. CBD highlights the logical fact that high heritability does not prevent changing the trait in question by changing the environmental conditions. At the same time, critics have worked hard to expose flaws in studies purporting to show high heritability, even though by their own logic nothing should hang on the magnitude of the heritability. Critics need not contest every high heritability estimate, however, once it is recognized that RC models allow high heritability *and* a significant role for environmental conditions. Instead, the emphasis in CBD could shift to exposing and contesting the self-fulfilling quality of the matching. After all, RC models do not dictate that "cognitively enriching experiences" must be channeled only to those whose slightly higher IQ might lead them to seek out such experiences on their own.

12.5 From commentary to engagement

This chapter's introduction described the three approaches as *complicating* the persistent contrasts: inborn and unchangeable vs. environmental and changeable; and biological vs. social. In the previous section I wrote of the approaches *challenging* each other. In both cases I might have better said having the *potential* for complicating and challenging; further steps are needed to realize such potential.

One direction for these steps would involve commentators on science delving more deeply than my brief overviews have into the research involved in the different approaches. Writing about their intellectual history and current concerns – their questions, concepts, methods, and findings – should help readers see that much is overlooked when scientists and boosters of biotechnology claim that understanding of diseases or the development of normal traits will flow from sequencing DNA and determining the traits' genetic bases. A comparison of contrasting approaches, such as GP, LED, and RC, would enrich the discussion of the complexities of environmental influences on developing organisms, and would do so without setting environmental influence in opposition to genetic control.

Another direction would involve commentary on the dynamics of science, exposing more of the relation between the approaches and their social context: How is the research funded and organized? How are the methods and findings received in the scientific community and in society more broadly? How do the scientists respond to or resist criticisms of their work? How do they use metaphors and rhetoric to sway their audiences? Writing about such matters should help readers see that much is overlooked when accounts of science portray advances as a matter of scientists uncovering, to the extent of their current technical capacity, the workings of nature. Again, a comparison of contrasting approaches could be helpful, allowing researchers in the social studies of science and technology to affirm that the workings of science are not simply or directly driving towards increased understanding of reality. (The possibilities of comparison entered strongly into my choosing to study the three approaches introduced in this essay – LED has a long history in the UK, but has never become well established in the USA; GP also originated in the UK, but took off in the USA in the late 1990s; and RC models involve a new contribution by a key figure in the area of IQ and heritability, and responses are mostly yet to emerge.)

A third direction would involve commentators on science and its social dynamics engaging in those dynamics with a view to influencing the future path of the science. Of course, writing commentaries is in itself a form of engagement. But it is possible to become a more "reflexive" agent of scientific change, to delve more self-consciously into the social dynamics with a view to identifying points of potential engagement for oneself and others (Scott *et al.* 1990). Such points of engagement are often specific to the individual in their context and provisional, subject to ongoing modification in light of responses to engagement and other changes in the individual's context. Nevertheless, writing about engagements that link science and its social dynamics may help readers to draw and explore analogies in their own situations. In this spirit, let me mention three modest examples from my own work in progress as a teacher and researcher of science in its social context.

In a course on "biology and society" I have used the LED case on depression to move discussion beyond the genes–environment dichotomy. After presenting Bowlby's reconstruction of Brown and Harris's analysis (Figure 12.2), I add a hypothetical genetic–biochemical strand, in which the developing individual is more susceptible to the biochemical shifts that are associated with depression. Although early diagnosis and lifelong treatment with prophylactic antidepressants could reduce the chances of onset of severe depression, there are many other readily conceivable engagements, such as quick action to ensure a reliable caregiver when a mother dies or is hospitalized, contraceptive education for adolescents, increasing state support for single mothers, and so on (Taylor 1995). In a follow-up activity I have students take the idea of multiple points of engagement in cross-connecting strands and extend it to their own development as future scientists or health care professionals. They try their hand at diagramming the life course that brought them to attend this kind of course, and I invite them to reflect on the contingent intersections of outside influences and their own agency. In short, at the same time as I stimulate students to think about more complex causal accounts in science, I ask them to think more deliberately about the complex causal connections that may shape their future work in science.

In my research I have begun to bring GP, LED, and RC to the attention of exponents of the other approaches. My plan is to note their immediate responses during interviews and to keep track of subsequent developments, if any, in their analyses and discussions of gene–environment interactions. This material should flesh out the potential challenges identified in the previous section at the same time as it reveals more of the dynamics of scientific change – or resistance to change. When I have enough material to present, I hope I can use it to stimulate more complex thinking, not only about the development of behavioral and medical conditions over any individual's lifetime, but also about multiple points of potential engagement that can affect those developmental processes.

In my teaching and my writing, I try to illustrate ways that close examination of concepts and methods within any given natural or social science can motivate or animate interpretations of the social influences on those sciences, and vice versa. In some accounts prepared for nonspecialist audiences I draw correlations between scientists' ideas and the actions that the ideas facilitate; that is, I interpret the science as building in the social action favored or privileged by the scientists (Taylor and Buttel 1992; but see Taylor 1995 for a more complex account). In this vein, after reflecting on some remarks made by Richard Lewontin during the course of a recent discussion around a brief account of RC models of IQ that I had sent him, I have begun to see curious commonalities between BG and CBD. Both sides, it seems, build into their analyses a view of social action as overarching change effected by some superintending agency. Although there is not space here to provide the full conceptual–social reconstruction behind this assertion, let me say a little more, which will allow me to return in closing to some issues raised in the preamble.

Plant breeding research and its recommendations to farmers can be effective without knowledge of biophysical causes involved in the pathways of plant growth and development and in the ways these pathways are affected by the different treatments (e.g., levels of fertilizer applied) studied. All that is required to cause the desired yields is control over which varieties to interbreed or plant and ability to replicate environmental conditions. This control/replication model of causes underlies the analysis of variance and related tools of quantitative genetics. Yet, the genetic and environmental control that makes this model of causes useful in agriculture is not possible for humans and their environments. The control/replication model, nevertheless, still shapes debates between BG and CBD. In CBD the conceptual point that heritability does not mean unchangeable is often illustrated with thought experiments that involve well-defined varieties of plants all grown in the same nutrient-deficient environment before they are all shifted to a uniform nutrient-rich environment. Moreover, when CBD addresses the conservative policy implications drawn from heritability research, it posits intelligence-boosting environments that American society has not yet explored (Lewontin 1970b). Likewise, when BG searches (unsuccessfully) for environmental factor-Xs that correlate with differences among means for racial groups, it also proceeds as if there could be something about American society that treats each racial group uniformly and differently from other groups.

Statistical analysis can be used with other models of causes that, unlike the analysis of variance, do not assume control and replication. Significant patterns can be considered to be invitations to search for underlying causes, which in agriculture would mean the biophysical causes involved in the pathways of plant growth and development under various treatments. In research on human behavior the search for underlying causes would mean piecing together the biosocial pathways of growth and development of the persons in particular conditions. Because BG researchers and their critics cannot – except in eugenic or revolutionary scenarios – have the control available to agricultural researchers, the alternative causal model might yield more insight, not only about current observations, but about potentialities for change.

The contrasting model is evident in LED research, to an increasing extent in GP, and potentially in RC. Recall the LED method: subjects are interviewed in depth to produce detailed case histories of life events and difficulties; the seriousness of events in the transcripts is rated with reference to context (but without knowledge of whether the subject became ill); and statistical analyses are used to identify combinations of proximate and background factors that distinguish ill from healthy subjects. The results are used to raise further questions and inquiry. For example, Brown and Harris (1978, p. 271) found that a supportive, confiding relationship with a partner had an effect in protecting a woman from developing depression after experiencing a severe event. They suggest, however, that the effect "might have little to do with confiding as such but with, say, the way she is able to think about the marriage and value it." Suggestions of this kind have led to an active interplay, characteristic of LED research, between statistical analysis of past case histories and design of interviews and rating schemes for new situations. More recently, LED has included interventions, such as home visits by volunteer conversation partners, in their research designs (Harris 2000).

LED and GP research suggest to me that it is possible for sociologists and biologists to measure and model interesting things about the complexities of biosocial development of human traits. To do so requires an approach that looks not for society-wide factors, but attends to contingency of development and its meaning in particular contexts. This suggests a lesson for evolutionary biology. I proposed in an earlier Lewontin festschrift (Taylor 2001) that "[i]n the center of any historical account we should see a lineage of active organisms, organisms that construct their responses to [dynamic ecological] situations that earlier responses in their lifetime and their lineage's earlier responses have helped construct." My essay was critical of population genetic models for "compressing organism–organism and organism–environment relationships into the fitness conferred on an organism by its characters." I did not, however, propose replacement models or methods for investigating evolution in a dynamic ecological context. I can now imagine a program applicable at least for microevolutionary research, which, like LED, integrates quantitative "epidemiology" and case history. To paraphrase my summary above: organisms could be observed in depth to produce detailed case histories of life events and difficulties; the events in the observations could be rated with reference to ecological context (but without knowledge of whether the organism survived and reproduced); and statistical analyses could be used to identify combinations of proximate and background factors that distinguish organisms that contribute to the next generation from those that did not. The results would be used to raise further questions and inquiry.

Of course, it is one thing for a commentator on science to suggest that something might be interesting to study; it is quite another to engage in the social dynamics of a science with a view to influencing its future path. I hope, at least, to provoke some further discussion of gene–environment complexities.

12.6 Acknowledgments

This chapter has developed in response to comments of Les Levidow, Susan Oyama, and audiences at the International Society for History, Philosophy, and Social Studies of Biology meetings in Oaxaca, Mexico, and at the University of California, San Francisco. David Gray, Amita Sudhir, and Suzanne Clark provided valuable research assistance. Peter Nathanielsz encouraged my early interest in Barker's research. George Brown, Tirril Harris, and Richard Lewontin were generous in the time they gave for interviews.

REFERENCES

Aldhous, P. (1992). The promise and pitfalls of molecular genetics. *Science* 257: 164–5.

Barker, D. J. P. (1994). *Mothers, Babies, and Diseases in Later Life.* London: BMJ Publishing Group.

Barker, D. J. P. (1995a). The fetal origins of adult disease. *Proc. R. Soc. Lond. Ser. B* 262:37–43.

Barker, D. J. P. (1995b). Fetal origins of coronary heart disease. *Brit. Med. J.* 311:171–4.

Barker, D. J. P., Osmond, C., Rodin, I., Fall, C. H. D., and Winter, P. D. (1995). Low weight gain in infancy and suicide in adult life. *Brit. Med. J.* 311:1203.

Block, N. (1995). How heritability misleads about race. *Cognition* 56:99–128.

Bouchard, T. J., and Propping, P. (eds) (1993). *Twins as a Tool of Behavioral Genetics.* Chichester: Wiley.

Bouchard, T. J., Lykken, D. T., McGue, M., Segal, N. L., and Tellegen, A. (1990). Sources of human psychological differences: the Minnesota study of twins reared apart. *Science* 250:223–8.

Bowlby, J. (1988). *A Secure Base.* New York: Basic Books.

Brown, G. W. (1989). Life events and measurement. In G. W. Brown and T. O. Harris (eds) *Life Events and Illness,* pp. 3–45. New York: Guilford.

Brown, G. W., and Harris, T. O. (1978). *Social Origins of Depression.* New York: Free Press.

Brown, G. W., and Harris, T. O. (eds) (1989a). *Life Events and Illness.* New York: Guilford Press.

Brown, G. W., and Harris, T. O. (1989b). Depression. In G. W. Brown and T. O. Harris (eds) *Life Events and Illness,* pp. 49–93. New York: Guilford Press.

Brown, G. W., and Moran, P. M. (1997). Single mothers, poverty and depression. *Psychol. Med.* 27:21–33.

Devlin, B., Resnick, D. P., and Fienberg, S. E. (eds) (1997). *Intelligence, Genes and Success: Scientists Respond to the Bell Curve.* New York: Springer-Verlag.

Dickens, W. T., and Flynn, J. R. (2001). Heritability estimates versus large environmental effects: the IQ paradox resolved. *Psychol. Rev.* 108:346–69.

Ewart, C. (1994). Non-shared environments and heart disease risk: concepts and data for a model of coronary-prone behavior. In E. M. Hetherington, D. Reiss and R. Plomin (eds) *Separate Social Worlds of Siblings: the Impact of Nonshared Environment on Development,* pp. 175–204. Hillsdale, NJ: Lawrence Erlbaum Associates.

Falconer, D. S. (1989). *Introduction to Quantitative Genetics,* 3rd edition. Harlow, Essex: Longman.

Flynn, J. R. (1987). Massive IQ gains in 14 nations: what IQ tests really measure. *Psychol. Bull.* 101:171–91.

Forsén, T., Eriksson, J. G., Tuomilehto, J., Teramo, K., Osmond, C., and Barker, D. J. P. (1997). Mother's weight in pregnancy and coronary heart disease in a cohort of Finnish men: follow up study. *Brit. Med. J.* 315:837–40.

Frankel, S., Elwood, P., Sweetnam, P., Yarnell, J., and Smith, G. D. (1996). Birthweight, body-mass index in middle age, and incident coronary heart disease. *The Lancet* 348:1478–80.

Gilbert, S. (1997). *Developmental Biology*. Sunderland, MA: Sinauer.

Gottesman, I. I. (1991). *Schizophrenia Genesis: the Origins of Madness.* New York: W. H. Freeman.

Harris, T. (ed.) (2000). *Where Inner and Outer Worlds Meet.* London: Routledge.

Herrnstein, R. J., and Murray, C. (1994). *The Bell Curve: Intelligence and Class Structure in American Life.* New York: Free Press.

Hetherington, E. M., Reiss, D., and Plomin, R. (eds) (1994). *Separate Social Worlds of Siblings: the Impact of Nonshared Environment on Development.* Hillsdale, NJ: Lawrence Erlbaum Associates.

Jensen, A. R. (1969). How much can we boost IQ and scholastic achievement? *Harvard Educational Rev.* 39:1–123.

Jensen, A. R. (1970). Race and the genetics of intelligence: a reply to Lewontin. *Bull. Atom. Sci.* 26:17–23.

Kamin, L. J. (1974). *The Science and Politics of I.Q.* New York: John Wiley.

Kramer, M. S., and Joseph, K. S. (1996). Enigma of fetal/infant origins hypothesis. *Lancet (North Am. edition)* 348:1254–5.

Lewontin, R. C. (1970a). Race and intelligence. *Bull. Atomic Scientists* 26:2–8.

Lewontin, R. C. (1970b). Further remarks on race and the genetics of intelligence. *Bull. Atom. Sci.* 26:23–5.

Lewontin, R. C. (1974a). *The Genetic Basis of Evolutionary Change.* New York: Columbia University Press.

Lewontin, R. C. (1974b). The analysis of variance and the analysis of causes. *Am. J. Hum. Genet.* 26:400–11.

Lewontin, R. (1982). Mental traits. In R. Lewontin (ed.) *Human Diversity*, pp. 88–103. New York: W. H. Freeman.

Loehlin, J. C. (2002). The IQ Paradox: Resolved? Still an Open Question (unpublished ms.)

McGue, M., and Lykken, D. T. (1992). Genetic influence on risk of divorce. *Psychol. Sci.* 3:368–73.

Neilson, E., Brown, G. W., and Marmot, M. (1989). Myocardial infarction. In G. W. Brown and T. O. Harris (eds) *Life Events and Illness*, pp. 313–342. New York: Guilford.

Neisser, U., Boodoo, G., Bouchard, T. J., Boykin, A. W., Brody, N., Ceci, S. J., Halpern, D. F., Loehlin, J. C., Perloff, R., Sternberg, R. J., *et al.* (1996). Intelligence: knowns and unknowns. *Am. Psychol.* 51:77–101.

Nelkin, D., and Lindee, M. S. (1995). *The DNA Mystique: the Gene as a Cultural Icon.* New York: W. H. Freeman.

Paneth, N. (1994). The impressionable fetus? Fetal life and adult health. *Am. J. Public Health* 84:1372–4.

Paneth, N., and Susser, M. (1995). Early origin of coronary heart disease (the "Barker hypothesis"). *Brit. Med. J.* 310:411–12.

Paul, D. (1997). Appendix 5. The history of newborn phenylketonuria screening in the U.S. In N. A. Holtzman and M. S. Watson (eds) *Promoting Safe and Effective Genetic Testing in the United States*, pp. 137–59. Washington, DC: NIH-DOE Working Group on the Ethical, Legal, and Social Implications of Human Genome Research.

Plomin, R. (1990). *Nature and Nurture: an Introduction to Behavioral Genetics.* Pacific Grove, CA: Brooks/Cole.

Plomin, R., Defries, J. C., and McClearn, G. E. (1990). *Behavioral Genetics.* New York: W. H. Freeman.

Schiff, M., and Lewontin, R. C. (1986). *Education and Class.* New York: Oxford University Press.

Science (1994). Behavioral genetics in transition (special section). *Science* 264:1686–97.

Scott, P., Richards, E., and Martin, B. (1990). Captives of controversy: the myth of the neutral social researcher in contemporary scientific controversies. *Sci. Technol. Hum. Values* 15:474–94.

Scrimshaw, N. (1997). The relation between fetal malnutrition and chronic disease in later life. *Brit. Med. J.* 315:825–6.

Stewart, J. (1979). Scientific findings that look awkward for socialists: how are we to respond? *Radical Sci.* 8:121–3.

Sulloway, F. J. (1996). *Born to Rebel: Birth Order, Family Dynamics, and Creative Lives.* New York: Pantheon.

Sultan, S. E. (1992). Phenotypic plasticity and the Neo-Darwinian legacy. *Evol. Trends Plants* 6:61–71.

Taylor, P. J. (1995). Building on construction: an exploration of heterogeneous constructionism, using an analogy from psychology and a sketch from socio-economic modeling. *Perspectives on Science* 3:66–98.

Taylor, P. J. (2001). From natural selection to natural construction to disciplining unruly complexity: The challenge of integrating ecological dynamics into evolutionary theory. In R. Singh, K. Krimbas, D. Paul and J. Beatty (eds) *Thinking About Evolution: Historical, Philosophical and Political Perspectives*, pp. 377–93. Cambridge: Cambridge University Press.

Taylor, P. J., and Buttel, F. H. (1992). How do we know we have global environmental problems? Science and the globalization of environmental discourse. *Geoforum* 23:405–16.

van der Weele, C. (1995). *Images of Development: Environmental Causes in Ontogeny.* Amsterdam: University of Amsterdam, Ph.D. dissertation.

Woodhead, M. (1988). When psychology informs public policy. *Am. Psychol.* 43: 443–54.

13

Genus-specific diversification of mating types

MARCY K. UYENOYAMA, NAOKI TAKEBAYASHI
Department of Biology, Duke University, Durham

13.1 Introduction

Self-incompatibility (SI) discourages self-fertilization in many hermaphroditic plants by causing the rejection of pollen that express genetically encoded specificities in common with the pistil. In the best-studied systems, a single region (*S*-locus) determines mating specificity in both pollen and pistil. Because pollen that express rarer specificities encounter rejection at lower rates, SI engenders intense balancing selection, maintaining dozens of *S*-alleles within local populations.

While the estimation of genealogical relationships among taxa is paramount in many phylogenetic studies, our primary objective here is to elucidate the evolutionary history of SI. This objective is served through the analysis of topology (pattern and order of branching) and genealogical shape (relative sizes of various features of the tree). Several likelihood- or moment-based methods for the estimation of demographic parameters under specified models have recently been developed (for example, Takahata *et al.* 1995, Yang 1997, Wakeley and Hey 1997, Nielsen 1998, Bahlo and Griffiths 2000, Beerli and Felsenstein 2001). Along a different tack, Hey (1992) and Nee *et al.* (1994) have based the inference of the rate of origin and extinction of taxa on the pattern of intervals between nodes in gene genealogies. Indeed, as Holmes *et al.* (1999) indicate, these methods seek to examine tree shape without having to estimate fundamental quantities such as Nu (product of effective population size and mutation rate). The method of Uyenoyama (1997) was explicitly designed to analyze tree shape apart from tree size, by examining ratios of sums of branch lengths from which Nu cancels out.

This chapter addresses the diversification of SI mating types. We describe the fundamental Wright (1939) model of SI, which determines the expected rate of bifurcation of *S*-allele lineages and number of segregating *S*-alleles as functions of effective population size and mutation rate. Our analysis indicates

The Evolution of Population Biology, ed. R. S. Singh and M. K. Uyenoyama. Published by Cambridge University Press.

that *S*-allele diversification has proceeded under different evolutionary processes in two solanaceous genera.

13.2 Determination of specificity

Under gametophytic SI (GSI), pollen specificity is determined in the gametophyte stage, by the genotype of the pollen, while under sporophytic SI (SSI), it is determined in the sporophyte stage, by the genotype of the parent plant that produced the pollen. In the form of SSI expressed in *Brassica*, proteins borne in the pollen coat determine pollen specificity (Stephenson *et al.* 1997, Schopfer *et al.* 1999, Takayama *et al.* 2000). Recognition of these proteins by a receptor protein kinase bound in the plasma membrane of stigma epidermal cells blocks the production by the stigma of hydrating factors required for germination of pollen grains (Takasaki *et al.* 2000). In the form of GSI expressed in species of the Solanaceae, Rosaceae, and Scrophulariaceae, an extracellular ribonuclease (*S*-RNase) produced in the stylar transmitting tract inhibits the growth of incompatible pollen tubes (McClure *et al.* 1990, Lush and Clarke 1997).

In the SSI system expressed in *Brassica*, a pollen coat protein encoded by a gene originally designated *SP11* by Suzuki *et al.* (1999) determines pollen specificity (Schopfer *et al.* 1999, Takayama *et al.* 2000). In the *S*-RNase-based GSI system of the Solanaceae, the determinant of pollen specificity has not yet been identified. Leading hypotheses for the mechanism of pollen inhibition in the *S*-RNase system include the receptor/gatekeeper model and the inhibitor model (see Golz *et al.* 1999, McCubbin and Kao 2000). Under the former model, a highly specific receptor permits entry into the pollen tube of only the *S*-RNase that matches the pollen specificity, while under the latter, all *S*-RNases may enter a pollen tube but only the matching specificity escapes inhibition. Detection of an *S*-RNase of a different specificity in the cytoplasm of pollen tubes lends support to the inhibitor model (Luu *et al.* 2000), although the precise mechanism is still to be determined (Luu *et al.* 2001, Golz *et al.* 2001).

Analyses of the pattern of sequence variation have contributed towards the identification of candidates for nucleotide sites involved in the determination of pistil specificity. Reasoning that specificity-determining sites likely differ between but not within *S*-specificities, Ioerger *et al.* (1991) proposed the involvement of two regions, designated HVa and HVb, which appear to show hypervariability among solanaceous *S*-RNases. Balancing selection is expected to accelerate substitution at specificity-determining sites (Hughes and Nei 1988, Takahata 1990, Sasaki 1992). In their analysis of *S*-RNases from the three plant families that express this form of GSI, Ishimizu *et al.* (1998) detected a significant excess of nonsynonymous over synonymous substitutions in four regions, which include HVa and HVb.

Definitive identification of specificity-determining sites awaits direct experimental analysis. The region of the *S*-RNase gene on which we based our estimates of genealogical relationships among *S*-alleles spans a number of these candidate regions.

13.3 Pattern of divergence

Figure 13.1 shows a maximum-likelihood (ML) genealogy reconstructed under a molecular clock from nucleotide sequences of *S*-RNases derived from the Solanaceae. From the nearly 160 solanaceous sequences available in GENBANK, we excluded those less than 200 base pairs in length, confining our analysis to 131 sequences at least 318 base pairs in length. Modeltest 3.04 (Posada and Crandall 1998) indicated that base substitution in this data set has proceeded under the Hasegawa, Kishino, and Yano (Hasegawa *et al.* 1985) model with gamma-distributed mutation rates among sites and the presence of invariable sites. Figure 13.1 shows the ML genealogy generated by Paup* 4.0b10 (Swofford 1999) under these specifications. A full description of the genealogical reconstruction will appear elsewhere (Takebayashi and Uyenoyama, in preparation).

From three species of the genus *Physalis* (*P. cineracens*, *P. crassifolia*, and *P. longifolia*), 62 distinct *S*-RNase sequences have been derived from 70 individual plants (Richman *et al.* 1995, 1996, Richman and Kohn 1999, Lu 2001). This rate of discovery of new *S*-alleles provides an estimate of 72 for the

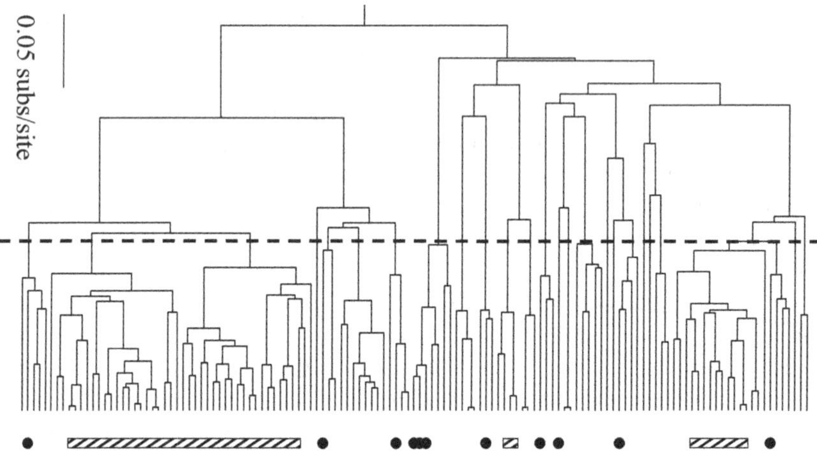

Figure 13.1. Maximum-likelihood genealogy, estimated under the Hasegawa *et al.* (1985) model with gamma-distributed mutation rates, of 131 *S*-alleles derived from the Solanaceae. Bars indicate sequences derived from species of *Physalis* (*P. cineracens*, *P. crassifolia*, and *P. longifolia*) and circles are *Solanum carolinense*. The broken line indicates the most recent node from which *S*-alleles derived from both *Physalis* and *Solanum* descend.

total number of *S*-alleles segregating within the genus (Paxman 1963). In contrast, the estimate of 13 *S*-alleles for *Solanum carolinense* is the lowest among all GSI systems for which this quantity has been determined (Lawrence 2000). Excluding short sequences, our analysis comprises 52 *Physalis* and 11 *Solanum* sequences. Figure 13.1 indicates that the many *Physalis* *S*-alleles fall in a few main clusters (bars), while the much fewer derived from *S. carolinense* (circles) span nearly the entire range of variation observed in the Solanaceae.

13.4 Process of divergence

In order to explore whether the patterns of diversity among *S*-allele lineages in *Physalis* and *Solanum* reflect the operation of the same evolutionary process under different parameter values or fundamentally different evolutionary processes, we first address the process of diversification of *S*-allele lineages.

13.4.1 Population and sample genealogies

A gene genealogy reconstructed on the basis of a sample from a population estimates a subset of the gene genealogy of the entire population. A given node in the population tree can be captured in the sample tree only if the sample includes descendants of both sister lineages. Modifying the approach of Takahata *et al.* (1992) and Hey (1992), we characterize the rate at which nodes appear in the population tree and are captured in the sample tree.

Turnover among mating types reflects invasion of new specificities and extinction of existing specificities. Denote the number of common specificities segregating in the population at steady state by n and the number of distinct specificities observed in the sample by m. The quantity n corresponds to the effective number of alleles (Kimura and Crow 1964), defined as the inverse of steady-state homozygosity

$$F = \sum x_i^2 = 1/n, \tag{13.1}$$

for x_i, the frequency of the ith *S*-allele in the population.

On average, the rise of each new specificity to common frequencies implies the extinction of one existing common specificity. At stochastic equilibrium, new specificities attain common frequencies within the population at rate λ. With probability $1 - 1/n$, bifurcation of an *S*-allele lineage in the population tree accompanies such an event only if the new specificity displaces one of the $n-1$ segregating types other than its immediate parent.

Consider state j of the sample tree, corresponding to the segment of the tree that contains j lineages ancestral to the m sampled specificities. A coalescence event in the population tree terminates state j of the sample tree only if both coalescing lineages occur among the j lineages of state j, implying a rate of termination of state j of

$$\frac{\lambda(1-1/n)\binom{n-2}{j-2}}{\binom{n}{j}} = \frac{\lambda j(j-1)}{n^2} \tag{13.2}$$

(compare Takahata *et al.* 1992, Hey 1992). This expression indicates that the rate of coalescence in the sample tree depends upon the number of lineages (j), the effective number of S-alleles maintained in the population (n), and the rate at which new specificities rise to common frequencies (λ). Wright's (1939) diffusion approximation of GSI evolution provides expressions for the number of common S-alleles (n) and the rate at which new S-alleles become common (λ) in terms of the effective population size and rate of mutation to new S-specificities.

13.4.2 Wright's diffusion model of GSI

Wright (1939) constructed a diffusion approximation to explore the joint operation of genetic drift, mutation, and the intense selection engendered by self-incompatibility. Zygotes of S-locus genotype $S_i S_j$ may receive S-allele S_i through the egg or pollen cell:

$$P'_{ij} = \left[q_i \sum_{k\neq i,j} \frac{P_{jk}}{1-q_j-q_k} + q_j \sum_{k\neq i,j} \frac{P_{ik}}{1-q_i-q_k} \right] /2,$$

in which P_{ij} denotes the frequency of $S_i S_j$ zygotes, the prime frequencies in the next generation, and q_k the frequency of S_k among pollen. The denominators ensure equal seed set among genotypes after rejection of incompatible pollen.

Stylar expression of GSI operates at the genotypic level, through rejection of pollen that express specificities in common with the pistil. Nevertheless, Wright succeeded in reducing the analysis to a one-dimensional diffusion approximation of allele frequency change by imposing symmetry assumptions. Because GSI distinguishes only between identical and nonidentical pollen specificities, S-alleles are exchangeable: the probability distributions of allele frequencies are unchanged under permutation of allele names. For $q(=q_i)$ the frequency of S-allele S_i, the assumption of symmetry among the $n-1$ other S-alleles segregating in the population implies

$$q_j = (1-q)/(n-1),$$

for j different from i. Symmetry among genotypic frequencies implies

$$P_{ij} = P,$$
$$P_{jk} = \frac{1 - P(n-1)}{\binom{n-1}{2}},$$
$$q = P(n-1)/2, \tag{13.3}$$

for j and k representing any specificity different from i.

Imposition of these symmetry assumptions and substitution of homozygosity for the inverse of allele number ($F = 1/n$) permits reduction of the system to a single dimension

$$\Delta q = \frac{q(F - q)}{1 - 3F + 2qF},$$

for Δq the change in frequency of the focal *S*-allele between generations. Replacement of q in the denominator by F (see Wright 1964) permits further simplification:

$$\Delta q = \frac{q(F - q)}{(1 - F)(1 - 2F)}.$$

Wright's (1939, 1969) diffusion approximation of GSI has drift and diffusion coefficients given by

$$\mu(q) = -aq(q - F) - uq,$$
$$\sigma^2(q) = q(1 - 2q)/(2N), \tag{13.4}$$

for

$$a = \frac{1}{(1 - F)(1 - 2F)},$$

with N representing the effective population size, u the rate of mutation to new *S*-alleles, and F the steady-state homozygosity (see Equation 13.1). The variance coefficient (Equation 13.4) reflects random sampling of zygotes (P_{ij}, see Equation 13.3), rather than of alleles (Fisher 1958, Chapter IV).

Virtually unique among diffusion models of selection, Wright's includes no parameter representing selection intensity. Existence of the diffusion limit as N grows arbitrarily large requires that the intense selection engendered by GSI operate on a time scale comparable to that of genetic drift and mutation:

$$aq(q - F) \sim O(1/(2N)).$$

This expression indicates that the diffusion approximation is valid only for frequencies of the focal allele either close to zero or close to the deterministic equilibrium value ($F = 1/n$). At other frequencies, selection overwhelms genetic drift and mutation, inducing virtually instantaneous jumps between the two regions in which the diffusion approximation applies (compare Sasaki 1989, 1992).

Wright's model shares a number of properties with the SSWM (strong-selection weak-mutation) models studied by Gillespie (1983, 1994): exchangeability, diffusion-scale changes near the extinction boundary and the deterministic equilibrium, and jump behavior between those regions. This approach has permitted the analysis of various aspects of symmetrically overdominant selection (e.g., Slatkin and Muirhead 1999, Slatkin 2000) and has been applied to GSI (Muirhead 2001).

13.4.3 Bifurcation rate and homozygosity

Wright's (1939, 1969) diffusion approximation model provides expressions for the rate at which new *S*-alleles become common (λ) and the steady-state number of common *S*-alleles (n), determinants of the rate of coalescence among *S*-allele lineages.

13.4.3.1 Rate of origin

New specificities become common at a rate proportional to the rate at which new *S*-alleles arise ($2Nu$):

$$\lambda = v[1/(2N)]2Nu,$$

for $v[1/(2N)]$ the probability that an allele introduced in a single copy (frequency $1/(2N)$) increases to frequency $1/n$ before declining to extinction. Substitution of the drift and diffusion coefficients for the Wright model into the expression for $v[1/(2N)]$ (see Karlin and Taylor 1981, Chapter 15) produces

$$\lambda = 2Nu(a - 2b) = \frac{4NuF}{(1 - F)(1 - 2F)} \tag{13.5}$$

(compare Vekemans and Slatkin 1994), for

$$b = \frac{1}{2(1 - F)} + u$$

(see Yokoyama and Hetherington 1982). This expression differs from that of Muirhead (2001), who substituted an approximate probability of fixation for $v[1/(2N)]$.

13.4.3.2 Rate of extinction

A fundamental result from diffusion theory provides $t[x]$, the expected time for an allele frequency initiated at x to reach either endpoint of the interval (α, β) (see Karlin and Taylor 1981, Chapter 15). Strong expression of

GSI maintains the number of *S*-alleles above two, with ultimate extinction the only possible fate of an *S*-allele. To determine the expected time for an *S*-allele to decline to extinction, we take the limits of the interval endpoints ($\alpha \to 0, \beta \to 1/2$). Substitution of the drift and diffusion coefficients of the Wright model yields

$$t[x] = \frac{2a}{(a-2b)^2} e^{N(a-2b)} \left(\frac{2b}{a}\right)^{2Nb} \sqrt{\frac{\pi}{Nb}}.$$

The rate of extinction of any particular *S*-allele corresponds to the reciprocal of $t[x]$, giving a total rate of extinction among the n common *S*-alleles segregating in the population of

$$\mu = n/t[x]. \qquad (13.6)$$

Equating the rates of origin and extinction ($\lambda = \mu$) implicitly determines F, the steady-state homozygosity (compare Sasaki 1989, Slatkin and Muirhead 1999):

$$2ue^{N(a-2b)} \left(\frac{2b}{a}\right)^{2Nb} \sqrt{\frac{N\pi}{b}} = 1. \qquad (13.7)$$

This equation corresponds to the solution derived by a different method by Yokoyama and Hetherington (1982).

Vekemans and Slatkin (1994) studied the evolution of GSI by adapting Takahata's (1990) approach to symmetric overdominance, incorporating the homozygosity determined from Equation 13.7. Their derivation followed Takahata's in replacing the diffusion coefficient $\sigma^2(q)$ (Equation 13.4) with $q/(2N)$. While this approximation likely introduces negligible error for small q, it is inconsistent with Equation 13.7, which uses the full coefficient. In her SSWM treatment of GSI, Muirhead (2001) inferred the rate of extinction by inserting the Yokoyama–Hetherington solution into her expression for the rate of increase in n.

13.4.4 Scaling factor

Takahata (1990) suggested that genealogical relationships among symmetrically overdominant alleles in a population of size N resemble those among neutral alleles in a population of size Nf, for f a scaling factor representing the expansion of coalescence times due to balancing selection. Coalescence among j neutral lineages in a population of effective size Nf occurs at rate

$$\frac{\binom{j}{2}}{2Nf}.$$

Equating this expression with the rate of bifurcation of *S*-allele lineages (Equation 13.2) determines the scaling factor for the Wright model:

$$f = \frac{n^2}{4N\lambda} = \frac{1}{16N^2uaF^2(F - u/a)}. \tag{13.8}$$

The latter expression differs from Vekemans and Slatkin's (1994) by the factor of $\sqrt{2}$ which Slatkin and Muirhead (1999) noted as equivocal.

13.5 Divergence among haplotypes

Using results derived from Wright's canonical model of GSI evolution, we explore whether the contrasting patterns of *S*-allele diversification in *Physalis* and *Solanum* reflect the operation of the same evolutionary process under different parameters or fundamentally different processes.

13.5.1 Spectrum of descendant number

Figure 13.1 illustrates transgeneric sharing, typical of *S*-allele lineages: several nodes give rise to descendants in both *Physalis* and *Solanum*. The vertical line marks the youngest node of this kind, with descendants of all more recent nodes restricted to one genus. This node provides an upper bound for divergence between the genera, preceding that event by the time to coalescence of two lineages within the common ancestor. We designate this upper bound as the boundary between the pre- and post-divergence eras. For group i (*Physalis*, $i = 1$; *Solanum*, $i = 2$), we assign each of the m_i specificities presently observed to one of the k_i ancestral lineages at the divergence boundary. We examine the process of descent of the present joint state $\{m_1, m_2\}$ from the ancestral joint state $\{k_1, k_2\}$.

We adopt a minimal model of *S*-locus evolution, assuming only equal rates of bifurcation among lineages. Under the Pólya urn model (see Feller 1957, Chapter II), the k ancestral *S*-allele lineages at the divergence boundary correspond to k urns, each containing exactly one ball. Representing bifurcation of an *S*-allele lineage by the random selection of a ball and its replacement in the same urn together with a new ball, we grow the genealogy to the present-day sample by adding $m - k$ new balls. Each arrangement occurs with equal probability, which corresponds to the reciprocal of the number of ways $m - k$ indistinguishable balls can be distributed among k distinguishable urns:

$$\binom{m-1}{k-1}.$$

We describe a spectrum of descendant numbers by $\{a_1, a_2, \ldots, a_{m-k+1}\}$, for a_j the number of urns that contain exactly j balls, without regard to urn identity. The probability of a given spectrum corresponds to the product of the probability of the arrangement and the number of ways the contents of the urns can be assigned to the urns:

$$\frac{k!}{\binom{m-1}{k-1}\prod_{j=1}^{m-k+1} a_j!} \tag{13.9}$$

Fu (1995) has noted that Expression 13.9 corresponds to Kingman's (1982, Equation (2.3)) distribution, for both balls and urns indistinguishable.

We adopt the Shannon–Weaver information measure as an index of evenness in descendant number among ancestral lineages:

$$H = -\sum_{j=1}^{m-k+1} a_j(j/m)\log[j/m].$$

Smaller values of H denote more unevenness. We characterize an observed spectrum of descendant numbers by determining the probability under Expression 13.9 of all spectra with equally or more deviant H values.

Our genealogical reconstruction indicates that the 52 *Physalis* S-alleles descend from 4 ancestral lineages, and 11 from 9 for *Solanum* ($\{m_1 = 52, m_2 = 11\}, \{k_1 = 4, k_2 = 9\}$). For *Solanum*, the descendant spectrum reflects the distribution of two new balls among nine urns. We observed the placement of both new balls into the same urn, an outcome with expected probability 2/10 (one of the two daughter lineages produced by the first bifurcation is chosen for the second bifurcation). For *Physalis*, the descent of 48 new lineages from the four ancestral lineages also shows nonsignificant departures from expectation (exact two-tailed probability: $P = 0.6$). While crude, this test suggests that S-allele lineages in the three *Physalis* species have diversified in a homogeneous manner. This homogeneity appears to reflect that the *Physalis* species have diverged recently compared with the S-allele lineages.

In contrast with the spectra of the two genera considered separately, the combined spectrum shows significant unevenness ($H = 1.83$, $P = 0.001$). Figure 13.2 shows the expected distribution of Shannon–Weaver indices and the observed value. Figure 13.3 compares the combined spectrum to a spectrum with Shannon–Weaver index close to the mode ($H = 2.31$).

Clearly, S-allele diversification has proceeded differently in the two genera. That *Physalis* maintains many more S-alleles than *Solanum* ($n_1 > n_2$) suggests a larger effective population size for *Physalis* ($N_1 > N_2$). We now address whether a difference in effective population size alone can account for the distinct patterns observed.

13.5.2 Process of bifurcation

13.5.2.1 Relative rate of bifurcation

We explore the joint process of S-allele diversification in the two genera by examining the order of bifurcations. Proceeding backwards from the present

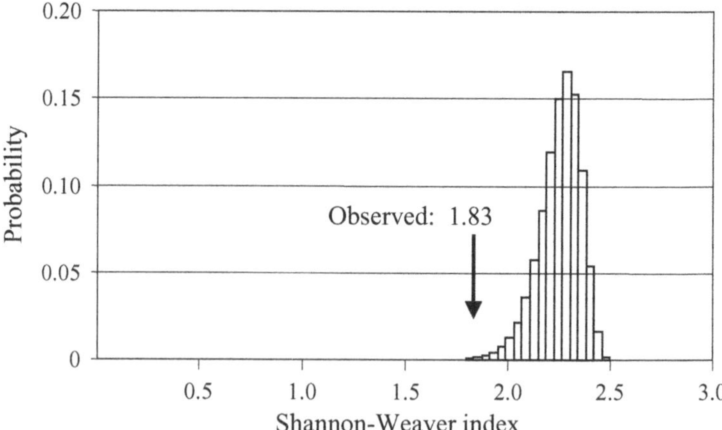

Figure 13.2. Expected distribution of Shannon–Weaver indices for descendant number spectra for 63 descendants of 13 ancestors, as indicated in Figure 13.1 for *Physalis* and *Solanum*. The arrow indicates the Shannon–Weaver index of the observed combined spectrum for *Physalis* and *Solanum* *S*-alleles.

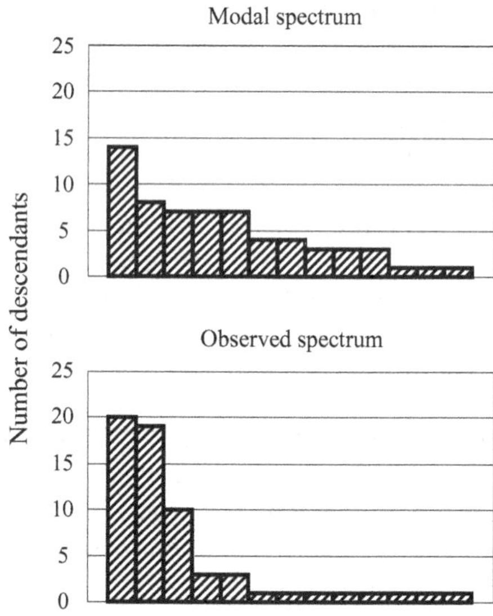

Figure 13.3. Spectra of numbers of descendants derived from ancestral lineages. Each bar indicates the number of descendants of a particular ancestral lineage at the divergence boundary. The upper panel corresponds to a spectrum with Shannon–Weaver index close to the mode ($H = 2.31$) and the lower panel corresponds to the combined spectrum for *Physalis* and *Solanum* *S*-alleles ($H = 1.83$).

to the divergence boundary, the joint state evolves from $\{m_1, m_2\}$ to $\{k_1, k_2\}$. At rate $\mu_1\{j_1, j_2\}$, the next coalescence event occurs among the lineages of group 1, taking the process to state $\{j_1 - 1, j_2\}$; at rate $\mu_2\{j_1, j_2\}(= 1 - \mu_1\{j_1, j_2\})$ the process moves to state $\{j_1, j_2 - 1\}$. For group i, λ_i (Equation 13.2) determines the birth rate of lineages proceeding forward in time and μ_i (Equation 13.6) the death rate proceeding backward in time. The duality of these two processes implies

$$\mu_1\{j_1, j_2\} = \frac{\lambda_1 j_1 (j_1 - 1)/n_1^2}{\lambda_1 j_1 (j_1 - 1)/n_1^2 + \lambda_2 j_2 (j_2 - 1)/n_2^2} = \frac{\alpha \binom{j_1}{2}}{\alpha \binom{j_1}{2} + \binom{j_2}{2}}, \tag{13.10}$$

in which

$$\alpha = \frac{\lambda_1 n_2^2}{\lambda_2 n_1^2} = \frac{N_2 f_2}{N_1 f_1},$$

for f_i, the scaling factor for group i (Equation 13.8). Coalescence rates depend upon the relative numbers of lineages in the groups and the rates at which new *S*-alleles arise and become common. Using the expected rate of invasion (Equation 13.5), we find that the relative rate of coalescence corresponds to

$$\alpha = \frac{N_1 u_1 n_2 (n_2 - 1)(n_2 - 2)}{N_2 u_2 n_1 (n_1 - 1)(n_1 - 2)}. \tag{13.11}$$

Under our null hypothesis, that the groups differ with respect to effective population size alone, the groups share a common rate of mutation to new specificities ($u_1 = u_2$), permitting reduction of α to a function of only allele numbers and the ratio of effective population sizes.

Figure 13.4 represents the initial, terminal, and intermediate states of the process as a matrix, with μ_1 and μ_2 representing rates of transition between cells and $Q\{j_1, j_2\}$, the probability of joint state $\{j_1, j_2\}$, the element in the j_1th row and j_2th column. Given the initial state $Q\{m_1, m_2\} = 1$, recursive determination of the state probabilities from Equation 13.10 is straightforward. $Q\{k_1, k_2\}$, the entry in the terminal cell, represents the total probability of reaching the terminal state from the initial state through all possible paths. Maximization of $Q\{k_1, k_2\}$ with respect to α determines $\hat{\alpha}$, a maximum-likelihood estimate of the rate of coalescence in group 1 relative to group 2. Further, the probability of any particular path between the initial and terminal cells corresponds to the product of the transition probabilities (μ_1 and μ_2) joining the constituent cells of the path. Division of this product by $Q\{k_1, k_2\}$ gives the probability of the path conditioned on both the terminal and initial states.

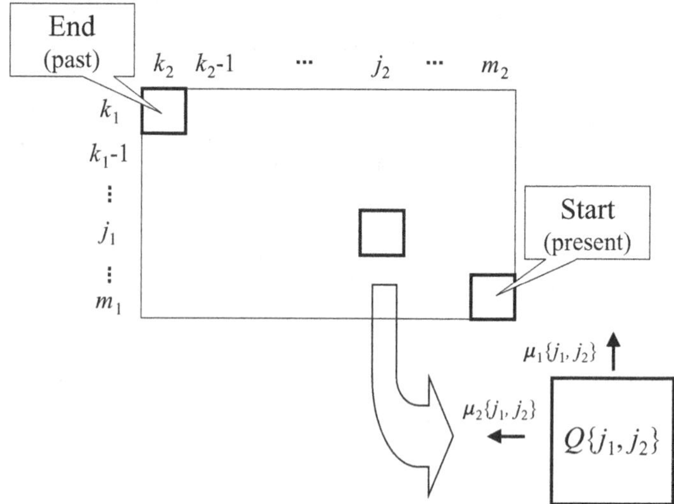

Figure 13.4. Joint ancestral states of genealogies of *S*-alleles from two groups. Joint state $\{m_1, m_2\}$ represents the m_1 present-day *S*-alleles observed in group 1 and the m_2 in group 2. The line of ascent from the present state to the ancestral state $\{k_1, k_2\}$ may proceed along various paths. $Q\{j_1, j_2\}$ represents the probability of state $\{j_1, j_2\}$, given the initial state ($Q\{m_1, m_2\} = 1$). Within state $\{j_1, j_2\}$, coalescence occurs among group 1 lineages at rate $\mu_1\{j_1, j_2\}$ and among group 2 lineages at rate $\mu_2\{j_1, j_2\}$.

13.5.2.2 Relative population sizes and allele numbers

Our analysis of the genealogy in Figure 13.1 indicates a value of 11.3 for $\hat{\alpha}$, the ML rate of coalescence of *Physalis* relative to *Solanum*. Together with estimates for the total numbers of segregating *S*-alleles ($n_1 = 72$, $n_2 = 13$), this value suggests a much higher rate of generation of new *S*-alleles in *Physalis*:

$$\frac{N_1 u_1}{N_2 u_2} = 2350$$

(from Equation 13.11). A rough calculation using a different approach indicated a value of about 100 for this ratio, reflecting a fourfold difference in population size and a 26-fold difference in mutation rate (Uyenoyama 2000).

To test the hypothesis that a difference in population size alone accounts for the observed difference in bifurcation rate, we restrict consideration to equal rates of mutation to new *S*-alleles ($u_1 = u_2$). Analysis of Equation 13.7 indicates that the ratio of population sizes (N_1/N_2), given the ratio of allele numbers (n_1/n_2), is largely insensitive to this common mutation rate. In the case at hand, the ratio of allele numbers ($n_1/n_2 = 72/13$) suggests that the effective population size of *Physalis* exceeds that of *Solanum* by 35- to 40-fold. Substitution of these quantities into Equation 13.11 produces an expected relative rate of coalescence of 0.2. Our ML estimate ($\hat{\alpha} = 11.3$) differs highly significantly from this expectation ($P = 1 \times 10^{-14}$), indicating that smaller effective population size cannot by itself account for the much lower rate of diversification of *S*-allele lineages in *Solanum*.

13.6 Conclusions

Our analysis indicates that under the canonical model of GSI, differences in effective population size alone cannot account for the different patterns of *S*-allele diversification observed in two solanaceous genera. In particular, diversification of *Solanum* *S*-allele lineages appears to have considerably pre-dated most of the bifurcation events in *Physalis*. Some genetic and demographic factors that might account in part for the empirical observations have been discussed by Uyenoyama (2000). Changes in demographic factors, particularly effective population size, may very likely have profound effects on the expected pattern of diversification. Such considerations demand extension of the theoretical description. Here we address the mechanism of origin of new specificities and the effect of demographic structure on the rate of origin.

Although early models of SI sought to conceive of mechanisms capable of generating a new *S*-allele through a single mutational event (Fisher 1961), virtually all recent analyses postulate the determination of pistil and pollen specificity by separate genes (Matton *et al.* 1999, Uyenoyama and Newbigin 2000, McCubbin and Kao 1999, 2000). Uyenoyama *et al.* (2001) explored conditions for the origin of new *S*-alleles through episodes of partial breakdown of GSI, reflecting the segregation of single-mutant haplotypes that express different specificities in pollen and style. This analysis indicated that a new *S*-allele can arise through this pathway only through the sequential invasion of a pollen-part mutation that specifies a new pollen specificity followed by a style-part mutation that permits recognition of the new pollen specificity. Because the invasion of the single and double mutants each entail the extinction of their immediate ancestral haplotypes, this mechanism would preclude lineage bifurcation, which requires coexistence of the parent and offspring specificities.

Uyenoyama *et al.* (2001) proposed a crucial role for population structure in permitting bifurcation of *S*-allele lineages. Balancing selection promotes the invasion of alleles introduced by migration into demes in which they are absent and their persistence, once established. As a consequence, the number of *S*-alleles maintained at steady state under the Wright model is expected to be fairly insensitive to population subdivision, except under very low migration rates (Wright 1939, Schierup 1998, Muirhead 2001). We conjecture that under the two-mutation mechanism studied by Uyenoyama *et al.* (2001), population subdivision might permit a significant increase in allele number through the independent derivation of *S*-alleles from the same ancestor in different subpopulations. Once reintroduced by migration into the same subpopulation, a new full-function *S*-allele could coexist with its full-function ancestors, even though the partial-function intermediate would have excluded its immediate full-function ancestor. However, bifurcation of *S*-allele lineages through this evolutionary pathway would appear to entail very low migration rates, sufficiently low to permit the rise of the double-mutant before the extinction throughout the metapopulation of the full-function ancestor as a consequence either of random turnover or of exclusion by the single-mutant. Whether such a low rate of gene flow is tantamount to virtual speciation among subpopulations remains to be determined. Exploration of these conjectures is currently under way.

13.7 Acknowledgments

We thank Subodh Jain and Rama Singh for constructive suggestions. US Public Health Service grant GM 37841 provided support for this study.

REFERENCES

Bahlo, M., and Griffiths, R. C. (2000). Inference from gene trees in a subdivided population. *Theor. Popul. Biol.* 57:79–95.

Beerli, P., and Felsenstein, J. (2001). Maximum likelihood estimation of a migration matrix and effective population sizes in *n* subpopulations by using a coalescent approach. *Proc. Natl. Acad. Sci. USA* 98:4563–8.

Feller, W. (1957). *An Introduction to Probability Theory and Its Applications*, Vol. I. Third Edition. New York: John Wiley.

Fisher, R. A. (1958). *The Genetical Theory of Natural Selection*, 2nd revised edition. New York: Dover.

Fisher, R. A. (1961). A model for the generation of self-sterility alleles. *J. Theor. Biol.* 1:411–14.

Fu, Y.-X. (1995). Statistical properties of segregating sites. *Theor. Popul. Biol.* 48:172–97.

Gillespie, J. (1983). Some properties of finite populations experiencing strong selection and weak mutation. *Am. Nat.* 121:691–708.

Gillespie, J. (1994). Substitution processes in molecular evolution. II. Exchangeable models from population genetics. *Evolution* 48:1101–13.

Golz, J. F., Oh, H.-Y., Su, V., Kusaba, M., and Newbigin, E. (2001). Genetic analysis of *Nicotiana* pollen-part mutants is consistent with the presence of an *S*-ribonuclease inhibitor at the *S* locus. *Proc. Natl. Acad. Sci. USA* 98:15372–6.

Golz, J. F., Su, V., Clarke, A. E., and Newbigin, E. (1999). A molecular description of mutations affecting the pollen component of the *Nicotiana alata S* locus. *Genetics* 152:1123–35.

Hasegawa, M., Kishino, H., and Yano, T. (1985). Dating the human–ape split by a molecular clock of mitochondrial DNA. *J. Mol. Evol.* 22:160–74.

Hey, J. (1992). Using phylogenetic trees to study speciation and extinction. *Evolution* 46:627–40.

Holmes, E. C., Pybus, O. G., and Harvey, P. H. (1999). The molecular population dynamics of HIV-1. In K. A. Crandall (ed.) *The Evolution of HIV*, pp. 177–207. Baltimore, MD: Johns Hopkins University Press.

Hughes, A. L., and Nei, M. (1988). Pattern of nucleotide substitution at major histocompatibility complex class I loci reveals overdominant selection. *Nature* 335:167–70.

Ioerger, T. R., Gohlke, J. R., Xu, B., and Kao, T.-H. (1991). Primary structural features of the self-incompatibility protein in Solanaceae. *Sex. Plant Reprod.* 4:81–7.

Ishimizu, T., Endo, T., Yamaguchi-Kabata, Y., Nakamura, K. T., *et al.* (1998). Identification of regions in which positive selection may operate in S-RNase of Rosaceae: implication for *S*-allele-specific recognition sites in S-RNase. *FEBS Lett.* 440: 337–42.

Karlin, S., and Taylor, H. M. (1981). *A Second Course in Stochastic Processes.* New York: Academic Press.

Kimura, M., and Crow, J. F. (1964). The number of alleles that can be maintained in a finite population. *Genetics* 49:725–38.

Kingman, J. F. C. (1982). The coalescent. *Stochastic Process. Appl.* 13:235–48.

Lawrence, M. J. (2000). Population genetics of the homomorphic self-incompatibility polymorphisms in flowering plants. *Ann. Bot.* 85, Suppl.:221–6.

Lu, Y. (2001). Roles of lineage sorting and phylogenetic relationship in the genetic diversity at the self-incompatibility locus of Solanaceae. *Heredity* 86:195–205.

Lush, M. W., and Clarke, A. E. (1997). Observations on pollen tube growth in *Nicotiana alata* and their implications for the mechanism of self-incompatibility. *Sex. Plant Reprod.* 10:27–35.

Luu, D.-T., Qin, X., Laublin, G., Yang, Q., Morse, D., *et al.* (2001). Rejection of S-heteroallelic pollen by a dual-specific S-RNase in *Solanum chacoense* predicts a multimeric SI pollen component. *Genetics* 159:329–35.

Luu, D.-T., Qin, X., Morse, D., and Cappadocia, M. (2000). S-RNase uptake by compatible pollen tubes in gametophytic self-incompatibility. *Nature* 407:649–51.

Matton, D. P., Luu, D. T., Xike, Q., Laublin, G., O'Brien, M., *et al.* (1999). Production of an S-RNase with dual specificity suggests a novel hypothesis for the generation of new *S* alleles. *Plant Cell* 11:2087–97.

McClure, B. A., Gray, J. E., Anderson, M. A., and Clarke, A. E. (1990). Self-incompatibility in *Nicotiana alata* involves degradation of pollen rRNA. *Nature* 347:757–60.

McCubbin, A. G., and Kao, T.-H. (1999). The emerging complexity of self-incompatibility (*S*-) loci. *Sex. Plant Reprod.* 12:1–5.

McCubbin, A. G., and Kao, T.-H. (2000). Molecular recognition and response in pollen and pistil interactions. *Annu. Rev. Cell Dev. Biol.* 16:333–64.

Muirhead, C. A. (2001). Consequences of population structure on genes under balancing selection. *Evolution* 55:1532–41.

Nee, S., May, R. M., and Harvey, P. H. (1994). The reconstructed evolutionary process. *Phil. Trans. R. Soc. Lond. B* 344:305–11.

Nielsen, R. (1998). Maximum likelihood estimation of population divergence times and population phylogenies under the infinite sites model. *Theor. Popul. Biol.* 53:143–51.

Paxman, G. J. (1963). The maximum likelihood estimation of the number of self-sterility alleles in a population. *Genetics* 48:1029–32.

Posada, D., and Crandall, K. A. (1998). MODELTEST: testing the model of DNA substitution. *Bioinformatics* 14:817–18.

Richman, A. D., and Kohn, J. R. (1999). Self-incompatibility alleles from *Physalis*: implications for historical inference from balanced genetic polymorphisms. *Proc. Natl. Acad. Sci. USA* 96:168–72.

Richman, A. D., Kao, T.-H., Schaeffer, S. W., and Uyenoyama, M. K. (1995). S-allele sequence diversity in natural populations of *Solanum carolinense* (horsenettle). *Heredity* 75:405–15.

Richman, A. D., Uyenoyama, M. K., and Kohn, J. R. (1996). S-allele diversity in a natural population of *Physalis crassifolia* (Solanaceae) (ground cherry) assessed by RT-PCR. *Heredity* 76:497–505.

Sasaki, A. (1989). *Evolution of Pathogen Strategies*, Ph.D. thesis, Kyushu University, Kyushu, Japan.

Sasaki, A. (1992). The evolution of host and pathogen genes under epidemiological interaction. In N. Takahata (ed.) *Population Paleo-genetics*, pp. 247–63. Tokyo: Japan Scientific Societies Press.

Schierup, M. H. (1998). The number of self-incompatibility alleles in a finite, subdivided population. *Genetics* 149:1153–62.

Schopfer, C. R., Nasrallah, M. E., and Nasrallah, J. B. (1999). The male determinant of self-incompatibility in *Brassica. Science* 286:1697–700.

Slatkin, M. (2000). Balancing selection at closely linked, overdominant loci in a finite population. *Genetics* 154:1367–78.

Slatkin, M., and Muirhead, C. A. (1999). Overdominant alleles in a population of variable size. *Genetics* 152:775–81.

Stephenson, A. G., Doughty, J., Dixon, S., Elleman, C., Hiscock, S., *et al.* (1997). The male determinant of self-incompatibility in *Brassica oleracea* is located in the pollen coating. *Plant J.* 12:1351–9.

Suzuki, G., Taki, N., Hirose, T., Fukui, K., Nishio, T., *et al.* (1999). Genomic organization of the *S* locus: identification and characterization of genes in *SLG/SRK* region of *S9* haplotype of *Brassica campestris* (syn. *rapa*). *Genetics* 153:391–400.

Swofford, D. L. (1999). *PAUP* Phylogenetic Analysis Using Parsimony (and Other Methods)*, Version 4.0b. Sunderland, MA: Sinauer.

Takahata, N. (1990). A simple genealogical structure of strongly balanced allelic lines and trans-species evolution of polymorphism. *Proc. Natl. Acad. Sci. USA* 87:2419–23.

Takahata, N., Satta, Y., and Klein, J. (1992). Polymorphism and balancing selection at major histocompatibility complex loci. *Genetics* 130:925–38.

Takahata, N., Satta, Y., and Klein, J. (1995). Divergence time and population size in the lineage leading to modern humans. *Theor. Popul. Biol.* 48:198–221.

Takasaki, T., Hatakeyama, K., Suzuki, G., Watanabe, M., Isogai, A., *et al.* (2000). The *S* receptor kinase determines self-incompatibility in *Brassica* stigma. *Nature* 403:913–16.

Takayama, S., Shiba, H., Iwano, M., Shimosato, H., Che, F.-S., *et al.* (2000). The pollen determinant of self-incompatibility in *Brassica campestris. Proc. Natl. Acad. Sci. USA* 97:1920–5.

Uyenoyama, M. K. (1997). Genealogical structure among alleles regulating self-incompatibility in natural populations of flowering plants. *Genetics* 147:1389–400.

Uyenoyama, M. K. (2000). The evolution of breeding systems. In R. S. Singh and C. Krimbas (eds) *Evolutionary Genetics: From Molecules to Morphology*, pp. 300–16. New York: Cambridge University Press.

Uyenoyama, M. K., and Newbigin, E. (2000). Evolutionary dynamics of dual-specificity self-incompatibility alleles. *Plant Cell* 12:310–12.

Uyenoyama, M. K., Zhang, Y., and Newbigin, E. (2001). On the origin of self-incompatibility haplotypes: transition through self-compatible intermediates. *Genetics* 157:1805–17.

Vekemans, X., and Slatkin, M. (1994). Gene and allelic genealogies at a gametophytic self-incompatibility locus. *Genetics* 137:1157–65.

Wakeley, J., and Hey, J. (1997). Estimating ancestral population parameters. *Genetics* 145:847–55.

Wright, S. (1939). The distribution of self-sterility alleles in populations. *Genetics* 24:538–52.

Wright, S. (1964). The distribution of self-incompatibility alleles in populations. *Evolution* 18:609–19.

Wright, S. (1969). *Evolution and the Genetics of Populations*, Vol. 2, *The Theory of Gene Frequencies.* Chicago: University of Chicago Press.

Yang, Z. (1997). On the estimation of ancestral population sizes of modern humans. *Genet. Res. Camb.* 69:111–16.

Yokoyama, S., and Hetherington, L. E. (1982). The expected number of self-incompatibility alleles in finite plant populations. *Heredity* 48:299–303.

PART IV

GENES, ORGANISMS, AND ENVIRONMENT: EVOLUTIONARY CASE STUDIES

14

Adaptation, constraint, and neutrality: mechanistic case studies with butterflies and their general implications

WARD B. WATT
Department of Biological Sciences, Stanford University
and
Rocky Mountain Biological Laboratory, Crested Butte

14.1 Introduction

I've known Dick Lewontin since the early 1960s. Hardly a month has passed in that time in which I've not made use of one or another of his contributions to evolutionary biology. That does not mean, as "the literature" bears witness, that he and I have always been in agreement! But we have been deeply interested in many of the same questions. So, it is a pleasant challenge for me to recount here some work on evolutionary processes that I and my associates have done. The mechanistic study of evolution, as we understand it, entails a central focus on adaptation, i.e., on organism–environment interactions, as a basis for predicting demographically based fitness outcomes (or their absence). Thus I begin by stating our views of the place of adaptation in the evolutionary process. I next turn to illustrative empirical examples from our work, and then to discussion of some general evolutionary questions of common interest, whose answers these examples illuminate.

14.2 The adaptation concept and its role in the study of evolution

Adaptation, correctly understood, refers to the suitedness of heritably varying phenotypes to the range of environments they occupy, and hence to these phenotypes' differential performance of biological functions in those environments (Watt 1994, 2000, 2001). As such, adaptation is the causal core of Darwin's argument for natural selection. *Natural* selection is not, despite claims to the contrary (e.g., Lewontin 1984), *sufficiently* defined by "variation, heritability, and differential reproduction of heritable variants." These three ideas suffice only to define *arbitrary* selection, of unspecified cause. Darwin's concept of differential adaptation as the nonpurposive, wholly natural cause of differential reproduction is required to distinguish natural from artificial

The Evolution of Population Biology, ed. R. S. Singh and M. K. Uyenoyama. Published by Cambridge University Press.

or other forms of selection (e.g., Darwin 1859, Hodge 1987, Brandon 1990, Watt 1994, Depew and Weber 1995).

The demographic results of adaptive performances accumulate, in ways which differ in algebraic detail depending on life-history variation (e.g., Charlesworth 1994, McGraw and Caswell 1996), into genotype-phenotype-environment-specific differences in organisms' net replacement rates and their ratios, i.e., absolute and relative fitnesses (Endler 1986, Feder and Watt 1992). To keep a definitionally sound view of "Darwin's theorem," one must not confuse adaptation with fitness (Watt 2001). Krimbas (1984, 2000) is right that defining adaptation in terms of fitness would be circular; he is wrong, however, to say that adaptation cannot be otherwise defined. It can and *must* be otherwise defined, as above, since the two phenomena are quite distinct. Indeed, adaptive differences may not always cause fitness differences. For example, if resources are plentiful, phenotypes may differ in foraging performance, i.e., state of feeding adaptation, yet still be fed to satiety and have the same resulting fitness (Grant 1986, Watt 1986, 2001). Yet in times of resource scarcity, those same adaptive performance differences may lead to large differences in fitness of their carriers.

Much uneasiness about adaptation stems from reaction against "adaptationism." This term does not properly refer to study of adaptation in general, but specifically to the a priori assumption of adaptiveness of phenotypic states. Adaptationism is needless at best, as the rigorous testing of adaptive hypotheses uses neutrality as usual null hypothesis, and can be misleading at worst (Watt 2001). But indictment of the adaptation concept as necessarily fostering adaptationism is just as misleading, and impedes evolutionary study (Watt 1994).

One rigorous strategy for the study of adaptation uses natural variants, in genes of known function, to probe mechanisms of organism–environment interaction – the "method of identified genes" (Watt 2000). Not all such variation is adaptive. Variants can be neutral to selection because they don't differ in function (e.g., Dykhuisen and Hartl 1983), or their effects are minimized by phenotypic organization (e.g., Dean 1989), or (as above), in a "relaxed" environment their performance differences, though measurable, have no fitness effects. Also, phenotypic reaction norms may be constrained in various ways (e.g., Kingsolver and Watt 1984, Watt and Dean 2000). The approach seeks general evolutionary principles by which to distinguish adaptive, constrained, or genuinely neutral variation in terms of processes and thus resulting patterns.

Some claim that study of adaptation at specific genes entails "atomistic" dissolution of phenotypes into myriad separate features. Beyond the deplorable tendency of some workers to speak inaccurately of "the genes for" complex traits, especially in human-behavioral context, this is a straw man: specific-gene studies do not imply "atomism." Some natural genetic variants may act independently, but natural epistatic interactions among variants are also often seen, for example among reactions within pathways or across branches of metabolism, and may be analyzed (e.g., Hartl and Dykhuisen 1981, Carter and Watt 1988, Watt and Dean 2000). Modularity and genetic hierarchy in developmental subsystems may likewise give analyzable order to the complexity of morphological evolution (e.g., Johnson and Porter 2001, Stern 2000, Wagner and Gauthier 1999).

Moreover, study of specific adaptive mechanisms in this way does not imply "Cartesian" or "strong" reductionism, i.e., the claim that phenomena at high organization levels are "nothing but" projections of simpler phenomena (rightly critiqued by Levins and Lewontin 1985). High-level properties, not existing at lower levels, may emerge from the integration of low-level parts – in chemistry as it arises from physics, and in biology as it arises from both. For example, constraint on thermal breadth of metabolic performance emerges from the assembly of multiple reaction steps into pathways (Watt 1991, Kohane and Watt 1999). To avoid the trap of strong reductionism, one follows the analysis of parts, necessary to mechanistic evolutionary study, with an equally necessary resynthesis – what Dobzhansky (1970) and Simpson (1964) called "compositionist" biology.

While we can achieve realistic generality about these complex evolutionary alternatives only by synthesizing from diverse case studies of putatively adaptive variation (Watt 2000), it is also true that those specific studies must be pursued before the synthesis can take place. I next discuss some of my own work to illustrate both the feasibility and the power of our approach.

14.3 Evolutionary generality via mechanistic specificity: bioenergetically based studies

To study evolution directly, one needs a suitable model system. A variety of organisms have been studied by mechanistic evolutionary approaches (e.g., Powers *et al.* 1991, Mitton 1997, Mangum and Hochachka 1998, Eanes 1999, Singh *et al.* 2000, Watt and Dean 2000). We use *Colias* butterflies (Lepidoptera, Pieridae) because of their variability, their suitedness to both laboratory and field study, and knowledge of their natural history. Bioenergetics offers a unifying context for case studies (e.g., Watt 1985a), so I begin with study of thermal adaptation and constraint in *Colias,* setting a context for study of adaptive genetic variation in mechanisms of energy processing.

14.3.1 Thermoregulatory adaptations and thermal constraints on butterfly flight

An afternoon's specimen sorting in Yale's Peabody Museum gave me, as a young graduate student, an entry to study of *Colias*' thermal adaptations and constraints: butterflies from cold habitats had darker hindwing undersides than those from warm ones, irrespective of taxonomic associations. The idea that this color difference might equate to greater absorptivity for sunlight, raising body temperature (T_b) and promoting flight in cold habitats, became a working hypothesis. From experimental tests of this, which sustained it against null hypotheses of no phenotypic and/or no fitness effects (Watt 1968, Roland 1982, Kingsolver 1983a,b), an unexpected constraint emerged: flight is confined to the same high and narrow range of T_b values throughout the genus. Contrary to some adaptationist expectations, *Colias* in cold habitats do not evolve to fly at lower or broader T_b (Watt 1968, 1997). Instead, they evolve locally specialized values of thermoregulatory parameters: solar absorptivity and anti-convection insulation (e.g., Watt 1968, Kingsolver 1983a, Tsuji *et al.* 1986).

That there are *two* constraints here, one of high T_b and one of a narrow range of T_b, was shown by Sherman and Watt (1973) while studying *Colias* larvae. These larvae are precluded from varying their absorptivity as a thermoregulatory device by an overriding selection pressure for camouflage coloration (first demonstrated by Gerould 1921; cf. also Hoffmann and Watt 1974). Despite this limitation, they fed and grew maximally in only a narrow T_b range, though that was lower by far than the T_b range for adult flight, and also, unlike flight T_b, varied among populations.

What could be the source of these constraints? I guessed initially that metabolic organization itself might impose them in some way. Indeed, later this guess became the explicit hypothesis, already mentioned, that metabolic pathways can only achieve high throughput over narrow T_b ranges, owing to interaction of the thermal dependencies of reactions in a pathway with the coadaptive organization of such pathways (Watt 1991). (This has been supported by experimental test against both null hypotheses and more conventional, single-reaction-oriented views of animal thermoregulation; Kohane and Watt 1999). To follow up at the time, I sought natural genetic variation in energy-processing reactions with which to probe this putative metabolic constraint.

14.3.2 Adaptation, constraint, and neutrality in variants of a specific gene: molecular mechanisms to fitness effects

The use of protein electrophoresis in population genetics, given major impetus by Jack Hubby and Dick Lewontin (1966, Lewontin and Hubby 1966), came to our aid in probing metabolic evolution. This means of identifying natural variants, though caught up in the unproductive dialectical maelstrom of the "neutralist–selectionist debate" (see below), gave us a new tool for probing the nature of functional phenotypes in energy processing. We began with natural variation in three enzymes at the metabolic branch-point which receives the input of nectar-derived sugars (Watt *et al.* 1974) as "flight fuel" for adult *Colias.* In our first few samples, we found strong suggestions of differential survivorship among genotypes of phosphoglucose isomerase, PGI (Watt 1977). This led to extended study of this polymorphism.

Central to this work has been the understanding of PGI's functional role as an "intervening" step in glycolysis, subject to selection for very high kinetic capacity in support of energy resupply under the extraordinarily high demands of insect flight (Watt 1977). This understanding is founded in theory of metabolic organization, especially in transient state (e.g., Easterby 1973, 1990), as applied to insect flight metabolism (Watt 1977, 1983, 1985a). Confusion about the meaning of this theory has beclouded its evolutionary interpretation in some quarters, but it is now clear (Watt and Dean 2000) that it is fully compatible with the finding of major phenotypic effects of single-gene genetic variants, or with the finding of different kinds of results in steady vs. transient state metabolism, etc. In this context, we could measure such functional differences as might exist among PGI genotypes, predict what flight performance differences should or should not result from them, and in turn predict presence or absence of fitness consequences in the wild. Importantly, this was done on a genotype-specific basis, not, for example, as vague predictions of general heterozygote superiority which might have other causes such as the segregation of deleterious recessives. We could, in short, address the problem posed by Lewontin (1974, p. 265): "Whether a 20 percent decrease in enzyme activity for an allozyme variant will or will not be detected by the physiological apparatus of the organism . . . ", and whether the "detection" has an impact on fitness in the wild.

We measured functional differences among the ten common PGI genotypes of lowland *Colias* (Watt 1977, 1983). We studied both thermal stability and kinetic capacity, the latter measured as the V_{max}/K_m ratio, which is the best single index of catalytic function with unsaturating *in vivo* metabolite concentrations, and is the basic performance measure for each reaction step in context of metabolic organization theory (review: Watt and Dean 2000). In summary, we found:

(a) extensive, but not universal, heterozygote advantage in V_{max}/K_m ratio, due primarily to up to threefold genotypic differences in values of the Michaelis constant K_m (a relative measure of substrate affinity, low K_m = tight binding), rather than differences in V_{max} (maximum saturated velocity, the product of the catalytic rate constant and enzyme concentration, $= k_{cat} \times [E]$);
(b) in homozygotes, a sharp tradeoff of V_{max}/K_m vs. a fourfold range of thermal stability values, so that functional disadvantages of homozygotes with respect to either parameter are asymmetrical around heterozygote values, while the most common heterozygotes may escape from this tradeoff.

Given PGI's functional role in support of flight, we could first translate these results into genotype-specific predictions of resulting differences in flight capacity:

(a) genotypes of higher V_{max}/K_m should support flight fuel resupply, hence flight, during a broader span of the daily thermal cycle than those of lower V_{max}/K_m;

(b) infrequent but recurrent heat stress should favor stable genotypes' better retention of enzyme activity; even if kinetically disadvantaged to start with (as is often so), they should suffer lesser diminution of flight capacity under heat stress than less stable genotypes.

All adult fitness components (survival, male mating success, female fecundity) depend on effective flight, so these flight predictions in turn directly predict genotypic fitness differences:

(a) in low- to moderate-temperature habitats, kinetically effective genotypes (mostly heterozygotes) should be favored in all fitness components in proportionate order of their $V_{\max}/K_m$ values;
(b) under heat stress, the most thermally sensitive genotypes lose their initial, kinetically based fitness advantages as their loss of enzyme concentration reduces their initial $V_{\max}/K_m$ values.

In other words, the interaction of genotypic kinetics and thermal stabilities with variation of thermal environments on diverse time scales yields predictable genotype-specific norms of reaction with respect to flight performance. These genotypic performance norms have equally predictable fitness consequences in those environments.

We then devised new lab and field protocols to test each of these genotype-specific predictions (Watt *et al.* 1983, 1985, 1986, Watt and Boggs 1987, and Watt 1992). *All* predictions, covering all adult fitness components (survivorship, male mating success, and female fecundity), have been confirmed in tests against null alternatives with high statistical significance. Recently (Watt 2003) we did the exercise of assembling the fitness component values for the three most common lowland genotypes (3/3, 3/4, and 4/4) into their relative fitnesses, W_{ij}, for each sex. Setting relative fitness of 1.0 for the 3/3 genotype in each case, the values for 3/4 males and females are 1.67 and 1.85 respectively, while for 4/4 the values for males and females are 0.17 and 0.19 respectively. Thresholds in the interactions of genotypic function with T_b, and the multiplicative relations among the fitness components l_x and m_x, combine to produce a tenfold range of W_{ij} values!

It is noteworthy that samples of newly emerging adults always show PGI frequencies as close to Hardy–Weinberg expectations as sample size would lead one to expect. This absence of any evidence for pre-adult selection fits with the idea that selection on this gene is dependent on the functional demands of high-throughput transient metabolism in support of flight. For larvae, moving slowly as they feed and thus spending most of their time in metabolic steady state, perhaps even the least effective of the PGI genotypes may have more than enough kinetic capacity.

Further, the balance of selective pressures maintaining this polymorphism is closely similar across broad expanses of the North American West (Watt *et al.* 2002). This is shown both by stability of the polymorphic genotypes' frequencies over distances up to 2000 km between California and the Rocky

Mountains (Watt 1983, Watt *et al.* 2003), and by closely similar values of flight performance (Watt *et al.* 1983) and fitness component (e.g., Watt 1977, 1983, Watt *et al.* 1985, 1986) differences measured among the genotypes at diverse locations over this geography.

14.3.3 Molecular aspects of the lowland *Colias* PGI polymorphism

New work extends our perspective on *Colias*' PGI variation to the levels of molecular sequence and structure (Wheat 2001, Wheat *et al.* 2003). One initial focus of this has been to address the protein-structural origins of the genotypic functional differences whose performance and fitness consequences maintain the variation. Beyond this, we can make process-based connections with the pattern-based subdiscipline of "molecular evolution."

Insight into heterozygotes' functional advantage in *Colias* PGI arises from the extensive sequence identity between this protein and rabbit PGI, for which there is a high-resolution crystal structure (Jeffrey *et al.* 2001). This allowed us to "thread", by computation, butterfly PGI sequences into the rabbit structure, compensating the structural representation for fixed sequence differences (Wheat 2001, Wheat *et al.* 2003). The resulting reconstruction shows that PGI's subunits, in native dimeric configuration, interpenetrate deeply. This establishes a basis in structural complementation (cf. Fincham 1966) for heterozygote properties different from those of homozygotes.

Turning to *Colias*' PGI variants, we see multiple substitutions, on average 3.5 amino acid differences between pairs of alleles (Wheat 2001, Wheat *et al.* 2003). These are found throughout the protein's structure, in contact with the surrounding hydration shell of water, and appear to alter catalytic properties by change of whole-protein folding patterns rather than by substitution in the catalytic center itself (Wheat *et al.* 2003); this is also seen in other systems (cf. Fields and Somero 1998, Watt and Dean 2000). Minor variation, i.e., substitution of very similar amino acids (e.g., valine for isoleucine), within electromorph allele classes is also seen (Wheat *et al.* 2003). We do not yet know, but will assess, what functional effects these minor substitutions may have. They may contribute to the variance around our successful predictions of mean performance and fitness component differences among the major PGI genotypes (e.g., Watt 1992). We will also study intron structure, intron variability, and silent-substitution nucleotide variability among these PGI alleles. These more probably neutral accompaniments to this largely strongly selected polymorphism may provide contrasting insight into the history of the polymorphism and its adjacent genetic regions.

This multiallelic molecular polymorphism also now offers a fertile field for evaluating the variation-multiplying evolutionary impacts of gene conversion and intragenic recombination. Our group has contributed to theoretical understanding of these "extra-Mendelian" phenomena arising from the molecular mechanism of recombination (Watt 1972, Leslie and Watt 1986), and their effects are already recognizable in other systems (e.g., Hall 1982,

Berry and Barradilla 2000, Hedrick and Kim 2000). We look forward to evaluating their role in *Colias* PGI's evolutionary history.

14.3.4 Adaptive complexities of PGI variation among taxa

A new line of work stems from one question about the generality of our PGI results: how does this gene vary in relatives of lowland *Colias*? Comparative studies of *Colias* PGI across taxa can now be pursued at all levels from molecular structure, through molecular and organismal function, to fitness effects. *Colias meadii* of the Rocky Mountains, the southernmost part of an arctic–alpine species complex which extends north to the Arctic Ocean and across Beringia into Asia, also has a multiallelic PGI polymorphism. Its genotypes' functional, hence performance and fitness, differences parallel those in lowland taxa in many ways, showing kinetics advantage in heterozygotes and kinetics/stability tradeoff in homozygotes (Watt *et al.* 1996). But these functionally based properties are not associated with identity or electrophoretic similarity among alleles or genotypes of the species complexes (Watt *et al.* 1996), putting a novel and complex twist on this parallel evolution.

Further, *C. meadii*'s PGI displays some remarkable biogeography. In this species' alpine–tundra habitats, PGI genotype frequencies are just as stable among geographically separated sites as those in the lowland taxa, though alpine geography sharply restricts exchange of migrants. But *C. meadii* also occurs in high montane steppe, and steppe populations' allele and genotype frequencies consistently differ from those in tundra (Watt *et al.* 2003). Where steppe and tundra populations make contact, exchanging many breeding individuals per generation (e.g., Watt *et al.* 1977), the frequency difference is stable year after year despite this migration (Watt *et al.* 2003). New field work will tease out how the insects' niche structure interacts with their PGI genotypes' norms of functional reaction to maintain this reproducible frequency difference across the treeline ecotone.

Broader comparisons are invited by the observation that PGI polymorphism seems to be ubiquitous among *Colias* species and in their sister group *Zerene* (Geiger, Descimon, Shapiro, Watt, Wheat, and Wright, unpublished). This suggests diversification by polymorphic allele replacement rather than by classic directional selection, as may also be so in plant self-incompatibility and vertebrate major-histocompatibility systems (e.g., Ioerger *et al.* 1990, Hedrick and Kim 2000). More study of this within and among species-complex clades of *Colias* may allow test of punctuated-equilibrium-based ideas about association between adaptive and phyletic differentiations (e.g., Eldredge 1995).

14.3.5 Epistasis in metabolism – a feasible new level of analyzable complexity

A second path to generality from the lowland *Colias* PGI case is to examine other variable enzyme genes involved in energy processing, to see what differentiation in function and in performance and fitness effects may occur among their variants. Glucose phosphate dehydrogenase (G6PD) and phosphoglucomutase (PGM) share a common substrate, glucose-6-phosphate, with PGI. PGM acts *en route* to or from glycogen storage; G6PD carries out the first step of the pentose shunt. The pentose shunt does not run in flight muscle, and muscle glycogen storage appears to play only a minor role in flight support (Watt and Boggs 1987), so PGM and G6PD variants *ought* not to impact flight. However, they could interact with one another, or with PGI variants, at the G6P branch point in fat body, a major site of insect biosynthesis and one in which glycolysis, the pentose shunt, and glycogen storage all run at once. Moreover, PGM and G6PD are loosely linked, showing about 40% recombination, though both assort independently of PGI (Carter and Watt 1988). Thus variants of PGM and G6PD have potential for both gametic disequilibrium and functional epistasis.

Five alleles are common at PGM, and four at G6PD, in lowland *Colias* (Carter and Watt 1988). While testing fitness predictions about PGI variants, we also scored PGM and G6PD. As expected, their genotypes had little effect on flight and survival. Heterozygotes were advantaged in male mating success at both genes, as much as for PGI variants but independent of them, showing the distinctness of effects resulting from the enzymes' metabolic roles. Epistasis occurs between PGM and G6PD: two of the most common two-locus genotypes (PGM 3/3 G6PD 1/1, and PGM 3/4 G6PD 1/2) were strongly favored, and the reverse combinations disfavored, in male mating success (Carter and Watt 1988). Functional study of these variants should clarify how epistasis among enzymes varies with their metabolic roles. Is it so, as may be theoretically expected, that epistasis mainly occurs among genes acting within single pathway branches, dividing metabolism into "modules" (cf. Savageau and Sorribas 1989, Wagner and Altenberg 1996)? We shall see.

14.4 Conceptual questions in evolution: some empirical answers

I now turn to showing how our specific studies of *Colias* variants, together with related specific studies on other systems, can illuminate some general issues that have been of concern to evolutionary biologists. The list is not exhaustive, and may be extended elsewhere.

14.4.1 The feasibility of mechanistic evolutionary study

We have shown with *Colias* that it is practical to disentangle analytically the organism–environment interactions of identified gene variants, then reassemble a comprehensive, predictive picture of their impacts on whole organism performance, thence on all major fitness components, in a range of natural environments. This can be done incrementally, building from one level of understanding to another, and testing for interactions with other genes or phenotypic subsystems, or with new habitat variables, as opportunities arise. Thus, setting aside truly rare catastrophic events which may negate the usual processes of evolution, evolutionists are not, despite some such claims, limited to fortuitous glimpses of those processes. It may well be easier to do such work on butterflies than on elephants, let alone (extinct) condylarths, but hope of developing mechanistic evolutionary approaches to a wide range of study systems is surely realistic (Feder and Watt 1992).

Bioenergetics provides an important unifying context of great power to assist this process, especially when metabolic organization theory and related derivations are used as a supporting framework (Watt 1985a, Watt and Dean 2000). It has been very useful not only for study of energy-processing genes standing alone, but for the beginnings of study of their epistatic interactions. It also contributes to study of other types of variation such as resource-related color polymorphisms, though other contexts such as evolved systems of mate-signalling may also affect such variants (e.g., Nielsen and Watt 1998, 2000). Such unifying contexts are central to extending the approach to a wide range of taxa in the search for mechanistically oriented general principles of adaptive change.

14.4.2 "Genic selectionism" vs. the genotypic level of selection

In haploid organisms, carrying single copies of their genes, only allelic variation among individuals is possible. Although diploids carry two copies of each gene, allowing allelic variation within individuals and hence genotypic variation among them, diverse authors (e.g., Dawkins 1989, Williams 1992) claim primacy for allelic, not genotypic, variants as the "fundamental" units of natural selection. Godfrey-Smith and Lewontin (1993) critique this claim, observing that quite aside from open questions of selection at levels above that of the individual, genotypes are units of selection. (The overall state of this dispute is well summarized by Lloyd 2000.)

Our results on *Colias* PGI strongly support Godfrey-Smith and Lewontin as regards the status of genotypic variation in diploids. The large advantage in kinetic function seen in most PGI heterozygotes, often alongside at least partial escape from the kinetics vs. thermal stability tradeoff which is the rule in homozygotes, proceeds directly from the molecular interaction of allelic gene products in native dimer structures, as noted earlier. It makes no functional sense to speak of "allelic contributions" here, when heterozygotes' properties are not linear combinations of the properties of related homozygotes, and present phenotypes to the test of organism–environment interactions which are not intermediate between those of homozygotes, but quite outside their ranges.

This case is no "exception": for example, in lactate dehydrogenase of the fish *Fundulus*, Powers and colleagues (e.g., 1991) found that heterozygotes, again via mediation of multimeric protein structure, alter their kinetics to approximate the function of each of two thermal-specialist homozygotes in the extreme water-temperature conditions to which the latter are alternatively best adapted. The homozygotes show no such thermally flexible kinetic norm-of-reaction (Place and Powers 1979). Thus, again, heterozygous genotypes determine fundamentally different phenotypes than do related homozygotes, and "genic" viewpoints on natural selection are inadequate to deal with such cases.

14.4.3 The continuity of adaptation from passive to constructionist emphases

Lewontin (1983) emphasized the active, "constructive" interaction of organisms with environments, modifying their habitats to increase their own adaptedness and hence resulting fitness therein. This idea was present in the *Origin of Species,* notably in discussion of pollinator–plant interactions, in which, by aiding the reproduction of their food sources, pollinators increase the suitability of their habitats to themselves. It is an important aspect of adaptation (Laland *et al.* 1996, 1999), but it is not a "substitute" for the adaptation concept. Philosophical aspects of the continuity between "passive" and "constructive" adaptations have been clarified by Godfrey-Smith (2000).

This continuity is clear in the variation of *Colias* butterflies' thermal ecology throughout their life cycle, and in comparison with other animals. The thermal balance of a new-laid egg on a leaf is at the mercy of its placement by its mother (done in context of *her* thermal balance, not its) and thermal conditions there until it hatches. A larva has only limited ability to change its thermal experience by moving in or out of shade, etc., and to set that experience for itself when choosing a pupation site (Sherman and Watt 1973). Behaviorally thermoregulating adults may change their experience of ecological thermal variables and their effects, for example by orienting to maximize or minimize absorption of sunlight so as to regulate their own T_b (Watt 1968, Tsuji *et al.* 1986).

However, no butterfly can alter habitat-scale thermal variables. Larger animals may have more opportunity to change their thermal balance by modifying external variables or at least by using their own metabolic heat as an important internal variable. Facultatively homeothermic insects such as bumblebees or hawkmoths, let alone vertebrates, can change their metabolic heat terms or their core-to-surface conductivities (e.g., Heinrich 1981, Harrison *et al.* 1996). Lizards can make burrow refuges from unfavorable thermal conditions, and birds or mammals may make nests or other shelters against them. At the extreme, *Homo sapiens* builds not only shelters but air conditioners, and thus may alter climate, hence its own survival prospects, for the worse by dissipating more heat to the global atmosphere while removing excess heat from its local niche. The continuum from passive to highly constructionist adaptations is especially clear in a thermal-ecological context.

14.4.4 The coincidence and coimportance of adaptation and constraint

Many have held that various agents may constrain the course of adaptive evolution (e.g., Gould and Lewontin 1979, Gould 1989, Feder and Watt 1992). Our work with *Colias* has found diverse evolutionary constraints or predispositions, ranging from molecular to ecosystem levels.

Gillespie (1991) asked: why is the PGI gene seemingly predisposed to adaptive variability in many diverse taxa? The interpenetrated state of PGI's subunits (above) may provide much of the answer. Heterogeneity of amino acid sequence between alleles and the influence of multimers in quaternary structure on one another's folding patterns (Fincham 1966) may lead to structures and functions impossible to achieve with subunits of identical sequence. Dissection by directed mutagenesis, amino acid by amino acid, of the structural differences among the *Colias* PGI alleles and their functional consequences in genotypic enzyme structures will allow us to test this explanation.

We've already seen that animal thermoregulation may be a collection of adaptive responses to a ubiquitously severe constraint placed by metabolic organization on the breadth of temperature over which metabolism can achieve high, controlled throughput (above; Watt 1991, Kohane and Watt 1999). Here is one of the clearest examples of "adaptationist" inadequacy: being a "thermal generalist", equally active over a wide range of body temperatures, would be a supremely adaptive phenotype, and yet no animal has achieved it. Modifying one's experience of one's habitat and/or one's own physiology to secure a narrow to invariant body temperature, and often a high one, may be a constructionist adaptive response, but its very necessity arises from a profound constraint.

Still other styles of constraint may be found in *Colias* thermal biology. Kingsolver and Watt (1984) found that *Colias* in cold habitats are neither as absorptive of sunlight, nor as insulated against convective cooling, as they should be to maximize flight time under average thermal conditions each day. This results from the great variance of microclimate in such habitats: for example, maximally absorptive *Colias* might make fuller flight use of mean alpine thermal conditions, but then literally cook themselves under uncommon but recurrent heat stress in these habitats. Constraints on adaptation may also arise from hierarchical selection priorities, as in the precluding of larval thermoregulatory color changes by larval needs for camouflage against visual predators (above). Organisms do not live in the best of all worlds, and "Darwin's Demon" is in no sense all-powerful.

14.4.5 Barring some holds: the dysfunction of "neutralism" and "selectionism"

The neutralist–selectionist dispute, provoked by findings of widespread natural variation in proteins assayable by gel electrophoresis, was marked by conflicting assertions of the neutrality or, *au contraire*, strong selectedness of this variation as a whole (reviews, e.g., Lewontin 1974, Ewens and Feldman 1976, Watt 1985b, 1995, Gillespie 1991). Discussion was often limited by selective focus on ideas supporting the position of each discussant. In this "no-holds-barred" atmosphere, many failed to consider that reality might be more complex than the most extreme views, and that some biologically realistic mixture of causes was probably at work.

The issue was not resolved by a "creative clash" of opposites leading to a new synthesis, as a dialectical view might hold (e.g., Levins and Lewontin 1985). Answers, in specific cases, to questions central to the dispute have owed little to the original controversy, but much to mechanistic study from its origins (e.g., Ingram 1963, Watt 1968, Koehn 1969, Clarke 1975) to the present. Many such efforts have focused on the variants' functional properties in the context of the organism–environment interactions surrounding the variation. In contrast, the neutralist–selectionist disputants were self-limited to analyzing statistical variant distributions, and often could not choose among abstract models with opposed (and often biologically implausible, Watt 1995) assumptions whose predictions proved indistinguishable in practice (e.g., Rothman and Templeton 1980, Gillespie 1991, Watt 1995).

The futility of the neutralist–selectionist clash is underscored by *Colias*' PGI variation. There are large functional and fitness differences among most PGI genotypes of the lowland taxa, but the homozygote 4/4 and the heterozygote 4/5 are indistinguishable in function (Watt 1983) and thus, predictably, in performance and fitness components across the range of these taxa (e.g., Watt 1992). Here is neutrality amid strong selection – not by a mystical "unity of opposites," but simply because functional differences and their fitness effects among variants range from zero to large. Thus theory of neutral genetic variants (Kimura 1983) does not "stand" or "fall" by whether or not it explains most observed patterns of genetic variation. It is the central null hypothesis and starting point for rigorous evidence-based testing of all evolutionary–genetic hypotheses, not just the extremes (cf. Platt 1964). The neutralist–selectionist dispute shows, indeed, how clash of extreme views, combined with a "no-holds-barred" atmosphere, may convert debate into unhelpful combat. Evolutionary biology, as a community of inquiry, must do better in future.

14.4.6 The evolutionary importance of strongly selected variants at major genes

Adherents of the amechanistic "Modern Synthesis" paradigm in evolutionary biology (cf. Watt 2000) commonly hold that genetic variants of large effect at "major" genes are unimportant in evolution, and if found must be "exceptions." This idea owes much to Haldane's cost-of-evolution argument (1957) and to the related concept of segregational load. Each view claims that if variation is selected upon strongly at too many genes at once (whether in directional or balancing mode), mean fitness will be so depressed that the population will go extinct. The belief also fits with the claim that "complex", "evolutionarily important" characters are necessarily polygenically determined, so that their variants will exhibit individually minute fitness differences, and thus will be sufficiently analyzable by quantitative genetics theory (e.g., Lande 2000). However, this belief persists despite severe flaws, repeatedly critiqued, in the cost-of-evolution and segregation-load concepts.

Brues, in an incisive but neglected analysis (1969), observed that both these ideas depend centrally on the convention of defining the maximum relative fitness in a population as 1.0. As soon as a new variant of superior fitness arises in a population, the "old" allele's homozygote, previously assigned relative fitness 1.0, is demoted to lower relative fitness. Brues pointed out that this seeming "depression of mean fitness" by the new variant's segregation is, in each case, an artifact of the fitness reassignment, lacking biological reality. These ideas also ignore the fact that population persistence is governed by, not relative, but absolute fitnesses, i.e., genotypic replacement rates (R or λ; Feder and Watt 1992, McGraw and Caswell 1996, Watt 2000) which can often be larger than 1.0. Thus, new variants can displace old ones in either directional sweeps or balanced polymorphism while actually *increasing* population sizes. Gillespie (1991) and Ewens (1993, 2000) focused on other flaws in cost-of-evolution or load models' fitness standardizations; in Ewens' summary phrase (1993, p. 12), "the immense genetic loads calculated . . . are artifacts of incorrect models." Thus, cost-of-evolution and segregation-load models may still apply to special cases, but they give no broadly valid reasons to regard large fitness differences among reasonable numbers of major-gene genotypes as "exceptional." The issue then comes down to the empirical question: what actually happens?

Empirically, our results on variants of energy processing in *Colias* (above), and diverse mechanistic results in other systems as well, show that major-gene variants of large effect may often (though not always) play important roles in evolution (e.g., Watt 1985a,b, 1994, Mitton 1997, Eanes 1999, Watt and Dean 2000). Such variants do contribute importantly to variation of many "complex characters" – locomotion as flight or swimming, osmoregulation, foraging behavior, heat stress resistance, detoxification of poisons, etc. – though polygene variants of small effect contribute to complex phenotypes as well. The study of "quantitative trait loci" (QTL; summarized by, e.g., Mackay 1995, Frankham and Weber 2000) now also shows that variants of large effect at major genes may often influence "complex characters," even though these characters are initially studied in polygenic terms. Frankham and Weber, indeed, remark (2000, p. 355) that "the infinitesimal model of a nearly infinite number of loci, each with a small effect, is clearly disproven." Thus our "method of identified genes" (Watt 2000) and the practice of QTL-oriented study may be converging to a view of a functionally characterizable spectrum of the nature and size of adaptive genetic changes. Mechanistic study of major-gene polymorphs and the quantitative genetics of "microvariants" would then become complementary parts of the working tool kit of post-"Synthesis" evolutionary biology.

14.5 Mechanistic evolutionary study in social context – initial suggestions and speculations

All peoples seek to understand the workings of the world and the meaning of their presence within it. We must respect nonscientific answers to such questions, from diverse cultures, as beliefs which members of those cultures have a right to hold. But the cultural independence of much of natural science, as seen in, for example, the convergence of folk taxonomies and formal systematics, or the contributions of persons of diverse origins to important progress in science, encourages us as to the genuine and humane power of science to increase our understanding of the world we live in.

A mechanistic approach to the study of evolutionary alternatives is a powerful complement to (*not* a replacement for) formal approaches and descriptive natural history alike in basic evolutionary biology (cf. Watt 2000, Watt and Dean 2000). It will also be important in applications to practical problems of agriculture, conservation, environmental management, and public health, for example in prudently constraining the uses of genetic-manipulation techniques. Each of these fields, central to human survival, is essentially a form of applied evolutionary biology, as Bennett *et al.* (1999) make clear. How the processes of evolution can be seen to work, in mechanistic and yet integrative fashion, in natural scenarios will be central to fostering the wide dissemination of public understanding, and preventing potentially disastrous mistakes, in these application areas.

Mechanistic evolutionary study may even shed light on humans' view of ourselves. Lewontin (1972) and others (e.g., Ruvolo and Seielstad 2000) emphasize that most human genetic variability is within population, not between population. A mechanistic view of this might observe that the most centrally human functional character – abstracting the habitat, manipulating the abstraction to solve problems "in principle," and then applying the solutions back to the habitat – is a universal feature of human organism–environment interaction (cf. Godfrey-Smith 1996). This is true wherever our species occurs – whether a person be an Inuit, a Dinka, or an Albanian. Whatever the complex neural underpinnings of this remarkable capacity may be, their variability is unlikely to differ among regional populations, given human migration and the parallelism of this organism–environment interaction in all human habitats. In contrast, mechanistic study of variation in human skin color identifies protection of blood folate levels against ultraviolet light as a source of local selection for skin melanization in UV-intensive latitudes, and *en passant* shows the lack of association between this character and other characters or human phylogeny as a whole (Jablonski and Chaplin 2000).

The sooner that insights from people of all cultural backgrounds, exercising their common skills of abstract problem-solving, can be brought to bear on human exploitationist and xenophobe tendencies, the sooner we may move toward a more humane, just, and environmentally sustainable global array of societies. Mechanistic evolutionary biology, cautiously and rigorously pursued, surely has a role to play in supporting those insights and fostering that problem-solving. We may well believe, along with Hutchinson (1965), that the pursuit of such studies in diverse systems, even far removed in phyletic terms from *Homo sapiens*, may contribute helpfully to this process.

14.6 Acknowledgments

I thank Carol Boggs, Egbert Leigh, Jessica Ruvinsky, Rama Singh, Jean Stamberger, and Chris Wheat for stimulating discussions, and the US National Science Foundation for research support (IBN 01-17754).

REFERENCES

Bennett, A. F., Brockmann, H. J., Feldman, M. W., Fitch, W. M., Futuyma, D. J., Godfrey, L. R., Jablonski, D., Lynch, C. B., Meagher, T. R., Real, L., Riley, M. A., Sepkoski, J. J., jr., and Smocovitis, V. B. (1999). *Evolution, Science, and Society: Evolutionary Biology and the National Research Agenda.* New Brunswick, NJ: Rutgers University, Office of University Publications.

Berry, A., and Barradilla, A. (2000). Gene conversion is a major determinant of genetic diversity at the DNA level. In R. S. Singh and C. B. Krimbas (eds) *Evolutionary Genetics: From Molecules to Morphology*, pp. 102–23. Cambridge: Cambridge University Press.

Brandon, R. N. (1990). *Adaptation and Environment.* Princeton, NJ: Princeton University Press.

Brues, A. M. (1969). Genetic load and its varieties. *Science* 164:1130–6.

Carter, P. A., and Watt, W. B. (1988). Adaptation at specific loci. V. Metabolically adjacent enzyme loci may have very distinct experiences of selective pressures. *Genetics* 119:913–24.

Charlesworth, B. (1994). *Evolution in Age-structured Populations*, 2nd edition. Cambridge: Cambridge University Press.

Clarke, B. (1975). The contribution of ecological genetics to evolutionary theory: detecting the direct effects of natural selection on particular polymorphic loci. *Genetics* 79:101–13.

Darwin, C. (1859). *The Origin of Species.* 6th edition, revised, 1872. New York: New American Library.

Dawkins, R. (1989). *The Selfish Gene*, revised edition, New York: Oxford University Press.

Dean, A. M. (1989). Selection and neutrality in lactose operons of *Escherichia coli.* *Genetics* 123:441–54.

Depew, D. J., and Weber, B. H. (1995). *Darwinism Evolving.* Cambridge, MA: MIT Press.

Dobzhansky, Th. (1970). *The Genetics of the Evolutionary Process.* New York: Columbia University Press.

Dykhuizen, D. E., and Hartl, D. L. (1983). Functional effects of PGI allozymes in *E. coli. Genetics* 105:1–18.

Eanes, W. F. (1999). Analysis of selection on enzyme polymorphisms. *Annu. Rev. Syst. Ecol.* 30:301–26.

Easterby, J. S. (1973). Coupled enzyme assays: a general expression for the transient. *Biochim. Biophys. Acta* 293:552–8.

Easterby, J. S. (1990). Integration of temporal analysis and control analysis of metabolic systems. *Biochem. J.* 269:255–9.

Eldredge, N. (1995). *Reinventing Darwin.* New York: John Wiley.

Endler, J. A. (1986). *Natural Selection in the Wild.* Princeton, NJ: Princeton University Press.

Ewens, W. (1993). Beanbag genetics and after. In P. P. Majumder (ed.) *Human Population Genetics: a Centennial Tribute to J. B. S. Haldane,* pp. 7–29. New York: Plenum Press.

Ewens, W. (2000). The mathematical foundations of population genetics. In R. S. Singh and C. B. Krimbas (eds) *Evolutionary Genetics: From Molecules to Morphology,* pp. 24–40. Cambridge: Cambridge University Press.

Ewens, W., and Feldman, M. W. (1976). The theoretical assessment of selective neutrality. In S. Karlin and E. Nevo (eds) *Population Genetics and Ecology,* pp. 303–37. New York: Academic Press.

Feder, M. E., and Watt, W. B. (1992). Functional biology of adaptation. In R. J. Berry, T. J. Crawford and G. M. Hewitt (eds) *Genes in Ecology,* pp. 365–92. Cambridge: British Ecological Society, Cambridge University Press.

Fields, P. A., and Somero, G. N. (1998). Hot spots in cold adaptation: localized increases in conformational flexibility in lactate dehydrogenase A_4 orthologs of Antarctic notothenioid fishes. *Proc. Natl. Acad. Sci. USA* 95:11476–81.

Fincham, J. R. S. (1966). *Genetic Complementation.* New York/Amsterdam: Benjamin.

Frankham, R., and Weber, K. (2000). Nature of quantitative genetic variation. In R. S. Singh and C. B. Krimbas (eds) *Evolutionary Genetics: From Molecules to Morphology,* pp. 351–68. Cambridge: Cambridge University Press.

Gerould, J. H. (1921). Blue-green caterpillars: the origin and ecology of a mutation in hemolymph color in *Colias (Eurymus) philodice. J. Exp. Zool.* 34:385–415.

Gillespie, J. H. (1991). *The Causes of Molecular Evolution.* Oxford: Oxford University Press.

Godfrey-Smith, P. (1996). *Complexity and the Function of Mind in Nature.* Cambridge: Cambridge University Press.

Godfrey-Smith, P. (2000). Organism, environment, and dialectics. In R. S. Singh, C. B. Krimbas, D. B. Paul, and J. Beatty (eds) *Thinking about Evolution,* pp. 253–66. Cambridge: Cambridge University Press.

Godfrey-Smith, P., and Lewontin, R. C. (1993). The dimensions of selection. *Phil. Sci.* 60:373–95.

Gould, S. J. (1989). A developmental constraint in *Cerion,* with comments on the definition and interpretation of constraint in evolution. *Evolution* 43:516–39.

Gould, S. J., and Lewontin, R. C. (1979). The spandrels of San Marco and the Panglossian paradigm. *Proc. R. Soc. Lond. Ser. B* 205:581–98.

Grant, P. R. (1986). *Ecology and Evolution of Darwin's Finches.* Princeton, NJ: Princeton University Press.

Haldane, J. B. S. (1957). The cost of natural selection. *J. Genet.* 55:511–24.

Hall, B. (1982). Evolution on a petri dish. *Evol. Biol.* 15:85–150.

Harrison, J. F., Fewell, J. H., Roberts, S. P., and Hall, H. G. (1996). Achievement of thermal stability by varying metabolic heat production in flying honeybees. *Science* 274:88–90.

Hartl, D. L., and Dykhuizen, D. E. (1981). Potential for selection among nearly neutral allozymes of 6-phosphogluconate dehydrogenase in *Escherichia coli. Proc. Natl. Acad. Sci. USA* 78:6344–8.

Hedrick, P. W., and Kim, T. J. (2000). Genetics of complex polymorphisms: parasites and maintenance of the major histocompatibility complex variation. In R. S. Singh and C. B. Krimbas (eds) *Evolutionary Genetics: From Molecules to Morphology*, pp. 204–34. Cambridge: Cambridge University Press.

Heinrich, B. (ed.) (1981). *Insect Thermoregulation.* New York: John Wiley.

Hodge, M. J. S. (1987). Natural selection as a causal, empirical, and probabilistic theory. In L. Kruger, G. Gigerenzer, and M. S. Morgan (eds) *The Probabilistic Revolution*, vol. 2, pp. 233–70. Cambridge, MA: MIT Press.

Hoffmann, R. J., and Watt, W. B. (1974). Naturally occurring variation in larval color of *Colias* butterflies: isolation from two Colorado populations. *Evolution* 28:325–8.

Hubby, J. L., and Lewontin, R. C. (1966). A molecular approach to the study of genic heterozygosity in natural populations. I. The number of alleles at different loci in *Drosophila pseudoobscura. Genetics* 54:577–94.

Hutchinson, G. E. (1965). *The Ecological Theater and the Evolutionary Play.* New Haven, CT: Yale University Press.

Ingram, V. I. (1963). *The Hemoglobins in Genetics and Evolution.* New York: Columbia University Press.

Ioerger, T. R., Clark, A. G., and Kao, T.-H. (1990). Polymorphism at the self-incompatibility locus in Solanaceae predates speciation. *Proc. Natl. Acad. Sci. USA* 87:9732–5.

Jablonski, N. G., and Chaplin, G. (2000). The evolution of human skin coloration. *J. Hum. Evol.* 39:57–106.

Jeffery, C. J., Hardre, R., and Salmon, L. (2001). Crystal structure of rabbit phosphoglucose isomerase complexed with 5-phospho-D-arabinonate identifies the role of Glu357 in catalysis. *Biochemistry* 40:1560–6.

Johnson, N. A., and Porter, A. H. (2001). Toward a new synthesis: population genetics and evolutionary developmental biology. *Genetica* 112/113:45–58.

Kimura, M. (1983). *The Neutral Theory of Molecular Evolution.* Cambridge: Cambridge University Press.

Kingsolver, J. G. (1983a). Thermoregulation and flight in *Colias* butterflies: elevational patterns and mechanistic limitations. *Ecology* 64:534–45.

Kingsolver, J. G. (1983b). Ecological significance of flight activity in *Colias* butterflies: implications for reproductive strategy and population structure. *Ecology* 64:546–51.

Kingsolver, J. G., and Watt, W. B. (1984). Mechanistic constraints and optimality models: thermoregulatory strategies in *Colias* butterflies. *Ecology* 65:1835–9.

Koehn, R. K. (1969). Esterase heterogeneity: dynamics of a polymorphism. *Science* 163:943–4.

Kohane, M. J., and Watt, W. B. (1999). Flight-muscle adenylate pool responses to flight demands and thermal constraints in individual *Colias* (Lepidoptera, Pieridae). *J. Exp. Biol.* 202:3145–54.

Krimbas, C. B. (1984). On adaptation, neo-Darwinism, tautology, and population fitness. *Evol Biol.* 17:1–57.

Krimbas, C. B. (2000). In defense of neo-Darwinism: Popper's "Darwinism as a metaphysical programme" revisited. In R. S. Singh, C. B. Krimbas, D. B. Paul, and J. Beatty (eds) *Thinking about Evolution*, pp. 292–308. Cambridge: Cambridge University Press.

Laland, K. N., Odling-Smee, F. J., and Feldman, M. W. (1996). The evolutionary consequences of niche construction: an investigation using two-locus theory. *J. Evol. Biol.* 9:293–316.

Laland, K. N., Odling-Smee, F. J., and Feldman, M. W. (1999). Evolutionary consequences of niche construction and their implications for ecology. *Proc. Natl. Acad. Sci. USA* 96:10242–7.

Lande, R. (2000). Quantitative genetics and phenotypic evolution. In R. S. Singh and C. B. Krimbas (eds) *Evolutionary Genetics: From Molecules to Morphology*, pp. 335–50. Cambridge: Cambridge University Press.

Leslie, J. F., and Watt, W. B. (1986). Some evolutionary consequences of the molecular recombination process. *Trends Genet.* 2:288–91.

Levins, R., and Lewontin, R. C. (1985). *The Dialectical Biologist.* Cambridge, MA: Harvard University Press.

Lewontin, R. C. (1972). The apportionment of human diversity. *Evol. Biol.* 6:381–98.

Lewontin, R. C. (1974). *The Genetic Basis of Evolutionary Change.* New York: Columbia University Press.

Lewontin, R. C. (1983). Gene, organism, and environment. In D. S. Bendall (ed.) *Evolution from Molecules to Men*, pp. 273–85. Cambridge: Cambridge University Press.

Lewontin. R. C. (1984). Adaptation. In E. Sober (ed.) *Conceptual Issues in Evolutionary Biology*, pp. 235–51. Cambridge, MA: MIT Press.

Lewontin, R. C., and Hubby, J. L. (1966). A molecular approach to the study of genic heterozygosity in natural populations. II. Amount of variation and degree of heterozygosity in natural populations of *Drosophila pseudoobscura. Genetics* 54:595–609.

Lloyd, E. A. (2000). Units and levels of selection: an anatomy of the units of selection debates. In R. S. Singh, C. B. Krimbas, D. B. Paul, and J. Beatty (eds) *Thinking about Evolution*, pp. 267–91. Cambridge: Cambridge University Press.

Mackay, T. F. C. (1995). The genetic basis of quantitative variation: numbers of sensory bristles of *Drosophila melanogaster* as a model system. *Trends Genet.* 11: 464–70.

Mangum, C. P., and Hochachka, P. W. (1998). New directions in comparative physiology and biochemistry: mechanisms, adaptations, and evolution. *Physiol. Zool.* 71:471–84.

McGraw, J. B., and Caswell, H. (1996). Estimation of individual fitness from life-history data. *Am. Nat.* 147:47–64.

Mitton, J. B. (1997). *Selection in Natural Populations.* Oxford: Oxford University Press.

Nielsen, M. G., and Watt, W. B. (1998). Behavioral fitness component effects of the "alba" polymorphism of *Colias* (Lepidoptera, Pieridae): resource and time budget analysis. *Func. Ecol.* 12:149–58. [also 1999. Erratum (for Table 2). *Func. Ecol.* 13:437.]

Nielsen, M. G., and Watt, W. B. (2000). Interference competition and sexual selection promote the maintenance of polymorphism in *Colias* (Lepidoptera, Pieridae). *Func. Ecol.* 14:718–30.

Place, A. R., and Powers, D. A. (1979). Genetic variation and relative catalytic efficiencies: lactate dehydrogenase B allozymes of *Fundulus heteroclitus. Proc. Natl. Acad. Sci. USA* 76:2354–8.

Platt, J. R. (1964). Strong inference. *Science* 146:347–53.

Powers, D. A., Lauerman, T., Crawford, D., and DiMichele, L. (1991). Genetic mechanisms for adapting to a changing environment. *Annu. Rev. Genet.* 25:629–59.

Roland, J. (1982). Melanism and diel activity of alpine *Colias* (Lepidoptera: Pieridae). *Oecologia* 53:214–21.

Rothman, E. and Templeton, A. M. (1980). A class of models of selectively neutral alleles. *Theor. Pop. Biol.* 18:135–50.

Ruvolo, M., and Seielstad, M. (2000). "The apportionment of human diversity" 25 years later. In R. S. Singh, C. B. Krimbas, D. B. Paul, and J. Beatty (eds) *Thinking about Evolution*, pp. 141–51. Cambridge: Cambridge University Press.

Savageau, M. A., and Sorribas, A. (1989). Constraints among molecular and systemic properties: implications for physiological genetics. *J. Theor. Biol.* 141:93–115.

Sherman, P. W., and Watt, W. B. (1973). The thermal physiological ecology of some *Colias* butterfly larvae. *J. Comp. Physiol.* 83:25–40.

Simpson, G. G. (1964). Organisms and molecules in evolution. *Science* 146:1535–8.

Singh, R. S., Eanes, W. F., Hickey, D. A., King, L. M., and Riley, M. A. (2000). The molecular foundation of population genetics. In R. S. Singh and C. B. Krimbas (eds) *Evolutionary Genetics: From Molecules to Morphology*, pp. 52–77. Cambridge: Cambridge University Press.

Stern, D. L. (2000). Perspective: evolutionary developmental biology and the problem of variation. *Evolution* 54:1079–91.

Tsuji, J. S., Kingsolver, J. G., and Watt, W. B. (1986). Thermal physiological ecology of *Colias* butterflies in flight. *Oecologia* 69:161–70.

Wagner, G. P., and Altenberg, L. (1996). Complex adaptations and the evolution of evolvability. *Evolution* 50:967–76.

Wagner, G. P., and Gauthier, J. A. (1999). 1,2,3 = 2,3,4: a solution to the problem of the homology of the digits in the avian hand. *Proc. Natl. Acad. Sci. USA* 96:5111–6.

Watt, W. B. (1968). Adaptive significance of pigment polymorphisms in *Colias* butterflies. I. Variation of melanin pigment in relation to thermoregulation. *Evolution* 22:437–58.

Watt, W. B. (1972). Intragenic recombination as a source of population genetic variability. *Am. Nat.* 106: 737–53.

Watt, W. B. (1977). Adaptation at specific loci. I. Natural selection on phosphoglucose isomerase of Colias butterflies: biochemical and population aspects. *Genetics* 87:177–94.

Watt, W. B. (1983). Adaptation at specific loci II. Demographic and biochemical elements in the maintenance of the *Colias* PGI polymorphism. *Genetics* 103:691–724.

Watt, W. B. (1985a). Bioenergetics and evolutionary genetics – opportunities for new synthesis. *Am. Nat.* 125:118–43.

Watt, W. B. (1985b). Allelic isozymes and the mechanistic study of evolution. *Isozymes: Curr. Top. Biol. Med. Res.* 12:89–132.

Watt, W. B. (1986). Power and efficiency as fitness indices in metabolic organization. *Am. Nat.* 127:629–53.

Watt, W. B. (1991). Biochemistry, physiological ecology, and population genetics – the mechanistic tools of evolutionary biology. *Func. Ecol.* 5:145–54.

Watt, W. B. (1992). Eggs, enzymes, and evolution – natural genetic variants change insect fecundity. *Proc. Natl. Acad. Sci. USA* 89:10608–12.

Watt, W. B. (1994). Allozymes in evolutionary genetics: self-imposed burden or extraordinary tool? *Genetics* 136:11–6.

Watt, W. B. (1995). Allozymes in evolutionary genetics: beyond the twin pitfalls of "neutralism" and "selectionism". *Rev. Suisse Zool.* 102:869–82.

Watt, W. B. (1997). Accuracy, anecdotes, and artifacts in the study of insect thermal ecology. *Oikos* 80:399–400.

Watt, W. B. (2000). Avoiding paradigm-based limits to knowledge of evolution. *Evol. Biol.* 32:73–96.

Watt, W. B. (2001). Adaptation, fitness, and the process of evolution. In M. W. Feldman and W. Durham (eds) *International Encyclopedia of Social and Behavioral Sciences: Evolutionary Sciences*, vol. 1, pp. 66–72. Oxford: Elsevier.

Watt, W. B. (2003). Mechanistic studies of butterfly adaptations. In C. L. Boggs, W. B. Watt, and P. R. Ehrlich (eds) *Butterflies: Ecology and Evolution Taking Flight*, pp. 319–52. Chicago: University of Chicago Press.

Watt, W. B., and Boggs, C. L. (1987). Allelic isozymes as probes of the evolution of metabolic organization. *Isozymes: Curr. Topics Biol. Med. Res.* 15:27–47.

Watt, W. B., and Dean, A. M. (2000). Molecular-functional studies of adaptive genetic variation in prokaryotes and eukaryotes. *Annu. Rev. Genet.* 34:593–622.

Watt, W. B., Carter, P. A., and Blower, S. M. (1985). Adaptation at specific loci. IV. Differential mating success among glycolytic allozyme genotypes of *Colias* butterflies. *Genetics* 109:157–75.

Watt, W. B., Carter, P. A., and Donohue, K. (1986). An insect mating system promotes the choice of "good genotypes" as mates. *Science* 233:1187–90.

Watt, W. B., Cassin, R. C., and Swan, M. S. (1983). Adaptation at specific loci. III. Field behavior and survivorship differences among *Colias* PGI genotypes are predictable from *in vitro* biochemistry. *Genetics* 103:725–39.

Watt, W. B., Chew, F. S., Snyder, L. R. G., Watt, A. G., and Rothschild, D. E. (1977). Population structure of pierid butterflies. I. Numbers and movements of some montane *Colias* species. *Oecologia* 27:1–22.

Watt, W. B., Donohue, K., and Carter, P. A. (1996). Adaptation at specific loci. VI. Divergence vs. parallelism of polymorphic allozymes in molecular function and fitness-component effects among *Colias* species (Lepidoptera, Pieridae). *Mol. Biol. Evol.* 13:699–709.

Watt, W. B., Hoch, P. C., and Mills, S. G. (1974). Nectar resource use by *Colias* butterflies: chemical and visual aspects. *Oecologia* 14:353–74.

Watt, W. B., Wheat, C. W., Meyer, E. H., and Martin, J. F. (2003). Adaptation at specific loci. VII. Natural selection, dispersal, and the diversity of molecular-functional variation patterns among butterfly species complexes (*Colias*: Lepidoptera, Pieridae). *Mol. Ecol.* 12: 1265–75.

Wheat, C. W. (2001). From DNA to differential reproductive success: an investigation of the phosphoglucose isomerase gene under selection in *Colias* butterflies. Ph.D. thesis. Stanford, CA: Stanford University.

Wheat, C. W., Watt, W. B., Pollock, D. D. and Schulte, P. M. (2003). From DNA to differential reproductive success: sequencing the adaptive polymorphic alleles of *Colias* phosphoglucose isomerase (Lepidoptera, Pieridae). (submitted).

Williams, G. C. (1992). *Natural Selection: Domains, Levels, and Challenges.* New York: Oxford University Press.

15

Evolution in hybrid zones

DANIEL J. HOWARD, SETH C. BRITCH, W. EVAN BRASWELL
Department of Biology, New Mexico State University, Las Cruces

JEREMY L. MARSHALL
Department of Biology, The University of Texas at Arlington

15.1 Introduction

Hybrid zones have been investigated for more than 70 years, yet the fascination that they hold for evolutionists shows no sign of diminishing. This fascination is easy to understand; not only are hybrid zones common evolutionary phenomena (Hewitt 1988) worthy of understanding on their own right, their study can provide insight into the nature of species (Harrison 1990), the genetic architecture of species differences (Harrison 1990, Rieseberg *et al.* 1999), the role of natural selection in maintaining species boundaries (Bigelow 1965, Barton and Hewitt 1981a) and the evolution of reproductive isolation (Howard 1993, Noor, 1999). Although studies of hybrid zones have a rich history, students entering the field have a strong tendency to focus on the most recent literature, with the consequence that through time the older literature becomes neglected and its findings lost. This is not a problem unique to the field of hybrid zone studies; it afflicts all of biology. The present chapter provides a brief introduction to hybrid zones and the evolutionary questions they present. It is written in the spirit that the old hybrid zone literature is worthwhile and many of its conclusions relevant to issues of current interest. We hope that it stimulates evolutionists to delve further into a body of work that deserves wider recognition.

A thorough history of hybrid zone studies remains to be written and we will not pretend that we can do the topic justice in the limited space allotted for this chapter. However, there is space enough to highlight some of the questions that motivated early studies of hybrids and hybrid zones, what those studies concluded, and to note that some of the questions and conclusions bear remarkable resemblances to questions being addressed and to conclusions being reached today.

The Evolution of Population Biology, ed. R. S. Singh and M. K. Uyenoyama. Published by Cambridge University Press.

15.2 The 1920s and 1930s

Although the literature on hybridization stretches back to the eighteenth century (Koelreuter 1766), it was not until the 1920s that field studies of hybrids and hybrid zones (defined as areas of interaction between genetically distinct groups of organisms resulting in at least some offspring of mixed ancestry; Harrison 1990) began to appear in the literature. Even with the gradual appearance of this new type of investigation, the vast majority of the literature through the 1930s documents the results of laboratory and greenhouse hybridization studies. Crosses between species and divergent populations were used to examine patterns of inheritance (McCluer 1892, Nilsson-Ehle 1930), meiotic patterns (Hoar 1927), the survival of hybrids (Appl 1928), and the fertility of hybrids (Appl 1928, Fukushima 1929). Motivating a good part of this work were applied considerations: could any of the hybrids produced in laboratories and greenhouses be used in agriculture? The finding that many species that remain distinct in nature give rise to fertile and viable hybrids in the laboratory led to the question of why more hybrids were not seen in nature. Ostenfeld (1927) argued that the reasons were twofold: (1) hybrids, although viable and fertile, were less fit than parental taxa; and (2) when hybrids crossed back to parental taxa, combinations that resembled parents most closely were most fit. Thus, hybrids were outcompeted by parentals and/or repeated backcrossing gave rise to individuals of mixed ancestry that could not be distinguished from parental taxa.

A more general question that attracted the attention of evolutionists can be framed in the following way: what is the role of interspecific hybridization in evolutionary change? Studies of chromosomes, particularly in plants, were generating intriguing findings and Winge (1917) had argued that hybridization followed by a doubling of chromosomes would give rise to polyploid offspring that could undergo normal meioses and would be isolated from the parental taxa. By the 1930s, enough polyploids had been developed in the laboratory (Marsden-Jones and Turrill 1930) and studied in nature (Darlington 1927, Huskins 1931) to convince biologists that hybridization followed by polyploidization occurred and could give rise to new species. What was not clear was whether this process was important in evolution. Anderson and Diehl (1932), after an extensive study of the plant genus *Tradescantia* in the field, argued against the importance of polyploidization as a factor in the evolution of new species. They noted that, in *Tradescantia,* there was little evidence of hybridization in the field and that species diversity was greater in areas where polyploids were absent.

In an influential paper, Wiegand (1935) also argued against an important role for hybridization in generating new species. He based his contentions on field studies of hybrids. First, according to Wiegand, the production of hybrids was local and seemed to be associated with disturbance. Second, hybrids did not persist in nature; they were outcompeted by parentals and/or absorbed back into the parentals by backcrossing. Finally, hybridization did not give rise to new characters of the sort associated with new species. Instead, hybrids tended to be blends of the parent taxa. Wiegand's point of view, although influential, was not universal. Hayata (1928) maintained that the crossing of species plays an important role in the origin of existing species and that the genealogy of species may well be represented by a net, rather than by a branching tree.

Although hybridization came under increasing scrutiny from evolutionists working in nature during the 1920s and 1930s, studies of areas of contact between largely allopatric, differentiated taxa were rare. Instead the focus was on the production of hybrids, something that might occur between any sympatric, closely related taxa. An exception was Meise (1928) who concentrated on areas of racial or subspecific intergradation, where, he argued, most hybrids occurred (Meise 1936). Influenced by his studies of the *Corvus* hybrid zone in central Europe, Meise (1928) observed that areas of intergradation were characterized by increased variation and that the location of these zones did not seem to be determined by the environment. He further noted that areas of intergradation could arise *in situ* through the advance of a wave of differentiation or through the meeting of two waves of differentiation. Alternatively, areas of intergradation could arise through the contact of two formerly isolated races. Meise (1928) attributed the relatively small width of these areas to the incomplete fertility of hybrids.

Perhaps the most influential evolutionist studying hybridization during the 1920s and 1930s was Edgar Anderson. In 1936, he introduced character index scoring as a means of analyzing patterns of interaction between hybridizing taxa. Hybrid zone workers use this approach to this day because it allows the visualization of patterns of interaction between two taxa within hybrid zones. In addition, Anderson and Hubricht (1938) introduced the term "introgressive hybridization" to describe what they felt was the principal result of hybridization: namely, a series of backcrosses to one or both parents which introduce new variation into the parental populations.

15.3 The 1940s and 1950s

In the 1940s and 1950s, evolutionists turned their attention *en masse* to the type of hybrid zones that had attracted Meise's attention years earlier. Clearly fueling this shift in interest was the publication of Dobzhansky's 1937 book *Genetics and the Origin of Species* and his 1940 paper "Speciation as a stage in evolutionary divergence". In the former, Dobzhansky argued that species are real entities isolated by reproductive isolating mechanisms, and in the latter, he argued that these mechanisms generally arise in zones of secondary contact as a consequence of selection against hybridization, a process that came to be known as reinforcement (Blair 1955). The clarity of Dobzhansky's thinking and the force of his personality galvanized studies of speciation and placed hybrid zones at the center of these studies. Zones of contact between differentiated taxa came under intensive study, motivated by questions about the evolution of reproductive barriers (Blair 1941, 1955, Thornton 1955, Sibley 1954, Brower 1959).

Commonly, these studies found that the distribution of character index scores within a zone of contact was bimodal: an indication that parental taxa remained distinct within the zone and that hybrids were generally backcrosses to one or both taxa (Hovanitz 1943, Baker 1951, Froiland 1952, Fassett and Calhoun 1952, Blair 1955, Thornton 1955). However, this finding of restricted gene flow did not appear to result from strengthening of isolating barriers. Hybrid zone workers rarely reported evidence of reproductive character displacement, the signature of reinforcement (although see Blair 1955). Instead, interacting taxa seemed to be at an impasse, neither fusing nor evolving stronger isolation, despite the limited production of hybrids. A different question came to the fore: how do taxa remain distinct in the face of many years of interbreeding (Hovanitz 1943, 1949, Baker 1951)? Clearly, natural selection had to play some role in this maintenance of species identity. Hovanitz (1943) postulated that some genes flow between species more easily than others and those genes responsible for giving races their individuality must be highly resistant to breakdown in hybrid crosses. These sentiments were echoed by Baker (1951), who noted that genes determining characters adapting a taxon to its ecological niche would tend to flow less freely than other genes.

A related question – why is natural hybridization largely limited to backcrosses that resemble parental species? – became the focus of Anderson's attention during this period. Anderson (1948) argued that most species had distinct habitat requirements and that hybrids required habitats intermediate to that of the parental species, intermediate habitats that were rarely available. Given the existing habitats, only backcross individuals that closely resembled one of the parental species could successfully compete with the parentals. Thus, in most situations in nature, the only hybrids found would be backcrosses. The exceptions, Anderson argued, were disturbed environments and recently opened environments (for example, those formed as glaciers receded at the end of an ice age). In the former situation, the external factors that isolate most closely related species would break down and lead to the production of a hybrid swarm. This hybrid swarm would find a footing in the environment because disturbance gives rise to a heterogeneous array of habitats. In the latter situation, competition would be reduced and new genotypes arising from hybridization could be molded by selection to fit into the developing assemblage of plants and animals. Thus, Anderson (1948) came to envision a creative role for hybridization in evolution beyond that of introducing new variation into populations. Studies of sunflowers (Smith and Guard 1958) and oaks (Muller 1952, Silliman and Leisner 1958) provided evidence of the importance of disturbance in the production of hybrid swarms.

As more hybrid zones were studied, certain regularities began to appear and to be noted by evolutionists. Again, Anderson made one of the more interesting observations. In a foreshadowing of the concept of suture zones, which would be more fully developed by Remington (1968), Anderson (1953) pointed out that species of plants in the eastern United States hybridized with closely related taxa that arose from five areas: Florida, the Rockies, Texas or Mexico, the north, and higher altitudes in the eastern mountains. Thus, hybrid zones in the eastern United States were concentrated in a limited number of regions.

Other noteworthy developments during the 1940s and 1950s were intensive studies of the herb *Gilia* by Grant (1950, 1952, 1953) and of the grasshopper *Moraba scurra* (later *Keyacris scurra*) by White and his colleagues (White 1957a,b, White and Chinnick 1957). Grant's work indicated that much of the taxonomic messiness of *Gilia* could be attributed to the origin of species through hybridization. These results and other considerations led him to argue that at least half of angiosperm species were of hybrid origin. White's work on the chromosome races of *Moraba* led him to emphasize that narrow, stable zones of contact could be maintained by selection against hybrids (White and Chinnick 1957).

15.4 The 1960s and 1970s

The influence of Dobzhansky continued to be felt in the 1960s and 1970s as questions about isolating barriers and their evolution remained a strong theme of hybrid zone studies. Littlejohn (1965) reported evidence of reinforcement (in this case, displacement of male calling song) in a study of the *Hyla ewingi* complex. The divergence found in one species, *H. verreauxi*, was so great that females from allopatric populations discriminated against songs of males from deep sympatry (Littlejohn and Loftus-Hills 1968). This finding led Littlejohn and Loftus-Hills to argue that the challenge from a related species can cause sufficient divergence within a zone of contact that a new species may develop, one isolated even from the parental population from which it is derived. Jones (1973) and Corbin and Sibley (1977) provided other compelling evidence that reproductive barriers had strengthened as a consequence of interactions within hybrid zones. In both hybrid zones, the number of hybrids decreased over the course of years.

Despite the discovery of patterns consistent with the action of reinforcement, the fortunes of the hypothesis suffered serious setbacks during the 1960s and 1970s. Many hybrid zones showed no evidence of the strengthening of reproductive barriers (Stebbins and Daly 1961, Levin 1963, Thaeler 1968, Gill and Murray 1972, Blackwell and Bull 1978, Woodruff 1979). Moreover, objections to the hypothesis that had begun to be voiced in the 1950s by J. A. Moore were amplified by Mayr (1963) and more fully developed by Bigelow (1965). Moore (1957) had noted that the hypothesis of reinforcement was unnecessary. His work on the *Rana pipiens* complex demonstrated that reproductive barriers between populations could arise simply as a byproduct of adaptation to distinct thermal regimes; contact was not necessary for the perfection of the barriers. Moreover, he observed that there were significant difficulties with the hypothesis of reinforcement; for example, isolating barriers would have an advantage within a zone but not outside a zone; how then do they spread through a species? Bigelow shared Moore's concerns and added others (1965), remarking that many hybrid zones were narrow and that gene flow from parental populations would swamp the effects of natural selection against hybridization within a zone. Perhaps most problematic for the hypothesis, the signature of reinforcement, reproductive character displacement, appeared to be a rare phenomenon in singing Orthoptera contact zones, where the obvious nature of the signaling by males (calling song) should have made detection easy (Walker 1974). Although the data reviewed by Walker were less comprehensive than some evolutionists realized (see Howard 1993), his paper had a major influence on thinking about reinforcement.

As reinforcement lost some of its luster, viewpoints on the age of hybrid zones began to shift. Although the dating of hybrid zones is fraught with difficulty, and the ages of few hybrid zones have been satisfactorily established (Harrison 1990), an increasing number of biologists came to see them not as ephemeral phenomena marking an intermediate stage on the path to fusion or reproductive isolation, but as stable phenomena demanding an understanding of the factors responsible for their maintenance. Three models for the maintenance of hybrid zones emerged during this period. The dynamic equilibrium model postulates that hybrids are intrinsically less fit than pure species individuals and that hybrid zones are maintained by the balance between selection against hybrids within the zone and dispersal into the zone (Bigelow 1965). Thus, hybrid zones of this sort will move under pressure from a more abundant species that sends out more migrants (Key 1968). The bounded hybrid superiority model, which owes much to Anderson, was developed in detail by W. S. Moore (1977). Moore argued that many hybrid zones occur in narrow ecotones and that in these intermediate environments hybrids are more fit than parentals. The third model, the ecological gradient model (Endler 1977), shares with the bounded hybrid superiority model the view that hybrid zones occur in ecological transition areas. However, unlike the bounded hybrid superiority model, this model presumes that hybrids are less fit than parentals.

As in earlier periods, hybrid zones were not seen as strong barriers to gene flow. Instead, they were viewed as semipermeable (Key 1968), with positively selected alleles relatively free to cross the zone and with the flow of neutral markers determined by their linkage to negatively selected loci. Also, introgression continued to be viewed by some biologists as an important source of genetic variation, a viewpoint tested in a remarkable study by Lewontin and Birch (1966). Through hybridization and selection experiments, Lewontin and Birch provided evidence that range expansion of *Dacus tryoni* could be attributed to introgression of genes from *D. neohumeralis*.

The 1960s and 1970s can be considered the age of allozymes in evolutionary studies, and hybrid zones were an obvious target for biologists interested in exploiting the power of this new technology. Allozymes provided evolutionists with a new suite of markers that could be used to describe allele frequency changes through, introgression across, and patterns of interaction within hybrid zones. Not surprisingly, a variety of different patterns were reported ranging from coincidence and concordance of allele frequency changes through a hybrid zone (Hall and Selander 1973, Blackwell and Bull 1978) to distinct differences in patterns of replacement and flow of alleles across zones (Hunt and Selander 1973, Gartside *et al.* 1979).

The most common outcome of interactions within hybrid zones, as revealed by studies of allozymes, chromosomes, and morphology, was not the formation of a hybrid swarm, but the maintenance of parental genotypes and phenotypes (Thaeler 1968, Gill and Murray 1972, Hall and Selander 1973, Blackwell and Bull 1978, Feder 1979, Woodruff 1979). Therefore, a major problem still confronting evolutionists was explaining how parental genotypes remain distinct in the face of gene flow and why barriers to gene flow do not break down or strengthen in areas of overlap and hybridization. One potential explanation was offered by Patton *et al.* (1979) who noted that terrestrial environments are unstable, thus contacts between differentiated taxa are likely to be ever shifting. In the absence of prolonged contact between two taxa in the same area, evolution within zones will be difficult.

15.5 The 1980s to present

Early during this period an influential review by Barton and Hewitt (1981a) contended that most hybrid zones are tension zones; that is, they are narrow clines maintained by the balance between dispersal into the zone and hybrid unfitness (due to intrinsic factors) within the zone. Although many previously described hybrid zones presented patterns inconsistent with this model, Barton and Hewitt's (1981a) viewpoints regarding the pattern of replacement of alleles across hybrid zones and the factors responsible for the maintenance of hybrid zones rapidly came to dominate the literature of evolutionary biology. Part of the attraction of this view were the analytical tools developed for such zones, tools that allowed for the estimation of dispersal rates, selection pressures, and the number of loci that maintain a zone (Barton and Hewitt 1981b, Szymura and Barton 1986, Barton and Gale 1993).

Despite the attractiveness of the tension zone model and its clear relevance to some real world hybrid zones (reviewed in Barton and Hewitt 1981a), detailed investigations continued to turn up patterns inconsistent with the model. In particular, spatial patterns of allele replacement across hybrid zones often appeared more mosaic than clinal. For example, Howard

(1986) reported that allele frequency change across the hybrid zone between the ground crickets *Allonemobius fasciatus* and *A. socius* is very erratic, with mixed populations and pure species populations forming a patchwork (clearly related to climatic factors; Howard and Waring 1991) across the landscape. Similarly, transects through the hybrid zone between *Gryllus pennsylvanicus* and *G. veletis* revealed abrupt reversals of allele frequencies (Harrison 1986). Once again, adaptation to environmental factors, in this case soil, accounts for the patchwork structure of the zone (Harrison and Rand 1989). Although mosaic hybrid zones can be maintained by a variety of different underlying mechanisms (Harrison and Rand 1989, Harrison 1990), the foregoing examples implicate an important role for the environment in determining the spatial structure of these zones. With regard to which is more frequent, a clinal structure or a mosaic structure, the most recent survey of the hybrid zone literature suggests that the two structures are about equally frequent (Braswell and Howard, unpublished data).

The view that hybrids are more unfit than parentals, another facet of the tension zone model, came under increasing fire, particularly from botanists, during the 1990s. Echoing Anderson and W. S. Moore, Arnold and his colleagues (Arnold *et al.* 1991, Arnold 1997, Arnold and Emms 1998, Arnold *et al.* 2001) have noted that hybrids are a heterogeneous assemblage of genotypes, and that while it is true that some of the genotypes are less fit than parentals, some of the genotypes may be more fit. Not only may they be more fit, they may be quite novel and therefore capable of exploiting new resources and new habitats (DePamphilis and Wyatt 1990, Arnold 1997). Thus, hybrids may give rise to new species, a contention that has existed in the literature for more than 70 years, and one consistent with new systematic studies indicating that many groups of plants and animals have a reticulate evolutionary history (Dowling and DeMarais 1993, Sang and Zhang 1999, van Oppen *et al.* 2001, Yoon and Boo 1999). Building on another theme initially developed by Anderson, Arnold (1997) has argued that introgression is an important source of variation that provides species with enhanced abilities to respond to changing conditions and to expand ranges.

Arnold's writing has had the salutary effect of forcing evolutionists to actively investigate the fitness of hybrids rather than simply assuming they are unfit. Unfortunately, no field studies have yet succeeded in comparing lifetime reproductive success of hybrids vs. parentals. Even the most rigorous have tended to focus on a single component of fitness – survival (e.g., Howard *et al.* 1993, Wang *et al.* 1997), a pre-meiotic component of fitness that often may not reflect post-meiotic components of fitness, such as fertility (Graham 1992, Alibert *et al.* 1994).

Ironically, even as biologists have begun to take seriously, once again, the possibility that hybrids and hybridization may be important in the creation of new species, reinforcement, a process dependent on selection against hybridization, made a comeback in the 1990s. Renewed interest in reinforcement was fueled by three factors. The first was the publication of three major reviews: one countered earlier criticisms of reinforcement (Howard 1993) and all three provided evidence that patterns consistent with reinforcement are relatively plentiful in nature (Coyne and Orr 1989, Howard 1993, Coyne and Orr 1997). The second was the publication of several case studies clearly documenting stronger reproductive isolation between taxa within areas of overlap and hybridization than outside (Noor 1995, Saetre *et al.* 1997, Cooley *et al.* 2001). The final factor was the development of new theory illuminating the conditions under which reinforcement was likely to occur (Liou and Price 1994, Kelly and Noor 1996, Servedio and Kirkpatrick 1997, Cain *et al.* 1999, Kirkpatrick and Servedio 1999, Kirkpatrick 2000, 2001). Among the factors enhancing the possibility of reinforcement are a mosaic structure to the zone of overlap and hybridization (Cain *et al.* 1999) and a larger number of loci contributing to the ecological trait that confers adaptation to local conditions (Kirkpatrick 2001).

The perception that hybrid zones are common (Hewitt 1988) helps explain the vigor with which they have been studied and raises the question of how they are initially formed. In 1928, Meise noted that hybrid zones could arise *in situ* or they could arise due to contact between previously isolated populations. These two possibilities for the origin of hybrid zones have come to be known as primary and secondary contact, respectively. The vast majority of zones have been interpreted as instances of secondary contact, even though Endler (1977) noted that primary and secondary contact would give rise to similar patterns of variation. Concordance of change at many loci across hybrid zones is seen as one argument for secondary contact (Nelson *et al.* 1987, Hewitt 1988, Harrison 1990). Another is the position of hybrid zones; in North America and Europe they seem to be concentrated in a relatively small number of areas. These areas can be interpreted as regions where closely related taxa would meet as ranges expanded following the last ice age (Anderson 1953, Remington 1968, Hewitt 1988). The ability to generate phylogenies of alleles provides evolutionists with new tools for tracing the movement and expansion of populations, thus allowing more rigorous testing of hypotheses regarding the origin of hybrid zones. By using such an approach, Cooper *et al.* (1995) documented the postglacial expansion of subspecies of *Chorthippus parallelus* and the secondary nature of the contact zone in the Pyrenees. The incorporation of geographic information science (GIS), which provides unprecedented power to map the association between genes and the environment, into studies of hybrid zones will also allow more thorough analyses of origins (Ritchie *et al.* 2001).

Whatever the interest in hybrid zones in their own right, many hybrid zone studies continue to be motivated by an interest in speciation and the insight that hybrid zones can provide into this process. The application of new genetic methodologies and attention to new molecules such as mtDNA have provided evolutionists with unprecedented insight into introgression across hybrid zones (Arnold *et al.* 1991, Young *et al.* 2001), directionality of crosses (Arntzen and Wallis 1991, Taylor and Hebert 1993), and the level of isolation between hybridizing taxa (Chu *et al.* 1995). Partial isolation of taxa within zones continues to be a common finding (see reviews by Harrison 1990, Howard 1993, Jiggins and Mallet 2000) and detailed analyses of hybrid zones have led to new insights into the nature of isolating barriers. In particular, hybrid zone studies have highlighted the importance of what has been termed conspecific sperm and pollen precedence (Howard 1999) in isolating closely related animals and plants (Bella *et al.* 1992, Howard and Gregory 1993, Carney *et al.* 1994, 1996, Gregory and Howard 1994, Rieseberg *et al.* 1995, Howard *et al.* 1998a,b).

Evolutionists have long noted that genetic studies of hybrid zones have the potential to provide insight into the genetic architecture of reproductive isolation (Baker 1951, Barton and Hewitt 1981a, Harrison 1990) and cline theory provides a coarse tool for ascertaining the number of loci that contribute to reproductive isolation (Barton and Hewitt 1981b, Szymura and Barton 1986). However, it was not until 1999 that hybrid zone studies truly began to fulfill their considerable promise in this area. That year, by documenting differential introgression of mapped markers across three hybrid zones, Rieseberg and his students (Rieseberg *et al.* 1999) identified 26 regions of the *Helianthus petiolaris* genome that contribute to reproductive isolation between *H. petiolaris* and *H. annuus*. The approach pioneered by Rieseberg and his students provides a very sensitive assay of genetic control of reproductive isolation by taking advantage of the many generations of recombination that occur within hybrid zones. The approach can be applied to any hybridizing taxa for which linkage maps are available and promises to keep hybrid zones at the center of genetic studies of speciation for many years to come.

Even as evolutionists gain a better understanding of hybrid zones through the application of new molecular and quantitative tools and through long-term studies (e.g., Britch *et al.* 2001), we are left with an old question, a question that we seem no closer to answering now than 50 years ago. What accounts for the evolutionary impasse that many hybrid zones seem to represent? That is, how do parental genotypes remain intact in the face of gene flow and why does it appear that, in many cases, isolating barriers neither strengthen nor weaken through time? One recent insight about bimodal hybrid zones is that partial isolation appears to be attributable to barriers to gene flow that operate prezygotically (Jiggins and Mallet 2000). While this is an important finding, the identification of the types of barriers that operate in bimodal hybrid zones still leaves the question of why these barriers appear to neither break down nor strengthen. The potential explanations are many and worthy of investigation by evolutionists. As noted by Woodruff (1981), hybrids may be ill-adapted and thus ineffective bridges for gene flow. Nevertheless, reinforcement may not occur because necessary variation does not exist in parental populations (Doherty and Howard 1996), gene flow into the zone from parental populations swamps the effects of selection within the zone (Bigelow 1965, Barton and Hewitt 1981a), or contact between taxa may be too ephemeral for selection to operate (Patton *et al.* 1979, Britch *et al.* 2001).

Interesting new explanations for lack of reinforcement have been raised by the discovery of conspecific sperm precedence (reviewed in Howard 1999, Marshall *et al.* 2002): namely, lack of selection against hybridization and/or interspecific sexual conflict. In cases of strong conspecific sperm precedence, matings between heterospecific males and females will result in few if any hybrid offspring, thus unless mating itself is costly to males and females, there is little or no selection to avoid mating with an individual of the "wrong" species. In cases where females benefit from mating, for example in situations in which males offer nuptial gifts or females obtain and eat a large spermatophore, interspecific sexual conflict may exist. In these situations, a female may benefit from mating with a heterospecific because she obtains all the benefits of a mating without producing costly (potentially) hybrid offspring. On the other hand, a heterospecific male provides valuable resources to a female and receives nothing in return. Thus, males that discriminate against heterospecific females should have a fitness advantage. The fact that the number of well-documented cases of reinforcement is small may signal that females have the upper hand in this conflict.

15.6 Concluding remarks

Hybrid zones have been under relatively intensive investigation by evolutionists for more than 70 years. Yet surprisingly, some of the major questions that motivated early studies of hybrid zones are with us still. How important are hybrids in serving as the foundation of new species? Is reinforcement important in the evolution of prezygotic barriers to gene exchange? What factors account for the maintenance of parental genotypes within hybrid zones in the face of gene flow? Surely, one reason for the durability of these questions is that no investigation of a single hybrid zone can answer them. They are "how important is" questions, the type of questions that require data from detailed studies of multiple situations. Thus, accumulating the data necessary to answer the questions is time consuming and requires continued focus. Maintaining this focus can be difficult as time passes, the literature ages, and new questions come to the fore. Yet, to allow the old literature to slip away means giving up knowledge that has been wrested at considerable cost from a natural world that does not yield its secrets easily.

We have written this brief review in the hope of stimulating interest in the old hybrid zone literature and to better frame some of the questions that currently hold the attention of evolutionists interested in hybrid zones. We recognize that much has been left out of this chapter and we hope that someone frustrated with its limitations will see fit to write a more complete history of hybrid zone studies.

15.7 Acknowledgments

We are grateful to two anonymous reviewers for their comments on an earlier draft of this chapter and to the National Science Foundation for its support of our research on hybrid zones and speciation, most recently through NSF grants DEB 9726502 and DEB 011613. This work was also supported, in part, by Texas Advanced Research Program Grant 003656-0067-2001.

REFERENCES

Alibert, P., Renaud, S., Dod, B., Bonhomme, F., and Auffray, J. C. (1994). Fluctuating asymmetry in the *Mus musculus* hybrid zone: a heterotic effect in disrupted co-adapted genomes. *Proc. R. Soc. Lond. Ser. B* 258:53–9.

Anderson, E. (1936). The species problem in *Iris*. *Ann. Miss. Bot. Gard.* 23:457–509.

Anderson, E. (1948). Hybridization of the habitat. *Evolution* 2:1–9.

Anderson, E. (1953). Introgressive hybridization. *Biol. Rev.* 28:280–307.

Anderson, E., and Diehl, D. G. (1932). Contributions to the *Tradescantia* problem. *J. Arn. Arb.* 13:213–31.

Anderson, E., and Hubricht, L. (1938). Hybridization in *Tradescantia*. III. The evidence for introgressive hybridization. *Am. J. Bot.* 25:396–402.

Appl, J. (1928). Uber einen Bastard von *Origanum majorana* F und *Origanum vulgare* M und dessen Aufspaltung in der F2 Generation. [On a cross between *Origanum majorana* F and *Origanum vulgare* M and the F2 progeny.] *Pres. Vest. Cesk. Bot. Spolec. Praze.* 6:3–13.

Arnold, M. L. (1997). *Natural Hybridization and Evolution.* New York: Oxford University Press.

Arnold, M. L., and Emms, S. K. (1998). Paradigm lost: natural hybridization and evolutionary innovations. In D. J. Howard and S. H. Berlocher (eds) *Endless Forms: Species and Speciation,* pp. 379–89. New York: Oxford University Press.

Arnold, M. L., Buckner, C. M., and Robinson, J. J. (1991). Pollen-mediated introgression and hybrid speciation in Louisiana (USA) irises. *Proc. Natl. Acad. Sci. USA* 88:1398–402.

Arnold, M. L., Kentner, E. K., Johnston, J. A., Cornman, S., and Bouck, A. C. (2001). Natural hybridisation and fitness. *Taxon* 50:93–104.

Arntzen, J. W., and Wallis, G. P. (1991). Restricted gene flow in a moving hybrid zone of the newts *Triturus cristatus* and *Triturus marmoratus* in western France. *Evolution* 45:805–26.

Baker, H. G. (1951). Hybridization and natural gene-flow between higher plants. *Biol. Rev.* 26:302–37.

Barton, N. H., and Gale, K. S. (1993). Genetic analysis of hybrid zones. In R. G. Harrison (ed.) *Hybrid Zones and the Evolutionary Process,* pp. 13–45. New York: Oxford University Press.

Barton, N. H., and Hewitt, G. M. (1981a). Hybrid zones and speciation. In W. R. Atchley and D. S. Woodruff (eds) *Evolution and Speciation: Essays in Honor of M. J. D. White,* pp. 109–145. Cambridge: Cambridge University Press.

Barton, N. H., and Hewitt, G. M. (1981b). The genetic basis of hybrid inviability between two chromosomal races of the grasshopper *Podisma pedestris*. *Heredity* 47:367–83.

Bella, J. L., Butlin, R. K., Ferris, C., and Hewitt, G. M. (1992). Asymmetrical homogamy and unequal sex ratio from reciprocal mating-order crosses between *Chorthippus parallelus* subspecies. *Heredity* 68:345–52.

Bigelow, R. S. (1965). Hybrid zones and reproductive isolation. *Evolution* 19:449–58.

Blackwell, J. M., and Bull, C. M. (1978). A narrow hybrid zone between two western Australian frog species *Ranidella insignifera* and *R. pseudinsignifera*: the extent of introgression. *Heredity* 40:13–25.

Blair, A. P. (1941). Isolating mechanisms in tree frogs. *Proc. Natl. Acad. Sci. USA* 27:14–17.

Blair, W. F. (1955). Mating call and stage of speciation in the *Microhyla olivacea – M. carolinensis* complex. *Evolution* 9:469–80.

Britch, S. C., Cain, M. L., and Howard, D. J. (2001). Spatio-temporal dynamics of the *Allonemobius fasciatus – A. socius* mosaic hybrid zone: a 14-year perspective. *Mol. Ecol.* 10:627–38.

Brower, L. P. (1959). Speciation in butterflies of the *Papilio glaucus* group. I. Morphological relationships and hybridization. *Evolution* 13:40–63.

Cain, M. L., Andreasen, V., and Howard, D. J. (1999). Reinforcing selection is effective under a relatively broad set of conditions in a mosaic hybrid zone. *Evolution* 53:1343–53.

Carney, S. E., Cruzan, M. B., and Arnold, M. L. (1994). Reproductive interactions between hybridizing Irises: analyses of pollen-tube growth and fertilization success. *Am. J. Bot.* 81:1169–75.

Carney, S. E., Hodges, S. A., and Arnold, M. L. (1996). Effects of differential pollen-tube growth on hybridization in the Louisiana Irises. *Evolution* 50:1871–8.

Chu, J., Powers, E., and Howard, D. J. (1995). Gene exchange in a ground cricket hybrid zone. *J. Hered.* 86:17–21.

Cooley, J. R., Simon, C., Marshall, D. C., Slon, K., and Ehrhardt, C. (2001). Allochronic speciation, secondary contact, and reproductive character displacement in periodical cicadas (Hemiptera: *Magicicada* spp.): genetic, morphological, and behavioural evidence. *Mol. Ecol.* 10:661–71.

Cooper, S. J. B., Ibrahim, K. M., and Hewitt, G. M. (1995). Postglacial expansion and genome subdivision in the European grasshopper *Chorthippus parallelus*. *Mol. Ecol.* 4:49–60.

Corbin, K. W., and Sibley, C. G. (1977). Rapid evolution in orioles of the genus *Icterus*. *Condor* 79:335–42.

Coyne, J. A., and Orr, H. A. (1989). Patterns of speciation in *Drosophila*. *Evolution* 43:362–81.

Coyne, J. A., and Orr, H. A. (1997). "Patterns of speciation in *Drosophila*" revisited. *Evolution* 51:295–303.

Darlington, C. D. (1927). The behaviour of polyploids. *Nature* 119:390–1.

DePamphilis, C. W., and Wyatt, R. (1990). Electrophoretic confirmation of interspecific hybridization in *Aesculus* (Hippocastanaceae) and the genetic structure of a broad hybrid zone. *Evolution* 44:1295–317.

Dobzhansky, T. (1937). *Genetics and the Origin of Species*. New York: Columbia University Press.

Dobzhansky, T. (1940). Speciation as a stage in evolutionary divergence. *Am. Nat.* 74:312–321.

Doherty, J. A., and Howard, D. J. (1996). Lack of preference for conspecific calling songs in female crickets. *Anim. Behav.* 51:981–9.

Dowling, T. E., and DeMarais, B. D. (1993). Evolutionary significance of introgressive hybridization in cyprinid fishes. *Nature* 362:444–6.

Endler, J. A. (1977). *Geographic Variation, Speciation, and Clines.* Princeton, NJ: Princeton University Press.

Fasset, N. C., and Calhoun, B. (1952). Introgression between *Typha latifolia* and *T. angustifolia*. *Evolution* 6:367–79.

Feder, J. (1979). Natural hybridization and genetic divergence between the toads *Bufo boreas* and *Bufo punctatus*. *Evolution* 33:1089–97.

Froiland, S. G. (1952). The biological status of *Betula andrewsii* A. Nels. *Evolution* 6:268–83.

Fukushima, E. (1929). Preliminary report on *Brassica–Raphanus* hybrids. *Proc. Imp. Acad.* 5:48–50.

Gartside, D. F., Littlejohn, M. J., and Watson, G. F. (1979). Structure and dynamics of a narrow hybrid zone between *Geocrinia laevis* and *Geocrinia victoriana* (Anura: Leptodactylidae) in southeastern Australia. *Heredity* 43:165–78.

Gill, F. B., and Murray, B. G. (1972). Discrimination behavior and hybridization of the blue-winged and golden-winged warblers. *Evolution* 26:282–93.

Graham, J. H. (1992). Genomic coadaptation and developmental stability in hybrid zones. *Acta Zool. Fenn.* 191:121–31.

Grant, V. (1950). Genetic and taxonomic studies in *Gilia*. I. *Gilia capitata*. *Aliso* 2:239–316.

Grant, V. (1952). Genetic and taxonomic studies in *Gilia*. III. The *Gilia tricolor* complex. *Aliso* 2:375–88.

Grant, V. (1953). The role of hybridization in the evolution of the leafy-stemmed *Gilias*. *Evolution* 7:51–64.

Gregory, P. G., and Howard, D. J. (1994). A postinsemination barrier to fertilization isolates two closely related ground crickets. *Evolution* 48:705–10.

Hall, W. P., and Selander, R. K. (1973). Hybridization of karyotypically differentiated populations in the *Sceloporus grammicus* complex (Iguanidae). *Evolution* 27: 226–42.

Harrison, R. G. (1986). Pattern and process in a narrow hybrid zone. *Heredity* 56:337–50.

Harrison, R. G. (1990). Hybrid zones: windows on the evolutionary process. *Ox. Surv. Evol. Biol.* 7:69–128.

Harrison, R. G., and Rand, D. M. (1989). Mosaic hybrid zones and the nature of species boundaries. In D. Otte and J. Endler (eds) *Speciation and its Consequences*, pp. 111–33. Sunderland, MA: Sinauer.

Hayata, B. (1928). The succession and participation theories and their bearings upon the objects of the Third Pan-Pacific Science Congress. *Proc. 3rd Pan-Pac. Sci. Cong., Tokyo* 2:1869–75.

Hewitt, G. M. (1988). Hybrid zones – natural laboratories for evolutionary studies. *Trends Ecol. Evol.* 3:158–67.

Hoar, C. S. (1927). Chromosome studies in *Aesculus*. *Bot. Gaz.* 84:156–70.

Hovanitz, W. (1943). Hybridization and seasonal segregation in two races of a butterfly occurring together in two localities. *Biol. Bull.* 85:44–51.

Hovanitz, W. (1949). Interspecific matings between *Colias eurytheme* and *Colias philodice* in wild populations. *Evolution* 3:170–3.

Howard, D. J. (1986). A zone of overlap and hybridization between two ground cricket species. *Evolution* 40:34–43.

Howard, D. J. (1993). Reinforcement: origin, dynamics, and fate of an evolutionary hypothesis. In R. G. Harrison (ed.) *Hybrid Zones and the Evolutionary Process*, pp. 46–69. New York: Oxford University Press.

Howard, D. J. (1999). Conspecific sperm and pollen precedence and speciation. *Annu. Rev. Ecol. Syst.* 30:109–32.

Howard, D. J., and Gregory, P. G. (1993). Post-insemination signaling systems and reinforcement. *Phil. Trans. R. Soc. Lond., Ser. B* 340:231–6.

Howard, D. J., and Waring, G. L. (1991). Topographic diversity, zone width, and the strength of reproductive isolation in a zone of overlap and hybridization. *Evolution* 45:1120–35.

Howard, D. J., Gregory, P. G., Chu, J., and Cain, M. L. (1998a). Conspecific sperm precedence is an effective barrier to hybridization between closely related species. *Evolution* 52:511–6.

Howard, D. J., Reece, M., Gregory, P. G., Chu, J., and Cain, M. L. (1998b). The evolution of barriers to fertilization between closely related organisms. In D. J. Howard and S. H. Berlocher (eds) *Endless Forms: Species and Speciation*, pp. 279–88. New York: Oxford University Press.

Howard, D. J., Waring, G. L., Tibbets, C. A., and Gregory, P. G. (1993). Survival of hybrids in a mosaic hybrid zone. *Evolution* 47:789–800.

Hunt, W. G., and Selander, R. K. (1973). Biochemical genetics of hybridisation in European house-mice. *Heredity* 31:11–33.

Huskins, C. L. (1931). Origin of *Spartina townsendii*. *Nature* 127:781.

Jiggins, C. D., and Mallet, J. (2000). Bimodal hybrid zones and speciation. *Trends Ecol. Evol.* 15:250–5.

Jones, J. M. (1973). Effects of thirty years of hybridization on the toads *Bufo americanus* and *Bufo woodhousii fowleri* at Bloomington, Indiana. *Evolution* 27:435–48.

Kelly, J. K., and Noor, M. A. F. (1996). Speciation by reinforcement: a model derived from studies of *Drosophila*. *Genetics* 143:1485–97.

Key, K. H. L. (1968). The concept of stasipatric speciation. *Syst. Zool.* 17:14–22.

Kirkpatrick, M. (2000). Reinforcement and divergence under assortative mating. *Proc. R. Soc. Lond. Ser. B* 267:1649–55.

Kirkpatrick, M. (2001). Reinforcement during ecological speciation. *Proc. R. Soc. Lond. Ser. B* 268:1259–63.

Kirkpatrick, M., and Servedio, M. R. (1999). The reinforcement of mating preferences on an island. *Genetics* 151:865–84.

Koelreuter, J. G. (1766). Dritte Fortsetzung der vorläufigen Nachricht von einigen das Geschlecht der Pflanzen betreffenden Versuchen und Beobahtungen. Leipzig. [Reprinted in Ostwald's *Klassiker der exakten Wissenschaften*, No. 41, Leipzig, 1893.]

Levin, D. A. (1963). Natural hybridization between *Phlox maculata* and *Phlox glaberrima* and its evolutionary significance. *Am. J. Bot.* 50:714–20.

Lewontin, R. C., and Birch, L. C. (1966). Hybridization as a source of variation for adaptation to new environments. *Evolution* 20:315–36.

Liou, L. W., and Price, T. D. (1994). Speciation by reinforcement of premating isolation. *Evolution* 48:1451–59.

Littlejohn, M. J. (1965). Premating isolation in the *Hyla ewingi* complex (Anura: Hylidae). *Evolution* 19:234–43.

Littlejohn, M. J., and Loftus-Hills, J. J. (1968). An experimental evolution of premating isolation in the *Hyla ewingi* complex (Anura: Hylidae). *Evolution* 22:659–63.

Marsden-Jones, E. M., and Turrill, W. B. (1930). Hybridization in certain genera of the British flora. *Gard. Chron.* 87:210–11.

Marshall, J. L., Arnold, M. L., and Howard, D. J. (2002). Reinforcement: the road not taken. *Trends Ecol. Evol.* 17:558–63.

Mayr, E. (1963). *Animal Species and Evolution.* Cambridge, MA: Belknap Press.

McCluer, G. W. (1892). Corn crossing. *Ill. Agr. Exp. Sta. Bull.* 21:82–101.

Meise, W. (1928). Rassenkreuzungen an den Arealgrenzen. *Zool. Anz., Suppl. (Verhandl. Deutsch. Zool. Ges. E. V.)* 3:96–105.

Meise, W. (1936). Uber Artenstehung durch Kreuzung in der Vogelwelt. *Biol. Zentralbl.* 56:590–604.

Moore, J. A. (1957). An embryologist's view of the species concept. In E. Mayr (ed.) *The Species Problem,* pp. 325–88. Washington, DC: American Association for the Advancement of Science.

Moore, W. S. (1977). An evaluation of narrow hybrid zones in vertebrates. *Quart. Rev. Biol.* 52:263–77.

Muller, C. H. (1952). Ecological control of hybridization in *Quercus*: a factor in the mechanism of evolution. *Evolution* 6:147–61.

Nelson, K., Baker, R. J., and Honeycutt, R. L. (1987). Mitochondrial DNA and protein differentiation between hybridizing cytotypes of the white-footed mouse, *Peromyscus leucopus. Evolution* 41:864–72.

Nilsson-Ehle, H. (1930). Racial crosses from the viewpoint of general biology. *Genetica* 11:213–24.

Noor, M. A. (1995). Speciation driven by natural selection in *Drosophila. Nature* 375:674–5.

Noor, M. A. F. (1999). Reinforcement and other consequences of sympatry. *Heredity* 83:503–8.

Ostenfeld, C. H. (1927). The present state of knowledge of hybrids between species of flowering plants. *J. R. Hort. Soc.* 53:31–44.

Patton, J. L., Hafner, J. C., Hafner, M. S., and Smith, M. F. (1979). Hybrid zones in *Thomomys bottae* pocket gophers: genetic, phenetic and ecologic concordance patterns. *Evolution* 33:860–76.

Remington, C. L. (1968). Suture-zones of hybrid interaction between recently joined biotas. *Evol. Biol.* 2:321–428.

Rieseberg, L. H., Desrochers, A. M., and Youn, S. J. (1995). Interspecific pollen competition as a reproductive barrier between sympatric species of *Helianthus* (Asteraceae). *Am. J. Bot.* 82:515–19.

Rieseberg, L. H., Whitton, J., and Gardner, K. (1999). Hybrid zones and the genetic architecture of a barrier to gene flow between two sunflower species. *Genetics* 152:713–27.

Ritchie, M. G., Kidd, D. M., and Gleason, J. M. (2001). Mitochondrial DNA variation and GIS analysis confirm a secondary origin of geographical variation in the bushcricket *Ephippiger ephippiger* (Orthoptera: Tettigonioidea), and resurrect two subspecies. *Molec. Ecol.* 10:603–11.

Saetre, G. P., Moum, T., Bures, S., Kral, M., Adamjan, M., and Moreno, J. (1997). A sexually selected character displacement in flycatchers reinforces premating isolation. *Nature* 387:589–92.

Sang, T., and Zhang, D. M. (1999). Reconstructing hybrid speciation using sequences of low copy nuclear genes: hybrid origins of five *Paeonia* species based on ADH gene phylogenies. *Syst. Bot.* 24:148–63.

Servedio, M. R., and Kirkpatrick, M. (1997). The effects of gene flow on reinforcement. *Evolution* 51:1764–72.

Sibley, C. G. (1954). Hybridization in the red-eyed towhees of Mexico. *Evolution* 8:252–90.

Silliman, F. E., and Leisner, R. S. (1958). An analysis of a colony of hybrid oaks. *Am. J. Bot.* 45:730–6.

Smith, D. M., and Guard, A. T. (1958). Hybridization between *Helianthus divaricatus* and *H. microcephalus. Brittonia* 10:137–45.

Stebbins, G. L., and Daly, K. (1961). Changes in the variation pattern of a hybrid population of *Helianthus* over an eight-year period. *Evolution* 15:60–71.

Szymura, J. M., and Barton, N. H. (1986). Genetic analysis of a hybrid zone between the fire-bellied toads, *Bombina bombina* and *Bombina variegata*, near Krakow in southern Poland. *Evolution* 40:1141–59.

Taylor, D. J., and Hebert, P. D. N. (1993). Habitat-dependent hybrid parentage and differential introgression between neighboring sympatric *Daphnia* species. *Proc. Natl. Acad. Sci. USA* 90:7079–83.

Thaeler, C. S. (1968). An analysis of three hybrid populations of pocket gophers (genus *Thomomys*). *Evolution* 22:543–55.

Thornton, W. A. (1955). Interspecific hybridization in *Bufo woodhousei* and *Bufo valliceps. Evolution* 9:455–68.

van Oppen, M. J. H., McDonald, B. J., Willis, B., and Miller, D. J. (2001). The evolutionary history of the coral genus *Acropora* (Scleractinia: Cnidaria) based on a mitochondrial and a nuclear marker: reticulation, incomplete lineage sorting, or morphological convergence? *Mol. Biol. Evol.* 18:1315–29.

Walker, T. J. (1974). Character displacement and acoustic insects. *Am. Zool.* 114:1137–50.

Wang, H., McArthur, E. D., Sanderson, S. C., Graham, J. H., and Freeman, D. C. (1997). Narrow hybrid zone between two subspecies of big sagebrush (*Artemisia tridentata*: Asteraceae): IV: Reciprocal transplant experiments. *Evolution* 51:95–102.

White, M. J. D. (1957a). Cytogenetics of the grasshopper *Moraba scurra.* I. Meiosis of interracial and interpopulation hybrids. *Aust. J. Zool.* 5:285–304.

White, M. J. D. (1957b). Cytogenetics of the grasshopper *Moraba scurra.* II. Heterotic systems and their interaction. *Aust. J. Zool.* 5:305–37.

White, M. J. D., and Chinnick, L. J. (1957). Cytogenetics of the grasshopper *Moraba scurra.* III. Distribution of the 15- and 17-chromosome races. *Aust. J. Zool.* 5: 338–47.

Wiegand, K. M. (1935). A taxonomists's experience with hybrids in the wild. *Science* 81:161–6.

Winge, O. (1917). The chromosomes. Their numbers and general importance. *Compt. Rend. Trav. Lab. Carlsberg* 13:131–275.

Woodruff, D. S. (1979). Postmating reproductive isolation in *Pseudophryne* and the evolutionary significance of hybrid zones. *Science* 203:561–3.

Woodruff, D. S. (1981). Toward a genodynamics of hybrid zones: studies of Australian frogs and West Indian land snails. In W. R. Atchley and D. S. Woodruff (eds) *Evolution and Speciation: Essays in Honor of M. J. D. White*, pp. 171–97. Cambridge: Cambridge University Press.

Yoon, H. S., and Boo, S. M. (1999). Phylogeny of Alariaceae (Phaeophyta) with special reference to *Undaria* based on sequences of the rubisco spacer region. *Hydrobiologia* 399:47–55.

Young, W. P., Ostberg, C. O., Keim, P., and Thorgaard, G. H. (2001). Genetic characterization of hybridization and introgression between anadromous rainbow trout (*Oncorhynchus mykiss irideus*) and coastal cutthroat trout (*O. clarki clarki*). *Mol. Ecol.* 10:921–30.

16

Nine relatives from one African ancestor: population biology and evolution of the *Drosophila melanogaster* subgroup species

DANIEL LACHAISE, PIERRE CAPY, MARIE-LOUISE CARIOU, DOMINIQUE JOLY, FRANÇOISE LEMEUNIER, JEAN R. DAVID

Laboratoire Populations, Génétique et Evolution, Centre National de la Recherche Scientifique, Gif-sur-Yvette

16.1 Introduction

Evolutionary studies have long come across the snag of that elusive evolutionary event called "speciation." Understanding speciation is understanding species boundaries, and hence the source of biodiversity. Therefore the genetics of speciation has been a major focus of evolutionary biology thinking over the last decade (Coyne 1992, Coyne and Orr 1998, Singh 2000, Wu 2001). However, major questions about speciation have remained unresolved, for instance whether splitting is instantaneous or whether species emerge gradually. Generally there have been two ways of approaching the speciation process: the *mapping of speciation phenotypes* and "*divergence population genetics*" as labeled by Kliman *et al.* (2000).

The *gene mapping* studies address the genetic architecture of phenotypes that are thought to be instrumental in driving speciation. However, these maps do not indicate which selective factors have actually caused speciation. Among these phenotypes are species-specific adaptations such as the resistance to a host-plant toxin, or "speciation" traits including species-specific mate detection pheromones affecting mating (Coyne 1996, Coyne and Charlesworth 1997), courtship songs affecting sexual isolation, or sperm diversity affecting sperm competition. As genes involved in male fertility are potential targets for sexual selection, it could be that those genes with a sex-related function play a role in reproductive isolation between species, and evolve more rapidly than genes with developmental or metabolic function (Civetta and Singh 1998, 1999, Swanson and Vacquier 2002, Kulathinal *et al.* 2003). It has been suggested that adaptive evolution may occur at the protein level, and that advantageous mutations contribute little to polymorphism, although they may contribute substantially to the divergence between species (Smith and Eyre-Walker 2002). Such a difference between polymorphism and divergence is limited to only a fraction of the genes, which evolve more rapidly, suggesting

The Evolution of Population Biology, ed. R. S. Singh and M. K. Uyenoyama. Published by Cambridge University Press.

that positive selection is involved (Fay *et al.* 2002). Among speciation traits are also those which arise after speciation that is primarily caused by selection on other phenotypes, including hybrid sterility and inviability. They can be genetically studied only in species where postzygotic isolation is incomplete.

The "*divergence population genetics*" approach addresses the history of species divergence as it is revealed in the polymorphism pattern at randomly selected genes (Kliman *et al.* 2000). In contrast to the gene map approach, divergence studies can focus directly on evolutionary forces, particularly those demographic factors that affect all the genes in the genome. Comparative DNA sequence data have been used to study relatedness of close sister taxa and populations.

To address which of the various mechanisms of speciation are consistent with empirical patterns, and can provide more general interpretations, an appropriate biological model is needed, and *Drosophila* is one such model. In 1830, Johann Wilhelm Meigen described an insignificant fruitfly that he named *Drosophila melanogaster.* One hundred and seventy years later, more than 80 000 works have been devoted to the *Drosophila* model and some 3500 species of Drosophilidae have been described (Ashburner 1989, Powell 1997, Bächli 1999). Since the overwhelming majority of these works deal with the now famous *D. melanogaster,* that species has become the first insect to have its genome completely sequenced (Adams *et al.* 2000) and increasingly attention will be paid to the species closely related to *D. melanogaster.* The reason why *D. melanogaster* relatives are so popular is that studies have been based on an equal knowledge of genetics, biogeography and ecology, a balance which is rarely respected in other groups of organisms (Turelli *et al.* 2001). Substantial progress in the study of speciation has indeed been driven by empirical results on *D. melanogaster* relatives, all of which originate from the Afrotropical regions.

16.2 The discoveries

Finding new species of drosophilid is quite usual whenever field surveys are made in tropical areas, but discovering a new species of the *melanogaster* species subgroup is undoubtedly a rare event. Only nine species have been discovered within 172 years, the last one being in 1998 (*D. santomea*). Table 16.1 summarizes the history of these findings and shows that the 1970s, with four new species (*D. teissieri, D. erecta, D. mauritiana, D. orena*), were the most productive years. It was also when it was realized that the Afrotropical region was the ancestral home range of the *melanogaster* species subgroup.

16.3 Morphology and taxonomy

The nine *D. melanogaster* subgroup species are distinguished on the basis of unequivocal morphological characters (Figure 16.1). These include various structures of male terminalia (genitalia) like the aedeagus, posterior

Table 16.1. *Summary of the origin of the nine relatives of the* Drosophila melanogaster *species subgroup*

Drosophila	Country of first record	Distribution	Habitat and resource	Discoverer	Description
melanogaster[a]	North Germany: Kiel, Hamburg	Cosmopolitan	Human commensal (domestic)	J. Meigen, 1830	Meigen, 1830
simulans	Lakeland, Florida	Cosmopolitan	Semi-domestic	Sturtevant, 1919	Sturtevant, 1919
yakuba	Ivory Coast	Afrotropical mainland, Madagascar, Zanzibar, São Tomé, Príncipe Islands	Savannahs and woodlands	H. Burla, 1951	Burla, 1954
teissieri	Zimbabwe, Mt Silinda, Chirinda	Afrotropical mainland	Rainforest *Parinari excelsa* (Chrysobalanaceae)	H. Paterson, 1970	Tsacas, 1971
erecta	Ivory Coast	West–West Central Africa, Gulf of Guinea	Screw pine marshes *Pandanus* spp. (Pandanaceae)	D. Lachaise, 1971	Tsacas & Lachaise, 1974
mauritiana	Mauritius	Mauritius, Rodrigues	Semi-domestic and forests	J. R. David, 1973	Tsacas & David, 1974
orena	Cameroon Volcanic Line: Mt Lefo	Bafut N'Guemba, Mt Lefo 2100 m	*Syzygium staudtii* submontane forest	L. Tsacas, 1975 CNRS expedition: Tsacas, David, Lachaise 1975	Tsacas & David, 1978
sechellia	Seychelles	Mahé, Praslin, Cousin, Frégate, Aride Islands	*Morinda citrifolia* (Rubiaceae)	Susan North & M. Nigel Varty Oxford Seychelles Expedition 1980	Tsacas & Bächli, 1981
santomea	Cameroon Volcanic Line: São Tomé Island	São Tomé 1150–1600 m	*Ficus chlamydocarpa fernandesiana* (Moraceae) Submontane forest	D. Lachaise, 1998	Lachaise & Harry, 2000

[a] syn *D. ampelophila* Loew, 1862.

Note: References can be found in Coyne and Orr 2000.

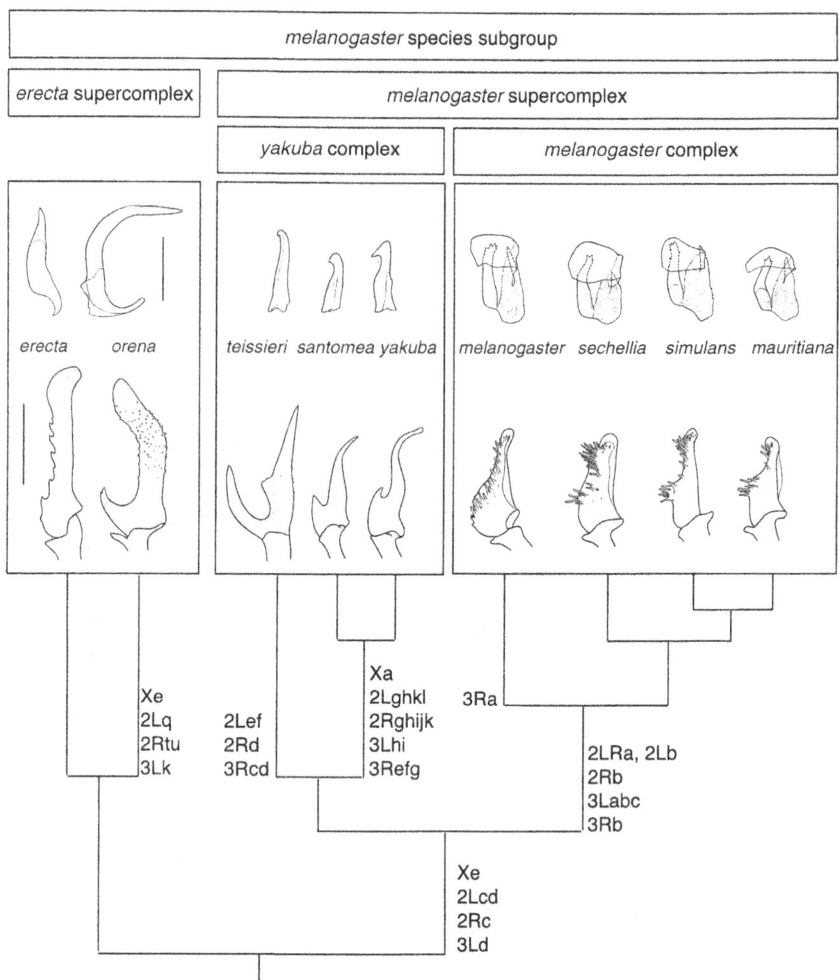

Figure 16.1. Consensus phylogenetic tree showing the relatedness between the nine species of the *D. melanogaster* subgroup (the branch lengths are not proportional to evolutionary time). A congruent topology is obtained for the phylogenetic tree regardless of the nuclear gene (*period, Amylase, Amyrel*) and algorithm (Neighbor-joining or Maximum likelihood) used (Cariou *et al.* 2001). The phylogenetic relationships within the *simulans* species subcomplex are nonetheless generally unresolved (Kliman *et al.* 2000) even though we favor here the emergence of *D. sechellia* first. Superimposed on the tree are fixed chromosomal inversions (Lemeunier and Ashburner 1984, Lemeunier and Aulard 1992) and diagnostic morphological features: (lower line) aedeagus; (upper line) posterior parameres (morphological structures are shown at the same scale all along each of the two lines; both scale bars: 0.01 mm). Although it is uncertain whether the totality of the intricate continuous structure of the parameres in the *melanogaster* complex is homologous to the simple structure in the *yakuba* complex, the two species complexes appear clearly distinguished on a morphological basis (Lachaise and Chassagnard, unpublished).

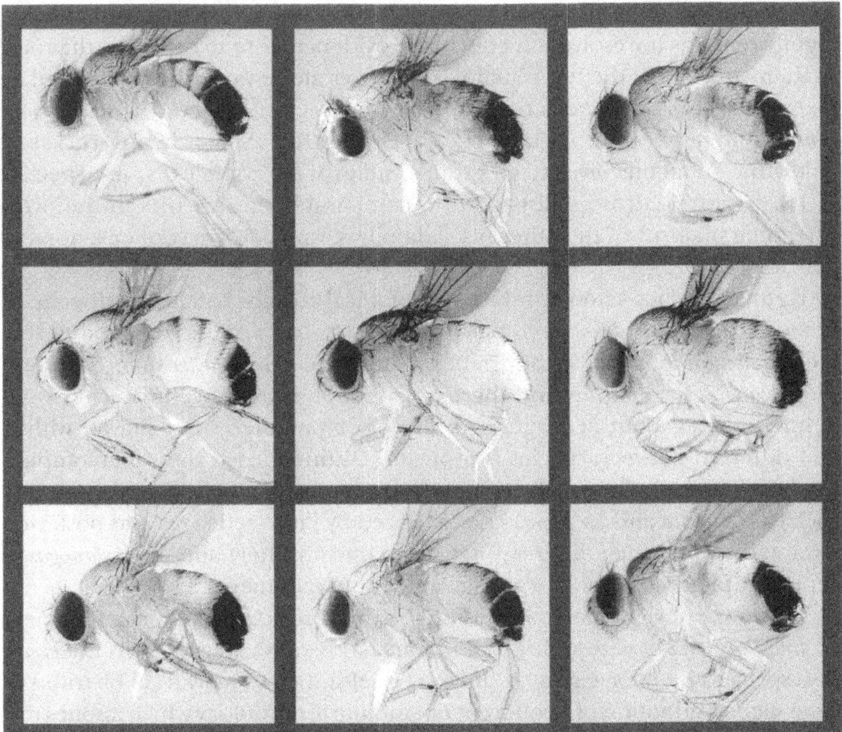

Figure 16.2. The males of the nine closely related species of the *Drosophila melanogaster* subgroup. Unlike its eight relatives, including its sister species *D. yakuba, D. santomea* almost completely lacks dark abdominal pigmentation in both males and females – evidence that the species difference in pigmentation in this species pair is polygenic (Llopart *et al.* 2002a). From left to right: (upper line) the *simulans* species subcomplex: *D. sechellia, D. simulans, D. mauritiana*; (middle line) the *yakuba* species complex: *D. yakuba, D. santomea, D. teissieri*; (lower line) *D. orena, D. erecta, D. melanogaster* (Lachaise *et al.* 2000).

parameres and cercus, which are most reliable for assessing diagnostic differences (but also species affinities). Other traits can also be used reliably, such as the tarsal sex combs in males (that of *D. mauritiana* is twice as long as that of *D. orena*) and the pigmentation pattern in both males and females (Figure 16.2). Natural hybrids, if any, can easily be identified on the basis of chimeric morphological patterns which match those observed in experimental hybrids, more especially in *santomea–yakuba* hybrids.

The various terms, such as "sister species" or "siblings," widely used to describe any of the closely related species of the *melanogaster* subgroup, may sometimes be misleading or confusing. Under the cladistic approach the terms should be used more restrictively. Strictly speaking, only *D. erecta* and *D. orena*, and *D. santomea* and *D. yakuba* are true sister species. Also, *D. mauritiana*,

D. sechellia and *D. simulans* will be seen as sister species so long as their phylogeny remains unresolved. If conclusive evidence were to be found that one of them arose first, the two remaining relatives alone would be considered as sister species *sensu stricto*. In connection with this, *D. melanogaster* and *D. simulans* should not be seen as sister species (Figure 16.1). Similarly, all terms used below the subgenus level (e.g., groups, subgroups, complexes) are specific to the jargon of drosophilists and no comparable entity exists in the overwhelming majority of the other insect families, with a few exceptions, notably culicids.

Figure 16.1 shows how tightly morphological affinity matches phylogenetic relatedness. In the *D. melanogaster* species subgroup there is a good agreement between morphology and genes. Within the *melanogaster* supercomplex, the *melanogaster* complex and the *yakuba* complex are morphologically well-defined clades: both aedeagus and posterior parameres are shared within and different between species complexes. Distinctions at the supercomplex level are more elusive, unless the *erecta* supercomplex were to comprise two species (*D. erecta* and *D. orena*) characterized by large aedeagus and posterior parameres (but which differ strongly from one another) and the *melanogaster* supercomplex species with small aedeagus and parameres.

The monophyletic unit comprising the three *simulans*-like species, namely *D. simulans, D. sechellia,* and *D. mauritiana,* has been variously called the *simulans* species complex, clade, or lineage, or else the *simulans* triad or trio. We here suggest adoption of a coherent taxonomic terminology which applies the term *simulans* subcomplex to the three-species cluster. Thus, *D. melanogaster* is a sister species to the *simulans* species subcomplex, as *D. teissieri* is a sister species to the *yakuba* subcomplex, composed of *D. yakuba* and *D. santomea.* Accordingly, the classification proposed in Table 16.2 could be adopted.

16.4 Ecology and biogeography

16.4.1 The western endemics: *D. erecta, D. orena, D. santomea*

D. erecta has a restricted geographical range along the Gulf of Guinea coast (Lachaise *et al.* 1988). This western endemic was recorded abundantly in the Ivory Coast and Gabon, and a few individuals were collected in Benin, Nigeria, Cameroon and Congo (Figure 16.3). In Cameroon, *D. erecta* has been reported from both the Bamileke and Adamawa Plateaux, that is all along the mainland Cameroon Volcanic Line (CVL). *D. erecta* is strictly specialized for breeding on the syncarps of a diversity of species (monosyncarpic and polysyncarpic) of the genus *Pandanus* (Pandanaceae). As a result, *D. erecta* is represented, like its host plant, by subdivided populations throughout the Gulf of Guinea mainland. New unpublished data in Gabon nevertheless suggest that a shift from one *Pandanus* to another displaying out-of-phase fruiting phenologies

Table 16.2. *Proposed classification of* D. melanogaster *subgroup*

- Family Drosophilidae
  - Subfamily Drosophilinae
    - Genus *Drosophila*
      - Subgenus *Sophophora*
        - *melanogaster* group
          - ***melanogaster* subgroup (the nine species)**
            - *erecta* supercomplex (new terminology)
              - 1. *D. erecta* Tsacas & Lachaise, 1974
              - 2. *D. orena* Tsacas & David, 1978
            - *melanogaster* supercomplex (new terminology)
              - *yakuba* complex
                - 3. *D. teissieri* Tsacas, 1971
                - *yakuba* subcomplex (new terminology)
                  - 4. *D. yakuba* Burla, 1954
                  - 5. *D. santomea* Lachaise & Harry, 2000
              - *melanogaster* complex
                - 6. *D. melanogaster* Meigen, 1830
                - *simulans* subcomplex (new terminology)
                  - 7. *D. sechellia* Tsacas & Bächli, 1981
                  - 8. *D. simulans* Sturtevant, 1919
                  - 9. *D. mauritiana* Tsacas & David, 1974

may occur more often than has generally been thought. The syncarps produce an offensive and pungent smell when ripening. Very few species other than *D. erecta* can breed in screwpine syncarps and it is therefore most likely that *Pandanus* contains a toxin to which *D. erecta* has evolved a resistance. But, neither the host toxin nor the genetic basis of *D. erecta*'s resistance to it has been studied as yet.

D. orena is only reported from the High Valley (2100 m) of Bafut N'Guemba on Mount Lefo in the Northwestern Province of Cameroon. Until contrary evidence is given, it should be regarded as endemic to a single volcano of the mainland CVL. *D. orena* is certainly the rarest and most enigmatic species of the *melanogaster* subgroup. All in all, less than ten individuals were caught in the wild when it was discovered in 1975, and none has been found since. One of them was a female that could be bred, and this isofemale line has provided all the material used in laboratories since then. Beyond localization to the *Syzygium staudtii* (Myrtaceae) submontane forest of Bafut N'Guemba, the ecology of *D. orena* is completely unknown. Given that similar submontane forest relicts exist on different uplands of the CVL, it is likely that the geographic range of *D. orena* extends over some of them.

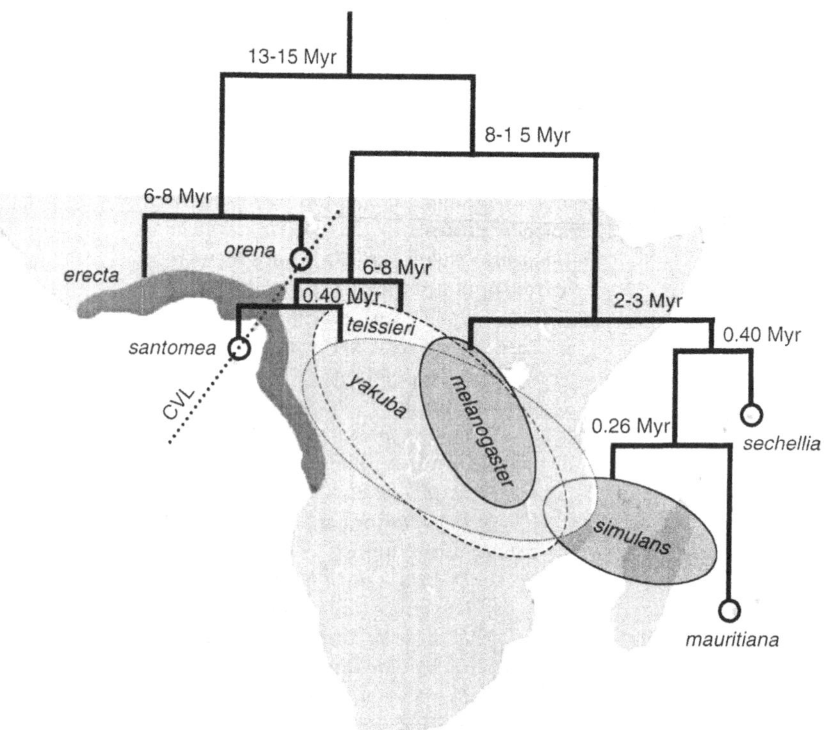

Figure 16.3. The phylogenetic tree of the nine *Drosophila melanogaster* subgroup species is superimposed onto their centers of origin in the Afrotropical region. The branch lengths are constrained by biogeography and are therefore not proportional to time. Estimations of splitting events are given in million years (Myr). Insular (*D. santomea, D. sechellia, D. mauritiana*) and local (*D. orena*) endemic species are indicated by circles. The geographic range of *D. erecta* is its current distribution. Ellipses represent putative ancestral home ranges (*D. teissieri, D. yakuba, D. melanogaster, D. simulans*). The biogeographic pattern shown here is a picture of the situation prior to the expansion of the two neocosmopolitans (recently evolved as cosmopolitans) out of their zone of origin, and of *D. yakuba* into neighboring western and eastern Afrotropical islands. CVL, the Cameroon Volcanic Line.

D. santomea is endemic to the remote submontane mist rainforests covering the higher volcanic slopes of São Tomé Island off the Cameroon and Gabon coastlines. This species is the first insular *melanogaster* relative found in the eastern equatorial Atlantic Ocean, and the first insular endemic not belonging to the *simulans* subcomplex (Lachaise *et al.* 2000). In the São Tomé submontane forest, *D. santomea* has so far been found exclusively on fallen figs of the endemic hemiepiphyte fig tree *Ficus chlamydocarpa* subsp. *fernandesiana*, between 1150 m and 1450 m in elevation. Of interest is that three subspecies of that fig species can be recognized along the CVL. On the mainland CVL the relevant fig tree subspecies is rare, being limited in distribution to a few submontane forests at altitudes between 1300 m and 2000 m, but in São Tomé the endemic subspecies is quite common. In fact, the authors' first field trip to São Tomé was planned with the idea that this endemic fig tree might yield an interesting endemic *Drosophila*, and this expectation turned out to be right. Whether the endemic *Ficus* species is the sole breeding site for *D. santomea* is, however, questionable, for the altitudinal range of *D. santomea* extends to at least 1600 m (probably more) while that of *F. chlamydocarpa* seems limited to 1450 m.

16.4.2 The eastern endemics: *D. mauritiana, D. sechellia*

D. mauritiana is endemic to two Mascarene islands, namely Mauritius and Rodrigues, to the east of Madagascar and Réunion islands. Mauritius is entirely volcanic and the basal brecciated series was roughly dated 8–10 Myrs. Population substructuring of *D. mauritiana* within Mauritius is plausible but the rapidly changing environment of the island has completely obscured the pattern. The current semi-domestic ecological status of *D. mauritiana* is certainly very different nowadays from what it was formerly (David *et al.* 1989).

D. sechellia, endemic to the Seychelles granitic archipelago, is a most interesting species because of its specialization on the fruit of *Morinda citrifolia.* This fruit is a highly toxic resource for all other *Drosophila* species so far investigated. The relationship between *D. sechellia* and morinda is, however, not restricted to a mere tolerance to toxic products. Complex adaptations have evolved in *D. sechellia,* which have been identified by a comparison with its relative *D. simulans* and an analysis of interspecific hybrids (R'Kha *et al.* 1991, 1997). *D. sechellia* adults are specifically attracted by the smell of morinda; females prefer to oviposit on morinda while those of *D. simulans* are repelled; and oogenesis is stimulated by morinda in *D. sechellia,* but inhibited in *D. simulans.* More than 150 different chemicals have been identified in ripe morinda fruit, among which hexanoic and octanoic acids, present in large amounts, are responsible for the toxicity (Farine *et al.* 1996, Amlou *et al.* 1997). These acids also induce oviposition in *D. sechellia* females but an aversion in *D. simulans* (Amlou *et al.* 1998), although the surprising oviposition preference for morinda in *D. mauritiana* is probably due to other compounds (Moreteau *et al.* 1994). The diversity of the traits which are involved in the adaptation of *D. sechellia* to its host plant raises a major problem: explaining the large number and nature of the genes which have changed during the adaptation process. The diversity of the traits which are involved suggests that several independent functions, and different sets of genes, have diverged from the ancestral state, although a pleiotropic action of a few genes cannot be completely excluded. The identification of the responsible genes will be a difficult challenge. Attempts to introgress genes mediating tolerance for morinda from *D. sechellia* into *D. simulans* were unsuccessful in spite of strong selection (Amlou *et al.*

1997). A QTL (quantitative trait loci) approach (Jones 1998) using 15 markers has suggested that all major chromosomes were involved in the tolerance. However, genetic results may also depend on the experimental technique implemented. Using a short-term technique, Jones (1998) found that the *D. sechellia* tolerance was dominant in F_1 hybrids while Amlou *et al.* (1997), measuring tolerance after a two-day exposure, found genetic additivity in hybrids. A similar discordance (dominance vs. additivity) was also observed in oviposition choice analyses (Amlou *et al.* 1998). For the moment a suggestion is that a fairly large number of QTL have been modified during the specialization process of *D. sechellia.*

16.4.3 The widespread mainland endemics: *D. teissieri, D. yakuba*

Although endemic to the Afrotropical region, *D. teissieri* and *D. yakuba* are widespread throughout sub-Saharan Africa where they are represented by numerous large populations.

D. teissieri, although widespread from Guinea to Zimbabwe, is strictly confined to the rainforest habitat (*sensu stricto*) and its geographical range strikingly matches the present-day distribution of the rainforest, including the scattered and remote rainforest patches of eastern and southeastern Africa (Usambara and Uzungwa montane forests in the Eastern Arc mountains of Tanzania; Chimanimani and Chirinda montane forests in the Eastern escarpments of Zimbabwe). However, the species distribution does not extend to the south beyond the classical biogeographic limit represented by the Limpopo River, and *D. teissieri* has not been reported from South Africa. More accurately, the distribution of *D. teissieri* tightly matches the geographic range of the Chrysobalanaceae *Parinari excelsa,* and evidence has been provided that *D. teissieri* breeds mostly in the *Parinari* fruits in localities as much distant as Mt Nimba in Guinea and Mt Usambara in northeastern Tanzania (Lachaise, unpublished). *D. yakuba* is widely distributed throughout the Afrotropical mainland from Sahel to Swaziland and has spread to inland Madagascar, Zanzibar island off Tanzania in the Indian Ocean, and São Tomé and Príncipe Islands off Gabon in the Atlantic Ocean. Its capacity to colonize new insular environments is likely related to its ability to cope with open, even semi-arid, habitats.

D. teissieri and *D. yakuba* live in markedly contrasted habitats, the rainforest for the former species, and open habitats, such as savannahs, open woodlands or secondary vegetation, for the latter. The two relatives thus show a striking ecological divide within the forest–savannah mosaic. They nonetheless happen to coexist whenever and wherever their habitats overlap, as for instance in the Marantaceae forest of Middle Ogooué in Central Gabon. There the Okoumé forest is tending to recolonize the open savannah habitat that has prevailed for thousands of years. As a consequence, *D. teissieri* is gradually tending to replace *D. yakuba.* In fact, it could be that the ratio between *D. yakuba* and *D. teissieri* is a way to estimate the age of the regenerating forest

(Bridle and Lachaise, unpublished). As the history of tropical Africa has mostly been a history of droughts interrupted by rainforest transgressions, it is possible that *D. teissieri* and *D. yakuba* have been alternately favored throughout the Pleistocene.

16.4.4 The neocosmopolitans: *D. melanogaster, D. simulans*

Wherever it has been recorded across the world, including within the Afrotropical region, *D. melanogaster* is a strict human commensal, and this peculiar and restricted ecology is by itself an evolutionary riddle. *D. melanogaster* natural populations show considerable genetic variation across several geographic regions and therefore the species is thought to have undergone recurrent worldwide expansions. How many such discrete events have occurred? When did they occur? Have all these events occurred through the agency of humans? All these points have been the subject of longstanding controversial debates.

D. simulans is generally seen as cosmopolitan, like *D. melanogaster.* However, there are major biogeographic distinctions between the two relatives. *D. simulans* is curiously lacking from Western Africa to the west of the Cameroon volcanic line (given that very few data are available from Nigeria, the limit might instead be the west of the Niger or Volta rivers). No less intriguing, *D. simulans* is rare or absent from the Southeast Asia and West Pacific regions, and immigrated only recently to Japan. Moreover, *D. simulans* contrasts with *D. melanogaster* in its ecology since it is not as strict a human commensal as its relative. In Eastern Africa, Madagascar, Seychelles and Mascarene islands, *D. simulans* thrives in wild highland habitats. It is, however, uncertain if *D. simulans* invaded highlands or evolved *in situ* there. In general the worldwide dispersal of *D. simulans* is thought to be more recent than that of *D. melanogaster,* but this point is still controversial.

16.5 Population substructuring

16.5.1 Rainforest fragmentation and *D. teissieri* population subdivision

Population substructuring in *D. teissieri* was first evidenced through the geographic variation of male terminalia, a trait that is remarkably invariant in most other species (Lachaise *et al.* 1981). Such differentiation, which has nonetheless not been accompanied by incipient reproductive isolation, is most likely a direct consequence of recurrent rainforest fragmentation during Pleistocene times. Extant populations in Eastern Africa are confined to the submontane rainforests covering remote isolated mountains. As these rainforests are currently separated from each other by semi-arid habitats or *Brachystegia* woodland unsuitable to *D. teissieri,* it is most likely that gene flow is considerably reduced among extant populations despite large local population size. In the western Guineo-Congolese rainforest block, the extant *D. teissieri* populations are more continuous, but fragmentation may also have occurred there during the reiterated drought periods that affected West Central Africa during the Pleistocene. A major consequence of the Rift mountain uplift and rainforest shrinkage on *D. teissieri* evolution is a marked molecular differentiation between East and West African populations (Cobb *et al.* 2000). Molecular data also indicate that the Eastern Arc mountain forests were formerly continuous (Harry, Lachaise, Lemeunier, Solignac, unpublished). Also, *D. teissieri* is the first organism where intron-present/intron-absent polymorphism was evidenced (here using the *jingwei* gene), and it was shown that such a polymorphism was maintained by Darwinian selection (Llopart *et al.* 2002b).

16.5.2 *D. melanogaster* populations with different status

D. melanogaster is a most efficient colonizing species: we are not aware of any place in the world with suitable climatic conditions where *D. melanogaster* has not been found. This is likely related to its domesticated status, which avails it of both convenient ecological niches and many opportunities for human-mediated transportation. It has been a longstanding *a priori* idea that *D. melanogaster*, the "garbage species," was genetically homogeneous all over the world. Empirical investigations have progressively revealed a contrary conclusion. *D. melanogaster* is highly differentiated into divergent geographic populations and races (David and Tsacas 1981, David and Capy 1988). It is now generally admitted that, with respect to their history, three kinds of populations can be identified: ancestral-stock derived (ancestral SD), ancient-stock derived (ancient SD) and newly introduced (NI) populations (David and Capy 1988). What differentiates them is the time elapsed since their arrival in the place where they are currently found. Ancestral-SD populations are those that arose from populations that have evolved *in situ* within the African ancestral home range. Ancient-SD populations are those that have evolved out of Africa in the first areas colonized by some ancient African migrants. Newly introduced populations are recent immigrants. Moreover, after a new place (e.g., America or Australia) was colonized, subsequent introductions have succeeded, obscuring the local population history.

The geographic structure of extant natural populations, in spite of gene flow, reflects that numerous traits have been subject to strong selective pressures and exhibit local adaptations. Two major environmental factors have been identified: climate (mostly temperature) and alcoholic resources. Both may vary according to latitude and are believed to be responsible for latitudinal clines. Indeed, latitudinal clines have been observed for a diversity of traits including morphometry, physiology, behavior, allozymes, and chromosome inversions (see David and Capy 1988, Capy *et al.* 1993, David *et al.* 2003, Gibert *et al.* 2003).

Ancestral SD populations are found in the Afrotropical region, south of the Sahara, and they are generally more polymorphic than derived populations.

These populations are far from being a single panmictic entity, as shown for instance by the structural pattern observed for amylase variants, which differ considerably among populations, not only in frequency but also in presence or absence (Dainou *et al.* 1993). Population structuring between western and eastern African populations is also suggested by the pattern of molecular and allozyme variation of *Adh* (Bénassi and Veuille 1995). Further evidence is provided by chromosome inversions, whose frequencies are highly variable among different countries, and more related to longitude than latitude (Aulard *et al.* 2002).

Ancient-SD populations are thought to be the extant populations that evolved out of Africa millennia ago. Non-African populations are very different from African ones in many respects. For instance, the prevalence of *Amy1* variant out of Africa is invariably correlated with low diversity at the amylase locus (Dainou *et al.* 1993).

The Asiatic Far East populations are also quite different from the European ones, and they remain an evolutionary challenge. Investigations of laboratory strains revealed that these populations were characterized by slow development, high body weight and low ovariole number (David *et al.* 1976), low fecundity and long lifespan (unpublished results). Recent collections in Far East countries failed, however, to find natural populations still showing these unusual life-history traits. It is possible that new immigrants, with a higher proliferative capacity, outcompeted the native populations. In other words it is possible that the Far East race, as previously defined, has disappeared. In contrast, recent investigations on mtDNA suggest that Far East populations are markedly divergent from western populations (Solignac 2003).

The newly introduced populations are found in America, Australia, and many oceanic islands, but their geographic origin is generally unclear. As regards North America, there is some evidence of a double colonization, first from tropical Africa and second from Europe (David and Capy 1988). Such between-continent migrations are evidenced by diffusion of transposable elements such as *P*, *hobo* and *I*. As old laboratory strains are free of these elements or do not contain active copies, it is assumed that natural populations worldwide acquired these elements during the twentieth century, and this is the best evidence that gene flows still exist among allopatric populations (Capy *et al.* 1997). As an example, the *P* element was apparently acquired in America after 1940, and owing to the P–M system of hybrid dysgenesis, has spread all over the world since (Kidwell 1994). There is now evidence that the *P* element is present in Japanese populations which are thought to be ancient-SD populations (Anxolabéhère *et al.* 1990). This indicates that contacts between populations have occurred. A similar phenomenon was described for the invasion of *D. melanogaster* by the *hobo* element (Bonnivard *et al.* 1997, 2000). In this case, the TPE tandem repeat of *hobo* was used as a marker. The authors proposed a scenario based on a two-step invasion: first, a complete invasion by elements with three TPE repeats, followed by the beginning of a new invasion involving *hobo* elements with five or seven repeats. In fact, it is difficult to characterize the extant populations owing to the admixture of ancestral-SD or ancient-SD with newly introduced populations, a situation that has probably grown in importance in the last hundred years or so.

16.5.3 The *D. simulans* enigma: lack of nuclear DNA population structuring conflicting with mtDNA haplotype geographic subdivision

Studies of phenotypic variation and of allozymes in *D. simulans* have suggested that there is almost no population subdivision within the species (Choudhary and Singh 1987), even though the adaptation of *D. simulans* to temperate regions has resulted, as in *D. melanogaster*, in latitudinal clines for many traits (see David *et al.* 2003, Gibert *et al.* 2003). As a rule, however, the clines in *D. simulans* are less marked than those in its relative, and this difference is not well explained. In contrast, DNA data suggest population differentiation in *D. simulans* mostly between Africa and the New World (Begun and Aquadro 1995).

A major breakthrough concerning population substructuring in *D. simulans* was the discovery of the three mainly allopatric mitochondrial haplotypes, designated as *si*I, *si*II and *si*III by Solignac (Solignac *et al.* 1986, Solignac 2003). The *si*II and *si*III types are more related to one another than either is to *si*I (Solignac and Monnerot 1986). Intriguingly, *si*I is found in the granitic Seychelles archipelago in the Indian Ocean and in New Caledonia, Moorea, Tahiti and Hawaii in the Pacific Ocean. *Si*III is endemic in the Afrotropical region, being confined to Madagascar, Réunion Island and also on Mt Kilimanjaro's lower slopes in Tanzania (Charlat *et al.* 2002). The *si*II type has a worldwide distribution (African continent, Middle East, America, Australia, Japan), but is curiously lacking in the Seychelles and Pacific islands. A marked 2–3% divergence exists between these mtDNA types, suggesting a separation time of 1–2 Myr (Solignac and Monnerot 1986, Monnerot *et al.* 1990, Satta and Takahata 1990).

In contrast with the high divergence among the major haplotypes, restriction site variability within each haplotype is very low or even absent (Solignac and Monnerot 1986, Baba-Aïssa *et al.* 1988, Nigro 1988). Complete sequences of the genome on more restricted samples of flies confirmed this low level of variability (Ballard 2000). Until recently, poor attention had been paid to the comparison of nuclear and mitochondrial variability of populations belonging to the three mitochondrial types. But, most recently, Ballard (2000) and Ballard *et al.* (2002) provided ample consistent evidence that the genetic subdivision observed in *D. simulans* mtDNA is not corroborated by nuclear DNA sequences, morphology of the genital arch, or assortative premating behavior. Ballard (2000) first addressed this question by comparing nucleotide variation of multiple complete mitochondrial genomes (excluding the A+T rich region) within the three distinct *D. simulans* mtDNA haplotypes with variation at intron 1 of the *alcohol dehydrogenase-related* locus (*Adhr*). Second,

Ballard *et al.* (2002) compared the genetic structure in each of the three mtDNA haplotype groups with data from another locus, *NADH: ubiquinone reductase 75kD subunit precursor* (*ND75*) locus. The *ND75* gene is located on the X-chromosome and encodes a mitochondrial polypeptide, and was chosen in an attempt to find an association between the mitochondrial and nuclear genealogies. The genealogies of *Adhr* and *ND75* consistently indicate extensive gene flow between the geographic regions and the lack of population structuring, data which conflict with the remarkable geographic subdivision of the mtDNA haplotypes.

Yet there is some evidence from two genes on the X-chromosome (*vermilion* and *G6pd*) for suggesting ancient subdivision of four African populations and recent admixture among African and derived populations (Hamblin and Veuille 1999). Also, analysis of two major components (7-tricosene and 7-pentacosene) of male cuticular hydrocarbons indicates a clear-cut distinction in male 7T/7P polymorphism on the African mainland between west equatorial and eastern–southern populations (Rouault *et al.* 2001). But more evidence is needed to understand the magnitude and antiquity of the genetic differentiation of *D. simulans* populations within the African mainland. Until contrary evidence is conclusively provided we will consider the lack of significant population structuring as the characteristic of *D. simulans.*

Two alternative hypotheses have been proposed to explain the genetic diversity of *D. simulans* and how the three extant *D. simulans* cytoplasmic "races" may have arisen: (1) a single very large ancestral population served as the source of three recent independent founder events; or (2) several highly structured ancestral populations existed (see Solignac 2003). Assuming three founder events (one for each mitochondrial type), the size of the single ancestral population is supposed to be large enough to maintain mtDNA haplotypes showing a 2–3% divergence in nucleotide sequence. Assuming several highly structured populations, the expectation is that high divergence in the mitochondrial genome would be consistently accompanied by high divergence at the nuclear level.

But the question then arises as to what evolutionary mechanism can influence the mtDNA so dramatically but yet have no detectable effect on autosomal DNA (Solignac 2003). In other words, how can one reconcile the conflicting findings that current populations of *D. simulans* are highly structured for their mtDNA types but not (or weakly) for their nuclear genome (Aquadro *et al.* 1988, Kliman *et al.* 2000, Solignac 2003)?

The most plausible explanation comes from the associated *Wolbachia,* a maternally inherited intracellular symbiont, whose strains appear to be structured consistently with mitochondrial types. As the endosymbiotic bacteria generate cytoplasmic incompatibility in their hosts, they increase the frequency of infected females. Crosses between infected males and uninfected females produce nonviable embryos, whereas the reciprocal cross is compatible (see review in Weeks *et al.* 2002). The endosymbiont is known to directly influence mitochondrial evolution but have a less direct influence on autosomal loci (Rousset *et al.* 1992, Merçot and Charlat 2003).

An alternative explanation for the conflict (a high structuring for mtDNA types *vis-à-vis* a weak structuring for nuclear genome) could be an asymmetrical pattern of dispersion between the sexes. If males leave their natal group before first mating, whereas females remain in their natal group throughout their lives, nuclear genes can flow solely by the transfer, and subsequent mating success, of males from one group to another. The rate and pattern of male-limited migration should affect the distribution of genetic variation within and between local populations. Nondispersing *simulans* females, on their side, would then make "mitochondrial matrilines," a pattern that is observed with *si*I, *si*II, *si*III. However, this hypothesis will be supported only if evidence is given that males are better dispersers than females, evidence that is presently lacking.

16.6 Reproductive relationships

16.6.1 The Zimbabwe and Brazza sexually diverging populations of *D. melanogaster*: novel races or new contacts?

An interesting debate arose from Brazzaville (Congo) and Zimbabwe *D. melanogaster* strains which revealed various degrees of sexual isolation with sympatric (Brazza) or allopatric (Zimbabwe) conspecific populations. The marked sexual isolation observed between the Zimbabwe and non-African "races" of *D. melanogaster* is clearly determined by many genes spread over the autosomal genome (Wu *et al.* 1995, Hollocher *et al.* 1997). The original series of 50 isofemale lines from Zimbabwe was isolated in 1990 from Sengwa Wildlife Research Area, at the junction of the Sengwa and Lutope rivers (Carson 2000) near Lake Kariba in northwestern Zimbabwe. According to Wu *et al.* (1995) these behavioral data reveal the early stages of speciation and suggest that sexual selection plays a major role in the process. But, while Carson (2000) agreed that sexual selection may serve as a driver that results in novel evolutionary divergence between populations, he considered the alternative possibility that Wu *et al.* had discovered a previously unrecognized older endemic population or sibling species of *D. melanogaster* from East Africa. A sibling species could be a possibility, but there are no "sibling species" strictly speaking (that is sister species without morphological distinction) in this species subgroup, and no – not even weak – morphological changes have been reported for the alleged "novel races." In fact, we concur with Carson in the possibility of a mixing between ancestral-stock derived and recently introduced *D. melanogaster* populations. It could be that such a situation is quite common in intertropical Africa. One more example is the remarkable co-occurrence of an ancestral-SD and a NI population of *D. melanogaster* within the urban area of Brazzaville in Congo in West Equatorial Africa (Vouidibio *et al.* 1989, Capy *et al.* 2000, Haerty *et al.*

2002). In Brazzaville, two populations living in different habitats can be distinguished: one on fruits and vegetables (the field population) and the other on alcoholic resources such as those found in the Kronenbourg brewery. Mating choice experiments clearly show the existence of a strong homogamy. Experiments between the field population and a European one show exactly the same phenomenon and no reproductive isolation was observed between European and brewery populations. Moreover, the brewery population is more similar to European than to African ones for several traits, including some allozyme frequencies, morphological and physiological traits, microsatellites and cuticular hydrocarbons (pheromones). Therefore, the brewery population was probably reintroduced in tropical Africa and maintained due to habitat choice.

While the Zimbabwe and Brazzaville cases seem to be similar (asymmetrical reproductive isolation), it remains difficult to determine whether or not they derive from the same phenomenon. In both cases, strong premating isolation is described while no postmating isolation can be detected (Hollocher *et al.* 1997, Haerty *et al.* 2002). However, it is difficult to compare these two situations because few analyses allowing comparison have so far been published. In Brazzaville, one of the major factors involved in the isolation could be the cuticular hydrocarbon (Haerty *et al.* 2002). Urban populations in Brazzaville show cuticular hydrocarbon profiles consistent with those of European populations, while rural populations show CHC profiles similar to those of African populations. The prevailing role of CHCs in reproductive isolation was previously proposed by Takahashi *et al.* (2001) for the Zimbabwe "race" but the suggestion requires conclusive evidence. In this case, it remains difficult to choose between the incipient speciation vs. the restored contact scenarios, inasmuch as they are not mutually exclusive.

16.6.2 Postmating isolation

Within the *melanogaster* species subgroup, various levels of postmating isolation are observed (Cariou *et al.* 2001). Table 16.3 shows that interspecific crosses involving two of the four mainland endemics (*D. orena, D. erecta, D. teissieri, D. yakuba*) are unsuccessful, whereas crosses involving an insular endemic species (from the Gulf of Guinea or the Indian Ocean) and a continental or insular allopatric sister (*sensu stricto*) species result in partial infertility, generally showing sterile male and fertile female hybrids in both reciprocal crosses. Most generally, sterility appears first in hybrids of the heterogametic sex in agreement with Haldane's rule (Coyne *et al.* 1991). However, crosses between *D. melanogaster* and any of the three *simulans* subcomplex species give sterile unisexual progeny of the sex of the *D. melanogaster* parent. Thus, if the direction of cross implicating *D. melanogaster* females and *simulans* subcomplex males obeys Haldane's rule, the reciprocal direction does not.

In contrast to the other interspecific crosses within the *simulans* subcomplex, the cross between *D. sechellia* and *D. simulans* shows a strongly

Table 16.3. *Reproductive relationships of the nine closely related species of the* Drosophila melanogaster *species subgroup as tested in standard "no-choice" conditions (generally five males of species A and five females of species B confined over two to four weeks together, with every two-day replicates)*

	melanogaster *subgroup*								
			melanogaster *supercomplex*						
	erecta *supercomplex*			yakuba *complex*		melanogaster *complex*			
Drosophila ♂	*orena*	*erecta*	*teissieri*	*yakuba*	*santomea*	*melanogaster*	*sechellia*	*simulans*	*mauritiana*
♀ *orena*	+	*no* F_1	*no* F_1	*no* F_1	*no* F_1	*no* F_1	*no* F_1	*no* F_1	*no* F_1
erecta	*no* F_1	+	*no* F_1	*no* F_1	*no* F_1	*no* F_1	*no* F_1	*no* F_1	*no* F_1 ♂ *ster.* F_1 ♀
teissieri	*no* F_1^d	*no* F_1	+	*no* F_1	*no* F_1	*no* F_1	*no* F_1	*no* F_1	ster. F_1 ♂ ster. F_1 ♀
yakuba	*no* F_1	*no* F_1	*no* F_1	+	**ster. F_1 ♂** **fert. F_1 ♀**	*no* F_1	*no* F_1	*no* F_1	ster. F_1 ♂ ster. F_1 ♀
santomea	*no* F_1	*no* F_1	*no* F_1	**ster. F_1 ♂** **fert. F_1 ♀**	+	*no* F_1	*no* F_1	*no* F_1	ster. F_1 ♂ ster. F_1 ♀
melanogaster	*no* F_1	*no* F_1	*no* F_1	*no* F_1	*no* F_1	+	*no* F_1 ♂ *ster.* F_1 ♀	*no* F_1 ♂ *ster.* F_1 ♀[a,c]	*no* F_1 ♂ *ster.* F_1 ♀
sechellia	*no* F_1	*no* F_1	*no* F_1	*no* F_1	*no* F_1	*ster.* F_1 ♂ *no* F_1 ♀	+	**ster. F_1 ♂** **fert. F_1 ♀** (rare)	**ster. F_1 ♂** **fert. F_1 ♀**
simulans	*no* F_1	*no* F_1	*no* F_1	*no* F_1	*no* F_1 ♂ *ster.* F_1 ♀	*ster.* F_1 ♂ *no* F_1^b ♀	**ster. F_1 ♂** **fert. F_1 ♀**	+	**ster. F_1 ♂** **fert. F_1 ♀**
mauritiana	*no* F_1	*no* F_1	*no* F_1	*no* F_1	*no* F_1	*ster.* F_1 ♂ *no* F_1 ♀	**ster. F_1 ♂** **fert. F_1 ♀**	**ster. F_1 ♂** **fert. F_1 ♀**	+

[a]Hybrid males die as late larvae. However, the *Lethal hybrid rescue* (*Lhr*) mutation from *D. simulans* (Watanabe 1979) and the *Hybrid male rescue* (*Hmr*) and *In(1)AB* mutations from *D. melanogaster* rescue larval hybrid viability (Hutter and Ashburner 1987, Hutter *et al.* 1990).

[b]Hybrid females die as embryo. However, embryonic inviability is rescued by two different mutations: *Zygotic hybrid rescue* (*Zhr*) from *D. melanogaster* (Sawamura *et al.* 1993b), and *maternal hybrid rescue* (*mhr*) from *D. simulans* (Sawamura *et al.* 1993a).

[c]One mutation rescues, albeit weakly, the fertility of female hybrids (Davis *et al.* 1996).

[d]One sterile F_1 hybrid male obtained once (Lemeunier *et al.* 1986).

asymmetrical mating preference. *D. simulans* females and *D. sechellia* males produce sterile male and fertile female hybrids, but the reciprocal cross rarely generates progeny, presumably due to behavioral causes. That crosses in both directions between *D. santomea* and *D. yakuba* result in fertile female and sterile male hybrids indicates that this reproductive pattern is not unique to the *simulans* subcomplex. However, this hybridization is also markedly asymmetrical: crossing between *D. yakuba* females and *D. santomea* males occurs easily and produces many F_1 hybrids of both sexes, while the reciprocal cross rarely produces offspring. Thus, it appears that males of insular endemic species (*D. mauritiana, D. santomea* and *D. sechellia*) that arose from recent (approximately half a million years) speciation events on offshore islands are more inclined to mate readily with heterospecific females of the *melanogaster* subgroup than are continental males.

16.6.3 Hybrid rescue genes

A major breakthrough in studies of postzygotic isolation was the discovery of "hybrid rescue" mutations (see Davis *et al.* 1996, Coyne and Orr 2000). These mutations are alleles that, when introduced singly into *Drosophila* hybrids, rescue the viability or fertility of normally nonviable larvae or sterile male individuals. All of the rescue mutations hitherto analyzed involve hybridization between *D. melanogaster* and *D. simulans.* In one direction of cross, that is female *simulans* × male *melanogaster,* hybrid males die as late larvae; in the other, hybrid females die as embryos (Table 16.3). All surviving hybrids are completely sterile. Several mutations, summarized in Table 16.3, are known that rescue the hybrid larva or embryo. That complementary sets of mutations rescue larval and embryonic viability strongly suggests that these forms of isolation have different developmental bases. The discovery of hybrid rescue genes has several important implications. First, it suggests that postzygotic isolation may have a simple genetic basis. Second, it may also permit precise delineation of reproductive isolation genes. Thus, using fertility rescue between *D. melanogaster* and *D. simulans* Sawamura *et al.* (2000) first produced fertile F_1 hybrid females between these two relatives, and discovered at least six genes of hybrid male sterility in a region of 5% of the *D. simulans* genome introgressed into *D. melanogaster.*

16.6.4 An insular altitudinal hybrid zone between *D. santomea* and *D. yakuba*

D. yakuba is widespread across western, eastern and southern Africa, but *D. santomea* is endemic to the island of São Tomé, a 860 km^2 volcanic island 320 km west of Gabon (Lachaise *et al.* 2000). *D. yakuba* also inhabits São Tomé, but preliminary molecular evidence points to *D. santomea* originating allopatrically after a colonization event by some inseminated females originating from the *yakuba* subcomplex ancestral population or from an ancient *D. yakuba* population, with *D. yakuba* subsequently invading the island a second time in recent times (Cariou *et al.* 2001).

On São Tomé, *D. yakuba* is limited to lower elevations, while *D. santomea* occurs in the mist forests at higher elevations. A hybrid zone occurs between 1150 m and 1450 m elevation on the volcanic island; here across the hybrid zone one finds a low frequency (*c.* 1%) of natural hybrids (Lachaise *et al.* 2000). However, the hybrid zone is nonuniform and the ratio between the two sister species shifts gradually to *D. santomea* from bottom to top of the hybrid zone. Given that the mating preference (behavioral) is markedly asymmetrical, the chance of finding natural interspecific hybrids is strongly uneven within that zone. Moreover, as hybrid females are fertile, there may also be some natural backcross hybrids coexisting with the parental species in that zone.

16.7 Discussion and questions arising

Why does the evolutionary biology of the nine *melanogaster* subgroup species have a general interest? Contrasting the ecologies of sister species within a diversity of species pairs enables us to understand the genetic mechanisms that has led some of these closely related species toward specialization and endemism (e.g., genetic resistance to host-plant toxin as assessed in *D. sechellia* in insular environments and as is suspected for *D. erecta* in continental environments), and some others toward human commensalisms and rapid dispersal (*D. melanogaster, D. simulans*). Moreover, within four sister species pairs (*yakuba–santomea, simulans–sechellia, simulans–mauritiana, sechellia–mauritiana*), hybridization is partially successful, generally resulting in sterile male but fertile female hybrids. Female hybrid fertility makes possible interspecific introgressions and may result in the propagation of endosymbiotic bacteria, which in turn may eventually generate mitochondrial hitchhiking across the species barrier.

16.7.1 Why do the frequencies of chromosomal inversions differ greatly among closely related species?

Many aspects remain unclear about chromosomal variants (fixed vs. polymorphic inversions), notably the markedly uneven amount of inversions on the various chromosomes: there is for instance no inversion on the chromosome X of all species, except a few in *D. melanogaster* (Lemeunier and Ashburner 1984, Lemeunier *et al.* 1986). The chromosome 2 is more rearranged (fixed inversions) and bears more polymorphic inversions than the chromosome 3 in all species of the subgroup except *D. melanogaster.* Why are there so many inversions in *D. melanogaster* populations (about 500 inversions described, among them 480 endemic to one single population), while few or no inversions exist in the three homosequential species of the *D. simulans* subcomplex?

Such rigidity of the *D. simulans* subcomplex compared with the flexibility of *D. melanogaster* is not clearly understood (Lemeunier and Aulard 1992). Curiously, this duality (flexibility vs. rigidity) exists in each of the two supercomplexes, *melanogaster* and *erecta*. In the latter, no inversions have been detected in *D. erecta* (but very few lines have been hitherto analyzed), while two inversions were observed in the single *D. orena* isofemale line available.

Furthermore, the number of inversions differs within species between western and eastern Africa, with west equatorial rainforest harboring populations with higher levels of chromosomal polymorphism. In *D. yakuba*, the most polymorphic species within the *melanogaster* subgroup, a number of West African populations exhibit four inversions (Kounden and N'Koemvone, Cameroon; Madibou, Congo), and some others three inversions (Banco and Lamto, Ivory Coast). In contrast, East African populations so far analyzed have only one inversion, such as Limbe in Malawi and Kampala in Uganda (Lemeunier and Ashburner 1976). Similarly, *D. teissieri* yielded five inversions in the Lopé population in central Gabon (most individuals having four inversions), and four inversions in the Lamto population in the preforest belt of the Ivory Coast (most individuals having four inversions). In contrast, in eastern Africa, there are rarely more than two inversions, for instance one to two in Tanzania (East and West Usambara Mts, Uzungwa Mts; and only one (rarely three) in Zimbabwe (Silinda Mts, Chirinda forest, Chimanimani Mts) (Lemeunier, unpublished).

Finally, it is still unclear whether the overall and average numbers of inversions per region and per individual reflect the antiquity and size of the local population.

16.7.2 Why *either* reduced *or* sustained genetic diversity in insular endemics?

The population genetics of the origin and divergence of the *D. simulans* subcomplex species (*D. simulans, D. mauritiana* and *D. sechellia*) were examined by Kliman *et al.* (2000) using the patterns of DNA sequence variation found within and between species at 14 different genes. *D. sechellia* consistently revealed low levels of polymorphism, and genes from *D. sechellia* have accumulated mutation at a rate that is nearly 50% higher than the same genes from *D. simulans*. At synonymous sites, *D. sechellia* has experienced a significant excess of unpreferred codon substitutions. As an endemic species on a small island, *D. mauritiana* is presumed to have a long history of small population size, but surprisingly high genetic diversity was found. *D. simulans* and *D. mauritiana* are both highly polymorphic and the two species share many polymorphisms, probably maintained since the time of common ancestry (both species have higher levels of polymorphism than reported for *D. melanogaster*).

Why has an insular endemic species like *D. mauritiana* retained genetic variation, while another (*D. sechellia*) has not? The reason may merely reflect differences in the sizes and histories of the islands. The Seychelles Bank was completely emergent 16 000 years ago and was reduced to its present condition of scattered islands about 10 000 years ago (see Lachaise and Silvain 2003). The populations of *D. sechellia* have then undoubtedly suffered dramatic reduction in size. If these islands have harbored small refugial *D. sechellia* populations, one would expect the archipelago to maintain considerable genetic variation. But the genetic diversity observed in *D. sechellia* for mtDNA, allozymes (Cariou *et al.* 1990), and sequences of nuclear genes in *D. sechellia* is exceedingly low (Hey and Kliman 1993, Kliman *et al.* 2000). In fact, *D. sechellia* shows very low polymorphism levels at nearly all loci, almost certainly due to a small effective population size. Therefore, the submersion over the last 10 000 years is thought to have reduced *D. sechellia* genetic diversity through extinction and genetic bottleneck, rather than set the stage for increased diversity by fragmenting a once continuous population into separate islands. The overall paleogeography of Mauritius has remained unchanged during and after the glacial period, and therefore it is most likely that *D. mauritiana* was not affected by a bottleneck as was *D. sechellia.*

16.7.3 Do genes with a sex-related function evolve more rapidly than others?

D. melanogaster subgroup species have provided interesting consistent data addressing the "faster male" hypothesis (Wu and Davis 1993). Civetta and Singh (1998) found a higher ratio of nonsynonymous to synonymous substitution rates in genes with a sex-related function than in genes with developmental or metabolic function. Consistent data were also obtained by Ting *et al.* (2000) who analyzed DNA polymorphism at the Odysseus (*OdsH*) locus of hybrid sterility between *D. mauritiana* and *D. simulans.* Ting *et al.* stressed that in order to reveal the differences in the phylogenetic pattern provided by *OdsH* vs. all other loci which do not appear to be directly relevant to speciation, we must distinguish between variant nucleotide sites that are phylogenetically ambiguous and unambiguous for the three species. Phylogenetically ambiguous sites designate shared variations across species, presumably resulting from ancient polymorphisms and/or subsequent introgressions. Ting *et al.* (2000) showed that a large majority of sites from other loci are ambiguous (30 of 31 sites), whereas, at *OdsH,* seven of the nine sites are unambiguous with six of them supporting the close kinship between *D. mauritiana* and *D. simulans.* Clearly, a strong disparity exists in the effect of ancient polymorphism and/or introgression on the current variations at speciation loci vs. others. Consistent data were provided by Parsch *et al.* (2001) who described a new *D. melanogaster* gene, *ocnus* (*ocn*), that encodes a protein abundant in testes. *Ocn* shares homology with another testis-specific gene, *janusB* (*janB*), and is located just distal to *janB* on chromosome 3. The two genes also share homology with the adjacent *janusA* (*janA*) gene, suggesting that multiple duplication events have occurred within this region of the genome. Parsch *et al.* sequenced these three genes from eight species of the *D. melanogaster* species subgroup. Their data suggest that diversification of gene function followed each duplication event and that each gene evolved under different selective constraints. Interestingly, all three genes showed faster rates of evolution than genes encoding proteins with metabolic function.

16.7.4 Conclusion

An explanation for the discordance observed between the "reproductive" and "molecular" phylogenies is that genomes may be mosaics with respect to molecular genealogy (Ting *et al.* 2000). Most loci, chosen without regard to their roles in reproductive differentiation, may not reflect the species divergence in their sequence polymorphism because of either shared ancient polymorphism or gene introgression through secondary contact. Indeed, it has been reported that 88% of *D. mauritiana* lines carry a *D. simulans* mtDNA (Solignac and Monnerot 1986, Solignac 2003). The one-base-pair difference over 15 kb of the molecules from the two species unequivocally indicates that the similarity is due to introgression (Ballard 2000).

Similarly, the difference between haplotypes taken in pairs among *D. santomea, D. yakuba* and *D. teissieri* does not exceed 0.6% (Lachaise *et al.* 2000), a level of similarity which confirms the negligible differentiation of the mitochondrial genome that was earlier detected between *D. yakuba* and *D. teissieri* (Monnerot *et al.* 1990). This unusually low polymorphism and the similarity of the mtDNAs between species imply a recent and common coalescence time for all haplotypes. A factor likely to clone the mtDNA could be an invasion of the cytoplasm by *Wolbachia.* This endosymbiont was found commonly in the *D. teissieri* lines (69 out of 72), rarely in *D. yakuba* (five out of 54) and at an intermediate frequency in *D. santomea* (13 out of 46). Sequencing of the hypervariable gene of a surface protein (*wsp*) in those lines analyzed for mtDNA showed that the current populations of the three species share the same bacteria, curiously not associated with cytoplasmic incompatibility. In contrast to nuclear markers which indicate a marked divergence, the subtle differences observed in mtDNA and in the endosymbiont are definitely more recent than the cladogeneses and likely reflect the history of the invasion of their cytoplasm by *Wolbachia.* If *Wolbachia* infection pre-dated speciation in the *D. simulans* subcomplex (Rousset and Solignac 1995), propagation of the bacteria occurred after speciations in the *yakuba* complex, indicating that the species barrier can be overcome (Lachaise *et al.* 2000). In *D. simulans Wolbachia* alone is probably not sufficient to account for the major conflict resulting from the inconsistency between the lack of population structuring reflected by the nuclear genome and the marked geographic subdivision of the mtDNA haplotypes.

In conclusion, the nine closely related species of the *melanogaster* subgroup have allowed us to address some of the major general questions of evolutionary biology and ecology, but a number of outstanding questions have remained unresolved. Do all nine species follow the same molecular clock? Can we rely on the calibration, assuming that the split between *D. melanogaster* and *D. simulans* occurred around 2–3 Myr ago, and on which a number of inferences rely?

REFERENCES

Adams, M. D. (and 194 others) (2000). The genome sequence of *Drosophila melanogaster*. *Science* 287:2185–95.

Amlou, M., Moreteau, B., and David, J. R. (1998). Genetic analysis of *Drosophila sechellia* specialization: oviposition behavior toward the major aliphatic acids of its host plant. *Behav. Genet.* 28:455–64.

Amlou, M., Pla, E., Moreteau, B., and David, J. R. (1997). Genetic analysis by interspecific crosses of the tolerance of *Drosophila sechellia* to major aliphatic acids of its host plant. *Genet. Sel. Evol.* 29:511–22.

Anxolabéhère, D., Hu, D. K., Nouaud, D., and Périquet, G. (1990). *PM* system: a survey of *Drosophila melanogaster* strains from the People's Republic of China. *Genet. Sel. Evol.* 22:175–88.

Aquadro, C. F., Lado, K. M., and Noon, W. A. (1988). The rosy region of *Drosophila melanogaster* and *Drosophila simulans*. I. Contrasting levels of naturally occurring DNA restriction map variation and divergence. *Genetics* 119:875–88.

Ashburner, M. (1989). *Drosophila: a Laboratory Handbook*. Cold Spring Harbor, N.Y.: Cold Spring Harbor Press.

Aulard, S., David, J. R., and Lemeunier, F. (2002). Chromosomal inversion polymorphism in ancestral Afrotropical populations of *Drosophila melanogaster*. *Genet. Res.* 79:49–63.

Baba-Aïssa, F., Solignac, M., Dennebouy, N., and David, J. R. (1988). Mitochondrial DNA variability in *Drosophila simulans*: quasi absence of polymorphism within each of the three cytoplasmic races. *Heredity* 61:419–26.

Bächli, G. (1999). TaxoDros: the new internet database. Program and Abstracts, 16th European Drosophila Research Conference, September 29–October 2, 1999, Zürich, Switzerland, p. 67 (http://taxodros.unizh.ch).

Ballard, J. W. (2000). Comparative genomics of mitochondrial DNA in *Drosophila simulans*. *J. Mol. Evol.* 51:64–75.

Ballard, J. W., Chernoff, B., and James, A. C. (2002). Divergence of mitochondrial DNA is not corroborated by nuclear DNA, morphology, or behavior in *Drosophila simulans*. *Evolution* 56:527–45.

Begun, D. J., and Aquadro, C. F. (1995). Molecular variation at the *vermilion* locus in geographically diverse populations of *Drosophila melanogaster* and *D. simulans*. *Genetics* 140:1019–32.

Bénassi, V., and Veuille, M. (1995). Comparative population structuring of molecular and allozyme variation of *Drosophila melanogaster Adh* between Europe, West Africa and East Africa. *Genet. Res.* 65:95–103.

Bonnivard, E., Bazin, C., Denis, B., and Higuet, D. (2000). A scenario for the *hobo* transposable element invasion, deduced from the structure of natural populations of *Drosophila melanogaster* using tandem TPE repeats. *Genet. Res.* 75:13–23.

Bonnivard, E., Higuet, D., and Bazin, C. (1997). Characterization of natural populations of *Drosophila melanogaster* with regard to the *hobo* system: a new hypothesis on the invasion. *Genet. Res.* 69:197–208.

Capy, P., Bazin, C., Higuet, D., and Langin, T. (1997). *Dynamic and Evolution of Transposable Elements.* Austin, TX: R. G. Landes Company.

Capy, P., Pla, E., and David, J. R. (1993). Phenotypic and genetic variability of morphometrical traits in natural populations of *Drosophila melanogaster* and *D. simulans.* I. Geographic variations. *Genet. Sel. Evol.* 25:517–36.

Capy, P., Veuille, M., Paillette, M., Jallon, J.-M., Voudibio, J., and David, J. R. (2000). Sexual isolation of genetically differentiated sympatric populations of *Drosophila melanogaster* in Brazzaville, Congo: the first step towards speciation. *Heredity* 84:468–75.

Cariou, M.-L., Silvain, J.-F., Daubin, V., Da Lage, J.-L., and Lachaise, D. (2001). Divergence between *Drosophila santomea* and allopatric or sympatric populations of *D. yakuba* using paralogous amylase genes and migration scenarios along the Cameroon volcanic line. *Mol. Ecol.* 10:649–60.

Cariou, M.-L., Solignac, M., Monnerot, M., and David, J. R. (1990). Low allozyme and mtDNA variability in the island endemic species *Drosophila sechellia* (*Drosophila melanogaster* complex). *Experientia* 46:101–4.

Carson, H. L. (2000). Sexual selection in populations: the facts require a change in the genetic definition of the species. In R. S. Singh and C. Krimbas (eds) *Evolutionary Genetics: From Molecules to Morphology*, vol. 1, pp. 495–512. Cambridge: Cambridge University Press.

Charlat, S., Le Chat, L., and Merçot, H. (2002). Characterization of non-cytoplasmic incompatibility inducing *Wolbachia* in two continental African populations of *D. simulans. Heredity* 90:49–55.

Choudhary, M., and Singh, R. S. (1987). A comprehensive study of genetic variation in *Drosophila melanogaster.* III. Variations in genetic structure and their causes between *Drosophila melanogaster* and its sibling species *Drosophila simulans. Genetics* 117:697–710.

Civetta, A., and Singh, R. S. (1998). Sex-related genes, directional sexual selection, and speciation. *Mol. Biol. Evol.* 15:901–9.

Civetta, A., and Singh, R. S. (1999). Broad-sense sexual selection, sex-gene pool evolution, and speciation. *Genome* 42:1033–41.

Cobb, M., Huet, M., Lachaise, D., and Veuille, M. (2000). Fragmented forests, evolving flies: molecular variation in East and West African populations of *Drosophila teissieri. Mol. Ecol.* 9:1591–7.

Coyne, J. A. (1992). Genetics and speciation. *Nature (London)* 355:511–15.

Coyne, J. A. (1996). Genetics of differences in pheromonal hydrocarbons between *Drosophila melanogaster* and *D. simulans. Genetics* 143:353–64.

Coyne, J. A., and Charlesworth, B. (1997). The genetics of a pheromonal difference affecting sexual isolation between *Drosophila mauritiana* and *D. sechellia. Genetics* 145:1015–30.

Coyne, J. A., and Orr, H. A. (1998). The evolutionary genetics of speciation. *Phil. Trans. R. Soc. Lond. Ser. B* 353:287–305.

Coyne, J. A., and Orr, H. A. (2000). The evolutionary genetics of speciation. In R. S. Singh and C. Krimbas (eds) *Evolutionary Genetics: From Molecules to Morphology*, vol. 1, pp. 532–69. Cambridge: Cambridge University Press.

Coyne, J. A., Charlesworth, B., and Orr, H. A. (1991). Haldane's rule revisited. *Evolution* 45:1710–13.

Dainou, O., Cariou, M.-L., Goux, J.-M., and David, J. R. (1993). Amylase polymorphism in *Drosophila melanogaster*: haplotype frequencies in tropical Africa and American populations. *Genet. Sel. Evol.* 25:133–51.

David, J. R., and Capy, P. (1988). Genetic variation of *Drosophila melanogaster* natural populations. *Trends Genet.* 4:106–11.

David, J. R., and Tsacas, L. (1981). Cosmopolitan subcosmopolitan and widespread species: different strategies within the Drosophilid family (Diptera). *C. R. Soc. Biogéogr.* 57:11–26.

David, J. R., Bocquet, C., and Pla, E. (1976). New results on the genetic characteristic of the Far East race of *Drosophila melanogaster. Genet. Res.* 28:253–60.

David, J. R., McEvey, S., Solignac, M., and Tsacas, L. (1989). *Drosophila* communities on Mauritius and the ecological niche of *D. mauritiana* (Diptera, Drosophilidae). *Revue Zool. Afr. – J. Afr. Zool.* 103:107–16.

David J. R., Allemand, R., Capy, P., Chakir, M., Gibert, P., Pétavy, G., and Moreteau, B. (2003). Comparative life histories and ecophysiology of *Drosophila melanogaster* and *D. simulans.* In P. Capy and P. Gibert (eds) Special issue of *Genetica* "Drosophila melanogaster, D. simulans*: so similar, so different*" (in press).

Davis, A. W., Roote, J., Morley, T., Sawamura, K., Herrmann, S., and Ashburner, M. (1996). Rescue of hybrid sterility in crosses between *D. melanogaster* and *D. simulans. Nature (London)* 380:157–9.

Farine, J.-P., Legal, L., Moreteau, B., and Le Quere, J.-L. (1996). Volatile components of ripe fruits of *Morinda citrifolia* and their effects on *Drosophila. Phytochemistry* 41:433–8.

Fay, J. C., Wyckoff, G. J., and Wu, C.-I. (2002). Testing the neutral theory of molecular evolution with genomic data from *Drosophila. Nature (London)* 415: 1024–6.

Gibert, P., Capy, P., Imasheva, A., Moreteau, B., Morin, J.-P., Pétavy, G., and David, J. R. (2003). Comparative morphometry in *Drosophila melanogaster* and *D. simulans.* In P. Capy and P. Gibert (eds) Special issue of *Genetica* "Drosophila melanogaster, D. simulans: *so similar, so different*" (in press).

Haerty, W., Jallon, J.-M., Rouault, J., Bazin, C., and Capy, P. (2002). Reproductive isolation in natural populations of *Drosophila melanogaster* from Brazzaville (Congo). In W. J. Etges and M. A. F. Noor (eds) Special issue of *Genetica* Vol. 116, "*Genetics of mate choice: from sexual selection to sexual isolation*".

Hamblin, M. T., and Veuille, M. (1999). Population structure among African and derived populations of *Drosophila simulans*: evidence for ancient subdivision and recent admixture. *Genetics* 153:305–17.

Hey, J., and Kliman, R. M. (1993). Population genetics and phylogenetics of DNA sequence variation at multiple loci within the *Drosophila melanogaster* species complex. *Mol. Biol. Evol.* 10:804–22.

Hollocher, H., Ting, C.-T., Wu, M.-L., and Wu, C.-I. (1997). Incipient speciation by sexual isolation in *Drosophila melanogaster*: extensive genetic divergence without reinforcement. *Genetics* 147:1191–201.

Hutter, P., and Ashburner, M. (1987). Genetic rescue of inviable hybrids between *Drosophila melanogaster* and its sibling species. *Nature (London)* 327:331–3.

Hutter, P., Roote, J., and Ashburner, M. (1990). A genetic basis for the inviability of hybrids between sibling species of *Drosophila*. *Genetics* 124:909–20.

Jones, C. D. (1998). The genetic basis of *Drosophila sechellia*'s resistance to a host plant toxin. *Genetics* 149:1899–908.

Kidwell, M. G. (1994). The Wilhelmine E. Key 1991 Invitational Lecture. The evolutionary history of the *P* family of transposable elements. *J. Hered.* 85:339–46.

Kliman, R. M., Andolfatto, P., Coyne, J. A., Depaulis, F., Kreitman, M., Berry, A. J., McCarter, J., Wakeley, J., and Hey, J. (2000). The population genetics of the origin and divergence of the *Drosophila simulans* complex species. *Genetics* 156:1913–31.

Kulathinal, R. J., Skwarek, L., Morton, R. A., and Singh, R. S. (2003). Rapid evolution of the sex-determining gene, transformer: structural diversity and rate heterogeneity among sibling species of *Drosophila*. *Mol. Biol. Evol.* 20:44–52.

Lachaise, D., Cariou, M.-L., David, J. R., Lemeunier, F., Tsacas, L., and Ashburner, M. (1988). Historical biogeography of the *Drosophila melanogaster* species subgroup. *Evol. Biol.* 22:159–225.

Lachaise, D., Harry, M., Solignac, M., Lemeunier, F., Bénassi, V., and Cariou, M.-L. (2000). Evolutionary novelties in islands: *Drosophila santomea*, a new *melanogaster* sister species from São Tomé. *Proc. R. Soc. Lond. Ser. B* 267:1487–95.

Lachaise, D., Lemeunier, F., and Veuille, M. (1981). Clinal variation in male genitalia in *Drosophila teissieri* Tsacas. *Am. Nat.* 117:600–8.

Lachaise, D., and Silvain, J.-F. (2003). How two vicariant Afrotropical endemics made two human commensals: the *Drosophila melanogaster–D. simulans* palaeogeographic riddle. In P. Capy and P. Gibert (eds) Special issue of *Genetica* "Drosophila melanogaster, D. simulans: *so similar, so different*" (in press).

Lemeunier, F., and Ashburner, M. (1976). Relationships within the *melanogaster* species subgroup of the genus *Drosophila* (*Sophophora*). II. Phylogenetic relationships between six species based upon chromosome banding sequences. *Proc. R. Soc. Lond. Ser. B* 193:275–94.

Lemeunier, F., and Ashburner, M. (1984). Relationships within the *melanogaster* species subgroup of the genus *Drosophila* (*Sophophora*). *Chromosoma* (Berlin) 89: 343–51.

Lemeunier, F., and Aulard, S. (1992). Inversion polymorphism in *Drosophila melanogaster*. In C. B. Krimbas and J. R. Powell (eds) *Drosophila Inversion Polymorphism*, pp. 339–405. Boca Raton, FL: CRC Press.

Lemeunier, F., David, J. R., Tsacas, L., and Ashburner, M. (1986). The *melanogaster* species group. In M. Ashburner, and E. Novitski (eds) *The Genetics and Biology of Drosophila* (Vol. 3_e), pp. 147–256. New York: Academic Press.

Llopart, A., Elwyn, S., Lachaise, D., and Coyne, J. A. (2002a). Genetics of a difference in pigmentation between *Drosophila yakuba* and *D. santomea*. *Evolution* 56: 2262–77.

Llopart, A., Comeron, J. M., Brunet, F., Lachaise, D., and Long, M. (2002b). Intron-present/absent polymorphism in *Drosophila* driven by Darwinian selection. *Proc. Natl. Acad. Sci. USA* 99:8121–5.

Merçot, H., and Charlat, S. (2003). *Wolbachia* infections in *Drosophila melanogaster* and *D. simulans*: polymorphism and levels of cytoplasmic incompatibility. In P. Capy and P. Gibert (eds) Special issue of *Genetica* "Drosophila melanogaster, D. simulans: *so similar, so different*" (in press).

Monnerot, M., Solignac, M., and Wolstenholme, D. R. (1990). Discrepancy in divergence of the mitochondrial and nuclear genomes of *Drosophila teissieri* and *Drosophila yakuba. J. Mol. Evol.* 30:500–8.

Moreteau, B., R'Kha, S., and David, J. R. (1994). Genetics of non-optimal behavior: oviposition preference of *Drosophila mauritiana* for a toxic resource. *Behav. Genet.* 24:433–41.

Nigro, L. (1988). Natural populations of *Drosophila simulans* show great uniformity of mitochondrial DNA restriction map. *Genetica* 77:133–6.

Parsch, J., Meiklejohn, C. D., Hauschteck-Jungen, E., Hunziker, P., and Hartl, D. L. (2001). Molecular evolution of the *ocnus* and *janus* genes in the *Drosophila melanogaster* species subgroup. *Mol. Biol. Evol.* 18:801–11.

Powell, J. R. (1997). *Progress and Prospects in Evolutionary Biology. The* Drosophila *Model.* New York: Oxford University Press.

R'Kha, S., Capy, P., and David, J. R. (1991). Host-plant specialization in the *Drosophila melanogaster* species complex: a physiological, behavioral, and genetical analysis. *Proc. Natl. Acad. Sci. USA* 88:1835–9.

R'Kha, S., Moreteau, B., Coyne, J. A., and David, J. R. (1997). Genetics and evolution of a lesser fitness trait: egg production in the specialist *Drosophila sechellia. Genet. Res.* 69:17–23.

Rouault, J., Capy, P., and Jallon, J.-M. (2001). Variations of male cuticular hydrocarbons with geoclimatic variables: an adaptive mechanism in *Drosophila melanogaster. Genetica* 110:117–30.

Rousset, F., and Solignac, M. (1995). Evolution of single and double *Wolbachia* symbioses during speciation in the *Drosophila simulans* complex. *Proc. Natl. Acad. Sci. USA* 92:6389–93.

Rousset, F., Vautrin, D., and Solignac, M. (1992). Molecular identification of *Wolbachia,* the agent of cytoplasmic incompatibility in *Drosophila simulans,* and variability in relation with host mitochondrial types. *Proc. R. Soc. Lond. Ser. B* 247: 163–8.

Satta, Y., and Takahata, N. (1990). Evolution of *Drosophila* mitochondrial DNA and the history of the *melanogaster* subgroup. *Proc. Natl. Acad. Sci. USA* 87:9558–62.

Sawamura, K., Davis, A. W., and Wu, C.-I. (2000). Genetic analysis of speciation by means of introgression into *Drosophila melanogaster. Proc. Natl. Acad. Sci. USA* 97:2652–5.

Sawamura, K., Taira, T., and Watanabe, T. K. (1993a). Hybrid lethal systems in the *Drosophila melanogaster* species complex. I. The *maternal hybrid rescue* (*mhr*) gene of *Drosophila simulans. Genetics* 133:299–305.

Sawamura, K., Yamamoto, M.-T., and Watanabe, T. K. (1993b). Hybrid lethal systems in the *Drosophila melanogaster* species complex. II. The *zygotic hybrid rescue* (*Zhr*) gene of *D. melanogaster. Genetics* 133:307–13.

Singh, R. S. (2000). Toward a unified theory of speciation. In R. S. Singh and C. Krimbas (eds) *Evolutionary Genetics: From Molecules to Morphology*, vol. 1, pp. 570–604. Cambridge: Cambridge University Press.

Smith, N. G. C., and Eyre-Walker, A. (2002). Adaptive protein evolution in *Drosophila. Nature (London)* 415:1022–4.

Solignac, M. (2003). Mitochondrial DNA in the *Drosophila melanogaster* complex. In P. Capy and P. Gibert (eds) Special issue of *Genetica* "Drosophila melanogaster, D. simulans: *so similar, so different*" (in press).

Solignac, M., and Monnerot, M. (1986). Race formation, speciation, and introgression within *Drosophila simulans, D. mauritiana,* and *D. sechellia* inferred from mitochondrial DNA analysis. *Evolution* 40:531–9.

Solignac, M., Monnerot, M., and Mounolou, J.-C. (1986). Mitochondrial DNA evolution in the *melanogaster* species subgroup of *Drosophila. J. Mol. Evol.* 23:31–40.

Swanson, W. J., and Vacquier, V. D. (2002). The rapid evolution of reproductive proteins. *Nat. Rev. Genet.* 3:137–44.

Takahashi, A., Tsaur, S. C., Coyne, J. A., and Wu, C.-I. (2001). The nucleotide changes governing cuticular hydrocarbon variation and their evolution in *Drosophila melanogaster. Proc. Natl. Acad. Sci. USA* 98:3920–5.

Ting, C.-T., Tsaur, S.-C., and Wu C.-I. (2000). The phylogeny of closely related species as revealed by the genealogy of a speciation gene, *odysseus. Proc. Natl. Acad. Sci. USA* 97:5313–16.

Turelli, M., Barton, N. H., and Coyne, J. A. (2001). Theory and speciation. *Trends Ecol. Evol.* 16:330–43.

Vouidibio, J., Capy, P., Defaye, D., Pla, E., Sandrin, J., Csink, A., and David, J. R. (1989). Short-range genetic structure of *Drosophila melanogaster* populations in an Afrotropical urban area and its significance. *Proc. Natl. Acad. Sci. USA* 86:8442–6.

Watanabe, T. K. (1979). A gene that rescues the lethal hybrids between *Drosophila melanogaster* and *D. simulans. Jpn. J. Genet.* 54:325–31.

Weeks, A. R., Reynolds K. T., and Hoffmann A. A. (2002). *Wolbachia* dynamics and host effects: what has (and has not) been demonstrated? *Trends Ecol. Evol.* 17:257–62.

Wu, C.-I. (2001). The genic view of the process of speciation. *J. Evol. Biol.* 14:851–65.

Wu, C.-I., and Davis, A. W. (1993). Evolution of postmating reproductive isolation: the composite nature of Haldane's rule and its genetic basis. *Am. Nat.* 142:187–212.

Wu, C.-I., Hollocher, H., Begun, D. J., Aquadro, C. F., Xu, Y., and Wu, M.-L. (1995). Sexual isolation in *Drosophila melanogaster,* a possible case of incipient speciation. *Proc. Natl. Acad. Sci. USA* 92:2519–23.

PART V

APPLIED POPULATION BIOLOGY: BIODIVERSITY AND FOOD, DISEASE, AND HEALTH

17

Conservation biology: the impact of population biology and a current perspective

PHILIP HEDRICK
Department of Biology, Arizona State University, Tempe

17.1 A personal reflection

As a post-doctoral fellow with Dick Lewontin at the University of Chicago for the academic year 1968–1969, I attended the inspiring Tuesday and Thursday morning class on population biology he jointly taught with Dick Levins. With 20 to 30 post-docs, students, and other visitors listening intently to these lectures, we were all sure that population biology was an approach that would be widely adopted over the coming years. The lectures of this class were mainly on theoretical topics and generally focused on population genetics and somewhat less on population ecology. After my post-doc year at Chicago, I went to the University of Kansas where an upper-level undergraduate course in population biology was being developed. There I taught population biology over the next 19 years and wrote a text for the course entitled *Population Biology: the Evolution and Ecology of Populations* (Hedrick 1984). The connection of evolution and ecology using introductory theoretical population biology principles and illustrative biological examples worked well teaching undergraduates. The introduction to population biology 35 years ago at the University of Chicago greatly influenced my approach to science and was an important factor resulting in my concentration on conservation biology research in the past decade.

17.2 Brief history of conservation biology

In the latter half of the twentieth century, it was widely recognized that the rate of species extinction was increasing and that many species were in imminent danger of extinction. The major factors related to these extinctions and declines are thought to be (1) overharvesting from hunting, fishing, trapping, and other killing, (2) loss, degradation, and fragmentation of habitat, and (3) introduction of nonnative species such as pathogens, parasites, predators, and competitors (e.g., Diamond 1989). In other words, the problem and the

The Evolution of Population Biology, ed. R. S. Singh and M. K. Uyenoyama. Published by Cambridge University Press.

potential causes of the problem were widely recognized, but only in the 1970s and 1980s was a new approach, generally called conservation biology, introduced and developed to understand the processes influencing extinction by utilizing the principles of population biology. Even though conservation biology was not specifically discussed in the Lewontin–Levins population biology class, both Lewontin and Levins wrote about extinction of populations during this period and the theoretical core of population biology was central to the new conservation biology approach.

Much of the credit for the development of the discipline of conservation biology utilizing both ecological and genetic principles can be attributed to Michael Soulé (Frankel and Soulé 1981, Soulé and Wilcox 1980, Soulé 1986, 1987) and several other insightful researchers. For example, Shaffer (1981) wrote an influential review of the threats to endangered species and explicitly introduced the concept of the minimum viable population size, i.e., the minimum population size necessary for long-term persistence. He suggested that there were four major sources of uncertainty facing endangered species and the maintenance of a viable population: demographic stochasticity, environmental stochasticity, natural catastrophes, and genetic stochasticity (see Lande, 1993, 1994 for a theoretical evaluation).

Gilpin and Soulé (1986) set out the proximate factors which they felt would influence the minimum viable population size and species (or population) extinction. Two of the major extinction factors they identified were ecological – decline and variation in demographic values and fragmentation of the population distribution – and two were genetic – increase in inbreeding (with a consequent loss in fitness) and loss of adaptation (or adaptive potential). They emphasized that multiple causes may simultaneously impact small populations and that there may be positive feedback between these factors such that the overall impact increasing the probability of, and decreasing the time to, extinction was greater than expected from the individual components. For example, variation in population numbers due to demographic factors could result in higher inbreeding (and inbreeding depression) which could in turn result in further reduction in population numbers, and so on.

Although extinction of a species, demise of all its individuals, is by definition a demographic phenomenon, genetics has played an important role in much of conservation biology. This visibility resulted both from an emphasis by Michael Soulé and the early enlistment of the ideas from many leading evolutionary geneticists (see contributors to Schonewald-Cox *et al.* 1983). In addition, the documentation by Ralls *et al.* (1979) of the impact of inbreeding on the survival of captive endangered species demonstrated the need to manage, primarily for genetics reasons, captive populations of endangered species (see also Ralls *et al.* 1988). The importance of captive management, both for the long-term survival of these populations and for use of captive animals in the recovery of wild populations, has been a focus of a significant amount of conservation biology research (see Ballou *et al.* 1995).

Partly as a reaction to the emphasis on genetics, reviews by Lande (1988) and Simberloff (1988) attempted to give a reasoned perspective to the impact of genetics on extinction and suggested that ecological factors are of more immediate concern in conservation of wild populations. Suggesting that the focus on conservation biology may actually make important contributions to population biology, Lande (1988) stated that "The immediate practical need in biological conservation for understanding the interaction of demographic and genetic factors in the extinction of small populations therefore may provide a focus for fundamental advances at the interface of ecology and evolution."

A more provocative view was taken by Caughley (1994), just before he died, outlining his perspective of the scientific approaches used in conservation biology. He suggested that there were two paradigms in conservation biology: the small population paradigm, which endeavors to determine "the effects of smallness on the persistence of a population" and the declining population paradigm which deals with "the cause of smallness and its cure." However, this separation seemed artificial because "the factors under the small-population paradigm are the stochastic ones that may result in the proximate cause of extinction, and the ones under the declining population paradigm are the deterministic (or ultimate) ones that reduced the population size so that it becomes vulnerable to random events and phenomena" (Hedrick *et al.* 1996). As a result, Hedrick *et al.* (1996) suggested that it would be "more productive and accurate to cast the discussion in terms of an analysis of viability that considers both the generally anthropogenic ultimate causes and the stochastic proximate causes – an inclusive population viability analysis."

Shaffer (1981) and Soulé (1987) focused on the factors that may potentially influence the minimum viable population and cause extinction. To examine these causes in more detail, an individual-based simulation approach called VORTEX, encompassing all the relevant ecological and genetic factors, was developed by Robert Lacy in the late 1980s (Lacy *et al.* 1995). Workshops, using population viability analysis (PVA) have evolved to include all the factors influencing extinction, now including human populations and activities (Lacy and Miller 2002), so that examination of individual endangered species may be appropriately tailored. Using this approach, the factors potentially influencing extinction in hundreds of endangered species have been examined worldwide. There are somewhat differing views on what type of analysis should be used in PVA and the overall value of PVA (Beissinger and McCullough 2002). However, it seems that all researchers involved would agree that more ecological and genetic information on particular species would allow greater confidence in any PVA predictions and a greater understanding of the important factors influencing persistence in specific endangered species.

Recently, there seems to be a trend in some of the conservation biology literature, particularly in the journal *Conservation Biology*, to focus less on the biological factors that may be important in persistence. In fact, there seems to be an emphasis on nonscientific areas of conservation that impact extinction. Although no one would deny that economic, sociological, policy, and other arenas might have great impact on endangered species, it seems shortsighted to assume that all is known about scientific aspects of conservation biology and particular endangered species. Data from genome projects, advanced GIS analysis, and other new developments may make the level of biological information on endangered species more detailed and realistic and the context of conservation biology using population biological principles even more useful.

It would greatly facilitate conservation biology research if model organisms were useful in examining and generating principles of conservation biology. However, in contrast to the basic facets of genetics, some of the factors influencing extinction are intrinsically different between model laboratory organisms, such as *Drosophila*, and most endangered species. It is true that laboratory experiments with model organisms may often serve a useful heuristic purpose to illustrate evolutionary genetic principles or captive management options but in many ways they are not much different from computer simulations. It seems unlikely that laboratory experiments on insects with a history of very large population size will provide new insight into conservation of endangered species, most of which are vertebrates of small population size, have a history of declining numbers, may have important social and mating structures, etc. For example, if laboratory experiments give counterintuitive results, is it relevant to endangered species or only to the model organism being used?

Much of the theoretical (both verbal and mathematical) research in conservation biology has sought to develop general principles that can be used for application for different endangered species. Although this is an important goal, many endangered species have special attributes that must be considered and may be the ones that have made them endangered. As a result, it is often necessary to examine in detail a given endangered species in order to understand what has made it endangered and what might be done to facilitate recovery. In this overall vein, I will discuss below some areas of conservation biology in which recent studies have provided new insights. Although it may be possible to make some generalities, data on endangered species, or closely related surrogates, often suggest that cautious interpretation is necessary to understand the important factors influencing the dynamics of endangered species. Below I will briefly discuss four applications of principles of population biology (biased by my interest in population genetics) that are of important current interest in conservation biology. These examples serve to illustrate many of the thoughts I have expressed above (see also Hedrick 2001) but also show that to make a real impact in understanding the conservation of particular endangered species detailed examination and experimentation are often essential.

17.3 Examples of population biology in conservation biology

17.3.1 Genetic restoration of populations with low fitness

Populations of some endangered species have become so small that they have lost genetic variation and appear to have deleterious genetic variants at either very high frequency or fixed (Land *et al.* 2001, Westermeier *et al.* 1998, Madsen *et al.* 1999). To avoid extinction from this genetic deterioration, some populations may benefit from the introduction of individuals from related populations or even subspecies for genetic restoration, i.e., elimination of deleterious variants and recovery to normal levels of genetic variation. Recent research in nonendangered species (Ebert *et al.* 2002, Saccheri and Brakefield 2002, Richards 2000) has also shown that immigration can result in genetic rescue of populations from low fitness.

The last remaining population of the Florida panther (*Puma concolor coryi*) provides an extreme example of this phenomenon in an endangered species. This population has a suite of traits that suggests genetic drift has nearly fixed the population for previously rare traits, some of which appear to have severe deleterious effects. These traits, which are found in high frequency only in the Florida panther and are unusual in other puma populations, include high frequencies of cryptorchidism (unilateral undescended testicles), kinked tail for the last five vertebrae, cowlick on the back, and the poorest semen quality recorded in any felid (Roelke *et al.* 1993; Table 17.1). In addition, a large survey of microsatellite loci has shown that Florida panthers have much lower molecular variation than other North American populations of pumas (Culver *et al.* 2000).

The potential positive and negative genetic effects of introducing individuals from genetically diverse but geographically isolated populations into apparently inbred population was theoretically evaluated before the introduction of Texas pumas into Florida (Hedrick 1995). Assuming 20% gene flow from outside in the first generation (and 2.5% every generation thereafter), the fitness quickly improves. One concern about this approach is that any locally adapted alleles may be swamped by outside gene flow. However, with this level of gene flow, fitness from advantageous alleles is only slightly reduced in spite of gene flow.

A program to release females from the closest natural population from Texas was initiated in 1995 to genetically restore fitness in this population. Five of the eight introduced Texas females produced offspring with resident Florida panther males and now there are a number of F_1 and F_2 offspring (Land and Lacy 2000). At this point approximately 20% of the overall ancestry is from the introduced Texas pumas. Of the animals with Texas ancestry, only 7% have a kinked tail (compared with 88% before) and the ones with a kinked tail are progeny from backcrosses to Florida cats (Table 17.1). Similarly but not as dramatic, only 24% have a cowlick (compared with 93% before), two

Table 17.1. *The proportion of Florida panthers (sample size in parentheses) with a kinked tail, cowlick, or cryptorchidism with no Texas puma ancestry and those with Texas puma ancestry*

	No Texas ancestry	Texas ancestry				
		F_1	F_2	BC-TX	BC-FL	Total
Kinked tail	0.88 (48)	0.00 (17)	0.00 (7)	0.00 (3)	0.20 (15)	0.07 (42)
Cowlick	0.93 (46)	0.20 (10)	0.00 (5)	0.00 (1)	0.60 (5)	0.24 (21)
Cryptorchidism	0.68 (22)	0.00 (2)	0.00 (2)	–	0.00 (1)	0.00 (5)

BC-TX (backcross to Texas), and BC-FL (backcross to Florida).
Source: Land *et al.* 2001, Roelke *et al.* 1993.

in F_1s and three in backcrosses to Florida cats. Only five males with Texas ancestry have been evaluated for cryptorchidism and all have two descended testicles, in other words, a reduction from 68% to 0% cryptorchidism. Semen characteristics have been evaluated in only one F_1 male and it appears as good as or better than the average of Texas pumas, much better than Florida panthers. In other words, the introduction of Texas pumas has already resulted in a substantial reduction of the frequency of the detrimental traits that have accumulated in the Florida panther.

Overall, the introduction of individuals from outside the impacted population, and the consequent increase in "genetic health," in the example of Florida panthers, appears to provide a paradigm for other similar situations. However, many endangered species do not have related unimpacted populations, so that this option may not be available. Of course, if the cause of the low numbers is not recognized and remedied, then low numbers may continue to occur in the endangered population and eventually detrimental genetic effects could again accumulate.

17.3.2 Impact of the natural environment lowering fitness in inbred individuals

Inbreeding depression appears to be nearly universal; for example, Lacy (1997) stated that he was "unable to find statistically defensible evidence showing that any mammalian species is unaffected by inbreeding. Moreover, endangered species seem no less impacted by inbreeding, on average, than are common taxa." However, the effects of inbreeding on endangered species have generally been examined in captive populations for which the environment may be less harsh than natural environments (for reviews, see Hedrick and Kalinowski 2000, Keller and Waller 2002). Estimates of inbreeding depression from captivity or laboratory environments are thought to be an underestimate, or at least different from the effects in a natural environment. In addition, the last populations of many endangered species may exist in marginal habitats or be impacted by new (or even unknown) environmental stresses, such as introduced species or pollution.

Some general insights into these phenomena related to inbreeding depression can be obtained from several recent experimental studies. For example, the effect of inbreeding on male reproductive success in wild mice under laboratory conditions was minor, whereas in semi-natural conditions, inbred males had only about 20% of the success of outbred males (Meagher *et al.* 2000). In a comprehensive examination in *Drosophila*, the extent of inbreeding depression was greatly increased in stressful laboratory conditions (Bijlsma *et al.* 1999). Bijlsma *et al.* (1999) also found a low correlation between the fitness of genetic variants in different stressful environments, so that generalizations about the fitness of a variant over different stressful environments were not possible.

Jimenez *et al.* (1994) examined the survival of adult noninbred and inbred white-footed mice (*Peromyscus leucopus noveboracensis*). Stock for the experiment was captured from the natural study site near Chicago, Illinois, brought into the laboratory, and bred to produce individuals with inbreeding coefficients of 0.00 or 0.25 (from full-sib matings). Almost 800 mice, nearly equally split between noninbred and inbred, were released during three different periods. The area had a low number of mice during the release, suggesting that the environment was harsh because of some unknown cause. For the 10 weeks following release, noninbred individuals had a higher weekly survivorship at all census times than the inbred individuals. Using an approach that utilizes capture–recapture data, the weekly survival of the inbred mice was estimated to be 56% that of the noninbred mice.

To quantify the effects of inbreeding in nature, let us assume that the fitness, or a fitness component, in the captive or laboratory environment for noninbred individuals is standardized to be unity (Hedrick and Kalinowski 2000). Let the fitness relative to that in the inbred population be $\bar{w}_I$ and to that in the natural environment $\bar{w}_N$. If we assume that the effects of inbreeding and the environment are multiplicative, then the expected fitness of the inbred group in the natural environment is $\bar{w}_I\bar{w}_N$ (Table 17.2). In Table 17.2, the data from the white-footed mouse example (Jimenez *et al.* 1994) are given with the survival of noninbreds in the natural environment relative to the survival of noninbreds in captivity as 0.221 and the survival of inbreds in captivity relative to noninbreds in captivity as 0.935. Therefore, the predicted survival of inbreds in nature is 0.207 while the actual observed survival is only 0.046, about 22% of that predicted, suggesting a synergism such that inbreds in nature survive much more poorly than expected.

The population crash of the wild population of song sparrows (*Melospiza melodia*) on Mandarte Island, British Columbia, involves a documented case on inbreeding depression in nature (Keller *et al.* 1994). This population appears to undergo periodic crashes, probably due to severe winter weather: the decline in 1989 killed 89% of the adult animals. Because there had been an

Table 17.2. *A general way to predict the fitness of inbred individuals in a natural environment based on the relative fitness of inbreds in the captive environment and the relative fitness of noninbreds in a natural environment, and an example using the observed relative survival in the white-footed field mouse*

		Population	
	Environment	Noninbred	Inbred
General	Captive	1	$\bar{w}_I$
	Natural	$\bar{w}_N$	$\bar{w}_I \bar{w}_N$
Field mouse	Captive	1	$\bar{w}_I = 0.935$
	Natural	$\bar{w}_N = 0.221$	$\bar{w}_I \bar{w}_N = 0.207$
			Observed = 0.046

Source: Hedrich and Kalinowski 2000, Jimenez *et al.* 1994.

extensive program to mark individuals in this population over several generations, the inbreeding coefficient for most individuals was known before the crash. After the crash, the inbreeding coefficient for the 10 survivors was 0.0065 (only three had known inbreeding) while the inbreeding coefficient for the 206 birds that died was significantly higher at 0.0312. All the birds with inbreeding coefficients of 0.0625 or higher (13% of the population) died during the crash. Later, high inbreeding depression was shown in the population (Keller 1998), consistent with the differential survival between inbred and noninbred individuals. Recently, examination of the fitness of the descendants of new immigrants has shown that components of their fitness differed from nonmigrants in a complicated pattern (Marr *et al.* 2002).

These studies suggest that natural populations may experience much more inbreeding depression than estimated from captive or laboratory studies carried out in less extreme environmental conditions. Assuming that this is a general phenomenon, the inbreeding depression in natural populations may impact demographic parameters more than previously expected and consequently result in a positive feedback causing decline and extinction, perhaps even more quickly than suggested by Gilpin and Soulé (1986).

17.3.3 Genetic resistance to pathogens

In recent years, it has become widely recognized that endangered species often may be threatened by exposure to pathogens (e.g., Laurenson *et al.* 1998, Lafferty and Gerber 2002), many of them exotic and novel to the endangered species. Pathogens can be defined broadly to include viruses, bacteria, protozoa, etc., which can cause a reduction in fitness to the host. In endangered species, a pathogen introduced from another more common species, particularly if the endangered species has low resistance, may result in final decline in numbers to extinction.

Resistance of a host to pathogens often has an important genetic component. In particular, genes in the major histocompatibility complex (MHC) (Edwards and Hedrick 1998) play an important role in disease resistance to pathogens in vertebrates (Hedrick and Kim 2000). For example, the MHC in humans has been shown to be important in resistance to HIV, hepatitis, and malaria. Resistance to HIV and hepatitis in humans appears to be higher for MHC heterozygotes than for homozygotes (Thurz *et al.* 1997, Carrington *et al.* 1999). Further, susceptibility to HIV may be the result of a single amino acid change in a MHC molecule (Gao *et al.* 2001), suggesting that the role of small changes in the MHC may be important.

In endangered species, the level of genetic variation in the MHC, and other genes that may influence host resistance, may be lower because of past or present small population size than in more common species. This may make endangered species even more susceptible to the effects of pathogens than common species that have large, stable population sizes, which generally have greater genetic variation. In the small populations of endangered species, there may also be a higher frequency of inbreeding, a factor that results in an increase in homozygosity for all loci. Below I will summarize some of the results of an experiment to investigate both genetic resistance from an MHC gene and from inbreeding in the endangered winter-run chinook salmon, *Oncorhynchus tshawytscha*, to three very different pathogens (Arkush *et al.* 2002). To my knowledge, this is the largest and most comprehensive investigation of genetic resistance to disease in an endangered species ever carried out.

The annual number of spawning adults of winter-run chinook salmon from the Sacramento River, California, drainage has declined from over 100 000 in 1969 to about 200 in 1991. One response to this situation was to establish a refuge population at Bodega Marine Laboratory (BML) in case the natural run became extinct. As a result, it was possible to investigate the genetic resistance of winter-run chinook salmon produced at BML to three different pathogens; a bacteria, a virus, and a parasite (Arkush *et al.* 2002). The bacterium used, *Listonella (Vibrio) anguillarum*, is found worldwide and is a significant pathogen of salmonids. The virus used, infectious hematopoietic necrosis virus (IHNV), is considered the most important viral pathogen affecting salmonids in North America. The other pathogen, *Myxobolus cerebralis*, is a myxozoan parasite that causes whirling disease, which has been detected in salmonid populations throughout the United States.

To examine the impact of MHC genetic variation, parents of families were chosen so that they would be segregating for MHC genotypes with the expectation that 50% of the progeny would be heterozygotes and 50% homozygotes. To examine the impact of inbreeding, females were mated both to their full brothers and to unrelated males. As a result, progeny of these matings had either an inbreeding coefficient of 0.25 or 0.0.

An easy way to visualize the influence of genetic variation at a MHC locus is to examine the difference in survival rates for the MHC heterozygotes and

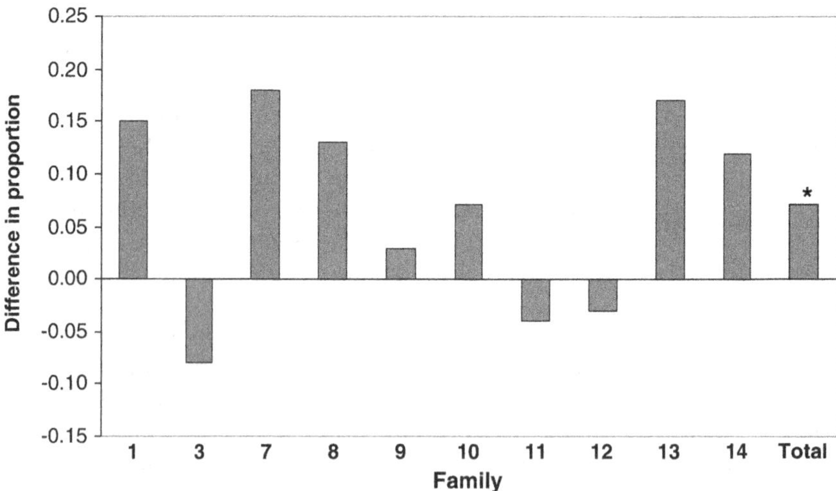

Figure 17.1. Difference in survival between major histocompatibility complex (MHC) heterozygotes and homozygotes after exposure to IHNV (infectious hematopoietic necrosis virus) for the ten families segregating for the MHC locus where * indicates $P < 0.05$ (from Arkush *et al.* 2002).

homozygotes. For the bacterium, there was no difference in overall survival of different MHC genotypes but there was a significant overall difference in survival of MHC heterozygotes and homozygotes upon exposure to IHNV with higher survival for heterozygotes than homozygotes in seven of the ten families (Figure 17.1) (only one family was examined for the whirling disease parasite). The overall survival upon exposure to IHNV of MHC heterozygotes and homozygotes were 0.82 and 0.75, respectively. The extent of differential selection of viability for IHNV can be estimated if these values are standardized so the relative survival of heterozygotes is unity and the relative survival of the homozygotes is $1 - s$, where s is the selection coefficient (Hedrick 2000). In this case, the estimate of s is 0.085.

Inbreeding affected the probability of infection (Figure 17.2) in the *M. cerebralis* trials (there was no significant effect for inbred vs. outbred for either the bacterium or IHNV). For the *M. cerebralis* trials, four of five families provided evidence that outbred fish had a higher resistance than inbred fish, as did the analysis of all families together. We can evaluate this effect by comparing the mean values of resistance to *M. cerebralis* for inbred ($\bar{w}_I$) and outbred ($\bar{w}_O$) fish and the ratio here is $\bar{w}_I/\bar{w}_O = 0.55/0.82 = 0.67$, a quite large reduction in the value for inbreds as compared with outbreds. This effect can be further estimated by determining a value analogous to the number of lethal equivalents ($2B$) (Hedrick and Kalinowski 2000). In this case, $2B = (-2\ln(0.55/0.82))/0.25 = 3.20$, or there are slightly over three genes (or a larger number of genes each with a smaller effect) that would result in no resistance when homozygous.

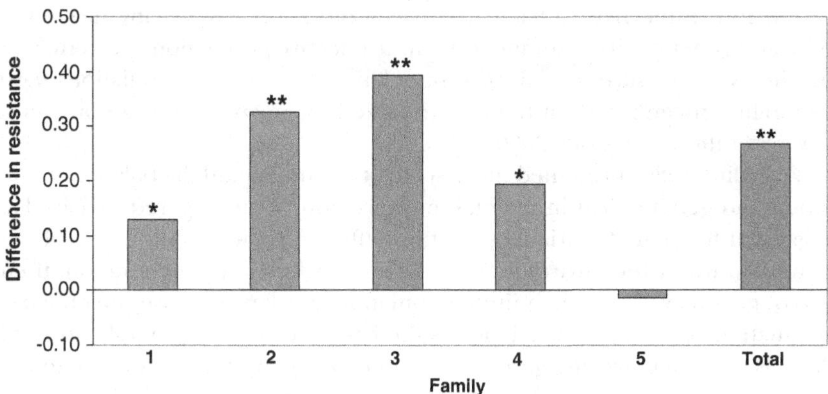

Figure 17.2. Difference in survival or resistance between inbred and outbred fish after exposure to the whirling disease parasite *M. cerebralis* for the five families with both inbred and outbred offspring where * indicates $P < 0.05$ and ** indicates $P < 0.005$ (from Arkush *et al.* 2002).

These results demonstrate that loss of genetic selection can result in increased susceptibility to disease and subsequent mortality. If there are further losses of genetic variation because of continued small population size in winter-run Chinook salmon, then one would predict both a higher level of homozygosity at MHC genes and higher levels of inbreeding. As a result, these known pathogens, and potentially others that may infect winter-run, may cause a decline in the number of winter-run chinook salmon and an additional extinction threat to this critically endangered species.

For species with a history of small population size, such as many endangered species, there also may be both low MHC variation and high inbreeding, suggesting that susceptibility to pathogens may result in a further reduction of fitness. In addition, endangered species may exist in stressful or marginal environments, potentially causing greater susceptibility to infectious disease. Finally, a number of pathogens appear to have host reservoirs in more common species, game species, livestock, or pets, and they may transmit pathogens to highly susceptible endangered species when they come in contact. For all these reasons, the low resistance of endangered species to pathogens may constitute a grave threat in coming decades.

17.3.4 Estimating long-term effective population size

Conserving genetic variation has been a major focus of recovery efforts for many endangered species. Retaining variation for adaptation to environmental change is of great concern, particularly because many imperiled taxa are in recently altered habitats and exposed to new biological threats, including nonnative predators, competitors, and pathogens. In general, the amount of genetic variation within a population available for future adaptation results from a balance between mutation introducing new variation, and genetic drift, resulting from finite effective population size, reducing it. Efforts to measure the effective population size in wild populations have generally concentrated on the effective size in a given generation or over a few generations (Hedrick 2000).

Franklin (1980) presented the first effort to understand the balance of mutation and genetic drift in maintaining variation of endangered species. He suggested for neutral variants that if the effect of new mutations was about a thousandth of the environmental variance in fitness per generation, then loss of genetic variation in a finite population is balanced when the effective population size (N_e) is 500. This was the basis for his very general choice of $N_e = 500$ for maintaining genetic variation. However, N_e is only equal to the adult breeding number if, from generation to generation, individuals at the same life stage are produced at random. For most organisms there is a higher variance in contribution than predicted from random because of unequal sex ratio, high variance in mating success, fecundity or progeny survival over individuals, and other factors. In addition, N_e over time (generations) depends on the harmonic mean for each generation, a value that may be far lower than the arithmetic mean over generations (Hedrick 2000).

Caution should be used in discussing N_e because many important parameters influencing it are not well understood and the actual N_e may be only a fraction of the total population of adults. For example, Frankham (1995), in a review of published estimates, suggested N_e is only about 10% that of adult population size. Within-generation estimates of the ratio of N_e to adult numbers often appear higher than 0.10 (Vucetich *et al.* 1997), but for maintaining genetic variation in the long term, variance in N_e over time should be included (as in many estimates by Frankham 1995; see also Kalinowski and Waples 2002). In other words, to maintain genetic variation in a population with an N_e of 500, even assuming that all variance generated by new mutations was potentially adaptive, would require a census population (N_c) size of ~5000 adults per generation.

In addition, Lande (1995) suggested as much as 90% of the increase in genetic variance by mutation over time may be caused by changes that unconditionally reduce fitness, so most new variation is unavailable for adaptive change. Based on this assumption, he thought $N_e = 5000$ may be required to maintain potentially adaptive genetic variation. Lynch and Lande (1998) also pointed out that the mutation rate for some traits, such as genes that may confer disease resistance, may be a thousandfold lower than for quantitative traits, making numbers necessary to maintain variation for these a thousandfold higher.

The amount of genetic variation potentially available for adaptation is determined by the long-term, or what could be called the evolutionary effective population size. The long-term effective size may not be reflected in estimates of contemporary effective population size or as some percentage of the contemporary census number because the ancestral effective population size, particularly for endangered species, could be much larger than contemporary estimates. Below, I will show how molecular data have been used to estimate the long-term effective population size in the endangered big fishes of the Colorado River and show that the contemporary effective size or census number are only a very small proportion of this estimated long-term effective size.

The four big fishes of the Colorado River, humpback chub, bonytail chub, razorback sucker, and Colorado pikeminnow (Colorado squawfish), are all endangered and have been in deep decline in recent decades (Minckley *et al.* 2003). The last wild Colorado pikeminnow in the lower Colorado River was caught in 1975. Bonytail chub are critically imperiled, persisting only in Lake Mohave, Arizona–Nevada, and perhaps Lake Havasu, Arizona–California, as a few wild fish augmented by hatchery reintroductions. Humpback chubs are represented by one viable population in the Little Colorado River–Grand Canyon complex. This population hovered near 10 000 adults into the early 1990s, but recently is thought to have declined substantially. Although annual spawning of razorback suckers occurs, the population consists mostly of large, very old adults and there is no evidence of recruitment success. A large population apparently formed when Lake Mohave filled in the early 1950s but annual estimates of adult fish have consistently declined to 9086 in 1999 (Minckley *et al.* 2003). Indications are that historically there were large populations of all four of these endangered big fishes in the lower Colorado River as late as the mid twentieth century (Minckley *et al.* 2003).

To estimate the long-term effective population size in three of these fishes (no samples of Colorado pikeminnow were available), Garrigan *et al.* (2002) examined mitochondrial DNA (mtDNA) sequence data and used the maximum likelihood approach in the program FLUCTUATE (Kuhner *et al.* 1998). Because the generation time is long for these fishes, and the decline is only over the past few generations, one would expect that estimates of the long-term effective population size from molecular data would represent the evolutionary important effective size. The maximum likelihood method assumes that new sequence variants appear by mutation and are eliminated by genetic drift. For a given mutation rate and N_e, a sample of mtDNA sequences should thus exhibit an appropriate pattern of pairwise differences. These long-term estimates are of the effective population size for a species throughout a substantial portion of its evolutionary history and do not necessarily reflect the historical or recent effective population sizes.

Examination of mtDNA sequence variation in samples of bonytail, humpback, and razorback showed substantial variation (Garrigan *et al.* 2002). For humpback chub, bonytail chub, and razorback sucker, there were 5, 3, and 10 haplotypes found in samples of 18, 16, and 49 fish, respectively (Table 17.3). Figure 17.3 depicts the maximum likelihood genealogies for the three species for these samples. For example for bonytail chub, the three haplotypes, Zx, Zz, and Yy are represented by 4, 7, and 5, individuals in the sample of 16.

Table 17.3. *Estimates of mtDNA variation in three endangered big fishes from the Colorado River*

Using these data, the maximum likelihood estimates are given of the long-term effective female population size, N_{ef}, if the population is assumed constant over evolutionary time and if the population is allowed to grow or contract and the direction of that change.

	Species		
	Humpback chub	Bonytail chub	Razorback sucker
Data			
mtDNA gene	ND2	ND2	cytb
No. of nucleotides	790	763	311
Sample size	18	16	49
No. of haplotypes	5	3	10
Estimates of N_{ef}			
N_{ef} (constant size)	97 500	89 500	669 000
N_{ef} (variable)	149 000	61 900	940 300
Population trend	Stable	Declining	Expanding

Source: Garrigan *et al.* 2002.

The humpback chub and razorback genealogies show similarity in that the rare haplotypes are the most divergent and the most common haplotypes are closely related. In addition, the humpback chub and razorback sucker show similar divergence over all sequences, about 1.5 nucleotides between all pairwise comparisons, while the bonytail had an average of 2.8 nucleotide differences.

If we conservatively assume that the mutation rate is 2×10^{-8} per nucleotide, then using the program FLUCTUATE, we can estimate the long-term female (mtDNA is maternally inherited in fish) effective population size from these data (Table 17.3). If the population size is assumed to be constant over evolutionary time, then estimates of female effective size are 97 500, 89 500, and 669 000 for humpback, bonytail, and razorback, respectively. Assuming that there are equal sex ratios for these species, then the overall effective population size should be about twice this value. When population growth is taken into account, the estimates suggest that bonytail chub has been declining and razorback sucker expanding in numbers over evolutionary time (Table 17.3). Overall, this analysis of genetic data suggests the three species existed in large numbers until only recently. Although there is less genetic variation for bonytail chub, and the estimates of effective population size are the smallest for it, the three remaining haplotypes are quite divergent, suggesting that present genetic variation still reflects a high degree of the variation present ancestrally. Variation in the other two species remains even more intact.

Many endangered species, such as these big fishes, may have evolved when their population size was much larger than presently or even known historically. As a result they may have substantial genetic variation remaining for

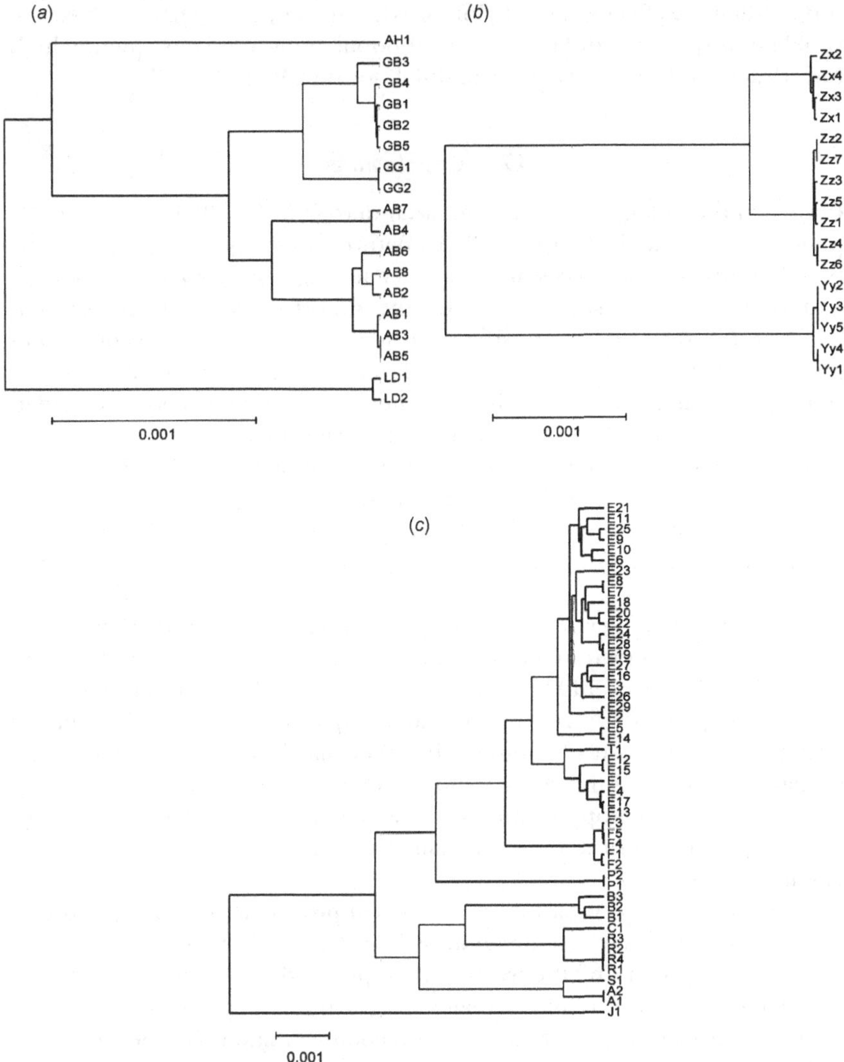

Figure 17.3. Coalescent genealogies for (*a*) humpback chub, (*b*) bonytail chub, and (*c*) razorback sucker that maximize the likelihood of the mtDNA data (from Garrigan *et al.* 2002). Branch lengths are scaled in terms of the number of substitutions per site. The letters of the tree branches represent the names of the haplotype and numbers represent individuals with those sequences. Unlike a phylogenetic tree, the distance between identical sequences represents the time in generations since a common ancestor.

potential adaptation but this variation could be quickly lost if the contemporary population size is small. Further, if the current effective population size is reduced, either naturally or through inappropriate management, genetic variation will be diminished and less new variation generated, increasing potential for reduced fitness due to fixation of detrimental alleles. Such reductions in fitness when the effective population size declines appear a particularly severe problem in species with large ancestral populations and consequently high historical genetic loads (Hedrick and Kalinowski 2000).

17.4 Conclusions

Conservation biology has been somewhat derogatorily called a "crisis" discipline or an "applied" discipline. By its nature, conservation biology may be both but this does not necessarily make it a less rigorous approach to biology than other areas of research. To make the important decisions in a crisis, it is important to utilize the most current ideas and research in ecology and genetics for endangered species. To apply techniques and theory from recent research in population biology, it is both necessary to understand them and to interpret their usefulness for endangered species. In some ways, research in conservation biology is more difficult than population biology in other species because there is less opportunity to experiment, sample sizes are smaller, less is known about the species, etc. Further, the future of an endangered species may critically depend upon any recommendations made from the research.

However, all of what is learned from population biology cannot be automatically used in conservation biology. For example, mathematical models from population biology should be used only as guidelines for decisions in conservation biology. Similarly, research in model organisms may provide heuristic examples of important phenomena but they may lack biological factors essential to consider in the conservation of endangered species. Alternatively, there may be unique opportunities to understand the population biology of small populations in the detailed examination of particular populations of endangered species.

The examples and topics I have discussed provide a personal glimpse of the important issues in conservation biology. Obviously, there is much more to understand, but using the framework of population biology with both traditional and new techniques, greater insight into the problems facing endangered species may be obtained and recommendations for their recovery achieved.

17.5 Acknowledgments

The context provided by population biology, and its architects Dick Lewontin and Dick Levins, is essential for conservation biology and the appropriate evaluation and examination of threats to endangered species. The inspiration for much of my research over the past decade was W. L. Minckley, a great scientist who dedicated his life to conserving the native fishes of the southwestern United States. I also wish to express my thanks to the colleagues who have contributed to the research that I have discussed here: in particular, Kristen Arkush, Dan Garrigan, and Steven Kalinowski.

REFERENCES

Arkush, K. D., Giese, A. R., Mendonca, H. L., McBride, A. M., Marty, G. D., and Hedrick, P. W. (2002). Resistance to three pathogens in the endangered winter-run Chinook salmon: effects of inbreeding and MHC genotypes. *Canad. J. Fish. Aquat. Sci.* 59:966–75.

Ballou, J. D., Gilpin, M., and Foose, T. J. (eds) (1995). *Population Management for Survival and Recovery.* New York: Columbia University Press.

Beissinger, S. R., and McCullough, D. R. (eds) (2002). *Population Viability Analysis.* Chicago, IL: University of Chicago Press.

Bijlsma, R., Bundgaard, J., and Van Putten, W. J. (1999). Environmental dependence of inbreeding depression and purging in *Drosophila melanogaster. J. Evol. Biol.* 12:1125–37.

Carrington, M., Nelson, G. W., Martin, M. P., Kissner, T., Vlahov, D., Goedert, J. J., Kaslow, R., Buchbinder, S., Hoots, K., and O'Brien, S. J. (1999). *HLA* and HIV-1: heterozygote advantage and *B*35-Cw*04* disadvantage. *Science* 238:1748–52.

Caughley, G. (1994). Directions in conservation biology. *J. Anim. Ecol.* 63:215–44.

Culver, M., Johnson, W. E., Pecon-Slattery, J., and O'Brien, S. J. (2000). Genomic ancestry of the American puma (*Puma concolor*). *J. Hered.* 91:186–97.

Diamond, J. (1989). Overview of recent extinctions. In D. Western and M. C. Pearl (eds) *Conservation for the Twenty-first Century,* pp. 37–41. New York: Oxford University Press.

Ebert, D., Haag, C., Kirkpatrick, M., Riek, M., Hottinger, J. W., and Pajunen, V. I. (2002). A selective advantage to immigrant genes in a *Daphnia* metapopulation. *Science* 295:485–8.

Edwards, S. V., and Hedrick, P. W. (1998). Evolution and ecology of MHC molecules: from genomics to sexual selection. *Trends Ecol. Evol.* 13:305–11.

Frankel, O. H., and M. E. Soulé (1981). *Conservation and Evolution.* Cambridge: Cambridge University Press.

Frankham, R. (1995). Effective population size/adult population size ratio in wildlife: a review. *Genet. Res.* 66:95–107.

Franklin, I. R. (1980). Evolutionary change in small populations. In M. E. Soulé and B. A. Wilcox (eds) *Conservation Biology: an Evolutionary–Ecological Perspective,* pp. 135–49. Sunderland, MA: Sinauer.

Gao, X., Nelson, G. W., Karacki, P., Martin, M. P., Phair, J., Kaslow, R., Goedert, J. J., Buchbinder, S., Hoots, K., Vlahov, D., O'Brien, S. J., and Carrington, M. (2001). Effect of a single amino acid substitution in the MHC class I molecule on the rate of progression to AIDS. *New England J. Med.* 344:1668–75.

Garrigan, D., Marsh, P. C., and Dowling, T. E. (2002). Long-term effective population size of three endangered Colorado River fishes. *Anim. Cons.* 5:95–102.

Gilpin, M. E., and Soulé, M. E. (1986). Minimum viable populations: processes of species extinction. In M. E. Soulé (ed.) *Conservation Biology: the Science of Scarcity and Diversity,* pp. 19–34. Sunderland, MA: Sinauer.

Hedrick, P. W. (1984). *Population Biology: the Evolution and Ecology of Populations.* Boston, MA: Jones and Bartlett.

Hedrick, P. W. (1995). Gene flow and genetic restoration: the Florida panther as a case study. *Cons. Biol.* 9:996–1007.

Hedrick, P. W. (2000). *Genetics of Populations,* 2nd edition Boston, MA: Jones and Bartlett.

Hedrick, P. W. (2001). Conservation genetics: where are we now? *Trends Ecol. Evol.* 16:629–36.

Hedrick, P. W., and Kalinowski, S. T. (2000). Inbreeding depression in conservation biology. *Annu. Rev. Ecol. Syst.* 31:139–62.

Hedrick, P. W., and Kim, T. J. (2000). Genetics of complex polymorphisms: parasites and maintenance of the major histocompatibility complex variation. In R. S. Singh and C. B. Krimbas (eds) *Evolutionary Genetics: From Molecules to Morphology*, pp. 204–34. Cambridge: Cambridge University Press.

Hedrick, P. W., Lacy, R. C., Allendorf, F. W., and Soulé, M. E. (1996). Directions in conservation biology: comments on Caughley. *Cons. Biol.* 10:1312–20.

Jimenez, J. A., Hughes, K. A., Alaks, G., Graham, L., and Lacy, R. C. (1994). An experimental study of inbreeding depression in a natural habitat. *Science* 266:271–3.

Kalinowski, S. T., and Waples, R. S. (2002). Relationship of effective to census size in fluctuating populations. *Cons. Biol.* 16:129–36.

Keller, L. F. (1998). Inbreeding and its fitness effects in an insular population of song sparrows (*Melospiza melodia*). *Evolution* 52:240–50.

Keller, L. F., and Waller, D. M. (2002). Inbreeding effects in wild populations. *Trends Ecol. Evol.* 17:230–41.

Keller, L. F., Arcese, P., Smith, J. M. N., Hochachka, W. M., and Stearns, S. C. (1994). Selection against inbred song sparrows during a natural population bottleneck. *Nature* 372:356–7.

Kuhner, M. K., Felsenstein, J., and Yamato, J. (1998). Maximum likelihood estimation of population growth rates based on the coalescent. *Genetics* 149:429–34.

Lacy, R. C. (1997). Importance of genetic variation to the viability of mammalian populations. *J. Mammal.* 78:320–35.

Lacy, R. C., and Miller, P. S. (2002). Incorporation of human populations and activities into population viability analysis. In Beissinger, S. R. and McCullough, D. R. (eds) *Population Viability Analysis*, pp. 490–510. Chicago, IL: University of Chicago Press.

Lacy, R. C., Hughes, K. A., and Miller, P. S. (1995). VORTEX: a stochastic simulation of the extinction process: version 7 user's manual. ICUN/SSC Conservation Breeding Specialist Group. Apple Valley, Minnesota.

Lafferty, K. D., and Gerber, L. (2002). Good medicine for conservation biology: the intersection of epidemiology and conservation theory. *Cons. Biol.* 16:593–604.

Land, D., Cunningham, M., Lotz, M., and Shindle, D. (2001). Florida panther genetic restoration and management: annual report 2000–2001. Florida Fish and Wildlife Conservation Commission.

Land, E. D., and Lacy, R. C. (2000). Introgression level achieved through Florida panther genetic restoration. *Endangered Species Update* 17:100–5.

Lande, R. (1988). Genetics and demography in biological conservation. *Science* 241:1455–9.

Lande, R. (1993). Risks of population extinction from demographic and environmental stochasticity and random catastrophes. *Am. Nat.* 142:911–27.

Lande, R. (1994). Risk of population extinction from fixation of new deleterious mutation. *Evolution* 48:1460–9.

Lande, R. (1995). Mutation and conservation. *Cons. Biol.* 9:782–91.

Laurenson, K., Sillero-Zubiri, C., Thompson, H., Shiferaw, F., Thirgood, S., and Malcolm, J. (1998). Disease as a threat to endangered species: Ethiopian wolves, domestic dogs and canine pathogens. *Anim. Cons.* 1:273–80.

Lynch, M., and Lande, R. (1998). The critical effective size for a genetically secure population. *Anim. Cons.* 1:70–2.

Madsen, T., Shine, R., Olsson, M., and Wittsell, H. (1999). Restoration of an inbred adder population. *Nature* 402:34–5.

Marr, A. B., Keller, L. F., and Arcese, P. (2002). Heterosis and outbreeding depression in descendants of natural immigrants to an inbred population of song sparrows (*Melospiza melodia*). *Evolution* 56:131–42.

Meagher, S., Penn, D. J., and Potts, W. K. (2000). Male–male competition magnifies inbreeding depression in wild house mice. *Proc. Natl. Acad. Sci.* 97:3324–9.

Minckley, W. L., Deacon, J. E., Dowling, T. E., Hedrick, P. W., Marsh, P. C., Matthews, W. J., and Mueller, G. (2003). A conservation plan for lower Colorado River fishes. *BioScience* 53:219–34.

Ralls, K., Ballou, J. D., and Templeton, A. R. (1988). Estimates of lethal equivalents and the cost of inbreeding in mammals. *Cons. Biol.* 2:185–93.

Ralls, K., Brugger, K., and Ballou, J. (1979). Inbreeding and juvenile mortality in small populations of ungulates. *Science* 206:1101–3.

Richards, C. M. (2000). Inbreeding depression and genetic rescue in a plant metapopulation. *Am. Nat.* 155:383–94.

Roelke, M. E., Martenson, J. S., and O'Brien, S. J. (1993). The consequences of demographic reduction and genetic depletion in the endangered Florida panther. *Curr. Biol.* 3:340–50.

Saccheri, I. J., and Brakefield, P. M. (2002). Rapid spread of immigrant genomes into inbred populations. *Proc. R. Soc. Lond. Ser. B* 269:1073–8.

Schonewald-Cox, C. M., Chambers, S. M., MacBryde, B., and Thomas, L. (eds) (1983). *Genetics and Conservation: A Reference for Managing Wild Animal and Plant Populations.* London: Benjamin-Cummings.

Shaffer, M. L. (1981). Minimum population sizes for species conservation. *BioScience* 31:131–4.

Simberloff, D. (1988). The contribution of population and community biology to conservation science. *Annu. Rev. Ecol. Syst.* 19:473–511.

Soulé, M. E. (ed.) (1986). *Conservation Biology: the Science of Scarcity and Diversity.* Sunderland, MA: Sinauer.

Soulé, M. E. (ed.) (1987). *Viable Populations for Conservation.* New York: Cambridge University Press.

Soulé, M. E., and Wilcox, B. A. (eds) (1980). *Conservation Biology: an Ecological–Evolutionary Perspective.* Sunderland, MA: Sinauer.

Thurz, M. R., Thomas, H. C., Greenwood, B. M., and Hill, A. V. S. (1997). Heterozygote advantage for HLA class-II type in hepatitis B virus infection. *Nat. Genet.* 17:11–2.

Vucetich, J. A., Waite, T. A., and Nunney, L. (1997). Fluctuation in population size and the ratio of effective to census population size. *Evolution* 51:2017–21.

Westemeier, R. L., Brown, J. D., Simpson, S. A., Esker, T. L., Jansen, R. W., Walk, J. W., Kershner, E. L., Bouzat, J. L., and Paige, K. N. (1998). Tracking the long-term decline and recovery of an isolated population. *Science* 282:1695–8.

18

The emergence of modern human mortality patterns

SHRIPAD TULJAPURKAR
Department of Biological Sciences, Stanford University

18.1 Introduction

In 1977, Dick Lewontin started me thinking about human mortality patterns with an introduction to Thomas McKeown's then new analysis of mortality declines in England (McKeown 1976). McKeown proposed three features of mortality decline that have become influential: there had been long-run steady progress against mortality since the late eighteenth century; medical technology did not play a central or even important role in mortality decline through the early twentieth century; and the principal driver of mortality decline in England had been a general increase in the standard of living. We now know that McKeown was wrong about his third point but right about his first two points, which demonstrate the importance in mortality studies of analyzing long-run change and of multifactorial explanation.

In recent years human mortality has become a popular topic because of economic concern that declining mortality and fertility may result in unprecedented percentages of older people. This concern and the resulting research funding have energized the study of mortality decline and aging. These subjects can be studied at several levels, from populations to genes. My own work has concerned mortality trends and mortality forecasts, the design of public pension systems, and the development of biodemography – an attempt to integrate demographic, evolutionary, and molecular information about mortality patterns. This chapter describes issues and problems that are important and intriguing in mortality analysis: those concerning trends, causes, forecasts, limits, and evolution. It is useful to start by describing trends in modern human mortality patterns and the questions that arise when one tries to understand them. Next, we discuss causes of mortality decline and demographic information about its limits. The chapter closes with a discussion of our limited understanding of the evolution of senescence. References are deliberately skimpy but the reader can find an extensive bibliography in the paper by Tuljapurkar and Boe (1999).

The Evolution of Population Biology, ed. R. S. Singh and M. K. Uyenoyama. Published by Cambridge University Press.

18.2 Modern human mortality

The rich industrialized countries of the world experienced a remarkable decline in mortality over the past century, notwithstanding two world wars and several lesser wars. The nature of this decline is illustrated by US mortality change from 1930 to 1996. Figure 18.1 shows that mortality decline favored the young: the top two lines in the figure show that the chance of survival from birth to age 45 increased from 79% in 1930 to 94% in 1996. Progress has been slower at higher ages: in 1996 there was an 85% probability of surviving from the "middle" age of 45 to a retirement age of 65 but only a 41% chance of making it from retirement to age 85. "Mortality" here is sexes-combined period (annual) mortality for the entire country. At any given time, there is much variation between sexes, between populations living in different states, and so on, but over the entire period there is an essentially parallel decline in mortality at such levels of disaggregation. Notice the slowing of the rate of change in those survival probabilities that have approached 1, whereas curves for the older age intervals look roughly linear even in recent years. This age pattern of mortality change is reflected in the expectation of life, which is just the number of years remaining until the average age at death. Figure 18.2 shows the gains in the expectation of life starting at birth, age 45 and age

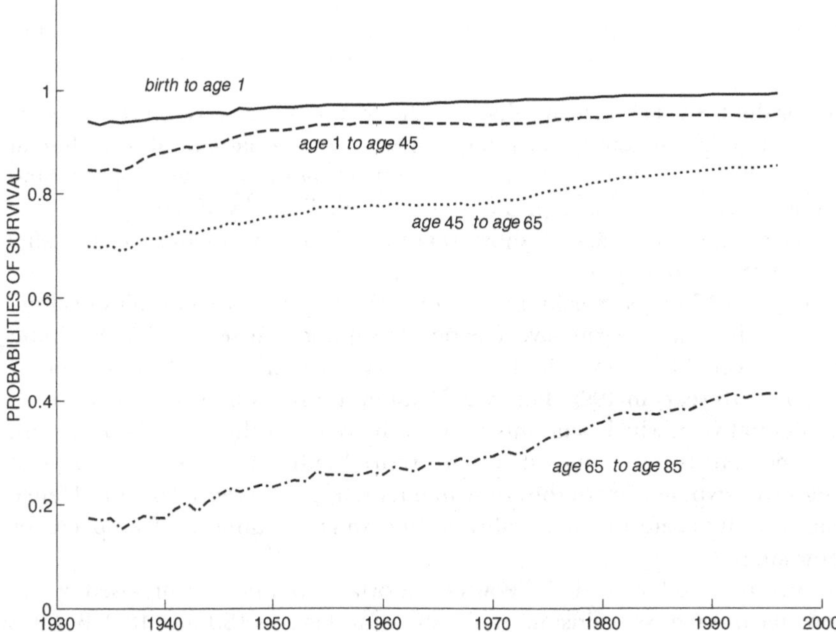

Figure 18.1. Probabilities of survival over particular age intervals corresponding to infants, young adults, middle age, and older ages. These are computed using period data, both sexes together, for the USA in the twentieth century.

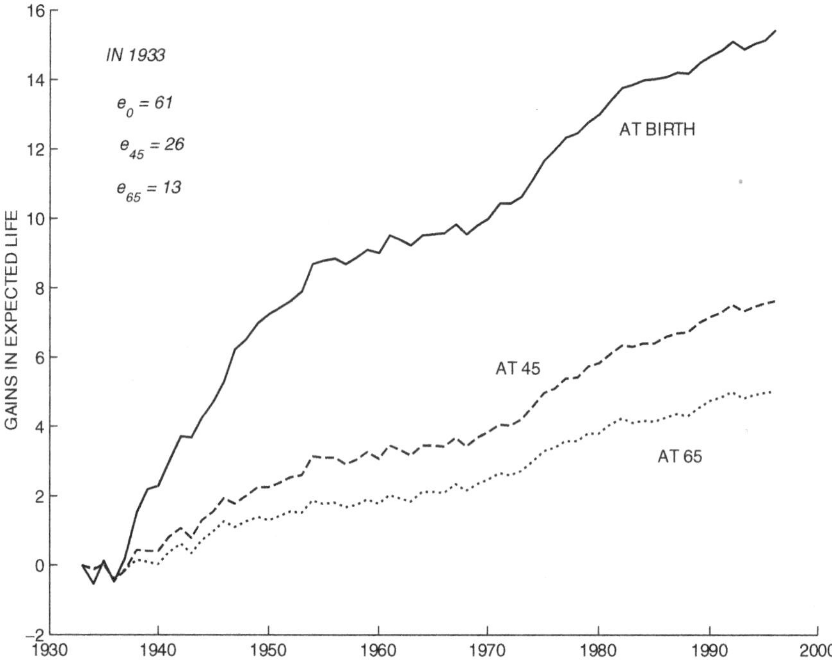

Figure 18.2. Gains in the remaining period life expectancy starting at birth, age 45 (for middle age), and age 65 (retirement) in the USA, both sexes combined.

65 for the USA. The concave shape of the first of these curves in recent years reflects the progressively declining scope for additional mortality decline at young ages. To some people this concavity implies that we are approaching a limit to lifespan – this implication is false: the linearity of the curve for the expectation of life at age 65 shows that there is plenty of scope for mortality reduction at older ages.

Figure 18.2 helps explain the economic consequences of mortality decline. If you retire at age 65 you have to expect to support yourself for the remaining expectation of life, shown in the lowest curve in Figure 18.2. That is a period of about 18 years in 1996, but by 2016 a new retiree will need to plan for an additional year, which amounts to a 5% increase in the cost of retirement. Five percent is real money if you are retired and is big money in terms of aggregate expenditure by the government in support of retired people. Hence the current debate about mortality decline: will it continue, at what speed, for how long?

Anyone who has read McKeown on mortality will not be surprised by the long-term progress against mortality shown in Figures 18.1 and 18.2. But it is surprising how steady that progress has been, a discovery made by Lee and Carter (1992). Their original approach to the problem is a classic example of good demographic analysis, but we will tell their story in a different way. The data on twentieth century US mortality consist of a set of age-specific central death rates $m(x, t)$ for each age x and year t in the period. If mortality decline is strongly correlated across all ages over time, there is an underlying time signal that captures the effect of all the drivers of mortality change. We can see if this is so by doing a singular value decomposition. We work with log $m(x, t)$ because of the huge variation in death rates over age. For the sample period (1930 to 1996) compute a time average, $a(x)$, for each age, as

$$a(x) = (1/T) \sum \log m(x, t).$$

Then there is an exact decomposition

$$\log m(x, t) = a(x) + s_1 u_1(x) v_1(t) + s_2 u_2(x) v_2(t) + \dots,$$

where the s_i, called singular values, are real numbers (greater than or equal to zero) arranged in declining value, and the $u_i(x)$, $v_i(t)$ are vectors with dimensions along age and time respectively. These vectors are known to form orthogonal sets, meaning that they are independent vectors within the sample. We may think of the successive $v_i(t)$ as independent "time signals" in the data, whose importance relative to the other $v_i(t)$ is measured by the singular values. If we find that the value of s_1 is much larger than that of all the other singular values, then the first vector $v_1(t)$ is the dominant signal. The proportion of temporal variance in the data explained by $v_1(t)$ is given by the ratio

$$s_1^2/(s_1^2 + s_2^2 + \dots).$$

What Lee and Carter found in the USA was that the first singular value accounts for over 97% of the variation in mortality rates over time, so we have an effectively one-dimensional pattern of mortality change. Thus we can represent mortality change to a good approximation by using just the first singular value and write

$$\log m(x, t) = a(x) + b(x)k(t) + e(t)$$

where $b(x)$ is an age schedule, $k(t)$ is a vector with as many entries as years in the data, and $e(t)$ is a residual error term. We normalize these so that the $b(x)$ sum to 1, and the $k(t)$ sum to zero. The first conclusion of this analysis is that the logarithm of mortality at each age changes in some fixed proportion to the temporal signal $k(t)$. The values of $k(t)$ are shown in Figure 18.3 (which was computed using US data from 1930 to 1996, with some years added to the original Lee–Carter data). Observe that $k(t)$ shows a strikingly regular, nearly linear decline over the period.

What is even more striking is that Tuljapurkar *et al.* (2000) found a similar one-dimensional pattern of mortality decline in all of the G7 industrialized countries from about 1955 through 1997. Thus the pattern – exponential decline in death rates at every age with a nearly constant rate of decline at each age – is not unique to the USA but holds for all the rich industrialized G7

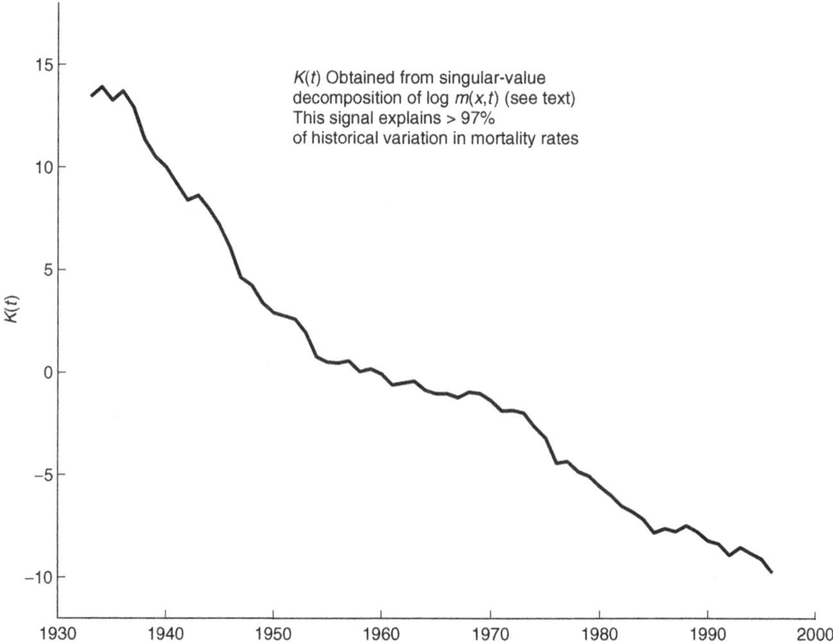

Figure 18.3. Mortality at every age declined at a roughly constant rate in the last century in the USA and elsewhere. The curve shown here is the common time factor of decline for the USA: an age-specific multiplier turns the quantity shown into an age-specific mortality.

countries. We have since found this pattern to hold for many other industrialized countries.

What can we make of this effectively one-dimensional, regular pattern of decline? As seen in Figure 18.3, the dominant time signal is roughly linear with time over the century; there are intervals of relatively slow or fast change but no systematic persistent deviation from linearity. One important conclusion is that we can and should use this linear regularity as a basis for making mortality forecasts, although such forecasts must be stochastic in order to include the short-term variability evident in Figure 18.3. It is straightforward to include stochastic variation around the linear trend in terms of a time-series model. The papers cited above make such forecasts, concluding that mortality decline is likely to be significantly larger than is anticipated by most official forecasts. History may not repeat, of course, and there is no way to tell in advance. A common argument against such forecasts is that the future will differ from the past because we are on the threshold of revolutionary change driven by genetic and other technologies. This is not a convincing argument – if anything, the twentieth century was surely revolutionary in all these ways. Retrospective analysis, in which we use a portion of the past data to forecast a subsequent observed span of years, supports forecasts based on a constant exponential trend. But this is not convincing to those who see the future as simply unknowable.

18.3 Causes of mortality decline

Forecasts that use the regularity of mortality change in the industrialized countries are simply a mechanical exploitation of a solidly established pattern. At a more fundamental level, why do we observe such regular mortality decline? Tuljapurkar *et al.* (2000) present an explanation for this based on the nature of mortality change and the roughly parallel socioeconomic development that has occurred in the industrialized countries since about 1950. Mortality declines are driven by a variety of factors – public health, living standards and arrangements, medical care, knowledge about health and illness, and so forth – that are in turn supported by a large and increasing social commitment of resources. Over time, these resources are deployed in order to reduce the most significant prevailing causes of death: childhood diseases in the early twentieth century, then infectious diseases, later still diseases of adulthood and chronic disease. Although the level of resources so deployed has risen and continues to rise rapidly, the complexity of the diseases we confront has also increased – heart disease, for example, demands far more expensive analysis and technology than did, say, bacterial infections in early childhood. As a result of this complexity the marginal effectiveness of additional resources, as measured by their impact on mortality levels, goes down over time. We argue that the twentieth century has been a period of balance between two trends – increasing resources, declining marginal effectiveness – resulting in a largely steady rate of mortality decline.

This hypothesis is persuasive (at least to us) but is difficult to test. More generally, it is notoriously difficult to identify the causes of mortality decline. To illustrate, recall McKeown's claim that a rise in living standards drove mortality decline in England from the late eighteenth to the early twentieth centuries. Sretzer (1988) showed that over long stretches of this period mortality fell even when real wages stagnated; the real driver of mortality decline in the late nineteenth century was effective public health management in urban areas. A similar pattern has been documented in the USA by Preston and Haines (1992). Yet the "living standards" argument refuses to die and has been resurrected most recently, in more sophisticated form, by Fogel and Costa (1997). This is not a surprise, given the secular nature of declining mortality and rising living standards and the inevitable correlations that must exist between the two.

A central difficulty in the study of causes is the absence of a theoretical model that captures the hierarchical nature of the factors (environments, interventions, genes) that determine mortality. Mosley and Chen (1984) developed a useful structural model for child mortality in developing countries, and a somewhat more general model of proximate determinants was developed by Landers (1992) for studies of preindustrial and early industrial England. Both frameworks were intended, in different ways, to cope with a limited amount of data. However, the identification of causes has not been made easier in more recent times by an explosion of data on putative risk factors, morbidity, mortality, and even genetics. For the industrialized countries, there is now something of a divide between studies that focus on longitudinal samples that contain extensive data on individual histories – these are analyzed using complex hazard models that are difficult to interpret, compare or generalize – and studies that use aggregated data on populations over time. The collection of ever more detailed data has become a preoccupation and I sometimes think it is a substitute for the development of good theory. A new wave of surveys that includes the gathering of DNA samples has now been launched, and these will undoubtedly muddy the waters with a host of variance partitioning studies. If these data are to be truly revealing, much more attention is needed to theory that can connect different levels of aggregation. Without them, analysis tends to be shaped by the nature of the data, by the appeal of complex statistical procedures, by some overriding organizational perspective such as economic utility theory, or by the investigator's strong attachment to one (or a few) driving factors.

18.4 Old age mortality and limits to lifespan

If we are right about a continued trend in mortality decline, how far will it go? One way to think about this is to consider a scenario. For example, Diane Wagener (another Lewontin alumna) and colleagues (Wagener *et al.* 2001) ask: if we eliminate mortality at ages up to 65 in the USA, what would be the expectation of life at birth? Their answer is 81 years, which seems a modest increase from 76 years in 1996. The increase they compute seems even more modest when we note that the expectation of life in Japan is already close to 81 years. But their answer depends on two assumptions: how mortality increases with age at ages 85+, and the correlation between mortality at 85+ and the level of mortality at younger ages. Wagener *et al.* do not say what they did, but these assumptions really determine the answer to their "what-if" question. Both are aspects of a debate about the limits to lifespan.

The age pattern of mortality at old ages is not easy to measure or to specify on a priori theoretical grounds. Someone who dies this year at, say, age 90 was born in a time when birth records were typically unreliable, certainly in the USA. In addition, mortality at younger ages has been high enough in the past that very few people lived to high ages, so even if we knew ages accurately we would have very small samples to work with. Actuaries have traditionally fallen back on models such as the Gompertz, which asserts that mortality increases exponentially fast with age from age 40 or so on up. We do know (see below) that mortality rises in roughly Gompertzian fashion between ages 65 and 85, though not necessarily at a constant exponential rate. If we assume that mortality continues to rise exponentially with age after age 85, the part of life lived at ages 85+ can contribute only modestly to the expectation of life at birth. To understand why, note that the ratio of the death rates at ages 85 and 65 in the USA in 1996 is 9.3, which means that if the probability of surviving from age 65 to 75 is L (less than 1) then the probability of surviving from age 85 to 95 is roughly L raised to the power 9.3 – this is going to be pretty small unless L is very close to one. There are only three ways in which ages over 85 can contribute much to the expectation of life at birth: mortality rises exponentially with age but the exponent becomes smaller, mortality rises slower than exponentially with age, or the level of mortality at these old ages falls. The last of these would imply a correlation in that old-age mortality must be driven down by some of the factors that act to reduce mortality at younger ages.

Recent work shows that all three things have happened in the latter part of the last century. Analyses of data from countries such as Sweden that have reliable age records and low mortality (Wilmoth *et al.* 2000, Thatcher *et al.* 1998) show that (1) mortality increases with age at a slower than exponential rate as one gets to the oldest ages, and (2) mortality at the oldest ages has been decreasing over time. With these analyses as a basis, it is straightforward to get a much larger increase in the expectation of life than did Wagener *et al.* in their scenario.

Even with good data in low mortality populations, we cannot say whether human mortality at even older ages, say 120 years or more, can or will continue to decline over time. This is one of those ghoulish problems in which you have to wait till enough subjects die in order to find an empirical answer. It is natural to ask if biology, either through observation or principle, provides an answer or useful parallels.

18.5 The evolution of senescence

Biologists have been interested in senescence and aging at least as long as demographers, using cross-species comparisons as well as evolutionary arguments. I consider the data first and then briefly discuss evolution. Comparisons between species have often relied on data of poor quality even though many an ecology text discusses "characteristic" mortality curves and allometric correlations between lifespan and other variables. The accurate measurement of survival rates has rarely been important to biologists, since small cohorts and short generation times have been the hallmark of good model species. The remarkable experiments of Carey *et al.* (1992) were the first to use massive cohorts (a billion flies) of Mediterranean fruit flies and accurately count deaths to measure the age pattern of survival. They found that mortality does not change with age in Gompertz (exponential) fashion. Instead the age pattern of old-age mortality can be very different, often with a significant deceleration of the rate of mortality increase with age. In further studies, Carey and his colleagues have shown that the age trajectory of mortality at old ages is strongly shaped by the environment over the life cycle, including nutrition and reproduction – not surprising in hindsight, perhaps, but not in keeping with the traditional view of biological senescence. The lessons for human mortality are that the shape of old-age mortality is not likely to have been fixed as some upper limit to life, that variability within and between populations in mortality pattern is to be expected, and that even what we consider very old-age mortality is responsive to environmental factors.

Readers will be familiar with the classical view of senescence due to Haldane and Medawar. Tuljapurkar (1997) and others have argued that this view does not adequately explain the age patterns we observe in recent studies of mortality, although there are plenty of people who would disagree. The classical view is a contingent explanation of the length of the life cycle. It says that if we were provided (on the proverbial tablet) the reproductive span of a species, then we would expect that mutations which have deleterious effects later in life could accumulate in higher frequency than those that have deleterious effects early in life. But here "early" and "late" are defined with respect to the reproductive span; in this view that span is life.

There are at least two problems here, which may be illustrated in terms of a quantitative genetics argument (Charlesworth 2001) in which the dynamics of the (mean) phenotypic values in a population evolve in the presence of weak selection – the argument is loosely based on an assumption that many alleles of small effect additively determine the phenotypes of interest. In this model, the phenotype consists of age-specific fertility $f(a)$ and survival $p(a)$, and the fitness function is taken to be total lifetime reproduction $R(f, p)$. Change in any phenotypic component is driven by a sum of two terms, the gradient of fitness R with respect to the phenotype and mutation. If we assume, for example, that the age of last reproduction is fixed at some A, then R is unaffected by changes in $p(a)$ beyond age A. Before age A, for any given assumption about mutation rates and their effects on $p(a)$, $f(a)$, it is possible to find a mutation-selection equilibrium for the phenotype components. A first problem arises if we suppose that reproductive span is itself affected by deleterious mutation. To see this, start with some reproductive span A and find the mutation-selection equilibrium, which will have $p(a) = 0$ for $a > A$. Next, allow mutations that shorten the span A to $B < A$ without reducing overall fitness. The mutation-selection process will now yield an equilibrium in which $p(a) = 0$ for $a > B$. Mutations that reduce the reproductive span will thus take us through a sequence of equilibria to ever smaller reproductive spans, until we are back to bacteria. A second problem, a limitation rather than a flaw, is that the classical theory ignores variation in reproductive span that can be maintained by, for example, temporal variability in the environment. Such variation actually changes the fitness measure relative to the one used in the classical analysis, and an analysis of this fitness leads to a variety of mortality patterns that can be maintained by selection. Examples of such patterns are given in a paper co-written with Steven Orzack, yet another Lewontin alumnus (Orzack and Tuljapurkar 1989).

The first problem above occurs because classical theory ignores the fundamentally coevolutionary nature of life cycles, that the age patterns of fertility and mortality must evolve together. It seems essential to incorporate into the theory mutations of positive effect that, for example, extend the reproductive span. Then we could explore a dynamic coevolutionary process whose equilibria correspond to the complex life-cycle phenotypes that we observe. The second problem occurs because the classical theory ignores the role of the environment in changing the nature of the selection regime and thus of the selective dynamics.

Progress here will come from several directions. To test the classical theory we need to extend and refine theoretical predictions about age-specific genetic covariance patterns (Charlesworth 2001). We also need experiments that measure and characterize the actual genetic covariance in aging cohorts – for example, from the sort of "shotgun" studies that are now possible. We need to develop predictions from theories that take environmental variation into account, and to confront those theories with data gathered on natural populations over time. We should begin to formulate and explore theories of the coevolution of the life cycle in terms of both positive and deleterious mutations. Finally, this kind of evolutionary theory needs to inform the analysis of large scale longitudinal data on humans. These are all elements of a research program in biodemography that is just getting off the ground.

18.6 Acknowledgments

My debt to Dick Lewontin may be larger than that of many, because he actually made it possible for me to switch from doing physics to doing biology and demography. I'm glad I made the switch.

REFERENCES

Carey, J. R, Liedo, P., Orozco, D., and Vaupel, J. W. (1992). Slowing of mortality rates at older ages in large medfly cohorts. *Science* 258:457–61.

Charlesworth, B. (2001). Patterns of age-specific means and genetic variances of mortality rates predicted by the mutation-accumulation theory of aging. *J. Theor. Biol.* 210:47–65.

Fogel, R. W., and Costa, D. L. (1997). A theory of technophysio evolution, with some implications for forecasting population, health care costs, and pension costs. *Demography* 34:49–66.

Landers, J. (1992). Historical epidemiology and the structural analysis of mortality. *Health Transition Rev.* (Supplementary Issue) 2:47–75.

Lee, R. D., and Carter, L. (1992). Modeling and forecasting the time series of U.S. mortality. *J. Am. Stat. Ass.*, 87:659–71.

McKeown, T. (1976). *The Modern Rise of Population.* New York: Academic Press.

Mosley, W. H., and Chen, L. C. (1984). An analytical framework for the study of child survival in developing countries. In W. H. Mosley and L. C. Chen (eds) *Child Survival: Strategies for Research*, pp. 25–48. Supplement to *Pop. Dev. Rev.* Vol. 10.

Orzack, S. H., and Tuljapurkar, S. (1989). Population dynamics in variable environments. VII. Demography and evolution of iteroparity. *Am. Nat.* 133:901–23.

Preston, S., and Haines, M. (1992). *Fatal Years: Child Mortality in Late Nineteenth Century America.* Princeton, NJ: Princeton University Press.

Sretzer, S. (1988). The importance of social intervention in Britain's mortality decline c1850–1914: a re-interpretation of the role of public health. *Social Hist. Med.* 1: 1–38.

Thatcher, A. R., Kannisto, V., and Vaupel, J. W. (1998). *The Force of Mortality at Ages 80 to 120.* Odense: Odense University Press.

Tuljapurkar, S. (1997). Theoretical perspectives on the evolution of senescence. In C. Finch and K. W. Wachter (eds) *From Zeus to the Salmon: Biodemography of Aging.* Washington, DC: National Academy Press.

Tuljapurkar, S., and Boe, C. (1999). Mortality change and forecasting: how much and how little do we know. *N. Am. Actuarial J.* 2:13–47.

Tuljapurkar, S., Li, N., and Boe, C. (2000). A universal pattern of mortality decline in the G7 countries. *Nature* 405:789–92.

Wagener, D. K., Molla, M. T., Crimmins, E. M., Pamuk, E., and Madans, J. H. (2001). *Summary Measures of Population Health: Addressing the First Goal of Healthy People 2010, Improving Health Expectancy.* Healthy People 2010, Statistical Notes No. 22. Hyatsville, MD: National Center for Health Statistics.

Wilmoth, J. R., Deegan, L. J., Lundström, H., and Horiuchi, S. (2000). Increase of maximum life-span in Sweden, 1861–1999. *Science* 289:2366–8.

19

Units of selection and the evolution of virulence

PAUL W. EWALD
Department of Biology, University of Louisville

GREGORY M. COCHRAN
Department of Anthropology, University of Utah, Salt Lake City

19.1 Introduction

In his classic paper on the units of selection Richard Lewontin, following Williams (1966), emphasized that natural selection for characteristics that benefited individual organisms was stronger than selection for characteristics that benefited groups of organisms. One of the most important generalizations from this insight is that natural selection will tend to favor characteristics that benefit the survival and reproduction of individuals (individual selection) over characteristics that benefit group survival or productivity (group selection) when these two benefits are contradictory (Williams 1966, Lewontin 1970). Lewontin went on to suggest that group selection might be particularly strong for parasites because the parasites within a host are a well-defined group and the transmission of such groups may be highly dependent on the characteristics of the group, such as the overall level of harm that the group of parasites imposes on the host; consequently if one wishes to evaluate whether group selection is important in nature, parasites would be appropriate objects for study (Lewontin 1970).

One reason for this expectation derives from the theory of inclusive fitness, which emphasizes how characteristics that benefit other group members can be favored even when they impose a cost on the individual possessing the characteristic as a result of the positive effects of the characteristic on other individuals that harbor the same genetic instructions (Hamilton 1964). Such selection for traits that are at a disadvantage within groups but increase the productivity of the group is often termed kin selection, but the range of situations in which it can act extends beyond groups of individuals that share alleles because they are members of the same family. Intermittent mixing of individuals and subdivision of individuals of the population into smaller groups can generate conditions that favor alleles that benefit the local group at a cost to the individuals within the group (Hamilton 1964, Wilson 1980, Sober and Wilson 1998).

The Evolution of Population Biology, ed. R. S. Singh and M. K. Uyenoyama. Published by Cambridge University Press.

Genetic variation among the parasites within a host can be generated by mutations within clones, inoculation with more than one parasite variant, or transfer of genetic material between clones of parasites that coexist within a host. The competitive success of a parasite may therefore depend on its success at competing with genetically different parasites within the host. But the competitive success of a parasite variant also depends on its success at transmission to new hosts, which in turn depends on the effects on the host of the entire group of parasites within the host. The overall success of a parasite variant therefore depends on the tradeoff between the competitive success of individual parasites within hosts and success of the group of parasites within a host at being transmitted between hosts: that is, a tradeoff between individual selection and group selection.

One of the most important attributes of parasites that can be analyzed from this perspective is parasite virulence, which for the purposes of this chapter is defined to be the inherent harmfulness of a parasite in a specified host population. Most of the analyses of the relationships between parasite virulence and levels of selection assume that increased strength of group level selection will favor decreased virulence. This line of argument emphasizes that variants inside of a host that exploit more of the host's resources will have more resources to invest in reproduction or other activities that increase their competitive advantage over other variants cohabiting the same host individual. The more exploitative variants will therefore tend to increase within the host at the expense of the less exploitative variants. Insofar as use of host resources causes harm to the host, such within-host competition is presumed to lead to increased virulence.

A tradeoff arises because the benefits of victory at within-host competition may be short-lived if the high overall level of exploitation in the host causes severe illness that inhibits transmission to the next host. The requirements for transmission may thus create a tradeoff between the within-host advantages and the transmission disadvantages incurred from intense exploitation of host resources (Levin and Pimentel 1981, Levin *et al.* 1982, Ewald 1983, May and Anderson 1983, Williams and Nesse 1991).

The experience with myxomatosis in Australian rabbits (Fenner and Ratcliffe 1965) has been advanced as the classic illustration of this tradeoff. Strains of myxoma virus that were highly lethal to the Australian rabbits were introduced in an effort to control the rabbit populations. Over a period of decades after this introduction the mortality induced by the myxoma viruses declined. In this case the tradeoff was cast in the context of host death weeding out groups composed of the most virulent clones of myxoma virus (Lewontin 1970, Levin and Pimentel 1981). But even when infections are not lethal the tradeoff between within-host and between-host competition occurs because pathogens will be less able to be transmitted from ill hosts if the illness makes the host less mobile when the pathogens depend on host mobility for transmission (Ewald 1983). This logic leads to the general hypothesis that the virulence to which pathogens evolve depends on the extent to which host mobility is important for transmission. On the basis of this general hypothesis several categories of transmission are predicted to be associated with increased virulence: (1) transmission by biting arthropod vectors, because such vectors transmit pathogens effectively from immobilized hosts; (2) waterborne transmission because attendants of sick hosts and the movement of water supplies can transport pathogens from immobile infected hosts to susceptibles; (3) persistence in the external environment, because durable pathogens can be transmitted from immobile hosts by "waiting" for susceptibles to move to the place in the external environment where they are located; (4) transmission via attendants in hospitals, because such attendants can transport pathogens from immobile infectious patients to susceptibles. In each of these cases comparative studies confirm these predictions (Ewald 1994, 2002, Ewald and Deleo 2002). Application of this tradeoff to the evolution of virulence has also been successful at explaining how virulence is positively associated with the importance of horizontal transmission relative to vertical transmission (Ewald and Shubert 1989, Bull *et al.* 1991, Turner *et al.* 1998, Messenger *et al.* 1999) and in the influence of host density and timing of transmission opportunities on virulence (Ebert 1994, Cooper *et al.* 2002).

19.2 Shared benefits

19.2.1 Virulence tradeoffs with shared benefits

Although the equating of within-host selection with selection for increased virulence and between-host selection as selection for decreased virulence has been a useful starting point for understanding the evolution of virulence, it is inadequate as a basis for a general theory to explain the evolution of virulence. Whether within-host selection leads to increased or decreased virulence depends on whether the benefit to the pathogen that is derived from its activities is shared among the group of conspecific pathogens within the host (Ewald 1995a). Such sharing of pathogen benefits opens the door for "cheater" variants that obtain the shared benefits but do not pay the price of generating the shared benefits. If pathogen products that generate these benefits negatively affect the host, that is, if they are virulence factors, the cheaters will be the benign variants within the group of pathogens. Because the cheaters obtain the benefits of the production of the virulence factor without paying the price they will tend to increase in frequency in a population composed of the more virulent individuals. Within-host selection in this case can favor decreased virulence.

19.2.2 An illustration: shared benefits of cholera toxin

Disease caused by *Vibrio cholerae,* the agent of cholera, illustrates the importance of considering whether benefits are shared among pathogens within a host. The virulence of *V. cholerae* results primarily from production of "cholera toxin" which is secreted from the bacterium and binds to receptors on cells that line the lumen of the small intestine. This binding causes the cells to transport chloride ions into the lumen of the small intestine, which increases the solute concentration in the lumen causing water from the epithelial cells to flow into the lumen. The increased fluid entry into the lumen flushes out the intestinal contents, thereby removing competing species of bacteria from downstream regions of the intestine. Because *V. cholerae* can swim and adhere to the epithelial cells it can remain in the intestinal tract when other bacteria are flushed out by the increased flow. Toxin production by *V. cholerae* in the small intestine thus clears the intestinal tract of a gauntlet of competing bacteria, transforming the intestinal contents into a growth chamber for the infecting group of *V. cholerae.* When humans are infected with a population of highly toxigenic *V. cholerae,* the fecal stools can have the appearance of cloudy water and consist of virtually pure culture of *V. cholerae.* The fluidity of the stools probably also enhances transmission by allowing billions of organisms to be rapidly dispersed in water supplies and on environmental surfaces, entering other people when contaminated water or food is ingested. Cholera toxin is not normally produced when *V. cholerae* is in the external environment and it has no other known function; it therefore appears to be selected specifically for this function. By this process, toxin production by some *V. cholerae* in the intestinal tract enhances both the takeover of the entire intestinal tract by the infecting group and the transmission of the progeny from the infecting group to other hosts. Because *V. cholerae* is often waterborne and watery diarrheal stools disperse readily, the inocula ingested will often be mixtures of progeny from different source individuals. This mixing and the differential regrouping of individuals are precisely the characteristics that predispose a population for group beneficial traits (Wilson 1980).

This argument proposes that cholera toxin is particularly important for the establishment of the invading group of *V. cholerae* in the intestinal tract. Inocula without at least some toxigenic strains would be unlikely to generate infections that are productively transmissible, an expectation consistent with experimental data using mice as subjects (Baselski *et al.* 1978, 1979, Sigel *et al.* 1980). However, once the takeover of the intestinal tract has occurred, strains that produce little or no toxin could accrue the benefits of the toxin production by others without paying the price of toxin production. One would therefore expect that during the course of an infection the strains that produce less toxin would tend to displace strains that produced more toxin. Although this prediction has not been prospectively tested, it is consistent with the outcome of an experiment in which strains with differing toxigenicities were introduced into the mouse intestinal tract: the less toxigenic strain tended to displace the more toxigenic strain (Baselski *et al.* 1978).

The variation in toxin production of isolates within regions also accords with the idea that cholera toxin contributes to the success of groups but is costly to individuals within the group. The toxin production of strains isolated at a particular time and place can vary by well over an order of magnitude. This polymorphism is expected when a characteristic is beneficial to the group as a whole but is disadvantaged within groups, because the advantage of the different morphs is frequency dependent, each being more successful when present in low frequency.

Simultaneous consideration of the different levels of selection may guide inquiry into the proximate mechanisms of virulence and the ways in which changes in environmental conditions may influence these mechanisms. When milder strains are favored by environmental conditions such as restricted opportunities for waterborne transmission (relative to opportunities for transmission that requires host mobility), a high frequency of low toxin producers may be favored in the population. Toxin production could be lowered through regulatory mechanisms that reduce amounts of toxin produced under particular environmental conditions or that restrict the range of environmental conditions permissive to toxin production.

The shift from the classical biotype of *V. cholerae* to the el tor biotype in South Asia accords with this explanation. The classical biotype was the predominant *V. cholerae* there during the first two-thirds of the twentieth century and persisted where improvements in water quality lagged (Ewald 1994). It produces toxin liberally over a range of culturing conditions, one of the few requirements being shaking of the culture. The el tor biotype, which became established much more broadly in areas with less opportunity for waterborne transmission tends to produce less toxin, and needs more restricted cues to initiate toxin production. A still period of about two hours must precede the period of shaking and the nutrient mix is more restrictive. These differences have been correlated with restriction of the period during which RNA that encodes a primary regulatory gene for toxin production (toxR) is transcribed. Presumably the more restricted set of *in vitro* conditions to which the el tor *V. cholerae* will respond with toxin production corresponds to a more restricted set of *in vivo* conditions (DiRita *et al.* 1996), which in turn corresponds to reduced tendencies to be the altruistically producing toxin that benefits other individuals in the group of *V. cholerae*.

19.2.3 The spectrum of shared benefits

A benefit associated with virulence may be shareable by various mechanisms.

1. *Systemic effect of local dissemination.* The virulence determinant could be secreted from the pathogen and have a localized effect on host tissue, which then has systemic effects, which in turn influence the group of the pathogens in a host, as is the case with cholera toxin.
2. *Systemic dissemination.* The virulence determinant could be secreted from a pathogen and travel broadly throughout the host causing the same effects on

other pathogens within the host that they do on the individual pathogen that produced it.

3. *Systemic effect of cellular pathology.* The virulence determinant could alter the pathogen's characteristics without being disseminated locally or systemically and still provide a shareable benefit if the cellular pathology affects the host's biology at the organismal level.

The toxin produced by the bacterium that causes anthrax, *Bacillus anthracis,* provides an example of the second category. The virulence attributed to anthrax results largely from the cell death caused by the anthrax toxin, which is disseminated throughout the victim's body. Host death allows the eventual liberation of anthrax spores, which, because of their extremely high durability outside the host, can wait for susceptibles to come to the place at which they are released in the external environment. The anthrax bacterium can thus exploit hosts to a great degree and still be transmitted from the host because the transmission benefits associated with tissue degradation and death of the host are generated in a shareable way by the anthrax toxin.

The third category of shareable effects can occur by a variety of mechanisms. Neuron function and thus behavior can be altered in such a way that the transmission of the pathogen population from the host is enhanced. Similarly, pathogens can alter circulating host cells or chemical messengers such as hormones or cytokines released from host cells so that growth, survival or transmission of the group of pathogens within the host is altered. The shared effect of such chemical messages may be on physiology, morphology, or behavior.

Shared effects through alteration of neuron function are illustrated by the trematode parasite, *Dicrocoelium dendriticum,* which is transmitted from ant to sheep when a sheep inadvertently eats an infected ant. The probability of this transmission is increased by the pathological effects that are generated by one or more of the *D. dentriticum* larvae; this increased probability of transmission is shared with the other *D. dentriticum* larvae inside the infected ant. During infection of the ant one or more larvae migrate into the brain of the ant and thereby cause the infected ant to migrate to the top of blades of grass, clamp onto the grass with its mandibles, and remain rigidly in that position throughout the day, thus increasing the chance that grazing sheep will ingest the ant. This transmission benefit is shareable because the larvae that have not migrated to the brain are transmitted even though they have not contributed to the manipulative pathology. This behavioral manipulation is therefore an example of increased virulence that is selected for through group benefit. The group benefit is especially germane to this example, because the individuals that migrate to the brain of the ant apparently are not able to infect the sheep. The behavioral pathology is therefore a characteristic that increases the chances for transmission of the other *D. dentriticum* in the ant at the expense of the transmission of the individual that is causing the pathology.

Its maintenance can therefore be understood only by taking into account the transmission benefits to the group of parasites within the host and is explainable by the high probability that trematodes within each ant are sharing alleles for the altruistic behavior.

Parasites with complex life histories are especially likely to generate such shareable benefits of virulence, especially when they are transmitted by predation. In such cases the prey hosts are typically severely affected, and the illness facilitates transmission to the predator host by making the prey more vulnerable to predation (Ewald 1995b). Parasitism by *D. dendriticum* represents a special case of this generalization because the behavioral manipulation of infected ants makes the sheep inadvertently into a predator of ants.

The life cycle of the protozoan *Toxoplasma gondii* illustrates a more normal state of affairs for a predator-borne parasite. *T. gondii* is transmitted from a rodent to a cat when the cat captures and eats the infected rodent. Infected cats shed cysts of *T. gondii* in their feces, and other rodents in the area complete the cycle when they eat or lick objects contaminated with cat feces containing *T. gondii* (Jewell *et al.* 1972). When ingested by a rodent the parasites burrow through its intestinal wall and encyst in muscles and brain tissue, which adversely affects the rodent's mobility and brain function. Some studies indicate that the parasite produces (or causes host cells to produce) an LSD-like compound, which may further compromise the accuracy of the rodent's mental function (Ledgerwood *et al.* 2003). These shared effects of virulence on the rodent's mental function probably facilitate transmission by making an infected rodent much easier for a cat to catch. Because the population of *T. gondii* within a rodent are often generated by more than one oocyst, each of which has subsequently divided clonally, these effects are often shared within genetically distinct clones within a host. The pathology associated with the invasive infections of *T. gondii* in intermediate hosts is often severe, demonstrating how behavioral manipulation of host behavior may apply to complex life histories in general whether the parasite is unicellular as in the case of *T. gondii* or multicellular as in the case of *D. dentriticum.* Vector-borne parasites may provide a further variation of this theme with severe effects of the parasites on intermediate hosts favoring transmission by facilitating the biting of the intermediate host by the vector. Vectors that transmit *Plasmodium* parasites, for example, are more able to bite sick hosts than healthy ones. The incapacitating illness of malaria may thus be partly a behavioral manipulation that enhances transmission of the group of parasites within the human host.

The same category of shared effects (systemic effects of cellular pathology) applies to virulence of some subcellular parasites. Rabies viruses reproducing in the limbic system of the brain, for example, cause the infected animal to bite, thus favoring the transmission of viruses in the saliva; viruses that have paralyzed the neurons that initiate swallowing similarly contribute to transmission by causing the accumulation of virally contaminated saliva in the oral cavity (Ewald 1980). The rabies virus thus illustrates shared benefits that derive from alterations of behavior and physiology that result from damage to neurons.

Viral infections may be transmitted as a result of damage that is restricted to particular host tissues and thus enhances the transmission of the group of organisms. Nuclear polyhedrosis viruses, for example, steadily destroy internal tissue of the abdomen and thorax of their caterpillar hosts. The infected caterpillars climb upwards in the vegetation, expand like a balloon, and then burst in response to slight disturbances. The bursting high in the vegetation liberates millions of virions. The triggering of the climbing behavior of the caterpillar and any other specific tissue tropisms that cause the caterpillar to continue to crawl upward (alteration of brain function while avoiding destruction of the brain) and then to burst high in the vegetation (rather simply causing general degradation of the caterpillar) provide shareable benefits of virulence that therefore need to be understood in the context of the selection for characteristics that benefit at the group level. The viruses within the brain are presumably less transmissible than those in the thorax and abdomen because the viruses in the brain are not released during the bursting of the carcass.

Manipulation of the endocrine system of a host can similarly create benefits associated with virulence that are shareable among all the parasites within a host. Microsporidians of the genus *Nosema*, for example, secrete a juvenile hormone analog that causes its host, larvae of the *Tribolium* beetles, to grow into ever larger larvae rather than into adults (Dawkins 1982). The effect on the beetle's fitness is tantamount to lethality, but the *Nosema* gain an ever larger and longer-lived resource base to exploit for their own reproduction. Any *Nosema* variant that did not produce the hormone analog in the midst of hormone producers would be at an advantage within the host because it could still share the benefits of being in a larger longer-lived host while using resources that would otherwise be spent on the hormone analog on its own reproduction. But if it were within a host that housed no producers of the hormone analog, its productivity would be relatively low. Thus, as is the case with *V. cholerae*, within-group competition probably favors variants of low virulence and between-group competition probably favors variants of high virulence.

Sexually transmitted pathogens of humans may similarly facilitate transmission by altering behavior. *Chlamydia trachomatis* causes infertility by having a tropism for the oviduct, which leads to formation of scar tissue there. This scarring can prevent pregnancy by prohibiting the migration of the ovum or fertilized egg to the uterus. Infertility is a major reason for divorce. Because infections with these organisms are long-lived, the breakup of sexual pairings that are attributable to this infertility probably enhances their transmission to new hosts. The individual organisms that cause the scarring are unlikely to be transmitted because they are not in a place where contact with a partner's genitals occurs. Their damage therefore may be analogous to that caused by

D. dentriticum, *T. gondii*, rabies viruses, and the nuclear polyhedrosis viruses: the damage facilitates transmission of genetically related parasites by altering host behavior.

These considerations could be dismissed by arguing that the parasites causing the systemic pathology and the parasites that are transmitted as members of the same clone are functioning like a subdivided organism, a single genetic individual. This argument misses the point, however, because a clonal structure explains how the altruistic behavior on the part of one individual could be favored by natural selection. The general argument is important because polyclonal inocula and mutations during infection tend to make pathogen populations within hosts deviate at least somewhat and perhaps greatly from being genetically identical, thus opening the door for within-host evolution of nonaltruistic characteristics (Bonhoeffer and Nowak 1994, Nowak and May 1994).

19.3 Virulence in uncooperative hosts

Neuronally sophisticated multicellular organisms can increase their fitness by identifying uncooperative individuals and then redirecting altruistic behavior away from them and toward organisms that are genetically related (thereby gaining inclusive fitness benefits; Hamilton 1964) or that will reciprocate with them (gaining benefits through reciprocal altruism (Trivers 1971, Axelrod and Hamilton 1981)). If the biological system allows for policing mechanisms to punish uncooperative individuals, then opportunities for cooperative behavior are enhanced. But policing is an altruistic activity when it is costly and when the benefits of controlling uncooperative individuals are shared by the other individuals in the group.

Most of the analyses of cooperation pertain to intraspecific interactions, but the principles apply to interspecific interactions as well, because individuals of different species can interact cooperatively and may do so depending on the potential for reciprocal altruism and exploitation. Associations between parasitic or mutualistic symbionts and their hosts may prove to be particularly strong examples of the influences on interspecific cooperation, because the intimacy of these associations creates a great potential for reciprocation or exploitation (Axelrod and Hamilton 1981). The principle of inclusive fitness is important for interspecific cooperation as well because benefits that may be obtained from damaging a nonreciprocating individual of another species (the benefits of policing) may sometimes be accrued through inclusive fitness benefits. For example, the death of a host that is unlikely to be of future use to a parasite will tend to decrease the local competition among hosts, and thus contribute to the survival of other genetically related parasites living in nearby hosts.

Some plasmids, for example, are able to harm bacterial cells that do not permit the plasmids to live within them. In "proteic killer gene systems" plasmids encode a toxin that kills uninfected cells and an antidote that makes the toxin ineffective so long as the plasmid persists (Jensen and Gerdes 1995). If the cell does not allow the plasmid to persist, the antidote is not produced and the toxin kills the cell. Destruction of the incompatible bacteria by plasmid-encoded factors may therefore favor the propagation of the plasmid alleles through inclusive fitness benefits – other copies of the plasmid in nearby bacteria may have increased success because they suffer less competition from the incompatible bacterium. Plasmid virulence in this case is a form of punishment against "uncooperative" bacterial hosts that is selected for on the basis of the benefit to the local group.

Punishment of uncooperative hosts may also be widespread among multicellular hosts. Microsporidian parasites of the genus *Amblyospora* provide an example (Ewald and Shubert 1989). When they infect female mosquito larvae from the water column they can persist in the mosquitoes and be transmitted to the offspring of the infected females. They tend to be benign in the female mosquitoes until this time. Infected adult male mosquitoes offer little if any opportunity for transmission and are therefore analogous to bacteria that are not permissive to a proteic killer plasmid. Accordingly, *Amblyospora* are much more lethal to male larvae than to female larvae. Although this death of male larvae does not liberate infective propagules into the water, it eliminates hosts that do not transmit the parasite into subsequent generations. By killing the males the parasites reduce competition with the female larvae and thereby increase the success of the parasites that are infecting these females. So long as the parasites in these females are genetically related to the parasites that kill male larvae, the genetic instructions for male-killing parasites may gain from the death of the males. One would expect that such inclusive fitness effects would occur most strongly where bodies of water are small (e.g., among tree-hole- or puddle-breeding mosquitoes), because parasites within such small bodies of water are likely to be closely related genetically owing to the reduced potential for multiple colonization that would occur in small bodies of water. This prediction has not yet been tested, however.

Similar explanations of virulence manifested as the killing of nonpermissive hosts apply to *Wolbachia*, a bacterial parasite of wasps, flies, and other arthropods. As is the case with *Amblyospora*, males are dead-end hosts for many *Wolbachia*. When infected males inseminate females, those females that do not already harbor the same *Wolbachia* are rendered infertile. The *Wolbachia* that do this damage do not benefit directly from the killing but may benefit indirectly because females in the vicinity that harbor the same *Wolbachia* experience reduced competition and may therefore have more (*Wolbachia*-infected) offspring (Hurst *et al.* 1996). Although once considered novelties, these associations now appear to be common over a wide range of hosts. The ciliate protozoan, *Paramecium aurelia*, for example, contains a bacterium that is distantly related to *Wolbachia* and kills the mate of the infected *P. aurelia* during the conjugation process if the mate is not susceptible to infection (Beale and Jurand 1966, Schmidt *et al.* 1988, Springer *et al.* 1993, Hurst *et al.* 1996).

As is the case with *Amblyospora* and proteic killer plasmids, the virulence inflicted by *Wolbachia* may benefit genetically related parasites in the vicinity, and can therefore be considered an altruistic act that evolves not by direct enhancement of the killer's competitive success within the group competition but rather by selection favoring the success of other individuals within the group. Whether one prefers to call the killing spite or altruism depends on the focus. The action is spiteful with regard to the host being harmed, but it is altruistic with regard to the individuals benefiting from the harm.

19.4 Horizontal transmission of virulence genes

The main selective pressure disfavoring altruism within groups results from the loss in direct fitness that the altruist incurs from the altruistic characteristic relative to nonaltruists within the group. This cost would be lessened if the altruistic allele could spread to new hosts. This possibility is feasible for prokaryotic hosts because genetic information encoding increased virulence can be readily transmitted horizontally by plasmids or viruses, which in turn can be controlled by proteins encoded in the bacterial genome. Thus a bacterium harboring genetic instructions for a increased virulence might be outcompeted within groups and across the population as a whole when the instructions are not horizontally transmitted, but horizontal transmission between bacteria may allow the instructions to persist because they can invade new hosts within groups.

Horizontal transmission of proteic killer plasmids has this effect. By eliminating bacteria that reject the plasmid, the killer plasmid increases the local density of infected bacteria and hence its own density. This system persists even though the plasmid represents a drain on cooperative bacteria due to the production of the toxin and its antidote. The high virulence of the plasmid in uncooperative hosts therefore can be understood only in the context of benefits to other group members (other plasmids in nearby bacterial hosts) and persists only because of horizontal transmission.

A similar argument applies to toxin production by bacteria that generate benefits that are shared by other bacteria within a host (e.g., the toxins secreted by *V. cholerae*). Because the positive effects of these toxins are shared among individuals that do not produce the toxin, one expects that within-host evolution will favor evolutionary reductions in toxin production and hence reduced virulence. The stability of relatively high virulence may be aided, however, by horizontal transfer of the gene that encodes the toxin. Because this gene is virally encoded, it can be transmitted to the bacteria that do not produce toxin, thus converting them from uncooperative group members to cooperative toxin producers, and enhancing the strength of the group selection favoring toxin production and hence high virulence.

19.5 Conclusions

Understanding the evolution of virulence requires consideration of how selection acts at the levels of genes, individuals, and groups. The interplay between these levels has generally emphasized two opposing effects: (1) within-host competition, which favors increased virulence because highly exploitative variants tend to be successful competitors within hosts; and (2) between-host competition, which favors decreased virulence because of the greater transmission from hosts that house the more benign groups. This chapter emphasizes that this characterization is incomplete, because consideration of within-group and between-group selection needs to consider whether the benefits associated with virulence are shareable among pathogens within a host and among pathogens within a local group of hosts. When such shareable benefits occur, within- and between-group selection may often influence virulence in directions that are opposite to the generally accepted influences mentioned above. Such shared virulence benefits may be widespread in nature. They are expected when virulence is caused by compounds that are released from parasites and benefit other parasites within the host. They are also expected when the virulence mechanism involves alteration of the host's organismal characteristics, such as behavior, in ways that improve transmission for the parasites within the host that are not generating the pathology.

Virulence can also arise as a result of attacks by parasites on uncooperative hosts. The feasibility of this strategy depends on how selection acts on local groups of pathogens and hosts, because the benefits of such attacks may be shared by the parasites in the local group of hosts.

Although this chapter has considered the evolution of virulence largely in the context of the tendency for the benefits associated with virulence to be shareable among the pathogens within a host and within a local group of hosts, casting tradeoffs according to the shared effects provides a general framework for understanding the full spectrum of effects of organisms that live in or on their hosts, from highly damaging parasitism to benign parasitism through commensalism and mutualism. The benefits of benignity, for example, require consideration of group structure and concepts such as inclusive fitness because these benefits are shareable. Specifically, if harm to the host inhibits transmission and one individual parasite in the host has suppressed reproduction, the positive effects of this suppressed reproduction on transmission are shared by the other conspecific parasites within the host. The conventional view of the tradeoff between within-host and between-host success therefore also rests implicitly on the concept of shared benefits.

REFERENCES

Axelrod, R., and Hamilton, W. D. (1981). The evolution of cooperation. *Science* 211:1390–6.

Baselski, V. S., Medina, R. A., and Parker, C. D. (1978). Survival and multiplication of *Vibrio cholerae* in the upper bowel of infant mice. *Infect. Immunol* 22:435–40.

Baselski, V. S., Medina, R. A., and Parker, C. D. (1979). *In vivo* and *in vitro* characterization of virulence deficient mutants of *Vibrio cholerae. Infect. Immunol* 24:111–16.

Beale, G. H., and Jurand, A. (1966). Three different types of mate killer (mu) particle in *Paramecium aurelia J. Cell Sci.* 1:31–4.

Bonhoeffer, S., and Nowak, M. A. (1994). Mutation and the evolution of virulence. *Proc. R. Soc. Lond. Ser. B.* 258:133–40.

Bull, J. J., Molineux, I. J., and Rice, W. R. (1991). Selection of benevolence in a host–parasite system. *Evolution* 46:882–95.

Dawkins, R. (1982). *The Extended Phenotype.* Oxford: W. H. Freeman.

DiRita, V. J., Neely, M., Taylor, R. K., and Bruss, P. M. (1996.) Differential expression of the ToxR regulon in classical and El Tor biotypes of *Vibrio cholerae* is due to biotype-specific control over toxT expression. *Proc. Natl Acad. Sci. USA* 93: 7991–5.

Ebert, D. (1994). Virulence and local adaptation of a horizontally transmitted parasite. *Science* 265:1084–6.

Ewald, P. W. (1980). Evolutionary biology and treatment of signs and symptoms of infectious disease. *J. Theor. Biol.* 86:169–76.

Ewald, P. W. (1983). Host–parasite relations, vectors, and the evolution of disease severity. *Annu. Rev. Ecol. Syst.* 14:465–85.

Ewald, P. W. (1994). *Evolution of Infectious Disease.* Oxford: Oxford University Press.

Ewald, P. W. (1995a). Response to "The scope for virulence management: a comment on Ewald's view on the evolution of virulence" by M. van Baalen & M. W. Sabelis. *Trends Microbiol.* 3:416–17.

Ewald, P. W. (1995b). The evolution of virulence: a unifying link between ecology and parasitology. *J. Parasitol.* 81:659–69.

Ewald, P. W. (2002). Virulence management in humans. In U. Dieckmann and H. Metz (eds) *Virulence Management: Mathematical Models of Virulence Evolution*, pp. 399–412. International Institute for Applied Systems Analysis: Vienna, Austria. Cambridge: Cambridge University Press.

Ewald, P. W., and DeLeo, G. (2002). Alternative transmission modes and the evolution of virulence. In U. Dieckmann and H. Metz (eds) *Virulence Management: Mathematical Models of Virulence Evolution*, pp. 10–25. International Institute for Applied Systems Analysis: Vienna, Austria. Cambridge: Cambridge University Press.

Ewald, P. W., and Schubert, J. (1989). Vertical and vectorborne transmission of insect endocytobionts, and the evolution of benignity. In W. Schemmler (ed.) *CRC Handbook of Insect Endocytobiosis: Morphology, Physiology, Genetics and Evolution*, pp. 21–35. Boca Raton, FL: CRC Press.

Fenner, F., and Ratcliffe, F. N. (1965). *Myxomatosis.* Cambridge: Cambridge University Press.

Hamilton, W. D. (1964). The genetical theory of social behavior (I and II). *J. Theor. Biol.* 7:1–52.

Hurst, L. D., Atlan, A., and Bengtsson, B. O. (1996). Genetic conflicts *Quart. Rev. Biol.* 71:317–64.

Jensen, R. B., and Gerdes, K. (1995). Programmed cell death in bacteria: proteic plasmid stabilization systems. *Mol. Microbiol.* 17:205–10.

Jewell, M. L., Frenkel, J. K., Johnson, K. M., Reed, V., and Ruiz, A. 1972. Development of *Toxoplasma* oocysts in neotropical felidae. *Am. J. Trop. Med. Hyg.* 21:512–17.

Ledgerwood, L., Ewald, P. W., and Cochran, G. M. (2003). Genes, germs, and schizophrenia: an evolutionary perspective. *Perspect. Biol. Med.* 46 (in press)

Levin, B. R., Allison, A. C., Bremermann, J., Cavalli-Sforza, L. L., Clarke, B. C., Frentzel-Beyme, R., Hamilton, W. D., Levin, S. A., May, R. M., and Thieme, H. R. (1982). Evolution of parasites and hosts. In R. M. Anderson and R. M. May (eds) *Population Biology of Infectious Diseases*, pp. 213–43. Berlin: Springer-Verlag.

Levin, S., and Pimentel, D. (1981). Selection of intermediate rates of increase in parasite–host systems. *Am. Nat.* 117:308–15.

Lewontin, R. C. (1970). The units of selection. *Annu. Rev. Ecol. Syst.* 1:1–18.

May, R. M., and Anderson, R. M. (1983). Epidemiology and genetics in the coevolution of parasites and hosts. *Proc. R. Soc. Lond. Ser. B* 219:281–313.

Messenger S. L., Molineux I. J., and Bull J. J. (1999). Virulence evolution in a virus obeys a trade-off. *Proc. R. Soc. Lond. Ser. B* 266:397–404.

Nowak, M. A., and May, R. M. (1994). Superinfection and the evolution of parasite virulence. *Proc. R. Soc. Lond. B* 255:81–9.

Schmidt, J. H, Gortz, H.-D, Pond, F. R., and Quackenbush, R. L. (1988). Characterization of *Caedibacter* endonucleosymbionts from the macronucleus of *Paramecium caudatum* and the identification of a mutant with blocked R-body synthesis. *Exp. Cell Res.* 174:49–57.

Sigel, S. P., Lanier, S., Baselski, V. S., and Parker, C. D. (1980). *In vivo* evaluation of pathogenicity of clinical and environmental isolates of *Vibrio cholerae. Infect. Immunol* 28:681–7.

Sober, E., and Wilson, D. S. (1998). *Unto Others: the Evolution and Psychology of Unselfish Behavior.* Cambridge, MA: Harvard University Press.

Springer, N., Ludwig, W., Amann, R., Schmidt, H. J., Gortz, H. D. and Schleifer, K. H. (1993). Occurrence of fragmented 16s rRNA in an obligate bacterial endosymbiont of *Paramecium caudatum. Proc. Nat. Acad. Sci. USA* 90:9892–5.

Trivers, R. L. (1971). The evolution of reciprocal altruism. *Q. Rev. Biol.* 46:35–57.

Turner, P. E, Cooper, V. S., and Lenski, R. E. (1998). Tradeoff between horizontal and vertical modes of transmission in bacterial plasmids. *Evolution* 52: 315–29.

Williams, G. C. (1966). *Adaptation and Natural Selection.* Princeton, NJ: Princeton University Press.

Williams, G. C., and Nesse, R. M. (1991). The dawn of darwinian medicine. *Q. Rev. Biol.* 66:1–22.

Wilson, D. S. (1980). *The Natural Selection of Populations and Communities.* Menlo Park, CA: Benjamin-Cummings.

20

Evolutionary genetics and emergence of RNA virus diseases

EDWARD C. HOLMES
Department of Zoology, University of Oxford

20.1 Introduction

RNA viruses are ubiquitous intracellular parasites responsible for many infectious diseases of humans. This is most publicly visible with the AIDS viruses, HIV-1 and HIV-2, which currently infect some 36 million people worldwide, including more than 30% of the adult population in parts of sub-Saharan Africa (Piot *et al.* 2001), and hepatitis C virus (HCV) which has ~175 million sufferers globally, many of whom will go on to develop serious liver diseases (WHO 1997). To these ailments can be added myriad other infections, from the benign to the lethal, including many that seem to have appeared only recently. Given the pace at which human ecology is changing, it is evident that more such "emerging diseases" will arise in the future, both in humans and wildlife species (Daszak *et al.* 2000, Morse 1994).

The success of RNA viruses may in large part be due to their remarkable genetic flexibility. In contrast to DNA-based life forms, the evolution of RNA viruses is fueled by extremely rapid rates of mutation, with around one error occurring during each round of genome replication (Drake and Holland 1999). When this mutational power is coupled to rapid rates of replication and large population sizes, immense amounts of genetic diversity can be produced, even within individual hosts. Although the end product of this evolutionary process will often be detrimental for our health by limiting the effectiveness of antiviral drugs and vaccines, the rapidity of RNA virus evolution also provides biologists with a unique opportunity to study evolutionary processes over the time-frame of human observation. It is not surprising, therefore, that RNA viruses are now commonly used in laboratory studies of evolutionary dynamics (Burch and Chao 2000, Miralles *et al.* 1999), have been used to test the accuracy of phylogenetic inference by reconstructing evolutionary pathways that have arisen in real time (Hillis *et al.* 1992), and have even been cited as exemplars of macroevolutionary processes such as punctuated equilibrium (Nichol *et al.* 1993).

The Evolution of Population Biology, ed. R. S. Singh and M. K. Uyenoyama. Published by Cambridge University Press.

Yet despite the wealth of gene sequence data now available from RNA viruses, there have been surprisingly few attempts to present a complete picture of their evolutionary biology, particularly the interaction between genetics and ecology. For example, RNA viruses have largely been excluded from the "Great Obsession" of population genetics – whether natural selection or genetic drift is the dominant evolutionary process (Lewontin 1974, Gillespie 1998). As this chapter will show, answering this question may have profound implications for how we treat viral infections. Further, while RNA viruses are the most common cause of emerging diseases (Cleaveland *et al.* 2001), little is known about the general evolutionary factors which allow viruses to cross species boundaries and establish productive infections in new hosts. It is likely that evolutionary genetics has an important role to play in studies of viral emergence as genome malleability may be a key reason why some viruses have catholic tastes in the hosts they infect, while others are more species specific. The aim of this chapter is to review the fundamental processes of virus evolution and then establish how they contribute to the emergence of new diseases.

Before examining the evolutionary genetics of RNA viruses, it is useful to remind ourselves of their salient biological features. As their name implies, RNA viruses have genomes composed of RNA arranged in a variety of ways (Table 20.1). The most common form of genome architecture comprises a single strand of RNA laid out in positive-sense polarity so that it corresponds to the mRNA of cellular genes. Other single-stranded RNA viruses have genomes with a negative-sense polarity, some have ambisense genomes of both positive and negative polarity and a small number have double-stranded genomes. In most cases, these genomes form a single replicatory unit. However, some viruses have genes located on independently replicating genomic segments, in some ways equivalent to chromosomes. Whatever their genome organization, all RNA viruses replicate using RNA-dependent RNA polymerase, an RNA-binding enzyme not found in cellular life forms. More complex are the retroviruses like HIV. Although these are usually classed as RNA viruses, and the genome that rests in the mature virus particle is composed of RNA, a DNA copy of the genome is also produced, which is then integrated into the host's cellular DNA. This DNA genome is created from the RNA template using reverse transcriptase, a virally encoded RNA-dependent DNA polymerase.

Not only do RNA viruses have very simple genome structures, but they possess tiny amounts of genetic material. Most have genome sizes ranging from just 7 kb to 10 kb, although in a few cases genome sizes may reach 30 kb in length. Finally, and as expected from their differing capacity for disease, RNA viruses interact with their hosts in a wide variety of ways. Most cause rapid, or acute, infections which only last a few days, while others, particularly the retroviruses, are able to infect their hosts for decades, often with continual replication, producing so-called chronic or persistent infections.

Table 20.1. *Biodiversity of families of animal RNA viruses*

Family	Examples
Positive-sense, single-stranded	
Picornaviridae	Poliovirus, hepatitis A, foot-and-mouth disease
Caliciviridae	Hepatitis E, human calicivirus, feline calicivirus
Astroviridae	Human astrovirus
Coronaviridae	SARS coronavirus, berne virus
Flaviviridae	Dengue, yellow fever, West Nile, hepatitis C
Togaviridae	Rubella, sindbis, Venezuelan equine encephalitis
Negative-sense, single-stranded	
Paramyxoviridae	Measles, mumps, parainfluenza
Rhabdoviridae	Rabies, vesicular stomatitis
Filoviridae	Ebola, Marburg
Orthomyxoviridae	Influenza viruses A, B and C
Bunyaviridae	Crimean-Congo hemorrhagic fever, hantaan
Arenaviridae	Lassa, guanarito, machupo, junin
Double-stranded	
Reoviridae	Reovirus, rotavirus, bluetongue
Birnaviridae	Infectious pancreatic necrosis
Reverse-transcribing	
Retroviridae	Human immunodeficiency virus
Subviral agents	
Genus *Deltavirus*	Hepatitis D

20.2 Do RNA viruses evolve like other organisms?

The most fundamental of all evolutionary questions relating to RNA viruses is whether these infectious agents evolve in a manner that differs qualitatively from DNA-based organisms. That this is raised at all stems from the concept of the quasispecies. Quasispecies theory was originally developed by Manfred Eigen and colleagues to describe the evolution of the first RNA replicators (Eigen and Schuster 1997). It was first applied to RNA viruses in the late 1970s following laboratory studies of the bacteriophage Qβ (Domingo *et al.* 1978) and since this time has come to dominate studies of RNA virus evolution (reviewed in Domingo and Holland 1997). As well as providing a basic description of evolutionary mechanism, quasispecies theory has major implications for other aspects of viral evolution such as the reconstruction of phylogenetic relationships – viruses will not possess independently evolving lineages so that network representations of phylogenetic history are more appropriate than bifurcating trees (Dopazo *et al.* 1993) – and for the acquisition of drug resistance in viruses like HIV, because in quasispecies theory fitness is a property assigned to the whole population, rather than to individual genotypes (Eigen 1996).

At the heart of quasispecies theory is extreme population heterogeneity, a function of the high mutation rate of RNA viruses. More fundamentally, high mutation rates mean that the frequency of any genotype is a function of its own replication rate *as well as* the probability that it will be produced by the erroneous replication of other genotypes in the population (Eigen 1987, 1996, Nowak 1992). Hence, there is a form of mutational coupling among genomes in the quasispecies, so that the entire population evolves as a single, interacting unit. The most important outcome of this population structure is that although there is a "master" sequence of highest individual fitness, natural selection acts on the quasispecies as a whole, which will evolve to maximize its average rate of replication. This sits in contrast to most population genetic models, aside from those based on group selection, in which individual genotypes are the unit of selection. Furthermore, because the quasispecies applies to organisms with small genomes, high mutation rates and large population sizes, the sequence space (that is, all possible combinations of mutants) surrounding the master sequence will be completely occupied, thereby preventing any random fluctuation due to genetic drift (Eigen 1987, 1996). The lack of genetic drift is critical to quasispecies theory as without it mutational coupling breaks down and natural selection loses its ability to act on the whole mutant distribution. It also means that despite the high error rate, the master sequence retains a stable frequency in the population, rather than diffusing through sequence space.

At face value, the quasispecies would appear an ideal model of RNA virus evolution. These organisms fit the bill in that they possess tiny genomes, are subject to intense mutation pressure, and have large populations. Moreover, nearly every study of RNA viruses in nature has shown that they are highly variable, but that population consensus sequences, assumed to represent the master sequence, are often stable through time. More powerful evidence for the quasispecies is that viral genotypes of relatively low individual fitness can outcompete those of higher fitness if they are surrounded by beneficial neighbors with which they evolve in a concerted fashion through mutational coupling. This has recently been depicted in simulations using digital organisms (Wilke *et al.* 2001) and may have also been observed in laboratory studies of vesicular stomatitis virus (VSV) (De la Torre and Holland 1990) and bacteriophage $\phi 6$ (Burch and Chao 2000). In the VSV experiments, for example, high fitness genotypes only came to dominate the population after they were seeded above a particular threshold level, suggesting that until this point they were outcompeted by low fitness genotypes which together had a higher mean fitness. If interactions of this type really do exist in quasispecies, then RNA viruses come with an inherent unpredictability.

Despite the ubiquity of the quasispecies concept in discussions of RNA virus evolution, and its potential importance for viral emergence, many of the observations said to support it can be equally well explained by standard population genetic models (Holmes and Moya 2002, Jenkins *et al.* 2001, Moya *et al.* 2000). More pertinently, there have been no definitive descriptions of quasispecies population structure in natural as opposed to laboratory populations of RNA viruses, so that whether the concept applies outside of an *in vitro* setting is unknown.

The key to whether the quasispecies reflects the true nature of RNA virus evolution lies in the strength of genetic drift. For example, genetic drift could explain why genotypes of lower individual fitness are seemingly more successful than those of higher fitness; the rarest genotypes, even if advantageous, will be lost by drift in small populations and both the VSV and ϕ6 experiments were performed with small population sizes. Although some have sought to bypass the problem of genetic drift by redefining a "phenotypic quasispecies" (Schuster and Stadler 1999), this will evidently only apply to a minority of the total genetic variation seen in virus populations.

As in any system, there are two key determinants of the strength genetic drift – the effective size (N_e) of populations and the distribution of selection coefficients (s) among nucleotides, with drift playing a major role in substitution dynamics when $N_e s < 1$. Most discussions of population size in RNA viruses have concentrated on the total viral load within an infected host. Such values will often be very large, especially in the absence of an immune response. For example, the viral load for foot-and-mouth disease virus (FMDV), a supposedly archetypal quasispecies, has been estimated at between 10^9 and 10^{12} (Domingo *et al.* 1992). In reality, however, these are estimates of the census population size, N, and not N_e. For RNA viruses, N_e may be roughly defined as that proportion of the total population that are able to seed progeny of the next generation. Unfortunately, it is difficult to measure this number using virological techniques so that estimates of N_e in viruses have utilized population genetic methods, most notably via estimates of neutral genetic diversity using $\theta(\theta = 2N_e\mu$; with μ the mutation rate). This is best documented in the case of HIV-1. All estimates of N_e in HIV-1 are in the region of 10^3 to 10^4, which are many orders of magnitude lower than total viral loads within patients which may range from 10^8 to 10^{10} (Leigh Brown 1997, Nijhuis *et al.* 1998, Zanotto *et al.* 1999). At face value, such low N_e estimates imply that genetic drift is an important force in HIV-1 evolution. Not only does this have important implications for quasispecies theory, but drift would then greatly influence the development of antiviral drug resistance. Indeed, randomness in the types of mutations that confer resistance, and the sequence in which they appear, has been cited as evidence for the action of genetic drift in HIV-1 (Leigh Brown and Richman 1997). As such, it is that clear that population genetic theory has a major role to play in predicting the nature of drug resistance in HIV-1 and other pathogenic organisms.

There is, however, a strong counter-argument that the low estimates of θ, and hence N_e, in HIV-1 are more likely to be due to periodic selective sweeps

(Zanotto *et al.* 1999). Other than drug resistance, the most compelling evidence for natural selection in HIV-1 involves the action of cytotoxic T CD8+ cells (CTLs), the cellular arm of the human immune system (McMichael and Phillips 1997, Ogg *et al.* 1998). If continual selective sweeps of viral mutants that escape CTL (or neutralizing antibody) recognition occur during the course of intrahost viral evolution, then levels of standing genetic variation, and hence N_e values, will be reduced. For other RNA viruses where immune selection is likely to be an important factor shaping intrahost evolution, including HCV (Farci *et al.* 2000) and influenza A virus (Fitch *et al.* 1991), θ and N_e values may likewise be low. Another important selectively driven process in this context is clonal interference, in which advantageous mutations become transiently common but do not achieve fixation because of interfering beneficial mutations (Miralles *et al.* 1999). Clonal interference is therefore another reason why high fitness variants do not always achieve fixation.

Measuring the strength of selection pressures in RNA viruses has proven equally complex. The most important number here is the proportion of the viral genome that can be considered truly neutral. The more neutral sites there are, the more genetic drift will disrupt the mutational coupling required for quasispecies formation. Data from VSV and FMDV suggest that the genomes of these viruses may contain sufficiently large numbers of neutral synonymous sites to prevent whole genomes forming quasispecies (Jenkins *et al.* 2001). Unfortunately, it is unclear whether the neutrality of synonymous sites applies to RNA viruses generally, or what proportion of nonsynonymous sites are free to change without constraints.

Two types of selective constraint could affect synonymous sites: those due to RNA secondary structure and those imposed by codon usage bias. Many RNA viruses possess short stretches of untranslated (nonprotein coding) sequence at their 5' and 3' ends which play critical roles in their life cycles. Flaviviruses, for example, which include such important human pathogens as dengue, yellow fever, and West Nile virus, have an ~500 nucleotide 3' untranslated region with strong RNA secondary structure. Critically, it is this RNA secondary structure, rather than specific sequence motifs, that seems to determine important phenotypic properties, including virulence (Leitmeyer *et al.* 1999, Proutski *et al.* 1997). More controversial is whether RNA secondary structures extend to the coding regions of genes. Studies of VSV and FMDV revealed long stretches of genome without clear RNA secondary structure, which also means that synonymous sites are unlikely to interact epistatically (Jenkins *et al.* 2001). In contrast, extensive secondary structure at synonymous sites has been detected in another flavivirus, GBV-C (also known as hepatitis G virus, HGV) (Simmonds and Smith 1999). Which pattern is the norm for RNA viruses remains to be determined.

A more tractable issue concerns the extent of codon usage bias in RNA viruses. We recently measured codon bias in 50 diverse human RNA viruses using the effective number of codons, N_C (Jenkins and Holmes 2003). N_C

can range from 20, when only one codon is used for each amino acid, to 61, when all codons are used equally (Wright 1990). Weak biases were observed, with an average N_C of 50.89. This means that there is less bias than in other well-studied species, including *E. coli* ($N_C = 45.0$), *S. cerevisiase* (48.3), *D. melanogaster* (46.2) and humans (range 30 to 61, mean ~45) (Powell and Moryama 1997, Wright 1990). More revealing was that codon bias generally matched that predicted by underlying base composition, which points to neutral mutation pressure and genetic drift as the principal causes of bias in RNA viruses, rather than natural selection for translational efficiency or accuracy. Such a conclusion would be surprising if RNA viruses do possess very large effective population sizes. Because of the small selection coefficients usually associated with synonymous substitutions, natural selection will only be able to control codon bias in species with high N_e, such as bacteria (reviewed in Mooers and Holmes 2000). In contrast, the much smaller N_e values of mammalian species preclude efficient selection for codon choice, so that mutation pressure is the most likely cause of any biases observed. Consequently, the weak codon biases observed in RNA viruses suggest that their effective population sizes may indeed be far lower than census numbers as suggested in HIV-1. Alternatively, it could be that mutation rates are so high and the selection coefficients associated with different codons so low that translational selection is unable to operate efficiently, or that successive population bottlenecks, which occur when viruses are transmitted between hosts, dissipate the action of natural selection (Bergstrom *et al.* 1999). Whatever the underlying cause, the evidence that many synonymous sites in RNA viruses are evolving neutrally argues for the general importance of genetic drift in viral evolution.

Even with the evidence for genetic drift in RNA viruses it is still possible that sequence space is in some cases limited enough for quasispecies formation. In particular, the fitness landscapes of viruses in nature may be extremely rugged, as appears to be true of ϕ6 (Burch and Chao 1999), so that variants become trapped in small regions of sequence space, which in turn will prevent widespread drift. As ever, determining the structure of fitness landscapes is notoriously difficult. To date, the only studies of fitness interactions in viral populations have used laboratory-adapted populations which may bear little resemblance to evolution *in vivo*.

The question of how many RNA viruses form quasispecies therefore remains unresolved. One possible research avenue for the future involves the wealth of data now being generated documenting resistance to antiviral therapy, particularly in HIV-1. As the fixation times of resistance mutations can be measured with some accuracy, so their selection coefficients can also be estimated (Goudsmit *et al.* 1996). Crucially, these data may be able to tell us whether genotypes of low fitness, such as drug-sensitive viral strains, are able to outcompete those of higher fitness if they act cooperatively with beneficial mutational neighbors.

20.3 What determines the rate of RNA virus evolution?

The analysis of rates of nucleotide substitution has long been used to investigate key processes of molecular evolution, including the roles played by selection and drift (Gillespie 1991, Kimura 1983). Such an approach has also been employed on RNA viruses, where their rapid and supposedly clock-like evolution has been cited as evidence that they conform to the predictions of the neutral theory (Gojobori *et al.* 1990).

It is now well established that the intrinsic rate of point mutation in RNA viruses is extremely high – in the region of 0.76 mutations per genome replication for lytic RNA viruses and 0.2 for retroviruses (Drake and Holland 1999). Such a high mutation rate is evidently due to the error-prone nature of replication using RNA polymerases, which lack any proof-reading or repair activity. This sits in contrast to DNA viruses replicated by DNA polymerases of greater fidelity which have mutation rates similar to those of bacteria – in the order of 0.0034 mutations per genome replication (Drake *et al.* 1998).

Perhaps of more interest to evolutionary geneticists is the proportion of these mutations that are deleterious. Determining the rate of deleterious mutation rate per replication, U, has become a *cause célèbre* of evolutionary biology as high values of U may constitute a key selection pressure for the maintenance of sexual reproduction (Kondrashov 1988). Although rarely studied in this context, RNA viruses represent highly informative model organisms since the negative side of their rampant mutagenesis is the production of many defective progeny. Although some of these defective viruses may still express enough surface protein to decoy immune defenses, and although others, the defective interfering (DI) viruses, will compete with fully functional viruses for replication materials, many will be quickly removed by purifying selection. Consequently, extreme mutation rates mean that RNA viruses are highly susceptible to fitness loss, especially when combined with transmission bottlenecks. In clonal viruses the endpoint of this cumulative loss of fitness will be Muller's ratchet (Chao 1990; see below). It is even possible that the fitness losses associated with unconstrained mutagenesis could be exploited in the development of new antiviral drugs; if the viral mutation rate can be increased artificially through the use of mutagens, then the reduction in fitness might be so severe that no genotype will be able to reproduce itself faithfully and the infection will be eradicated (Crotty *et al.* 2001, Loeb *et al.* 1999).

To date, the only direct estimate of U in a RNA virus – recorded at 1.2 mutations per genome replication – was obtained in a mutation-accumulation study of VSV, an apparently clonal RNA virus (Elena and Moya 1999). Broadly similar rates were provided by Drake and Holland (1999) who estimated the overall mutation rate in four RNA viruses (median = 0.76 mutations per genome replication) and suggested that most of these mutations would be deleterious. Such high rates must be close to the error threshold (Eigen 1996) – theoretically the maximum mutation rate tolerable before extensive fitness losses are encountered. However, that RNA polymerases are intrinsically error prone does not necessary mean that all viruses evolve both quickly and at roughly equivalent rates. In particular, it is possible that rates of replication vary extensively, particularly as viruses infect a broad range of host species and are transmitted in a wide variety of ways. Moreover, aside from well-studied viruses like HIV-1, little is known about replication rates *in vivo.*

The analyses of nucleotide substitution rates undertaken to date agree, with a few exceptions (see below), that RNA viruses evolve rapidly and that these rates are in the range 10^{-3} to 10^{-4}substitutions per site per year. One recent study considered substitution rates in 50 diverse RNA viruses, recording the numbers of changes that accumulated between viruses sampled at different times (Jenkins *et al.* 2002). Although for approximately one-third of the viruses studied the lower confidence limit of the substitution rate was zero, making accurate rate estimation impossible, the remaining substitution rates occupied a surprisingly narrow range of variation, from 0.16×10^{-3} substitutions per site per year (Western equine encephalitis) to 2.8×10^{-3} (swine vesicular disease virus). Intriguingly, there was some correlation between rate variation and other biological features of the viruses in question. Most notably, vector-transmitted viruses evolved at significantly lower rates than other viruses, which may be due to increased selective constraints imposed on viruses that need to replicate in very different sets of host species (Weaver *et al.* 1999, although this has been debated – Novella *et al.* 1999). Substitution rate also declined with increasing genome length, which in turn suggests that there is an inverse relationship between mutation rate and genome length, as expected if there is an error threshold. However, no relationship was found between substitution rate and codon usage bias, again supporting the idea that synonymous site evolution is predominantly neutral.

The most likely explanations for a substitution rate overlapping zero in some viruses were the use of short sequences or a lack of temporal distinction among samples. However, it is also possible that there is a subset of slowly evolving RNA viruses which have accumulated few or no sequence changes over the sampling period. This is best documented in GBV-C and the human T-lymphotropic retroviruses (HTLV-I and HTLV-II) where substitution rates in the range 10^{-6} to 10^{-7} have been estimated (Suzuki *et al.* 1999, Van Dooren *et al.* 2001). The prime evidence for a low substitution rate in GBV-C is its apparent cospeciation with various primate species (Charrell *et al.* 1999), as well as the supposed correspondence between viral genotypes and human population groups (Pavesi 2001). Conversely, studies of GBV-C sequence variability within individual patients and families have produced rate estimates in the range 10^{-3} to 10^{-4} substitutions per site per year, similar to those in other RNA viruses (Nakao *et al.* 1997). Hence, the true substitution rate of GBV-C, or why it might evolve anomalously slowly, are uncertain. Substitution rates within the range of other RNA viruses have also been estimated in HTLV-II strains sampled from injecting drug-users (Salemi *et al.* 1999).

In this case it is suggested that substitution rates vary according to rate and mode of transmission. Hence, in endemic situations where vertical transmission is common, both HTLV-I and HTLV-II maintain themselves within hosts through the clonal expansion of infected cells rather than active replication. However, in epidemics with high rates of transmission, such as those in injecting drug-users, most viral production is due to replication as the virus rapidly establishes itself in new hosts (Salemi *et al.* 1999).

Lastly, Jenkins *et al.* (2002) also observed that although substitution rates did not vary greatly among viruses, there was often enough rate variation *within* each virus that a molecular clock was rejected. Such local rate variation may be due to variable selective constraints among lineages, perhaps expected if a virus infects a variety of host species causing different disease syndromes. As a case in point, although the virulent HIVs and avirulent SIVs (simian immunodeficiency viruses which infect a wide range of African primates) seem to replicate and mutate at similar rates, SIVs accumulate proportionally fewer nonsynonymous substitutions (d_N) to synonymous substitutions (d_S) per site during intrahost viral evolution, suggesting that they are subject to purifying selection, whereas the HIVs experience strong immune-driven positive selection (reviewed in Holmes 2001).

20.4 How important is recombination in RNA virus evolution?

There is a growing awareness that RNA viruses recombine more frequently than was previously imagined, although usually far less often than most DNA-based organisms. Recombination has now been documented within particular RNA viruses, between viruses from different species groups, such as the different primate immunodeficiency viruses (Jin *et al.* 1994), between RNA and DNA viruses (Gibbs and Weiller 1999) and even between viral and cellular genes (Meyers *et al.* 1989). As such, recombination is clearly an important mechanism in the generation of new viruses. In particular, it has been proposed that viruses undergo a form of "modular evolution" in which recombination produces new combinations of key functional modules, such as those which determine the viral polymerase and capsid (Gorbalenya 1995). Quite clearly, recombination among diverse viruses also means that their evolutionary history cannot be depicted by simple bifurcating trees and perhaps that taxonomic units above the family level should be avoided altogether (Zanotto *et al.* 1996).

Two forms of recombination occur in RNA viruses. Reassortment describes the process in which those RNA viruses with segmented genomes exchange these segments. This is best documented in influenza A virus where genome segments containing the hemagglutinin (H) and neuraminidase (N) proteins are interchanged, including those from different host species, producing strains which can evade preexisting immunity. These novel recombinant forms are often associated with major influenza epidemics through an "antigenic shift." The second process is "true" RNA recombination. The most likely mechanism for this is "copy-choice" replication, in which the viral RNA-dependent RNA polymerase switches from one RNA molecule to another during replication, generating mosaic genomes (Nagy and Simon 1997). This process has now been documented in a variety of RNA viruses (reviewed in Worobey and Holmes 1999).

Although much is known about the precise molecular mechanics of RNA recombination, the broad evolutionary implications of this process are far less certain. In particular, it is unclear why rates of recombination vary so extensively among viruses, even between those that are closely related. As a case in point, HCV and GBV-C are both members of the Flaviviridae, both cause persistent infections and both are found at relatively high frequencies in human populations. Yet whereas HCV appears to be clonal, GBV-C recombines very frequently (Worobey 2001). There are also important practical benefits to understanding the determinants of recombination rate in RNA viruses. High levels of recombination indicate that multiple infections must be common, which in turn means that there is little immune protection against heterologous strains.

There are three main, although not mutually exclusive, theories for why rates of recombination vary among RNA viruses. The first, and most obvious, is that recombination produces advantageous combinations of mutations that increase fitness. The selective advantage of viral recombinants has now been documented on a variety of occasions. For example, recombination in HIV-1 has been cited as a means by which resistance to antiviral therapy can be rapidly generated (Moutouh *et al.* 1996). A second, rather more controversial hypothesis, is that recombination allows deleterious genomes to be purged from populations, thereby allowing viruses to escape Muller's ratchet. Indeed, RNA viruses represent one of the few organisms in which Muller's ratchet has been tested experimentally. These studies have generally supported its existence, at least if there are major bottlenecks at transmission (Chao *et al.* 1992), and revealed that sex in the form of reassortment can allow viruses to escape from deleterious mutation accumulation (Chao *et al.* 1997). A key question for the future is whether rates of recombination are in any way correlated with those of deleterious mutation.

Although it is tempting to invoke only adaptive explanations for variation in recombination rates in RNA viruses, a third theory is simply that it is a passive outcome of genome structure and ecological opportunity and not a selected entity at all. Constraints against recombination could arise in a variety of ways. First, coinfection of individual hosts might not occur because viral strains are ecologically or temporally separated or because strong immune responses prevent multiple infection. Similarly, viruses may be unable to coinfect a single cell because of cellular factors that block entry of more than one virus particle. This blocking may even be virally determined. Studies of bacteriophage $\phi 6$ found an optimal coinfection number of two to three viruses per cell, possibly because this balances the benefits of reassortment with the costs of intracellular competition (Turner *et al.* 1999). It is also clear that RNA viruses differ in their capacity to undergo copy-choice replication. Most notably, negative-strand RNA viruses, whose genomes are packaged into filamentous ribonucleoprotein (RNP) structures, may be less permissive to this process than other RNA viruses, and recombination in negative-strand RNA viruses has only been described on occasion. Furthermore, sequence dissimilarity will impose a physical threshold on which viruses are able to recombine as divergent viruses are less likely to form hybrids. For example, while recombination has been described within each of the four serotypes of dengue virus, it has never been documented among these serotypes which are far more genetically divergent (Worobey *et al.* 1999). Even if recombination does take place, the novel viruses produced must pass through the sieve of purifying selection, perhaps the strongest constraint of all.

20.5 The evolutionary genetics of viral emergence

Thus far I have discussed the evolutionary genetics of RNA viruses as they interact with their long-term hosts. An issue of equal importance in viral evolution is understanding how viruses establish themselves in new hosts; that is, the mechanisms by which new viruses emerge. Although the origins of some important RNA viruses of humans remain a mystery, most notably HCV, the observation that related viruses are often found in animal species points to cross-species transfer as the key determinant of viral emergence. This is clearly documented in HIV, where the related SIVs are commonly found in African primates, with no evidence that they cause disease (Hahn *et al.* 2000). In this case, SIVcpz from chimpanzees is the progenitor of HIV-1 and SIVsm from the sooty mangabey, the reservoir for HIV-2 (Figure 20.1).

In general, the factors that enable viruses to jump species boundaries are of two types, reflecting either genetic or ecological characteristics of the virus and host in question. The genetic factors that underpin viral emergence can be classed as those affecting either host genetic susceptibility to infection, such that the recipient species has a cellular or immunological constitution that determines whether a new virus will take hold, or viral infectiousness, so that a virus has particular characteristics that control its ability to successfully replicate in the cells of a new host species. In this context it is clear that there are phylogenetic constraints to the extent of cross-species transfer, with viruses that normally infect such distantly related host species as plants unable to infect humans despite the number we ingest on a regular basis. Furthermore, although there is no "universal" cell type preferred by emerging viruses, there is some correlation between the diversity of cells a virus can infect and the diversity of species it uses as hosts (R. T. Palframan and E. C. Holmes, unpublished data). Consequently, viruses that are "generalist" in the cells they infect may also be generalist in the species they infect. This is exemplified by the rabies viruses (genus *Lyssavirus*) which infect a wide range of mammalian species. Although

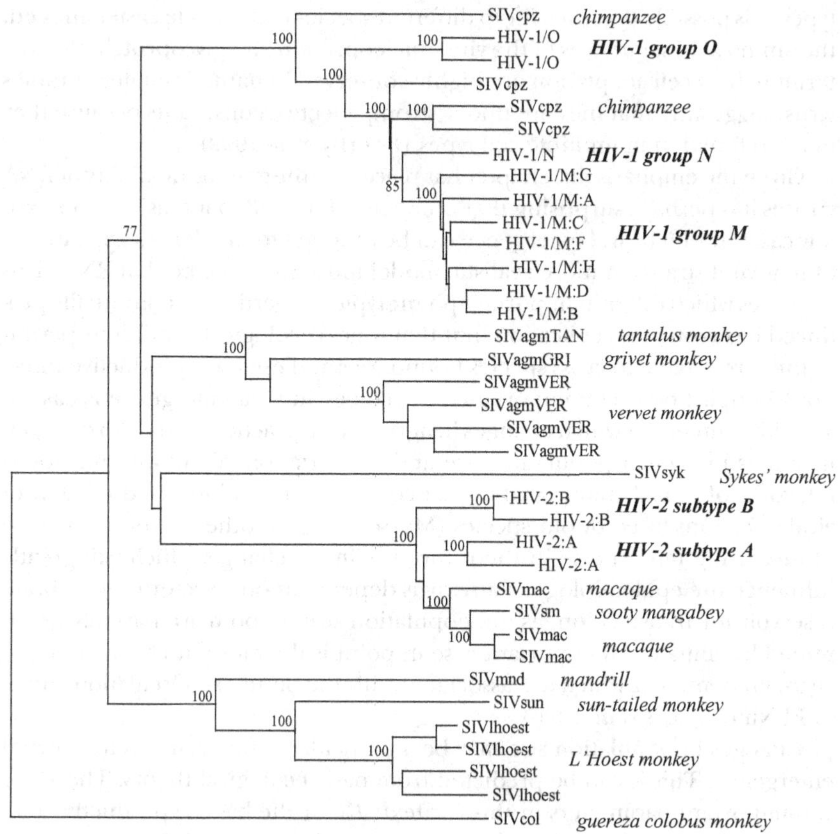

Figure 20.1. Phylogenetic tree of the major groups of primate immunodeficiency viruses. The tree was reconstructed using amino acid sequences from the polymerase gene under the neighbor-joining method (JTT distance statistic). The tree is mid-point rooted and bootstrap values for key nodes are shown. Three groups of HIV-1 have been identified, the globally distributed "M" ("main") group, and two groups of much lower prevalence from West-Central Africa – "O" ("outlier") and "N" ("new"). The M group can be further subdivided into a series of subtypes (denoted A–D, F–H, J, and K), some of which are shown here. The tree reveals that groups M, N, and O most likely represent independent transfers from chimpanzees (SIVcpz). HIV-2 seems to have been transferred from sooty mangabey monkeys, again on multiple occasions, and is generally restricted to people of West African origin. To date, seven subtypes of HIV-2 have been identified.

rabies virus exhibits a strong preference for neuronal cells, the virus also appears to replicate in muscle tissue at the site of inoculation before entering the central nervous system, the salivary glands, as well as other nonnervous tissues (Charlton *et al.* 1997). Hence, if there is strong selection pressure for broad tropism in rabies virus, so it is adapted to replicate in a diverse range of cell types, it is possible that transfer to different species is also made easier. Indeed, the amino acid sequences of the viral nucleoprotein and glycoprotein that determine host cell adaptation are highly conserved in natural isolates of rabies virus, suggesting that they are under strong selective constraints because they need to function in multiple cell types (Bourhy *et al.* 1999).

Given the emphasis this chapter has placed on the genetic flexibility of RNA viruses it is perhaps surprising that, some novel recombinants aside, there are few cases in which viral emergence can be assigned to the *de novo* appearance of new viral strains. A more realistic model must therefore be that RNA virus genomes which differ in important phenotypic properties are continually produced by erroneous replication, but that unless ecological conditions permit it, they are unable to cross species boundaries and establish productive infections in new hosts. The importance of ecology in viral emergence is easy to see with humans, in which changes in agricultural practices and deforestation, increases in human population size and urbanization, along with improved networks of global transportation, all contribute to an increased burden of viral infections faced by our species (Morse 1994). Another factor likely to be of increasing importance in the future is climate change, which will greatly influence the epidemiology of any virus dependent on a vector or vertebrate reservoir for transmission, as the population sizes of both are partially determined by climate. An important case in point is the mosquito-borne dengue virus, epidemics of which are associated with the Southern Oscillation Index of El Niño (Hales *et al.* 1996).

Changes in population size may be a particularly important factor in viral emergence. This is can be predicted from basic ecological theory. There are two important parameters in this context: R_0, or the basic reproductive rate of the pathogen, defined as the average number of new infections caused by a single index case in an entirely susceptible population (Anderson and May 1991), and N_T, the threshold density of hosts which must be breached if a viral epidemic is to be sustained (i.e., for $R_0 > 1$). The value of N_T required for any virus depends on various aspects of its biology, such as the duration of infectiousness, transmittability, and virulence. Hence, viruses of short duration, low transmittability and high virulence require the largest host populations to sustain their spread. Such a relationship explains why many small hunter–gatherer populations generally do not carry acute RNA virus infections endemically (Black 1975), and why disease burden in humans increased following the advent of agriculture, when population sizes rose but societies became less mobile, and with urbanization (Strassman and Dunbar 1999). Demographic changes will be of equal importance for disease emergence in wildlife species. For example, the importation of large numbers of raccoon dogs (*Nyctereutes procyonoides*) into northeast Europe for fur farming during the 1920s to 1950s provided a large pool of susceptible hosts for rabies virus, and this canid species is now an important virus reservoir in Europe (Bourhy *et al.* 1999).

While the rapidly expanding human population will mean that more viruses of increased virulence may eventually emerge in the future, it is important to remember that new viruses often first appear in small populations. This may have important consequences for the evolution of RNA virus diseases. Most notably, the genetic drift of viruses which occurs because of the random sampling of strains at transmission will be more pronounced in small host populations so that there is likely to be a strong stochastic element in the types of viruses that emerge and the phenotypic properties they possess. Indeed, chance sampling might be so strong that even highly transmittable strains will be lost by drift when they first emerge and the level of virulence attained may not be that which maximizes the reproductive success of the virus (Bergstrom *et al.* 1999).

An example of the potential importance of small populations in viral evolution is HIV-1. For sexually transmitted diseases like HIV-1, genetic drift will be particularly strong if there is great variability in rates of partner exchange, such as if most transmissions are due to a few core individuals. In an analogous manner to intrahost viral evolution, this means that the N_e of HIV-1 at the population level may be substantially less than its overall prevalence (Grassly *et al.* 1999). This will greatly limit the power of natural selection at the population level, which may explain why there is no strong evidence that different strains of HIV-1 (referred to as subtypes) are outcompeting each other, despite the phenotypic diversity produced on a daily basis. Further, even with a high population N_e, the bottleneck at HIV-1 transmission might be so great that substitution dynamics are mainly under the control of stochastic processes.

A second, more speculative but no less intriguing, impact small population size may have on HIV-1 concerns the evolution of virulence. It seems likely that HIV-1 was circulating in small rural populations in Africa for decades before finally entering far more dense urban populations. Because smaller and less dense host populations are only able to sustain infections of lower virulence, it follows that the virulence of HIV-1 may have been lower in the past than it is today, otherwise susceptible hosts would be lost too rapidly and the infection would burn out. This could explain why the emergence of HIV-1 in humans seems to pre-date World War II (Korber *et al.* 2000), but that the first well-documented cases of AIDS disease in Africa did not occur until the early 1980s. Although we may never know whether the pathogenic potential of HIV-1 has changed through time, the effect of demographic history on the evolution of virulence is clearly an area that should be explored in more detail for pathogens in general.

Although a terrible burden for human society, the example of HIV-1 illustrates how RNA viruses can be used to study the dynamic interplay between genetics and ecology. Not only will this greatly enhance our understanding of the basic mechanics of molecular evolution, but it may also allow us to predict, for the first time, the types and properties of the viral diseases that will inevitably affect us in the future.

20.6 Acknowledgments

I thank Dr David Robertson for assistance with the phylogenetic analysis of the primate immunodeficiency viruses.

REFERENCES

Anderson, R. M., and May, R. M. (1991). *Infectious Diseases of Humans*. Oxford: Oxford University Press.

Bergstrom, C. T., McElhany, P., and Real, L. A. (1999). Transmission bottlenecks as determinants of virulence in rapidly evolving pathogens. *Proc. Natl. Acad. Sci. USA* 96:5095–100.

Black, F. L. (1975). Infectious diseases in primitive societies. *Science* 187:515–18.

Bourhy, H., Kissi, B., Audry, L., Smreczak, M., Sadkowska-Todys, M., Kulonen, K., Tordo, N., Zmudzinski, J. F., and Holmes, E. C. (1999). Ecology and evolution of rabies virus in Europe. *J. Gen. Virol.* 80:2545–58.

Burch, C. L., and Chao, L. (1999). Evolution by small steps and rugged landscapes in the RNA virus ϕ6. *Genetics* 151:921–7.

Burch, C. L., and Chao, L. (2000). Evolvability of an RNA virus is determined by its mutational neighborhood. *Nature* 406:625–8.

Chao, L. (1990). Fitness of RNA virus decreased by Muller's ratchet. *Nature* 348:454–5.

Chao, L., Tran, T. R., and Matthews, C. (1992). Muller's ratchet and the advantage of sex in the RNA virus ϕ6. *Evolution* 46:289–9.

Chao, L., Tran, T. T., and Tran, T. T. (1997). The advantage of sex in the RNA virus ϕ6. *Genetics* 147:953–9.

Charlton, K. M., Nadin-Davis, S., Casey, G. A., and Wandeler, A. I. (1997). The long incubation period in rabies: delayed progression of infection in muscle at the site of exposure. *Acta Neuropathol.* 94:73–7.

Cleaveland, S., Laurenson, M. K., and Taylor, L. H. (2001). Diseases of humans and their domestic mammals: pathogen characteristics, host range and the risk of emergence. *Phil. Trans. R. Soc. Lond. B* 356:991–9.

Charrel, R. N., De Micco, P., and Lamballerie, X. (1999). Phylogenetic analysis of GB viruses A and C: evidence for cospeciation between virus isolates and their primate hosts. *J. Gen. Virol.* 80:2329–35.

Crotty, S., Cameron, C. E., and Andino, R. (2001). RNA virus error catastrophe: direct test by using ribavirin. *Proc. Natl. Acad. Sci. USA* 98:6895–900.

Daszak, P., Cunningham, A. A., and Hyatt, A. D. (2000). Emerging infectious diseases of wildlife – threats to biodiversity and human health. *Science* 287:443–9.

De la Torre, J. C., and Holland, J. J. (1990). RNA virus quasispecies populations can suppress vastly superior mutant progeny. *J. Virol.* 64:6278–81.

Domingo, E., and Holland, J. J. (1997). RNA virus mutations for fitness and survival. *Annu. Rev. Microbiol.* 51:151–78.

Domingo, E., Escarmis, C., Martínez, M. A., Martínez-Salas, E., and Mateu, M. G. (1992). Foot-and-mouth disease virus populations are quasispecies. *Curr. Top. Microbiol. Immunol.* 176:33–47.

Domingo, E., Sabo, D., Taniguchi, T., and Weissman, C. (1978). Nucleotide sequence heterogeneity of an RNA phage population. *Cell* 13:735–44.

Dopazo, J., Dress, A., and von Haeseler, A. (1993). Split decomposition: a technique to analyze viral evolution. *Proc. Natl. Acad. Sci. USA* 90:10320–4.

Drake, J. W., and Holland, J. J. (1999). Mutation rates among RNA viruses. *Proc. Natl. Acad. Sci. USA* 96:13910–13.

Drake, J. W., Charlesworth, B., Charlesworth, D., and Crow, J. F. (1998). Rates of spontaneous mutation. *Genetics* 148:1667–86.

Eigen, M. (1987). New concepts for dealing with the evolution of nucleic acids. *Cold Spring Harbor Symp. Quant. Biol.* 52:307–20.

Eigen, M. (1996). *Steps Towards Life.* New York: Oxford University Press.

Eigen, M., and Schuster, P. (1977). A principle of natural self-organization. *Naturwissenschaften* 64:541–65.

Elena, S. F., and Moya, A. (1999). Rate of deleterious mutation and the distribution of its effects on fitness in vesicular stomatitis virus. *J. Evol. Biol.* 12:1078–88.

Farci, P., Shimoda, A., Coiana, A., Diaz, G., Peddis, G., Melpolder, J. C., Strazzera, A., Chien, D. Y., Munoz, S. J., Balestrieri, A., Purcell, R. H., and Alter, H. J. (2000). The outcome of acute Hepatitis C predicted by the evolution of the viral quasispecies. *Science* 288:339–44.

Fitch, W. M., Leiter, J. M. E., Li, X., and Palese, P. (1991). Positive Darwinian evolution in human influenza A viruses. *Proc. Natl. Acad. Sci. USA* 88:4270–4.

Gibbs, M. J. and Weiller, G. F. (1999). Evidence that a plant virus switched hosts to infect a vertebrate and then recombined with a vertebrate-infecting virus. *Proc. Natl. Acad. Sci. USA* 96:8022–7.

Gillespie, J. H. (1991). *The Causes of Molecular Evolution.* Oxford: Oxford University Press.

Gillespie, J. H. (1998). *Population Genetics: a Concise Course.* Baltimore, MD: Johns Hopkins University Press.

Gojobori, T., Moriyama, E. N., and Kimura, M. (1990). Molecular clock of viral evolution, and the neutral theory. *Proc. Natl. Acad. Sci. USA* 87:10015–18.

Gorbalenya, A. E. (1995). Origin of RNA viral genomes; approaching the problem by comparative sequence analysis. In A. Gibbs, C. H. Calisher, and F. García-Arenal (eds) *Molecular Basis of Virus Evolution*, pp. 49–66. Cambridge: Cambridge University Press.

Goudsmit, J., De Ronde, A., Ho, D. D., and Perelson, A. S. (1996). Human immunodeficiency virus in vivo: calculations based on a single zidovudine resistance mutation at codon 215 of reverse transcriptase. *J. Virol.* 70:5662–4.

Grassly, N. C., Harvey, P. H., and Holmes, E. C. (1999). Population dynamics of HIV-1 inferred from gene sequences. *Genetics* 151:427–38.

Hahn, B. H., Shaw, G. M., de Cock, K. M., and Sharp, P. M. (2000). AIDS as a zoonosis: scientific and public health implications. *Science* 287:607–14.

Hales, S., Weinstein, P., and Woodward, A. (1996). Dengue fever epidemics in the South Pacific: driven by El Niño Southern Oscillation? *Lancet* 348:1664–5.

Hillis, D. M., Bull, J. J., White, M. E., Badgett, M. R., and Molineux, I. J. (1992). Experimental phylogenetics: generation of a known phylogeny. *Science* 255:589–92.

Holmes, E. C. (2001). On the origin and evolution of the human immunodeficiency virus (HIV). *Biol. Rev.* 76:239–54.

Holmes, E. C. and Moya, A. (2002). Is the quasispecies concept relevant to RNA viruses? *J. Virol.* 76:460–2.

Jenkins, G. M., and Holmes, E. C. (2003). The extent of codon usage bias in human RNA viruses and its evolutionary orgin. *Virus Res.* 92:1–7.

Jenkins, G. M., Rambaut, A., Pybus, O. G., and Holmes, E. C. (2002). Rates of molecular evolution in RNA viruses: a quantitative phylogenetic analysis. *J. Mol. Evol.* 54:152–61.

Jenkins, G. M., Worobey, M., Woelk, C. H., and Holmes, E. C. (2001). Evidence for the non-quasispecies evolution of RNA viruses. *Mol. Biol. Evol.* 18:987–94.

Jin, M. J., Hui, H., Robertson, D. L., Müller, M. C., Barré-Sinoussi, F., Hirsch, V. M., Allan, J. S., Shaw, G. M., Sharp, P. M., and Hahn, B. H. (1994). Mosaic genome structure of simian immunodeficiency virus from West African green monkeys. *EMBO J.* 13:2935–47.

Kimura, M. (1983). *The Neutral Theory of Molecular Evolution.* Cambridge: Cambridge University Press.

Kondrashov, A. S. (1988). Deleterious mutations and the evolution of sexual reproduction. *Nature* 336:435–40.

Korber, B., Muldoon, M., Theiler, J., Gao, F., Gupta, R., Lapedes, A., Hahn, B. H., Wolinksy, S., and Bhattacharya, T. (2000). Timing the ancestor of the HIV-1 pandemic strains. *Science* 288:1789–96.

Leigh Brown, A. J. (1997). Analysis of HIV-1 *env* gene sequences reveals evidence for a low effective number in the viral population. *Proc. Natl. Acad. Sci. USA* 94:1862–5.

Leigh Brown, A. J. and Richman, D. D. (1997). HIV-1: Gambling on the evolution of drug resistance? *Nat. Med.* 3:268–71.

Leitmeyer, K. C., Vaughn, D. W., Watts, D. M., Salas, R., Villalobos, I., de Chacon, Ramos, C., and Rico-Hesse, R. (1999). Dengue virus structural differences that correlate with pathogenesis. *J. Virol.* 73:4738–47.

Lewontin, R. C. (1974). *The Genetic Basis of Evolutionary Change.* New York: Columbia University Press.

Loeb, L. A., Essigmann, J. M., Kazazi, F., Zhang, J., Rose, K. D., and Mullins, J. I. (1999). Lethal mutagenesis of HIV with mutagenic nucleoside analogs. *Proc. Natl. Acad. Sci. USA* 96:1492–7.

McMichael, A. J. and Phillips, R. E. (1997). Escape of human immunodeficiency virus from immune control. *Annu. Rev. Immunol.* 15:271–96.

Meyers, G., Rumenapf, T., and Thiel, H. J. (1989). Ubiquitin in a togavirus. *Nature* 341:491.

Miralles, R., Gerrish, P. J., Moya, A., and Elena, S. F. (1999). Clonal interference and the evolution of RNA viruses. *Science* 285:1745–7.

Mooers, A. Ø. and Holmes, E. C. (2000). The evolution of base composition and phylogenetic inference. *Trends Ecol. Evol.* 15:365–9.

Morse, S. S. (1994). The viruses of the future? Emerging viruses and evolution. In S. S. Morse (ed.) *The Evolutionary Biology of Viruses*, pp. 325–35. New York: Raven Press.

Moutouh, L., Corbeil, J., and Richman, D. D. (1996). Recombination leads to the rapid emergence of HIV-1 dually resistant mutants under selective drug pressure. *Proc. Natl. Acad. Sci. USA* 93:6106–11.

Moya, A., Elena, S. F., Miralles, R., and Barrio, E. (2000). The evolution of RNA viruses: a population genetics view. *Proc. Natl. Acad. Sci. USA* 97:6967–73.

Nagy, P. D. and Simon, A. E. (1997). New insights into the mechanisms of RNA recombination. *Virology* 235:1–9.

Nakao, H., Okamoto, H., Fukuda, M., Tsuda, F., Mitsui, T., Masuko, K., Lizuka, H., Miyakawa, Y., and Mayumi, M. (1997). Mutation rate of GB virus C hepatitis G virus over the entire genome and in subgenomic regions. *Virology* 233: 43–50.

Nichol, S. T., Rowe, J. E., and Fitch, W. M. (1993). Punctuated equilibrium and positive Darwinian evolution in vesicular stomatitis virus. *Proc. Natl. Acad. Sci. USA* 90:10424–8.

Nijhuis, M., Boucher, C. A. B, Schipper, P., Leitner, T., Schuurman, R., and Albert, J. (1998). Stochastic processes strongly influence protease-inhibitor therapy. *Proc. Natl. Acad. Sci. USA* 95:14441–6.

Novella, I. S., Hershey, C. L., Escarmis, C., Domingo, E., and Holland, J. J. (1999). Lack of evolutionary stasis during alternating replication of an arbovirus in insect and mammalian cells. *J. Mol. Biol.* 287:459–65.

Nowak, M. A. (1992). What is a quasispecies? *Trends Ecol. Evol.* 7:118–21.

Ogg, G. S., Jin, X., Bonhoeffer, S., Dunbar, P. R., Nowak, M. A., Monard, S., Segal, J. P., Cao, Y., Rowland-Jones, S. L., Cerundolo, V., Hurley, A., Markowitz, A., Ho, D. D., Nixon, D. F., and McMichael, A. J. (1998). Quantitation of HIV-1 specific cytotoxic T lymphocytes and plasma load of viral RNA. *Science* 279:2103–6.

Pavesi, A. (2001). Origin and evolution of GBV-C/Hepatitis G virus and relationships with ancient human migrations. *J. Mol. Evol.* 53:104–13.

Piot, P., Bartos, M., Ghys, P. D., Walker, N., and Schwartländer, B. (2001). The global impact of HIV/AIDS. *Nature* 410:968–73.

Powell, J. R. and Moriyama, E. N. (1997). Evolution of codon usage bias in *Drosophila. Proc. Natl. Acad. Sci. USA* 94:7784–90.

Proutski, V., Gaunt, M. W., Gould, E. A., and Holmes, E. C. (1997). Secondary structure of the 3'-untranslated region of Yellow Fever virus: implications for virulence, attenuation and vaccine development. *J. Gen. Virol.* 78:1543–9.

Salemi, M., Lewis, M., Egan, J. F., Hall, W. W., Desmyter, J., and Vandamme, A. M. (1999). Different population dynamics of human T cell lymphotropic virus type II in intravenous drug users compared with endemically infected tribes. *Proc. Natl. Acad. Sci. USA* 96:13253–8.

Schuster, P. and Stadler, P. F. (1999). Nature and evolution of early replicons. In E. Domingo, R. Webster, and J. Holland (eds) *Origin and Evolution of Viruses*, pp. 1–24. London: Academic Press.

Simmonds, P. and Smith, D. B. (1999). Structural constraints on RNA virus evolution. *J. Virol.* 73:5787–94.

Strassman, B. I. and Dunbar, R. I. M. (1999). Human evolution and disease: putting the stone age in perspective. In S. C. Stearns, (ed.) *Evolution in Health and Disease*, pp. 91–101. Oxford: Oxford University Press.

Suzuki, Y., Katayama, K., Fukushi, S., Kageyama, T., Oya, A., Okamura, H., Tanaka, Y., Mizokami, Y., and Gojobori, T. (1999). Slow evolutionary rate of GB virus C/hepatitis G virus. *J. Mol. Evol.* 48:383–9.

Turner, P. E., Burch, C. L., Hanley, K. A., and Chao, L. (1999). Hybrid frequencies confirm limit to coinfection in the RNA bacteriophage ϕ6. *J. Virol.* 73: 2420–4.

Van Dooren, S., Salemi, M., and Vandamme, A. M. (2001). Dating the origin of the African human T-cell lymphotropic virus type-1 (HTLV-I) subtypes. *Mol. Biol. Evol.* 18:661–72.

Weaver, S. C., Brault, A. C., Kang, W., and Holland, J. J. (1999). Genetic and fitness changes accompanying adaptation of an arbovirus to vertebrate and invertebrate cells. *J. Virol.* 73:4316–26.

WHO (World Health Organization) (1997). *Weekly Epidemiological Record* 72:65.

Wilke, C. O., Wang, J. L., Ofria, C., Lenski, R. E., and Adami, C. (2001). Evolution of digital organisms at high mutation rates leads to survival of the flattest. *Nature* 412:331–3.

Worobey, M. (2001). A novel approach to detecting and measuring recombination: new insights into evolution in viruses, bacteria, and mitochondria. *Mol. Biol. Evol.* 18:1425–34.

Worobey, M. and Holmes, E. C. (1999). Evolutionary aspects of recombination in RNA viruses. *J. Gen. Virol.* 80:2535–44.

Worobey, M., Rambaut, A., and Holmes, E. C. (1999). Widespread intra-serotype recombination in natural populations of dengue virus. *Proc. Natl. Acad. Sci. USA* 96:7352–7.

Wright, F. (1990). The effective number of codons used in a gene. *Gene* 87:23–29.

Zanotto, P. M. de A., Gibbs, M. J., Gould, E. A., and Holmes, E. C. (1996). A re-assessment of the higher taxonomy of viruses based on RNA polymerases. *J. Virol.* 70:6083–96.

Zanotto, P. M. de A., Kallas, E. G., de Souza, R. F., and Holmes, E. C. (1999). Genealogical evidence for positive selection in the *nef* gene of HIV-1. *Genetics* 153:1077–89.

21

A scientific adventure: a fifty years study of human evolution

L. LUCA CAVALLI-SFORZA
Genetics Department, Stanford University

Fifty years ago I began dedicating my attention to the study of human evolution, and it gradually became my only interest. In this chapter, I summarize my personal experience in this venture. As a student and shortly after my degree I worked in bacteriology, bacterial genetics, immunology, *Drosophila* genetics, and plankton populations. Between 1948 and 1950 I went back to microbial genetics in Cambridge (UK), at the invitation of R. A. Fisher, but my earliest training with Adriano Buzzati-Traverso, and early contacts with N. W. Timofeeff-Ressowsky had already given me a taste for population genetics. Fisher was, with J. B. S. Haldane and S. Wright, one of the three fathers of the mathematical theory of evolution, so in my stay at Cambridge I divided my time between bacterial genetics and evolutionary theory.

21.1 The role of random genetic drift in human variation

My specific interest in human evolution began as a sideline when I was introduced to documents of the Roman Catholic Archives by Father Antonio Moroni, a student of my first course in Genetics at the University of Parma, Italy, in 1951–52. I was fascinated by consanguinity dispensations and parish registers of births, deaths, and marriages, thinking of the potential interest they offered for the study of some aspects of human evolution. I was not convinced by the idea, prevailing in England at the time, that random genetic drift was unimportant as an evolutionary factor. Drift is entirely determined by demographic factors, and it occurred to me that the demographic data available in parish records could be useful for a balanced analysis of the importance of drift. The expected amount of variation due to drift can be estimated from adequate demographic knowledge of populations. The idea was to evaluate how much of the genetic variation observed among a sample of the human population, the demography of which was known, could be due to drift. The only genes whose frequencies could be studied in human populations for

The Evolution of Population Biology, ed. R. S. Singh and M. K. Uyenoyama. Published by Cambridge University Press.

estimating their variation in the early 1950s were blood groups, and for only three of them was there already a fair amount of information in the literature: ABO, MNS and RH.

I started collecting blood samples from parishes of the Parma Valley, but it took three years before I had enough data. Population density decreased tenfold from the plain to the mountains, and it soon became very clear that genetic variation between villages became greater the lower the population density. In the plains nonrandom variation between villages could not be demonstrated, at least with the number of individuals we tested, while at the higher altitude variation of gene frequencies between villages was very significant. Studying demography of the last three centuries from records of the parishes for which we had collected blood samples, we estimated the theoretical variation expected under drift, and it matched quite well the observed one. It looked, in fact, as if drift would explain all the genetic variation that we found. Natural selection might be expected to alter the observed genetic variation, but it did not.

At the beginning, the expected variation on the basis of the demographic data of village size and migration was calculated rather roughly, with the help of existing elementary theories of isolation by distance. Later we found that computer simulation of the population applied to the region where most drift was observed, a population of about 5000 individuals and 22 villages, gave more satisfactory results, because the computer simulation imitated the real population much more closely than the theoretical models available (Cavalli-Sforza and Zei 1967).

I had not published the original data of the Parma Valley before, but in the last five years I spent a substantial amount of time in Italy and, thanks to the help of Gianna Zei, I could complete their analysis. We were also able to collect new genetic data for the same villages from telephone books, where the population could be genotyped for an unusual "gene" of the Y-chromosome: surnames. The enormous number of alleles surnames supplied, and their male-limited transmission, provided much supplementary knowledge compared with the old blood group data. We had already accumulated considerable experience on the study of surnames of Sardinia (Zei *et al.* 1983) and of China (Du *et al.* 1992). At a time when money for research was scarce, surnames allowed us to collect very cheaply a large set of data, practically the whole population. Until then we had not used surnames as if they were exactly a gene with many alleles. With a suitable modification of a formula by Fisher (1943) we could calculate the most comprehensive value of drift expected for every deme, *Nm*, the product of effective population size times migration. Comparisons of results with surnames and with genes in the same population were satisfactory. In the enthusiasm generated by having discovered how closely we could study drift thanks to the power of surnames, we also produced a geographic map of drift calculated from the surname distribution in each of the 8000 or so communities in the whole of Italy. This map of drift in Italy shows the effects of environment and of local socioeconomic developments with extraordinary detail (Cavalli-Sforza *et al.* 2003).

21.2 Tree analysis of population evolution

The suggestion that the importance of drift had been underestimated began to be appreciated by geneticists following the famous *Nature* 1968 paper by Kimura. My observations had indicated earlier, to me at least, that drift was often a dominant influence. This was a very useful finding also for another reason. I had been interested since my time at Cambridge in trying to reconstruct evolutionary trees on the basis of polymorphic gene frequencies in living populations. If natural selection was the only important cause of genetic variation, a tree would reflect mostly adaptation. If chance, that is drift, had an important role in determining genetic variation between populations, genetic trees were more likely to reconstruct correctly their history of separation and differentiation. But it was clear that it was necessary to have more gene systems than the three we used in Parma, and to study data on the same genes for many populations.

In the early 1960s I felt that as enough genetic polymorphisms were already known in humans, it might be possible to reconstruct an evolutionary tree of human living populations with their help. Anthony Edwards, another student of Fisher, collaborated in the development of mathematical methods and the use of computers, which were then becoming available to universities. We also obtained the cooperation of Arthur Mourant, who had a thorough knowledge of existing data on human polymorphisms. I wanted three geographically well-separated populations from each continent, and as many genes as available. In addition to the three genes we had already used in Parma, it was only possible to use two other blood group systems, Diego and Duffy. Altogether this meant 15 independent alleles, not as many as I hoped. We generated several methods of tree reconstruction and we also successfully used principal components displays, a classic statistical method that had never been used in genetics. Our tree put together, automatically, populations from the same continent, and gave to continents positions in agreement with their geographic locations. This looked like a success, but was the tree a true reconstruction of the history of separations and differentiations of populations? At the time, the history of human evolution was practically unknown, at least as regards the most recent part. There are people who deny even today that trees represent history. But when I gave at the International Congress of Genetics in the Netherlands, in 1963, our paper on the evolutionary analysis of human populations (Cavalli-Sforza and Edwards 1964), the response was very positive. Fisher, Haldane, Neel, Morton and others made favorable comments and gave suggestions.

In the lack of methods for establishing the root of the tree, we had given it a central location, and therefore it separated Africa and Europe on one side, and the rest of the world on the other. We know today that this is wrong: the root is not at all central, mostly because the origin of modern humans, in East Africa, was followed by an expansion eastwards, all the way through Asia to America, and a strongly asymmetric tree was the result. Today the root is more easily discovered because one can use the nearest Primates as "outgroup." The nonrecombinant Y (NRY) chromosome tree (Underhill *et al.* 2000) gives undoubtedly the clearest support to the root separating Africans from non-Africans, and other observations and analyses also show two major expansions of a small initial population: an early one, about 100 000 years ago, from East Africa to the rest of Africa, and a second one, about 50 000 years ago, from Africa to Asia and from Asia to the other continents. Naturally, the number of genes we could use at the beginning was too small for allowing such inferences. But even before DNA polymorphisms became available it was possible for Nei and Roychoudhury (1974, 1982), using many more protein markers that meanwhile became available, to establish that the root separated Africans from the rest of the world.

Another anomaly of our tree, which is repeated in many other world trees, was that Europe and to lesser extent some other populations had short branches. We believe, and already suspected at the time, that the anomaly is not due to a smaller evolutionary rate of Europeans, but to their mixed origin. A formal analysis of the hypothesis that both Asians and Africans contributed to Europeans was given much later, using 100 RFLPs (Bowcock *et al.* 1991).

21.3 Principal components analysis vs. trees

Anthony Edwards and I were able to carry out the very first analysis of 15 world populations with five gene systems, begun in 1961, because we had at Pavia in Italy an Olivetti Elea computer and one of the first FORTRAN compilers (Edwards and Cavalli-Sforza 1964, Cavalli-Sforza and Edwards 1967). This gave us a chance to try not only our new methods of tree analysis, but also principal components analysis (PCA). Not surprisingly, although it was invented in 1935, it had had only very few applications before computers became available. We now know there are two advantages to PCA compared with trees. When there have been major episodes of gene exchanges between two or more populations, tree analysis is misleading, but plots of the first two or three PCAs in two or three dimensions may sometimes give a more faithful graphic representation of a set of populations than a tree. Trees are very parsimonious models, as they need a number of parameters equal to the number of populations minus one (without root), which are sufficient to define the tree branches. But they use a very rigid hypothesis: equal rate of differentiation in all branches, sharp separations, and independence of evolution after separation (which includes absence of cross-migration between branches). Some of the constraints can be relaxed even when using trees, at the cost of additional labor.

PCAs use many more parameters than trees and are therefore more flexible. When data follow closely the simplest expectations generating a good tree, trees and PCAs tend to give identical results (Cavalli-Sforza and Piazza, 1975), with a one-to one correspondence of tree nodes and PCA eigenvalues. Genetic exchanges are almost always more likely among geographically adjacent populations, and then PCA displays in two dimensions tend to resemble the geographic map. Thus they come close to telling us about geography, but may lose information on history, which is more directly told by trees when all the major hypotheses are respected.

21.4 The spread of farming

PCAs are also useful for generating geographic maps of single components. These maps may help in detecting major internal migrations and expansions within continents, and also other factors of evolution. Our first geographic maps of single PCAs (Menozzi *et al.* 1978) were made in the hope that they might supply a way to dissect major population radiations. The radiation to be tested had been the subject of an earlier collaboration with archeologist Albert Ammerman, started in 1971 (see Ammerman and Cavalli-Sforza 1971, 1973, Cavalli-Sforza and Ammerman 1984), and was designed to study the spread to Europe of farming from its original area in the Middle East (the area known as the Fertile Crescent). Wheat was first domesticated in the Middle East and was practically nonexistent in Europe, prior to its appearance with neolithic farmers. We were specifically interested in understanding if the spread of agriculture, marked archeologically by the appearance of wheat, was effected by the expansion of farmers themselves, rather than of the farming technology, adopted by hunter–gatherers learning from neighboring farmers. We called the first type of diffusion "demic," and the second "cultural." We had no idea what the possible relative importance of the two mechanisms could be, but geographic distributions of genes like RH and HLA gave an impression that there might have been some demic diffusion from the Middle East. Mapping the radiocarbon dates of the earliest appearance of wheat in European archeological sites convinced Ammerman and me that the diffusion of neolithic farming was very slow. Population growth is slow, and a demic diffusion was likely to be slow compared with the cultural one, which can be rapid. But there is also an example of a true cultural diffusion built into the radiation of agriculture. For the first millennia in the Middle East and Turkey, farmers did not use pottery. The invention of ceramics came later to this area compared with other areas (e.g., Japan), and began in the Middle East only around 8500 years ago – the oldest known site with agropastoral economy, Abu Hureyra in northern Syria, is about 11 500 years old. Thus the farmers' culture remained aceramic for a long time, but after pottery became available in the Middle East it took very little time to reach the furthest outposts of the Mideastern neolithic, which were already in Greece.

From an analysis of radiocarbon dates of first arrival of wheat in Europe we estimated the observed rate of spread of farming to be about 1 km per year. Using a mathematical theory by Fisher (1937) that could be adapted to predict the rate of diffusion of a population from the geometric mean of the rate of population growth and its migration, we showed that this rate is compatible with the spread of farming by the demographic growth and expansion of farmers. But in the 1960s and 1970s, Anglo-American archeologists believed that people never moved, an attitude called indigenism, and would not accept the demic hypothesis. Fortunately one of their leaders, Renfrew, changed his mind (1987), and accepted demic diffusion in the case of the spread of agriculture from the Middle East. He also suggested that Mideastern farmers spread Indo-European languages to Europe, and to southwest Asia. This hypothesis of the diffusion of Indo-European languages is still fiercely debated, but I believe it is likely to be correct, with minor modifications.

Today the resistance of Anglo-American archeologists to demic farming has decreased, although some are still rather vociferous. Archeologists of continental Europe seem to be fully satisfied with the demic explanation (Ammerman and Biagi 2003). We developed genetic observations to test the demic hypothesis. A purely cultural diffusion was unlikely to change the gene frequencies, but if farmers moved into the territory of hunter–gatherers bringing their farming technology, and married in the hunter–gatherer society which was almost certainly genetically different, especially away from the place(s) of origin of agriculture, then a gradient of gene frequencies might be observed around the area of origin of agriculture. The same result would follow if, at the time of the arrival of farmers, some of the previous settlers joined the bandwagon (i.e., if there was cultural diffusion). The two phenomena: farmers/foragers marriages, and cultural diffusion accompanying demic diffusion, would inevitably cause, jointly or in isolation, a genetic gradient in the area of the spread of farming, if there was a sufficient genetic difference between the previous settlers at the extremes of the area.

The geographic map of the first PCA of 95 gene frequencies in Europe produced by us did show a gradient originating in the Middle East (Menozzi *et al.* 1978), which closely resembled the map of archeological radiocarbon dates of wheat arrival. We always insisted that no genetic gradient of PCAs could exist unless both mechanisms of diffusion, demic and cultural, participated. Even today there are misunderstandings of this phenomenon. Naturally, the gradient can be detected only for genes that showed a difference between foragers and farmers since the beginning, but it should be comparable. Sokal *et al.* (1991) showed this to be true for six autosomal genes they tested. One can always hypothesize that strong selection at the extremes of the gradient range, and not the radiation postulated, was responsible for the observed genetic gradients, but this is perhaps beyond the limits of coincidence. There are today three examples of haploid (NRY, nonrecombinant Y chromosome) genes, all in strong agreement with the first autosomal PCA (Semino *et al.* 2000).

We showed that the first PCA explains 26–28% of European genetic variation. Later tests with haploid markers, like NRY and mtDNA (Richards *et al.* 1998) indicated that around 20% of the European population seems to have originated in the Middle East. It would be strange if this concordance were due to selection. The estimate is based on mutants which arose too late for having been part of earlier radiation(s) from Africa.

21.5 The correlation of genes and languages

The gene–geographic study of Europe, carried out with Paolo Menozzi and Alberto Piazza (Menozzi *et al.* 1978) for the purpose of testing the demic hypothesis in the spread of agriculture, motivated us to extend it to the rest of the world. It took 16 years to collect the data from the literature, analyze them, write a book and bring it to publication, under the name of *History and Geography of Human Genes* (Cavalli-Sforza *et al.* 1994). Results of the analysis of some 1800 populations for over 110 polymorphic genes (for nearly 100 000 gene frequency data) convinced us that there was a good correspondence of genetics, archeology, and linguistics. The correlation between genetics and archeology can be summarized by saying that the ratio between the average genetic distance between Africa and the rest of the world, and that among non-Africans, was quite similar to the ratio of the archeological time of the first appearance of modern humans in Africa (*c.* 100 000 years ago) to the date of the expansion from Africa to the rest of the world (40 000–50 000 years ago). The comparison of genetic distances with archeological estimates of population separations was at the time the only possible approach. Data on protein alleles did not permit a direct dating based on knowledge of mutation rates, which is possible today with DNA polymorphisms. But these methods, leading to estimates of the time of a most recent common ancestor (TMRCA), identified by a specific mutation, have huge statistical errors, at least with the number of individuals usually employed. Moreover, the time separation in the gene genealogy is expected to precede the population separation. How long is the interval between the time at which the mutation occurred, and the time of the population separation? It may be estimated using rather restrictive assumptions, but it is even possible that migration may cause the mutation to postdate the separation (Rosenberg and Feldman, 2002).

Using the large number of protein data collected from the literature, we generated a tree of about 40 populations representing the world. This tree was published prior to the 1994 book (Cavalli-Sforza *et al.* 1988) to illustrate also a serendipitous finding: the existence of a strong correlation between the world genetic tree and what was then known of the linguistic tree. Languages were used by us when collecting the population genetic data from the literature, because they supplied the only practically complete, hierarchical list of human populations. The linguistic coding was very useful as a help to create a data base of the genetic information, available from the *c.* 100 000 gene frequency data of blood groups and other protein polymorphisms collected for generating the tree. The tree of languages used was based on the most satisfactory classification then available of the *c.* 5000 existing languages in 17 families, and fewer superfamilies connecting them. It was first published in 1987, and was kindly made available to us by its author, M. Ruhlen. It is almost entirely derived from the remarkable taxonomic work by J. Greenberg (1963). It was then, and still is, incomplete in its upper part, but several gaps have been filled by unpublished work.

There was at the time (1988) no simple and fully satisfactory method for measuring the correlation between trees. But the similarity of the genetic and linguistic trees was unmistakable, in spite of some obvious exceptions, all of which had clear historical explanations. Later we could show that the observed similarity could have arisen by chance only with a probability of 1/10 000 (Cavalli-Sforza *et al.* 1992). A different method by Penny *et al.* (1982) gave independently a very similar estimate.

Why should there be a correlation between the genetic and the linguistic tree? My interest in cultural transmission and evolution, which gave rise to a long collaboration with Marc Feldman (Cavalli-Sforza and Feldman 1981 and many papers) suggested the answer. We called "vertical" the cultural transmission from parents to children, or generally within the family, and we showed that it is much more conservative than the more common type of cultural transmission, called "horizontal," between unrelated people. The latter can also rapidly change a whole population, while vertical transmission has undoubted formal similarity with genetic transmission. Language replacement is obviously caused by what we called horizontal cultural transmission. It may take place in as few as three generations, and is responsible for the exceptions to a perfect correlation of genetic similarity and language families we noted and discussed in the 1988 paper. It occurred frequently in historical times, as a consequence of conquest. Armies conquering wide territories by a relatively small, well-armed group are relatively recent, but not every conquest involved language replacement. Examples from the recent European history: Latin replaced earlier languages in most of the western Roman empire, but later the Franks' and Lombards' conquests in France and Italy did not replace languages of Latin origin. Mandarin languages replaced a number of other languages in China, at the time of the conquest of China by the Qin and the Han dynasties, but many of the earlier languages are still in existence in the 55 ethnic minorities recognized by the Chinese government. Language replacement is not universal, however, and for instance Basque was possibly never replaced, but naturally it must have changed considerably from Cro-Magnon times. Many linguists acknowledge the existence of similarities between Basque and other language isolates in the Caucasus, in Pakistan (Burushaski) and even SinoTibetan languages, all possible examples of a common Eurasian family that were probably dominant in vast regions of Eurasia in much earlier times, but were later replaced by languages of the Indo-European, Altaic, Uralic and other families of the Eurasiatic–Nostratic superfamily, which spread in Eurasia in the last 10 000 years.

It is significant that other totally unrelated organisms show evolutionary patterns that, like languages, imitate those of modern humans: virus infections such as hepatitis B, papilloma and others, and also the bacterium *Helicobacter pylori*. At least for hepatitis B, transmission from the mother is a well-known common phenomenon (affecting about 50% of the progeny), and although direct contagion is also possible, this vertical transmission must be sufficient to dictate an evolutionary pattern of the virus similar to that of the host.

21.6 DNA vs. proteins

In the early 1980s, techniques of studying DNA polymorphisms became available as restriction fragment length polymorphisms or RFLPs (Botstein *et al.*). They gave an excellent chance of mapping disease genes, and also of studying DNA evolution directly. RFLPs of autosomal genes showed evolutionary patterns perfectly comparable to those of proteins (e.g., Bowcock *et al.* 1991 and many other papers), but mitochondrial DNA (mtDNA) genetics gave the first revolutionary advantages. Its almost haploid condition, and relatively high mutation rate, made it possible to generate uniparental genealogies of single individuals on the basis of a relatively short DNA segment (about 16 500 nucleotides). The 1987 paper by Cann *et al.* from Allan Wilson's laboratory published the tree of mtDNA of 135 individuals representing the whole world, and announced two major discoveries. One was the statement that the MRCA of mtDNA was located in Africa, because the first split in the tree was between a group of Africans and all the rest of the world (including many other Africans). The second was the date of this MRCA (the "mitochondrial Eve"), set at about 150 000 to 300 000 years ago. This estimate was based on the ratio between the difference of DNA among the descendants of the MRCA, and the rate of change per year calculated from comparisons among mtDNAs of many species, including chimps, of known approximate evolutionary separation times. An interesting observation is that these results were both widely criticized, and some of the criticisms were undoubtedly right from a statistical point of view, as the authors themselves acknowledged. But later, more abundant and satisfactory data on mtDNA and NRY from other laboratories have fully confirmed these two very important conclusions. The dates changed somewhat, and, for demographic reasons (higher male mortality and variance of reproductive success), the TMRCA of extant men is much more recent than that of extant women. We had earlier published the same tree with mtDNA, but with much fewer genes, and a large number of individuals, but fewer populations (Johnson *et al.* 1983). It was clear that our tree was not strong enough for making any such statement, but the similarity to the Cann *et al.* tree of 1987 is remarkable. The technique we were using for studying RFLPs (Southern blots) was also inferior to that introduced by Wilson (end labeling), which could also detect smaller restriction fragments, but needed large amounts of purified mtDNA. All these technical details are now surpassed by later developments that made DNA sequencing popular, but they emphasize both the limitations and the major role played by the quality of technology.

Ten years later, mtDNA was the source of another important conclusion (Krings *et al.* 1997). Neanderthals are human fossils common in Europe and West Asia until they disappeared about 30 000 to 40 000 years ago, shortly after the arrival in Europe and West Asia of modern humans. Many anthropologists considered them the ancestors of modern Europeans, but the study of the mtDNA of three Neanderthals from very different locations showed them to be descendants of an earlier human fossil found in Europe. The lineage leading to Neanderthals separated probably between 500 000 and 800 000 years ago from the ancestors of modern humans, today believed by many archeologists to have developed in Africa. MtDNA has thus played a key role in the understanding of human evolution, and keeps an important role in forensic, medical, and evolutionary applications. But its advantage, the high mutation rate, is also a disadvantage in that there is frequent back mutation and recurrent mutation, generating noise in trees. Its small size inevitably limits information, and another source of evolutionary noise is the presence in many cells, including egg cells, of a large number of mtDNA copies. Thus one finds in some subjects genetic variation within a single cell, or different cells of the same individual. Conversely, the large number of mtDNA in many cells is of major benefit, because there often remain enough undamaged copies of the segments under study even in relatively old bones. This made it possible to study Neanderthals and other fossils, whereas nobody has so far succeeded in studying any other DNA in really old bones.

Other DNA peculiarities have been bonuses for evolutionary research. Repetitions of very short segments (2–5 nucleotides), called microsatellites, have proved of considerable interest, as they are responsible for certain genetic diseases, and also provide excellent genetic markers for evolutionary studies (Bowcock *et al.* 1994). The number of repeats of a given microsatellite changes fairly frequently, and it is easy to count the number of repeats. The high mutation rate of repeat number (of order 1/1000) generates a large number of alleles, making microsatellites very good markers for certain applications. One of these is genetic dating (Goldstein *et al.* 1995). One can also thus estimate dates of surnames, as in the case of the Cohanim, a caste of priests whose origin supposedly goes back to Aaron, the brother of Moses, around 3000 years ago.

21.7 Single-nucleotide polymorphisms

The most common polymorphism is that for single nucleotides (SNPs), and there is now much knowledge accumulated about them because of their importance for medical and evolutionary genetics. Apart from mtDNA, their frequency in nuclear chromosomes is of the order of 1/1000. There was much interest in Y-chromosome SNPs, in the hope it would allow accumulatation of direct knowledge about male genealogies. Although RFLPs, microsatellites and grosser DNA variants had been found in the nonrecombinant portion of the Y-chromosome (NRY), SNPs were not available for a while. We found the first by brute force, through resequencing enough individuals and DNA (Seielstad *et al.* 1994). I was very fortunate that my laboratory colleague Peter Underhill patiently shopped for a new way to increase by a large factor the rate and probability of SNP discovery, and found one in collaboration with Peter Oefner (called DHPLC, denaturing high-performance liquid chromatography, Underhill *et al.* 1996). Today over 300 NRY SNPs have been detected almost entirely by them, with the new technique. The first 150 allowed us to generate a tree on about 1000 individuals from the whole world (Underhill *et al.* 2000, 2001). A new set of 121 mutations generated an almost identical tree (except for one mutation, M09, for which no independent other mutation was found occupying the same branch in the genealogy). The haplotypes are still grouped into the ten haplogroups described in the 2000 paper. Seven of them are monophyletic, four originally distinguished from the rest by a single mutation, three by chains of several mutations whose sequence within the branch of the genealogy they form can therefore not be reconstructed. The other three haplogroups are made up of five, eight, and five monophyletic subgroups respectively, each distinguished by at least one common mutation, or absence of it. They were at first assembled into the ten haplogroups by virtue of their closely related geography, but were kept distinct in an international classification submitted to *Genome Research*, which uses 25 haplogroups labeled with alphanumeric labels.

The NRY SNPs are all biallelic except three, and only two SNPs appeared twice in different haplogroups. The rarity of recurrences and apparent absence of back mutations make the NRY tree especially strong, and there is so far no reason not to consider it as a true genealogy. The MRCA was determined by comparison with the nearest Primates. It separates from all the rest eight haplotypes (forming the first haplogroup), all present in East Africa or among Khoisan (southern Africa). The second haplogroup originates after some time of the order of one-third of the total time, as measured in terms of mutations in the branch that generated it, and is again found in East Africa and among Khoisan, but also in Central African Pygmies.The first two haplogroups of NRY thus correspond at least approximately to the early African branch of the 1987 mtDNA tree (now distinguished into three branches). The third NRY haplogroup must be part of a later expansion, and is again separated by many mutations from the two earliest haplogroups. The mutations preceding it in the genealogy mark the expansion that involved all the rest of Africa, and all the other continents. There have probably been different expansions from Africa to the rest of the world. This is shown also by the analysis of mismatch distributions (Shen *et al.* 2000, and unpublished). The 1988 autosomes, mtDNA, and NRY trees of populations (as distinct from the genealogies of single individuals and haplotypes) agree between themselves in a general way, and indicate three major expansions or groups of expansions, the first limited to Africa, a later one to southern Asia, perhaps by the coastal route, which reached Oceania and even East Asia, and a numerically more important one towards central Asia, which continued to Europe, northeast Asia and America. But it is likely that each of these expansions never involved many individuals.

NRY can be especially informative because it shows more geographic detail and more reproducibility than any other set of markers, but we will need many more markers and individuals than currently available for giving a more detailed picture. There are some differences between the details of the picture given by NRY and mtDNA (Seielstad *et al.* 1998, and many others), in part because of differences in migration patterns, especially at marriage (70% of societies are patrilocal), in part because of differences in the mutation pattern, and also in the variance of reproductive success of males and females. The variance of reproductive success, and a major putative difference of mortality, which was probably greater in males (and probably greater than now), causes the age of the female TMRCA to be at least twice as old as that of the male TMRCA. This finding generates some doubt in the usefulness of MRCA dating for estimation of separation times, but a date of 100 000 years for that related to the first expansion within Africa is probably the most likely, with that/those from Africa to Asia having started between 60 000 and 40 000 years ago.

21.8 Understanding the last 100 000 years of human evolution

This summary is concentrated on our work, and a few major findings in other laboratories that I found especially stimulating for my work. Genetics has given an important contribution to the understanding of human evolution in the last 100 000 years, but without paleoanthropology and archeology the picture would be much poorer and less convincing, especially regarding dates and motivations of expansions. The contribution of linguistics to determining and understanding human evolution can be considerable. It is clear that languages can sometimes change dramatically, but I suspect that a major cause is language replacement due to foreign conquest, and it is likely it was frequent mostly or only in the last few thousand years, with the origin of state armies. There still exist pockets where older languages are spoken in many parts of the world, especially those whose history is less well known, which can be useful for understanding the past.

This and other facts seem to justify, at least in part, Darwin's prophecy, in chapter 14 of the second edition of *The Origin of Species*, that when the genetic tree is known we will also know the tree of languages. I find genetics can also be useful to linguistics in another way. The rate of linguistic evolution is so fast, that one may think, and most linguists seem to share this idea, that it is impossible to answer the question if there was a single, or many, original languages. A few linguists are less pessimistic and think they will find evidence that all living languages had a common origin. Communication may of course have taken many forms at the beginning, and still does: there is a language of gestures, a body language, and the expression of many emotions is often clear in mere intonation. But genetics has some relevant information. At present, the best interpretation from coalescence analysis of haploid genes (mtDNA and NRY) is that the origin of modern humans is from a single, small population, probably a single tribe. There is a frequent one-to-one relationship between language and tribe. This suggests that, even if there were many languages in Africa 100 000 years ago, only one or a few survived and differentiated into the many thousands that survive today.

There is other genetic support to this idea. The acquisition of language has certainly involved a number of gene changes, to effect the anatomic and physiological transformations that make language possible and made humans different from other close Primates. In order to have the most effective communication, it must work two ways, in the sense that everybody must be able to be speaker and listener. Each mutation favoring communication, and we do not know how many they were, must have been found in at least two people to become useful. It is unlikely that the same mutation occurred twice independently: it is more likely that once one occurred, it was transmitted genetically to other individuals who also gained the advantage of being better endowed speakers. There may have been many mutations involved in the process, but perhaps all mutations causing improved speaker–listener communication that happened to generate two equally endowed individuals, e.g., by transmission of the same mutation from parent to child, and/or to sibs, were those most likely to survive. Once this initial event happened, then the favorable trait may have rapidly expanded to the whole population, especially if this was small. Eshel and Cavalli-Sforza (1982) suggested that, for this reason, genetic traits favorable to cooperation (and communication is a special case) must have arisen and spread initially in the family. DNA analysis should help us to understand the genetic basis for the acquisition of modern language, which must obviously exist. But after the original language existed, and extended to all members of a small group that is genetically sufficiently homogeneous that all its members can participate in communication by language, its further evolution will be entirely cultural. Language may have been the major factor assuring success and expansion of the group of modern humans thus endowed.

Other cultural innovations thought to be responsible for modern humans, enabling their rapid spread to the whole world, including the development of a set of tools called Aurignacian, which is more refined and efficient when compared with the older one available to both Neanderthals and archaic, premodern *H. sapiens*, called Mousterian (Klein 1999). Simple nautical means must have been available at least in the passage from South East Asia to New Guinea and Australia, and, if already developed in East Africa at the beginning of the expansion, they would have also favored the settling of the entire southern Asian coast. Unfortunately this coast is submerged and hardly accessible to archeological research.

The modern human languages, from those of Bushmen and Hottentots (Khoisan) to those of the most economically successful populations of Europe and Asia, show no major differences except in technical terminologies, implying that the MRCAs of all living humans already spoke a sophisticated language. Although it is difficult to prove it, it is not difficult to understand that ability of communication by language must have been a precious asset in settling completely new territory, where changes of climate, flora and fauna, and possible dangers arising in encounters with earlier, *H. erectus* settlers must have generated a number of new difficulties, and the need to respond with appropriate innovations to cope with them. The ability to communicate by modern language must have been a very important, perhaps the most important, help to expanding modern humans.

Isaac (1972) noted that there is an increasing cultural fragmentation of archeological artifacts in the last 100 000 years in Africa, revealed by the increase in names of archeological cultures with time. He thought this was a proof of the increasing importance of language. Perhaps it was accompanied by an increasing population density, and consequent dialect differentiation. Be this as it may, it seems reasonable to expect that studies of language evolution will also help the understanding of genetic evolution, and vice versa. After all, linguists started making trees of languages before geneticists (Schleicher 1865), preparing the ground for methods of study of isolation by distance (called "wave theory" by Schmidt) and dating language separations by measurement of linguistic distance and rates of language change (glottochronology). I was amazed, in the course of a research with linguist Bill Wang on Micronesian dialects (Cavalli-Sforza and Wang 1986), to discover how useful is standard population genetics theory applied to language microdifferentiation. It is high time to reverse the trend in the separation of disciplines.

REFERENCES

Ammerman, A. J., and Biagi, P. (eds) (2003). *The Widening Harvest: the Neolithic Transition in Europe.* Boston, MA: Archaeological Institute of America; Oxford: Oxbow.

Ammerman, A. J., and Cavalli-Sforza, L. L. (1971). Measuring the rate of spread of early farming in Europe. *Man* 6:674–88.

Ammerman, A. J., and Cavalli-Sforza, L. L. (1973). A population model for the diffusion of early farming in Europe. In C. Renfrew (ed.) *The Explanation of Culture Change*, pp. 343–57. London: G. Duckworth.

Botstein, D., White, R. L., Skolnick, M., and Davis, R. W. (1980). Construction of a genetic linkage map in man using restriction fragment length polymorphisms. *Am. J. Hum. Genet.* 32(3):314–31.

Bowcock, A., Kidd, J., Mountain, J., Hebert, J., Carotenuto, L., Kidd, K., and Cavalli-Sforza, L. L. (1991). Drift, admixture, and selection in human evolution: a study with DNA polymorphisms. *Proc. Natl. Acad. Sci. USA* 88:839–43.

Bowcock, A. M., Ruiz-Linares, A., Tomfohrde, J., Minch, E., Kidd, J. R., and Cavalli-Sforza, L. L. (1994). High resolution of human evolutionary trees with polymorphic microsatellites. *Nature* 368(6470): 455–7.

Cann, R. L., Stoneking, M., and Wilson, A. C. (1987). Mitochondrial DNA and human evolution. *Nature* 325:31–6.

Cavalli-Sforza, L. L., and Ammerman, A. (1984). *The Neolithic Transition and the Genetics of Population in Europe.* Princeton, NJ: Princeton University Press; Turin: Boringhieri Publishers.

Cavalli-Sforza, L. L., and Edwards, A. W. F. (1964). Analysis of human evolution. In S. J. Geerts, (ed.) *Genetics Today*, pp. 923–33. Proc. XI Int. Cong. Genet., The Hague. New York: Pergamon Press.

Cavalli-Sforza, L. L., and Edwards, A. W. F. (1967). Phylogenetic analysis: models and estimation procedures. *Am. J. Hum. Genet.* 19:233–57.

Cavalli-Sforza, L. L., and Feldman, M. (1981). *Cultural Transmission and Evolution: A Quantitative Approach.* Princeton, NJ: Princeton University Press.

Cavalli-Sforza, L. L., and Piazza, A. (1975). Analysis of evolution: evolutionary rates, independence, and treeness. *Theor. Pop. Biol.* 8:127–65.

Cavalli-Sforza, L. L., and Wang, W. S-Y. (1986). Spatial distance and lexical replacement. *Language* 62:38–55.

Cavalli-Sforza, L. L., and Zei, G. (1967). Experiments with an artificial population. In Crow, J. F. and Neel J. V. (eds) *Proceedings III International Congress of Human Genetics*, pp. 473–8. Baltimore, MD: Johns Hopkins Press.

Cavalli-Sforza, L. L., Piazza, A., Menozzi, P., and Mountain. J. (1988). Reconstruction of human evolution: bringing together genetic, archeological and linguistic data. *Proc. Natl. Acad. Sci. USA* 85:6002–6.

Cavalli-Sforza, L. L., Minch, E., and Mountain, J. L. (1992). Coevolution of genes and languages revisited. *Proc. Natl. Acad. Sci. USA* 89:5620–4.

Cavalli-Sforza, L. L., Menozzi, P., and Piazza, A. (1994). *History and Geography of Human Genes.* Princeton, NJ: Princeton University Press.

Cavalli-Sforza, L. L., Moroni, A., and Zei, G. (2003). *Consanguinity, Inbreeding, and Drift in Italy.* Princeton, NJ (in press)

Du, R., Yuan, Y., Hwang, J., Mountain, J., and Cavalli-Sforza, L. L. (1992). Chinese surnames and the genetic differences between north and south China. Monograph. *J. Chinese Ling.* 5:1–93.

Edwards, A. W. F., and Cavalli-Sforza, L. L. (1964). Reconstruction of evolutionary trees. In V. H. Heywood and J. McNeill (eds) *Phenetic and Phylogenetic Classification*, pp. 67–76. London: Systematics Association Publication no. 6.

Eshel, I., and Cavalli-Sforza, L. L. (1982). Assortment of encounters and evolution of cooperativeness. *Proc. Natl. Acad. Sci. USA* 79:1331–5.

Fischer, R. A. (1936) Has Mendel's work been rediscovered? *Ann. Sci.* 1:115–37.

Fisher, R. A. (1937). The wave of advance of advantageous genes. *Ann. Eugenics,* 7:355–69.

Fisher, R. A. (1943). The relation between the number of species and the number of individuals in a random sample of an animal population. *J. Anim. Ecol.* 12:42–58.

Goldstein, D. B., Ruiz-Linares, A., Cavalli-Sforza, L. L., and Feldman, M. W. (1995) Genetic absolute dating based on microsatellites and the origin of modern humans. *Proc. Natl. Acad. Sci. USA* 92:6723–7.

Greenberg, J. H. (1963). *The Languages of Africa.* Bloomington: Indiana University.

Isaac, G. (1992). Chronology and the tempo of cultural change during the Pleistocene. In W. Bishop and J. A. Miller (eds) *Calibration of Hominoid Evolution.* Edinburgh: Scottish Academy Press.

Johnson, M. J., Wallace, D. W., Ferris, S. D., Rattazzi, M.C., and Cavalli-Sforza, L. L. (1983). Radiation of human mitochondria DNA types. *J. Mol. Evol.* 19:255–71.

Kimura, M. (1968). Evolutionary rate at the molecular level. *Nature* 217:624–6.

Klein, R. G. (1999). *The Human Career: Human Biological and Cultural Origins.* 2nd edition. Chicago: University of Chicago Press.

Krings, M., Stone, A., Schmitz, R. W., Krainitzki, H., Stoneking, M., and Paabo, S. (1997). Neanderthal DNA sequences and the origin of modern humans. *Cell* 90:19–30.

Kulathinal, R. J., Skwarek, L., Morton, R. A., and Singh, R. S. (2003). Rapid evolution of the sex-determining gene, transformer: structural diversity and rate heterogeneity among sibling species of *Drosophila. Mol. Biol. Evol.* 20:44–52.

Llopart, A., Elwyn, S. Lachaise, D., and Coyne, J. A. (2002a). Genetics of a difference in pigmentation between *Drosophila yakuba and D. santomea. Evolution* 56:2262–77.

Menozzi, P., Piazza, A., and Cavalli-Sforza, L. L. (1978). Synthetic maps of human gene frequencies in Europeans. *Science* 201:786–92.

Nei, M., and Roychoudhury, A. K. (1974). Genic variation within and between the three major races of man, Caucasoids, Negroids, and Mongoloids. *Am. J. Genet.* 26(4):421–43.

Nei, M., and Roychoudhury, A. K. (1982). Genetic relationship and evolution of human races. *Evol. Biol.* 14:1–59.

Penny, D., Foulds, L. R., and Hendy, M. D. (1982). Testing the theory of evolution by comparing phylogenetic trees constructed from five different protein sequences. *Nature* 297:197–200.

Penny, D., Hendy, M. D., and Steel, M. A. (1991). Testing the theory of descent. In M. Miyamoto, M. and J. Cracraft (eds) *Phylogenetic Analysis of DNA Sequences,* pp. 155–83. New York: Oxford University Press.

Renfrew, C. (1987). *Archaeology and Language: the Puzzle of Indo-European Origins.* London: Jonathan Cape.

Richards, M. B., Macaulay, V. A., Bandelt, H.-J., and Sykes, B. C. (1998). Phylogeography of mitochondrial DNA in western Europe. *Ann. Hum. Genet.* 62:241–60.

Rosenberg, N. A., and Feldman, M. W. (2002). The relationship between coalescence times and population divergence times. In M. Slatkin and M. Veuille (eds) *Modern Developments in Theoretical Population Genetics,* pp. 130–64, Oxford, Oxford University Press.

Ruhlen, M. (1987) *A Guide to the World's Languages.* Stanford, CA: Stanford University Press.

Schleicher, A. (1865). On the significance of language for the natural history of man. In E. F. K. Koerner (ed.) (1983) *Linguistics and Evolutionary Theory: Three Essays by August Schleicher, Ernst Haeckel, and William Bleek,* with an Introduction by J. Peter Maher. Amsterdam: John Benjamins.

Seielstad, M. T., Hebert, J. M., Lin, A. A., Underhill, P. A., Ibrahim, M., Vollrath, D., and Cavalli-Sforza, L. L. (1994). Construction of human Y-chromosomal haplotypes using a new polymorphic A to G transition. *Hum. Mol. Genet.* 3: 2159–61.

Seielstad, M. T., Minch, E., and Cavalli-Sforza, L. L. (1998) Genetic evidence for a higher female migration rate in humans. *Nat. Genet.* 20:278–80.

Semino, O., Passarino, G., Oefner, P. J. Lin, A. A., Arbuzova, S., and Beckman, L. E. (2000). The genetic legacy of Paleolithic *Homo sapiens sapiens* in extant Europeans: a Y chromosome perspective. *Science* 290:1155–9.

Shen, P., Wang, F., Underhill, P. A., Franco, C., Yang, W-H., Roxas, A., Sung, R., Lin., A. A., Hyman, R. W., Vollrath, D., Davis., R. W., Cavalli-Sforza, L. L., and Oefner, P. J. (2000). Population genetic implications from sequence variation in four Y chromososme genes. *Proc. Natl. Acad. Sci. USA* 97:7354–9.

Sokal, R., Oden, N., and Wilson, C. (1991). Genetic evidence for the spread of agriculture in Europe by demic diffusion. *Nature* 351:143–5.

Underhill, P. A., Passarino, G., Lin, A. A., Marzuki, S., Oefner, P. J., Cavalli-Sforza, L. L., and Chambers, G. K. (1996). A pre-Columbian Y-chromosome-specific transition and its implications for human evolutionary history. *Proc. Natl. Acad. Sci. USA* 93:196–200.

Underhill, P. A., Shen, P, Lin, A. A., Jin, Li, Passarino, G., Yange W. H., Kauffman, E., Bonne-Tamir, B., Bertranpetit, J., Francalacci, P., Ibrahim, M., Jenkins, T., Kidd, J. R., Mehdi, S. Q., Seielstad, M. T., Wells, R. S., Piazza, A., Davis, R. W., Feldman, M. W., Cavalli-Sforza, L. L., and Oefner, P. J. (2000). Y chromosome sequence variation and the history of human populations. *Nat. Genet.* 26: 358–61.

Underhill, P. A., Passarino, G., Lin, A. A., Shen, P., Mirazon Lahr, M., Foley, R. A., Oefner, P. J., and Cavalli-Sforza, L. L. (2001). The phylogeography of Y chromosome binary haplotypes and the origins of modern human populations. *Ann. Hum. Genet.* 65:43–62.

Zei, G., Guglielmino Matessi, R., Siri, E., Moroni, A., and Cavalli-Sforza, L. L. (1983). Surnames in Sardinia: I. Fit of frequency distributions for neutral alleles and genetic populations. *Ann. Hum. Genet.* 47:65–78.

22

Geneticists and the biology of race, 1900–1924

WILLIAM B. PROVINE
Department of Ecology and Evolutionary Biology, Cornell University

Attitudes about the kind and amount of genetic variation in populations, like all attitudes about unresolved scientific issues, reflect and are consistent with the intellectual histories of their proponents. People see the new problems mirrored in a glass that has been molded by their solutions to old problems. A scientist's present view of difficult questions is chiefly influenced by the history of his intellectual and ideological development up to the present moment, and the resolution of current difficulties will in turn precondition his view of future problems.

(Lewontin 1974, p. 29)

22.1 Introduction

The rediscovery of Mendel's theory of heredity in 1900 started the century of genetics. In the first quarter of the century, the new science was applied to many social questions, especially eugenics and the subject of this chapter, human race differences and race crossing.

Who would have guessed that a study of hybridization in peas could have such wide ramifications? Rediscovered by three men in 1900, Mendel's theory revolutionized the study of heredity and evolution within 15 years. Many of the long-standing puzzles and contradictions in the evidence about heredity and hybridization yielded to Mendelian analysis. Named "genetics" by William Bateson in 1908, the experimental study of heredity burgeoned in the United States, England, and the European continent. As geneticists successfully solved previously inscrutable problems, their confidence in the wide applicability and importance of genetics grew accordingly. The English geneticist Reginald C. Punnett expressed the excitement and optimism of many researchers in the new field when he wrote these words in the preface to the second (1907) edition of his little book, *Mendelism*:

Each new riddle solved propounds fresh riddles, and strengthens the hope of their solution. As year follows year, and experiment succeeds experiment, there

The Evolution of Population Biology, ed. R. S. Singh and M. K. Uyenoyama. Published by Cambridge University Press.

is forced upon us a sense of what it may all come to signify for ourselves, of the tremendous powers of control that a knowledge of heredity implies. To-day we are only at the beginning. The prologue is nearing completion; the drama is yet to be written – and played.

(Punnett 1907, p. vii)

Like most scientists, these early geneticists had a strong reductionist urge. They demonstrably possessed a major key to understanding the mysteries of heredity; thus if problems of society could be reduced to problems of heredity, geneticists might be able to solve them. In their first flush of success, many geneticists believed a wide range of social ills could be understood by genetic analysis. Punnett, for example, who began his book with the statement just quoted, closed it with this advice for those who were concerned with the social problems caused by the lower strata of society. "The facts of heredity," he said, "speak with no uncertain voice":

Education is to man what manure is to the pea. The educated are in themselves the better for it, but their experience will alter not one jot the irrevocable nature of their offspring. Permanent progress is a question of breeding rather than of pedagogics; a matter of gametes, not of training. As our knowledge of heredity clears, and the mists of superstition are dispelled, there grows upon us with ever-increasing and relentless force the conviction that the creature is not made but born.

(Punnett 1907, pp. 80–1)

Ten years later, Punnett was more skeptical about the possibility of genetic cures for social problems (see for example Punnett's refutation of eugenist's aim of eliminating feeblemindedness by genetic selection (Punnett 1917)). But in 1907 he expressed the hopes of many geneticists who found very appealing the prospect that genetics might provide scientific palliatives for some pressing social problems.

Nineteenth-century evolutionists all had theories of heredity, and despite great disagreements about these theories, they agreed easily that human races exhibited hereditary differences in physical characteristics and mentality (Provine 1986). The social problems associated with race differences and race crossing seemed obviously to have strong genetic components, and race issues inevitably drew the attention of some geneticists. Although admitting ignorance of the precise genetics of human race differences, these geneticists developed before 1924 a definite attitude about the biology of race differences and crossing, and offered guidelines for social policy.

The conclusion to this chapter will assess the question: Does the framework given by the Lewontin quotation that prefaces this chapter provide an enhanced historical understanding of geneticists' views on race differences and crossing?

22.2 Early successes of Mendelism

Mendel's hypothesis was that in the formation of the reproductive cells, all germinal elements from a parent segregate freely to produce as many kinds of egg and pollen cells as there are combinations of germinal elements. This hypothesis, Mendel stated, was "entirely adequate to explain the development of hybrids in separate generations if one could assume at the same time that the different kinds of germinal and pollen cells of a hybrid are produced on the average in equal numbers" (Stern and Sherwood 1966, p. 24). Every elementary genetics textbook illustrates how neatly Mendel's hypothesis fit his data.

The crossing of closely related varieties or races was the key to Mendelian analysis, and geneticists conducted a great number of controlled crosses of animals and plants between 1900 and 1912 (see Dunn 1965, Part II). The experimental subjects ranged from orchids, wheat, peas, beans, corn, and tomatoes in plants to butterflies, fruit flies, canaries, chickens, mice, guinea pigs, dogs, cattle, and horses in animals. Researchers immediately discovered that Mendel's results with peas did not hold in all particulars with other crosses. Many F_1 offspring showed less than complete dominance; others exhibited hybrid vigor, or heterosis, as geneticists called it. Nor did all germinal elements participate independently in the production of reproductive cells; some were linked together to varying degrees. But the principle of segregation was obvious to any researcher who bred F_2 or later generations of crosses between distinct varieties.

The literature on genetics expanded rapidly. William Bateson had trouble summarizing the work on genetics in his important book, *Mendel's Principles of Heredity*, published in March, 1909. By 1911 the amount of genetical work published on animals was so great that Arnold Lang, who attempted a comprehensive report on heredity in animals, found that the mammals alone required 427 pages. With the introductory material on Mendelism and biometry, the first volume of Lang's (1911) *Die experimentelle Vererbungslehre in der Zoologie seit 1900* ran to 892 large pages. He abandoned the project, and no one has ever tried to summarize the full range of zoological genetics again.

22.3 Human heredity, genetics, and eugenics

Knowledge of Mendelian heredity in humans was also expanding, but with limitations. In *Mendel's Principles of Heredity*, Bateson had a chapter entitled "Evidence as to Mendelian inheritance in man"; it began with the statement:

> Of Mendelian inheritance of normal characteristics in man there is as yet but little evidence. Only a single case has been established with any clearness, namely that of eye-colour. The deficiency of evidence is probably due to the special difficulties attending to the study of human heredity. Human families are small compared with those of our experimental animals and plants, and the period covered by each generation is so long that no observer can examine any.
>
> (Bateson 1909, p. 205)

Bateson added in the next paragraph that the inheritance of hair color and skin color in man was far more complex than the simple pattern followed by color or shape in peas. These were indeed substantial obstacles to the Mendelian analysis of human heredity; even simple 3:1 ratios could not be seen in families with three or fewer children.

Despite these difficulties, Bateson pointed out that a considerable amount was known about the inheritance of diseases and abnormalities in humans. Many of these conditions were inherited as dominants and were therefore much easier to trace than recessive or blending traits. Brachydactylous hands (with short, stubby fingers), cataracts, tylosis (thick skin on palms of hands and soles of feet), and the sex-linked traits hemophilia and color blindness were among the dominant characters described by Bateson. Albinism and alkaptonuria were two of the recessive traits. Bateson was confident that knowledge of both normal and abnormal Mendelian traits in humans would rapidly expand. But he did not pursue the study of human heredity himself, in part because of the experimental difficulties he outlined above.

With the rapidly increasing knowledge of heredity in general, and of the inheritance of abnormal traits in humans in particular, came the rise of the eugenics movement. Riding the crest of genetic discoveries, eugenists wished to apply the newfound knowledge to the genetical improvement of the human race. The possibilities for genetic improvement seemed both great and feasible. In addition to eliminating medical defects, many eugenists hoped to eliminate problems like criminality and feeblemindedness.

Especially in the United States and Germany, geneticists gave their prestige and support to the early eugenics movement. I will not here examine in detail the eugenics movement or geneticists' participation in it because several such studies are available (Haller 1963, Ludmerer 1972, Kevles 1985). The eugenics movement is, however, very important as the setting in which many geneticists spoke about race. Most geneticists who wrote about race in the period up to 1924 treated it as a subdivision of their larger interest in the eugenic improvement of humans.

The most influential early investigator of human heredity in the United States before 1912 was Charles Benedict Davenport, who was also the leading advocate and organizer of the eugenics movement. Unlike Bateson, Davenport was initially skeptical about the wide applicability of Mendelism. Once convinced, however, Davenport eagerly found simple Mendelian ratios in the most meager evidence. Bateson met Davenport on a trip to the United States in September, 1907. From Davenport's laboratory at Cold Spring Harbor, Bateson wrote his wife: "Mendel is a rage here, but I can see that Davenport is floored by any irregularity or difficulty. His mind must be quite devoid of invention" (Bateson to Beatrice Bateson, August 8, 1907, Bateson papers, Reel A, no. 2).

Perhaps Bateson's assessment was overly harsh, but certainly Davenport was uncritical in many of his assertions about Mendelian heredity in man, especially when he believed that reaching a firm conclusion might further the eugenics movement.

Davenport and his wife Gertrude began researches into human heredity in 1907. Their papers appeared in rapid succession: on eye color (1907), hair form (1908), hair color (1909), and skin pigment (1910). Although many of Davenport's genetic hypotheses in these papers were later modified by more exact research, the papers established him as a major student of human heredity. Every contemporary genetics textbook cited these papers; they stood out in a field where little research was being conducted.

Then in 1911 Davenport published his book, *Heredity in Relation to Eugenics*, which contained almost all that was then known of human genetics. The purpose of the book was to interest educated lay persons in eugenics. Realizing that eugenics had to be founded upon exact knowledge of human heredity, Davenport attempted to convince the reader that geneticists knew a great deal about human heredity. He included every possible hereditary trait, with little critical distinction. Thus he presented albinism, alkaptonuria, musical ability, and feeblemindedness as simple Mendelian recessives; compelling evidence was available for the first two, but unavailable for the second two. From the vantage point of modern human genetics, *Heredity in Relation to Eugenics* seems naive and overinflated; but in 1911 most geneticists found the aim of the book congenial, even if they disagreed about particulars.

Although he gained the position partially by default of accomplished geneticists who found *Drosophila* or maize superior to humans as experimental organisms, Davenport was clearly the leading student of human heredity in the United States for over 20 years. He was also respected as an administrator and teacher. Research geneticists Herbert Spencer Jennings and William Ernest Castle were among his more famous students, and though they often disagreed with Davenport on questions of genetics or eugenics, they retained much respect for him over the years. As director of the Cold Spring Harbor Biological Experiment Station, financed by the Carnegie Institution of Washington, Davenport held the purse strings for many research projects in genetics. Thus when he wrote about the genetics of race differences and race crossing, many geneticists listened.

Although human inheritance and eugenics were closely allied in England as in the United States, Davenport had no counterpart there. Galton's disciple Karl Pearson was the leading eugenist and student of human heredity, but he disliked Mendelism intensely. The arguments between the Biometricians, led by Pearson, and the Mendelians, led by Bateson, were ferocious. Consequently Bateson and his Mendelian followers generally steered clear of human heredity.

Bateson wanted to see rapid and clear-cut progress, but humans were terrible experimental animals. With Pearson and his workers critically scrutinizing every example of supposed Mendelian inheritance in man, Bateson thought it simplest to leave that field to Pearson until later, and to concentrate on more certain projects. Of Bateson's workers, only C. C. Hurst worked seriously on human inheritance, and he only on the problems of eye color and musical temperament. Like Davenport, Hurst was overeager to find 3:1 Mendelian ratios in his data. Unlike Davenport, he had Bateson looking critically over his shoulder. Hurst abandoned work on human heredity after 1912.

Beginning in 1909 the "Treasury of Human Inheritance" series began to appear from the Galton Laboratory of Eugenics, of which Pearson was the director. This series contained most of the work done on human heredity in England for the next two decades. The approach was almost entirely biometrical description rather than Mendelian analysis. During these same years Pearson and his workers published pamphlets on eugenics and related topics. Thus both in England and in the United States the study of human inheritance and eugenics were closely allied, but Davenport's Mendelian ratios were suspect, and Pearson despised Mendelism. Mendelian analysis of human heredity therefore proceeded even more slowly than the difficulty of the material dictated.

As the leading student of human heredity and advocate of eugenics in England, Pearson's ideas on race differences and race crossing were influential in educated circles. He took the same position as Galton had in the nineteenth century – that mental characters were inherited to the same extent as physical, and that a superior race should not allow itself to be lowered by crossing with an inferior race. In 1900, with the Boer War in mind, Pearson expressed clearly his ideas on race differences and race crossing in a lecture entitled "National Life from the Standpoint of Science":

How many centuries, how many thousands of years, have the Kaffir and the Negro held large districts in Africa undisturbed by the white man? Yet their inter-tribal struggles have not yet produced a civilization in the least comparable with the Aryan. Educate and nurture them as you will, I do not believe that you will succeed in modifying the stock. . . .

If you bring the white man into contact with the black, you too often suspend the very process of natural selection on which the evolution of a higher type depends. You get superior and inferior races living on the same soil, and that co-existence is demoralizing for both. They naturally sink into the position of master and servant . . . Frequently they intercross, and if the bad stock be raised, the good is lowered. Even in the case of Eurasians, of whom I have met mentally and physically fine specimens, I have felt how much better they would have been had they been pure Asiatics or pure Europeans.

(Pearson 1901, pp. 20–1)

Superior races should either stay away from areas occupied by inferior races, or should completely drive out the inferior race, as Europeans did with the Indians in America:

I venture to assert that the struggle for existence between white and red man, painful and even terrible as it was in its details, has given us a good far outbalancing its immediate evil. In place of the red man, contributing practically nothing to the work and thought of the world, we have a great nation, mistress of many arts, and able, with its youthful imagination and fresh, untrammelled impulses, to contribute much to the common stock of civilized man. Against that you have only to put the romantic sympathy of the Red Indian generated by the novels of Cooper and the poems of Longfellow, and then – see how little it weighs in the balance!

(Pearson 1901, pp. 22–3)

Pearson's views on race were unchanged by his further research into human heredity. Indeed, he believed his biometrical analysis of intelligence, which indicated that mental characters were inherited to the same extent as physical characters, supported his attitude about race. The fear of Pearson and Galton that high races would be lowered by crossing with inferior races would soon appear with a Mendelian analysis to support it.

Although genetical research on lower organisms was proceeding rapidly on the European continent, research on human genetics was even more limited than in the United States or England. By 1911 three major textbooks of genetics had appeared on the continent, all in German (Baur 1911, Goldschmidt 1911, Haecker 1911). Each referred primarily to research on human heredity done in England or the United States. William Bateson's observation that very little was known of the inheritance of normal characters in man was certainly accurate in 1909, and continued to be accurate for several more decades. His assessment of geneticists' understanding of human race crossing has remained largely accurate even to the present:

As to the results of inter-crossing between distinct races of mankind there exist, so far as I am aware, no records of that critical and minute sort which are alone of value for the adequate study of Mendelian inheritance.

(Bateson 1909, p. 208)

The scarcity of genetical evidence about human heredity, racial differences, and race crossing should be kept in mind during the analysis in this chapter of geneticists' ideas about race.

22.4 Eugen Fischer and the Rehoboth Bastards

Beginning in 1909 the first significant and influential genetical research on race differences and race crossing in humans began to appear in the work of Eugen Fischer. Trained in both anthropology and genetics, Fischer went in 1908 to South Africa to study a hybrid population supposedly descended from matings between Boer (Dutch) men and native Hottentot women. The original matings occurred in the Cape Colony in the late eighteenth century. In some areas these hybrids, or Bastards, tended to live apart from both whites and blacks. They mostly married among themselves; but occasionally some of the male Bastards married Hottentot or other black women, and some of the hybrid women married European men. Some of these Bastard populations were pushed northward by whites pouring into the south, and in 1870 a group of about 300 Bastards settled the old mission station of Rehoboth, situated in the center of what is now Namibia. This is the population that Fischer studied in 1908.

The Rehoboth Bastards had some advantages as experimental subjects. They felt superior to the Hottentots and the other black natives, whom they employed for menial jobs; whites in turn felt superior to the Bastards. Thus they were relatively, but not absolutely, reproductively isolated. This was a basic requirement for a hybridization experiment. The Bastards were proud of the family names of their white male ancestors and had retained them through the years. Utilizing the records of the Cape Colony, Fischer was able to isolate the names of 37 male ancestors, and from their names he deduced their national origin. Of the 37, 17 were Dutch, 11 German, and 9 of unknown origin. This information was, of course, useful for an analysis of the Bastards. Fischer published his detailed analysis in a 1913 book, entitled *Die Rehobother Bastards.*

He divided his subjects into four groups. The Eu-group (12 men and 15 women) had a preponderance of European features, the Mitt-group (32 men and 43 women) was intermediate, and the Hott-group (23 men and 22 women) had mostly Hottentot features. The fourth group (7 men and 11 women) contained those who did not fit the above categories. Fischer of course wanted to compare his hybrids with the original European and Hottentot populations. These measurements being unavailable, he used a series of measurements of 100 Low Germans from Baden for the "Boers," and for the "Hottentots" Fischer measured eight male and seven female Hottentots. The characters measured and recorded by Fischer were such things as stature, nasal breadth, cephalic index, nasal index, and eye color and shape.

For most characters the Bastards averaged, as Fischer expected, intermediate between the Badeners and Hottentots. But for stature he found that the average height for male Bastards was 168.4 cm, much greater than the 157.9 cm for the Hottentots, and slightly greater than the 167.5 cm for contemporary Dutch, and barely below the 171.0 cm for the 100 Badeners. Average height in the Eu-group was 174.4 cm. For the women the results were even more striking, the average of the Bastards being 157.0 cm, compared with 156.0 cm for the Badeners and 150.5 cm for the Hottentots. The Eu-group of women averaged 159.9 cm, or 3.9 cm greater than the Badeners. Clearly the hybrids had suffered no loss of height as a consequence of crossing. The figures for stature stood out particularly because in general the hybrids were intermediate with respect to a given characteristic.

The most important section of the book was Fischer's analysis of the Bastards in terms of Mendelian heredity. He found a considerable amount of segregation for many of the characters he had measured in the anthropological section, and some, including eye color, eye shape, hair color, and hair form (straight or curly) produced apparent 3:1 Mendelian ratios. Several other characters clearly segregated in F_2 offspring, but did not produce simple Mendelian ratios. Fischer concluded that his study "established – for the first time upon a broad base – that human races hybridized according to Mendel's laws, exactly like countless races of animals and plants" (Fischer 1913, p. 171).

When *Die Rehobother Bastards* was published in 1913 many, probably most, Mendelians already believed this conclusion, but had little evidence to support it. Thus they welcomed Fischer's book.

The fecundity of the Bastards was a crucial point. As the previous chapter showed, several nineteenth-century authors, especially Paul Broca, claimed that hybrids between divergent races suffered a significant loss of fertility. In contrast, Fischer found that the average number of children per Bastard family was 7.7 children. Fischer did not, however, generalize to all cases from this fact, concluding that:

> 1. All races are completely fertile in crossing with each other. 2. The hybrids thus produced are fertile with the parent races, but whether fertility is undiminished for all races is questionable. 3. The hybrids can be completely fertile inter se, as our case here demonstrates; but in many cases they appear to be less fertile or perhaps even infertile. 4. The factors upon which these surprising results depend are wholly unknown to us.
>
> (Fischer 1913, p. 183)

At least in the case of the Bastards, wide race crossing had not led to any obvious biological problems. But Fischer suggested that other wide race crosses might well lead to a loss of fertility.

Die Rehobother Bastards was influential among geneticists. For nearly 15 years it was the only available in-depth study of a wide race cross by a geneticist. Thus the book was almost always cited when geneticists spoke of race crossing. When Samuel J. Holmes, a University of California biologist, published his extensive bibliography of eugenics in 1924, Fischer's was the only genetical study of race crossing that he could recommend (Holmes 1924, pp. 465, 468).

But Fischer's work on the Rehoboth Bastards had very serious deficiencies. He gathered all his data in one four-month visit in 1908; thus he could scarcely conduct careful pedigree studies of the sort that Mendelians found so useful in studying other organisms. Although Fischer knew the nationalities of some of the white men (or at least of their names) who participated in the initial cross, he knew little about their physical features. Of the "Hottentot" women these men married, Fischer knew nothing. Thus the criticism of J. P. Lotsy and W. A. Goddjin in 1928 seems justified: "Very little is known with certainty about the origin of the Bastards, except that there was white and coloured blood in them" (Lotsy and Goddjin 1928, p. 206). Furthermore, the sample sizes Fischer utilized were much too small for any but the barest statistical validity. In short,

Fischer's study was scarcely even a minimal start on the genetical analysis of a human race cross. But it was the best available study until the late 1920s.

Lack of adequate data did not prevent geneticists from publishing on the topics of race differences and race crossing between 1913 and the late 1920s. Race was a popular topic, and with their strong interest in social problems, some geneticists simply could not refrain from hypothesizing about race in articles and books. These hypotheses often seemed to stem from irrefutable biological facts. Despite an almost total lack of adequate data, between 1913 and 1924 one may easily find in the genetics literature the development of a definite, and unopposed, position on the issues of race differences and race crossing.

22.5 Charles Benedict Davenport and race crossing in chickens and humans

Davenport was the first American geneticist to devote considerable attention to the problems of race crossing. Although aware of the small amount of genetic evidence available, he believed the growing evidence from other animals could be meaningfully extended to humans.

To understand Davenport's views on race crossing it is first necessary to investigate some aspects of his general view of Mendelian heredity. He thought that often Mendelian factors independently controlled very specific bodily traits. Thus eye color seemed independent of skin color in human crosses. In 1914 William Ernest Castle, who had earlier been one of Davenport's favorite students, argued that because of the high correlations between measurable parts of the body (bone lengths, etc.), the genetic factors which controlled size must be general factors affecting all parts of the body simultaneously. Davenport was unconvinced. In 1917 he published a long article on the inheritance of stature in man, arguing that the components of stature could to a great extent be inherited separately. For example, he thought an individual could inherit long arms from one parent and short legs from another (Davenport 1917a). This genetic proposition is fundamental for the theory of race crossing that Davenport published the same year.

On April 13, 1917, Davenport read to the American Philosophical Society in Philadelphia a paper entitled "The effects of race intermingling." He began the paper by saying how geneticists defined race. "A race is a more or less pure bred 'group' of individuals that differs from other groups by at least one character, or, strictly, a genetically connected group whose germ plasm is characterized by a difference, in one or more genes, from other groups" (Davenport 1917b, p. 364). Davenport's definition differs little from that of modern geneticists. In his important book *Heredity and Evolution in Human Populations* (1959), geneticist L. C. Dunn defined race like this: "a race is a population which differs from other populations in the frequency of some of its genes... We put together, as members of a race, populations which have many, perhaps most, of their genes in common" (Dunn 1959). Thus Davenport saw races as somewhat more genetically uniform than did Dunn. This is a difference, however, in degree rather than in kind. Davenport certainly realized that distinct races overlapped in many genetical traits, especially those governing mental abilities.

Having defined race, Davenport then stated what almost all geneticists believed at the time: that human races differed by hereditary physical and mental characters. He then attempted to evaluate the consequences of human race crossing. "I have to say," he began, "that this subject has not been sufficiently investigated; but we may, by inference from studies that have been made, draw certain conclusions." As an example of one of these studies, from which the consequences of human race crosses could be deduced, Davenport offered this:

> Any well-established abundant race is probably well adjusted to its conditions and its parts and functions are harmoniously adjusted. Take the case of the Leghorn hen. Its function is to lay eggs all the year through and never to waste time in becoming broody. The brooding instinct is, indeed, absent; and for egg farms and those in which incubators are used such birds are the best type. The Brahma fowl, on the other hand, is only a fair layer; it becomes broody two or three times a year and makes an excellent mother. It is well adapted for farms which have no incubators or artificial brooders. Now I have crossed these two races; the progeny were intermediate in size. The hens laid fairly well for a time and then became broody and in time hatched some chicks, and then began to roost at night in the trees and in a few days began to lay again, while the chicks perished at night of cold and neglect. The hybrid was a failure both as egg layer and as a brooder of chicks. The instincts and functions of the hybrids were not harmoniously adjusted to each other.

What did this engaging example have to do with the problem of human race crossing? Davenport believed the moral was clear. Each race had, through a long process of natural selection, developed genetical traits which were harmoniously adjusted both with each other and with the environment. When two races differing by a number of characters interbred, some new combinations of characters were formed in the hybrids. Mendelian segregation would produce many more new combinations in subsequent offspring of the hybrids.

Davenport thought many of these new combinations would be disharmonious, although some would be beneficial. For example, he said a large, tall race might breed with a small, short one, yielding, in the second generation, some offspring with "large frames and inadequate viscera" or "children of short stature with too large circulatory apparatus." Another example was the overcrowding or wide spacing of teeth probably caused by the "union of a large-jawed, large-toothed race and a small-jawed, small-toothed race." Nor were disharmonious combinations confined to physical characters. "One often sees in mulattos an ambition and push combined with intellectual inadequacy which makes the unhappy hybrid dissatisfied with his lot and a nuisance to others." In short,

> . . . miscegenation commonly spells disharmony – disharmony of physical, mental and temperamental qualities and this means also disharmony with environment. A hybridized people are a badly put together people and a dissatisfied, restless, ineffective people. One wonders how much of the exceptionally high death rate in middle life in this country is due to such bodily maladjustments; and how much of our crime and insanity is due to mental and temperamental friction.

The United States, he added, "is in for hybridization on the greatest scale that the world has ever seen" (Davenport 1917b, pp. 365–7).

What was the solution to this dilemma? The answer, said Davenport, was not prevention of all race crossing in the United States, but eugenics.

> The result of hybridization after two or three generations is great variability. This means that some new combinations will be formed that are better than the old ones; also others that are worse. If selective annihilation is permitted to do its beneficent work, then the worse combinations will tend to die off early. If now new intermixing is stopped and eugenical mating ensues, consciously or unconsciously, especially in the presence of inbreeding, strains may arise that are superior to any that existed in the unhybridized races. This, then, is the hope for our country; if immigration is restricted, if selective elimination is permitted, if the principle of the inequality of generating strains be accepted and if eugenical ideals prevail in mating, then strains with new and better combinations of traits may arise and our nation take front rank in culture among nations of ancient and modern times.
>
> (Davenport 1917b, pp. 367–8)

Race crossing combined with eugenic selection and judicious inbreeding was the best way to produce a biologically superior human race. Davenport was almost certainly correct in this conclusion. After about 1915, cross breeding of varieties accompanied by stringent selection was precisely the procedure recommended by geneticists for the production of genetically superior strains of domesticated livestock, such as cattle, horses, sheep, and dogs. The method works, as any modern day breeder will testify. A breeder, however, can select and inbreed his cows much more easily then a government can select and inbreed its people.

22.6 The color line

Davenport carefully avoided condemnation of entire races as inferior; he wanted the best individuals to be breeders of the future. Others had no such hesitancy. In 1918 Roswell H. Johnson, who had studied under Davenport at Harvard, and Paul Popenoe, editor of *The Journal of Heredity*, published their book *Applied Eugenics*, the most widely used textbook on this subject for over 15 years. The first six chapters outlined the current knowledge of heredity in man, and argued that by eugenic selection the human race could be significantly improved. The remaining 14 chapters analyzed the practical means by which society could encourage eugenic selection, with recommendations for social policy. The chapter on race, entitled "The Color Line," began with the recitation of a story:

> A young white woman, a graduate of a great university of the far north, where Negroes are seldom seen, resented it most indignantly when she was threatened with social ostracism in a city farther south with a large Negro population because she insisted upon receiving upon terms of social equality a Negro man who had been her classmate.
>
> (Popenoe and Johnson 1918, p. 280)

To Popenoe and Johnson, the story indicated the existence of a "color line" in the southern United States. Why should a color line exist? Racial antipathy, they said, was a biological mechanism to protect the white race from miscegenation; and there were good biological reasons to prevent miscegenation.

Negroes were inferior to whites, they argued. Their evidence was that Negroes had made no original contributions to world civilization; they had never risen much above barbarism in Africa; they did no better when transplanted to Haiti; and they failed to achieve white standards in America. Utilizing results from the newly developed intelligence tests, they presented evidence which showed that Negroes scored significantly worse than whites. Furthermore, the disease resistance of the Negro was inferior to that of the white in North America: "biologically, North America is a white man's country, not a Negro's country," but they admitted this relative fitness of the two races might be reversed in Africa. Popenoe and Johnson concluded that all these racial differences were largely hereditary:

> . . . the Negro race differs greatly from the white race, mentally as well as physically, and that in many respects it may be said to be inferior, when tested by the requirements of modern civilization and progress, with particular reference to North America.
>
> (Popenoe and Johnson 1918, pp. 291–2)

Having elucidated the problem of hereditary racial differences, Popenoe and Johnson turned to the problem of race crossing between Negroes and whites.

Mulattoes, they claimed, were intermediate between the two parent races in color and intelligence; thus "in general the white race loses and the Negro gains from miscegenation." For this reason they "unhesitatingly condemn miscegenation." But what of the argument that the surest way to elevate the Negro was through crossing with whites? They answered in Galtonian terms.

> To insure racial and social progress, nothing will take the place of leadership, of genius. A race of nothing but mediocrities will stand still, or very nearly so; but a race of mediocrities with a good supply of men of exceptional ability and energy at the top, will make progress in discovery, invention, and organization, which is generally recognized as progressive evolution.

If the level of the white races be lowered, it will hurt that race and be of little help to the Negro. If the white race be kept at such a level that its productivity of men of talent will be at a maximum, everyone will progress; for the Negro benefits just as the white does from every forward step in science and art, in industry and politics.

(Popenoe and Johnson 1918, pp. 292, 293)

Here was Galton's conclusion again, but now supported by the data and wording of twentieth-century genetics and quantitative psychology.

The conclusions which Popenoe and Johnson drew from this chapter they described as "in accordance with modern science." Their conclusions would warm the hearts of many white readers:

1. We hold that it is to the interests of the United States, for the reasons given in this chapter, to prevent further Negro–white amalgamation. 2. The taboo of public opinion is not sufficient in all cases to prevent intermarriage, and should be supplemented by law, particularly as the United States have of late years received many white immigrants from other countries (e.g., Italy) where the taboo is weak because the problem has never been pressing. 3. But to prevent intermarriage is only a small part of the solution, since most mulattoes come from extramarital miscegenation. The only solution of this, which is compatible with the requirements of eugenics, is not that of laissez faire, . . . but an extension of the taboo, and an extension of the laws, to prohibit all sexual intercourse between the two races. . . . 4. Miscegenation can only lead to unhappiness under present social conditions and must, we believe, under any social conditions be biologically wrong.

(Popenoe and Johnson 1918, pp. 296–7)

Applied Eugenics sold well. The first edition went through several printings and a second edition, revised by Popenoe, appeared in 1933. Among those who read or cited the book were geneticists W. E. Castle, Edward Murray East, and Sewall Wright. University of Chicago biologist Horatio Hackett Newman published a chapter of *Applied Eugenics* in his widely used textbook, *Readings in Evolution, Genetics, and Eugenics*, first published in 1921. Not one geneticist publicly challenged the use of genetics to support the racial policies found in *Applied Eugenics* when it first appeared. In 1971 Paul Popenoe said that he could "not recall that any geneticists disapproved of that chapter ['The Color Line']. It was definitely in line with the views of the majority, so far as I then knew" (Popenoe, pers. comm. 1971).

22.7 Edward Murray East and the genetics of race crossing

Although Popenoe and Johnson had both studied genetics, neither was a research geneticist of note. Davenport was respected by many of his colleagues and students, but his work on human genetics was clearly tentative when compared with other genetic research. Edward Murray East of Harvard's Bussey Institution was, on the other hand, one of the most highly respected research geneticists in America. He was among the first to clarify multifactorial Mendelian inheritance; he was also an expert on breeding and cross-breeding plants, and a pioneer in hybrid corn research. During World War I the government asked his assistance in agricultural planning, a responsibility that spurred his interest in the social significance of genetics. When East and his former student Donald F. Jones published *Inbreeding and Outbreeding* in 1919, they subtitled it *Their Genetic and Sociological Significance.* The book was a basic contribution to the Mendelian interpretation of breeding, and its significance was recognized by all experimental geneticists. It was published in a distinguished series edited by Jacques Loeb and Thomas Hunt Morgan.

The last two chapters of *Inbreeding and Outbreeding,* written by East alone, dealt with the sociological significance of genetics, particularly the problem of race mixture. East divided race mixture into two kinds, those between closely related races and those between distantly related races. The former, as between various white races of Europe, had produced the most civilized humans. But East cited two genetical objections to wide human race crosses, as between Negroes and whites. First, Mendelian segregation would "break apart those compatible physical and mental qualities which have established a smoothly operating whole in each race by hundreds of generations of natural selection." Second, it was "an unnecessary accompaniment to humane treatment, an illogical extension of altruism ... to seek to elevate the black race at the cost of lowering the white" because "in reality the Negro is inferior to the white. This is not hypothesis or supposition; it is a crude statement of actual fact" (East and Jones 1919, pp. 253–4).

East's first objection to race crossing was that promulgated by Herbert Spencer and Davenport. The second was a direct reiteration of the objection raised by Popenoe and Johnson a year earlier. These objections to race crossing now had the clear approval of a famous research geneticist who said he was examining the issue in accordance with the biological facts; and East was without question one of the world's experts on the biological facts of cross-breeding.

Geneticists reacted very favorably to *Inbreeding and Outbreeding.* They even reacted favorably to East's double-barreled objections to race crossing. Geneticist Raymond Pearl, who later boasted of his opposition to "Nordic enthusiasts," wrote in *Science* that the last two chapters dealing with race and society might "fairly be regarded as among the sanest and most cogent arguments for the integral incorporation of eugenic ideas and ideals into the conduct of social and political affairs of life.... There is refreshing absence of blind and blatant propaganda (Pearl 1920, p. 415).

As far as I can discover, no geneticist objected in print to East's views on race crossing. And there can be no doubt that *Inbreeding and Outbreeding* was required reading for geneticists. *Inbreeding and Outbreeding* marked only the beginning of East's interest in the human implications of genetics.

In the 1920s he published two large volumes, *Mankind at the Crossroads* (1923) and *Heredity and Human Affairs* (1927). The introduction to *Mankind at the Crossroads* explains why an expert on corn genetics or fruit flies could have the expertise to talk about human genetics:

Genetics have enticed a great many explorers during the past two decades. They have labored with fruit-flies and guinea-pigs, with sweet peas and corn, with thousands of animals and plants in fact, and they have made heredity no longer a mystery but an exact science to be ranked close behind physics and chemistry in definiteness of conception. One is inclined to believe, however, that the unique magnetic attraction of genetics lies in the vision of potential good which it holds for mankind rather than a circumscribed interest in the hereditary mechanisms of the lowly species used as laboratory material. If man had been found to be sharply demarcated from the rest of the occupants of the world, so that his heritage of physical form, of physiological function, and of mental attributes came about in a superior manner setting him apart as lord of creation, interest in the genetics of the humbler organisms – if one admits the truth – would have flagged severely. Biologists would have turned their attention largely to the ways of human heredity, in spite of the fact that the difficulties encountered would have rendered progress slow and uncertain. Since this was not the case, since the laws ruling the inheritance of the denizens of the garden and the inmates of the stable were found to be applicable to prince and potentate as well, one could shut himself up in his laboratory and labor to his heart's content, feeling certain that any truth which it fell to his lot to discover had a clear human interest, after all.

(East 1923, pp. v–vi)

East's reasoning here helps to explain why experts in plant genetics, or the genetics of protozoa, could feel qualified and be motivated to write books about the social consequences of human heredity.

Mankind at the Crossroads contained a long chapter entitled "Racial Prospects and Racial Dangers." Here East presented further evidence for the supposed genetic inferiority of the Negro, citing from Robert M. Yerkes' report on the psychological tests administered to United States recruits during World War I (Yerkes 1921). This wealth of often conflicting data and methods had already been mined by Paul Popenoe a year earlier (Popenoe 1922).

The basic facts were that Negro recruits on an average scored significantly lower than whites on the same intelligence tests; both East and Popenoe were certain this difference resulted primarily from genetic differences in the two races. That mulattoes scored intermediate between whites and dark-skinned Negroes they believed was an added proof for their conclusion. This conclusion would later be challenged by revisionists, mostly anthropologists, who pointed out that Negroes from Ohio scored on an average higher than whites from eleven Southern states, and that other facts indicated the importance of environmental influences (see M. F. Ashley Montagu 1945, pp. 138–41, for an account and bibliography). But in 1923 most geneticists had little reason to challenge East's conclusions because they agreed with them.

East was not a simple racist who argued that all Negroes were mentally inferior to whites. He was a population biologist, not a "typologist" of the sort emphasized by Ernst Mayr. East specifically ridiculed the biology of the popular racists such as Madison Grant, Seth Humphrey, and Lothropp Stoddard. Although these writers claimed to have based their assertions upon modern genetics, East denied this vehemently and classified them as "race dogmatists," whose belief that one race was completely superior to another was faulty biology. He sketched the difference between a biologist's point of view and that of a race dogmatist:

> The one forbids racial crossing because of an indefensible belief in the general superiority of all the individuals of one race over all the individuals of another; the other advises against racial crossing even between widely separated races of equal capacity simply because the operation of the heredity mechanism holds out only a negligible prospect of good results through disturbing the balanced whole of each component. Both recognize differences in racial levels or averages, but the biologist realizes what an immense amount of overlapping there is. He sees how small is the gap between the efficiency levels of each race as a whole, and how great is the chasm between the superior and inferior extremes within the race, even though each race may have exclusive possession of certain hereditary units.
>
> (East 1923, pp. 131–2)

Clearly East was looking at the problem of race differences as a population biologist, not a "typologist."

Furthermore, East was a staunch supporter of civil liberties for every individual. He was thoroughly indignant about discrimination against Negroes on trains and in theaters and restaurants. He exclaimed in *Heredity and Human Affairs* that such discriminatory actions were "the gaucheries of a provincial people, on a par with the guffaws of a troop of yokels who see a well-dressed man for the first time" (East 1927, p. 181). But East, the population biologist who believed in civil rights for all, is the same person who concluded that "the negro race as a whole is possessed of undesirable transmissible qualities both physical and mental, which seem to justify not only a line but a wide gulf to be fixed permanently between it and the white race" (East 1923, p. 133).

Many people now may find it easy to dismiss East as an evil racist; but this is too simple. He was also a dedicated scientist who tried to help society with the fruits of his labors in a young and exciting science.

Between 1919, when *Inbreeding and Outbreeding* appeared, and 1923, when *Mankind at the Crossroads* appeared, not one geneticist had published criticism of the views represented by East on race differences and race crossing.

In 1923, the English geneticist R. Ruggles Gates, whose major work was in botany, published the best available textbook on human heredity in England (Gates 1923). Gates clearly agreed with East's arguments on race, and cited Davenport's 1917 paper on race crossing.

University of California biologist Samuel J. Holmes, who had a keen interest in the social implications of genetics, agreed in 1923 that the crossing of races on different mental levels was inadvisable. But he argued that although studies of lower organisms indicated that disharmonies might occur in crosses between "equivalent" races, no clear evidence of these was as yet available. He counseled caution, emphasizing that "the argument from ignorance should not be used to defend race crossing because we cannot prove that it is bad; it should be used rather to counsel caution because we do not know that it is not bad (Holmes 1923, p. 223).

Holmes' position here contrasts sharply, as I have shown, with the 1951 UNESCO statement by geneticists and physical anthropologists which argued that race crossing should be permitted because there was no evidence it was harmful (Provine 1986).

Any educated layperson, who in 1923 was interested in the problem of human race crossing and who examined the genetics literature for guidance, could come to only one conclusion – that wide race crosses presented a clear biological danger and should be prevented by law if necessary. The available genetics literature contained not one dissent from this view. Soon that situation would change. One geneticist, William Castle, would speak out loud and clear, challenging the dominant view. He found a genuine battle on his hands. Castle, however, believed along with his colleagues that human races differed in genetically determined intelligence, and that blacks were mentally inferior to whites (Provine 1973, 1986).

22.8 Conclusion

The quotation from Lewontin that prefaces this chapter provides a strong historical thesis for explaining the views of geneticists about race differences and race crossing in the first quarter of the twentieth century. Geneticists knew almost nothing about hereditary race differences in intelligence, yet agreed to a person that races did in fact differ in this respect. Their attitude about human race crossing was equally free of genetic understanding, yet most geneticists rejected wide race crossing. All historians who have examined geneticists' views about race and race differences in the early twentieth century have found the same basic agreement that I find here in greater detail. The historians, however, have uniformly scolded and blamed the geneticists for holding such views, arguing that Davenport was a fool and others, better geneticists such as East and Jennings, should have known better (Haller 1963, Ludmerer 1972, Kevles 1985).

Lewontin offers a more robust historical interpretation. He encourages historians to find specific cultural influences (including, of course, from science) that determined geneticists to hold the observed views. I argue that geneticists adopted null hypotheses about race and race crossing from both scientists and the broader cultures, and that they saw no reason from the scientific evidence to change them. In the USA and Europe nearly everyone believed that races differed in intelligence and that wide race crosses had bad biological consequences. Related scientists from psychology and anthropology held the same views. Holding views of racial equality in intelligence and belief in beneficial consequences of race crossing would net a geneticist of the time charges of social motivation rather than scientific motivation.

Mere condemnation of the racial views of geneticists in the early twentieth century nets no historical understanding. Indeed, condemnation can obscure real historical connections. For example, branding Davenport a fool enables us to dismiss his influence on modern eugenics. Davenport vociferously argued for voluntary eugenics, eugenics education, and use of eugenics counseling centers. Were he able to see the profusion of genetic counseling education and centers of genetics counseling information now available (by fee) to upper and middle classes, he would have been thrilled with the success of his social aims. Eugenics has merely been renamed "genetic counseling," and modern eugenics is not only alive and well, but expanding at a great rate. Far from a mere fool, Davenport is the father of modern genetic counseling.

Lewontin has offered in his quote a profound challenge for historians of science. Easy assessments of blame and praise for scientific views are out, no matter how ugly or attractive to the historian. What counts is an historical analysis that incorporates all influences upon scientists, with special attention paid to those issues on which scientists know little, but say a lot.

Detailing geneticists' views of race differences and race crossing in the early twentieth century is easy compared with the much longer and difficult task of detailing influences upon individual geneticists to explain historically the views they hold.

REFERENCES

Bateson, W. (1909). *Mendel's Principles of Heredity.* Cambridge: Cambridge University Press.

Baur, E. (1911). *Einführung in die experimentelle Vererbungslehre.* Berlin: Borntraeger.

Davenport, C. B. (1911). *Heredity in Relation to Eugenics.* New York: Henry Holt.

Davenport, C. D. (1917a). The inheritance of stature. *Genetics* 2:313–89.

Davenport, C. D. (1917b). The effects of race intermingling. *Proc. Am. Phil. Soc.* 56:364–8.

Davenport, G. C., and Davenport, C. D. (1907). Heredity of eye color in man. *Science* 26:589–92.

Davenport, G. C., and Davenport, C. D. (1908). Heredity of hair form in man. *Am. Nat.* 42:341–9.

Davenport, G. C., and Davenport, C. D. (1909). Heredity of hair color in man. *Am. Nat.* 43:193–211.

Davenport, G. C., and Davenport, C. D. (1910). Heredity of skin pigment in man. *Am. Nat.* 44:642–72, 705–31.

Dunn, L. C. (1959). *Heredity and Evolution in Human Populations.* Cambridge, MA: Harvard University Press.

Dunn, L. C. (1965). *Short History of Genetics.* New York: McGraw-Hill.

East, E. M. (1923). *Mankind at the Crossroads.* New York: Charles Scribner's Sons.

East, E. M. (1927). *Heredity and Human Affairs.* New York: Charles Scribner's Sons.

East, E. M., and Jones, D. F. (1919). *Inbreeding and Outbreeding: Their Genetic and Sociological Significance.* Philadelphia: Lippincott.

Fischer, E. (1913). *Die Rehobother Bastards und das Bastardierungsproblem beim Menschen: anthropologische und ethnographische Studien am Rehobother Bastardvolk in Deutsch-Südwest-Afrika.* Jena: Gustav Fischer. Reprinted photomechanically in 1961 by Akademische Druck- und Verlagsanstalt, Graz, Austria, with new introduction by Hans Biedermann.

Gates, R. R. (1923). *Heredity and Eugenics.* London: Constable.

Goldschmidt, R. (1911). *Einführung in die Vererbungswissenschaft.* Leipzig: Wilhelm Engelmann.

Haecker, V. (1911). *Allgemeine Vererbungslehre.* Braunschweig: Vieweg und Sohn.

Haller, M. H. (1963). *Eugenics: Hereditarian Attitudes in American Thought.* New Brunswick, NJ: Rutgers University Press.

Holmes, S. J. (1923). *Studies in Evolution and Eugenics.* New York: Harcourt Brace.

Holmes, S. J. (1924). *A Bibliography of Eugenics.* University of California Publications in Zoology, vol. 25. Berkeley: University of California Press.

Hurst, C. C. (1912). Mendelian heredity in man. *Eugen. Rev.* 4:1–25.

Kevles, D. J. (1985). *In the Name of Eugenics.* New York: Knopf.

Lang, A. (1911). *Die experimentelle Vererbungslehre in der Zoologie seit 1900.* Jena: Gustav Fischer.

Lewontin, R. C. (1974). *The Genetic Basis of Evolutionary Change.* Columbia Biological Series, no. 25. New York: Columbia University Press.

Lotsy, J. P., and Goddjin, W. A. (1928). Voyages of exploration to judge of the bearing of hybridization upon evolution. *Genetica* 10:206.

Ludmerer, K. L. (1972). *Genetics and American Society: a Historical Appraisal.* New Brunswick, NJ: Rutgers University Press.

Montagu, M. F. A. (1945). *Man's Most Dangerous Myth: the Fallacy of Race.* 2nd edition New York: Columbia University Press.

Newman, H. H. (1921). *Readings in Evolution, Genetics, and Eugenics.* Chicago: University of Chicago Press. Many later editions.

Pearl, R. (1920). Review of *Inbreeding and Outbreeding* by E. M. East and D. F. Jones. *Science* 51:415.

Pearson, K. (1901). *National Life from the Standpoint of Science.* London: Adam and Charles Black.

Popenoe, P. (1922). Intelligence and race. *J. Hered.* 13:265–9.

Popenoe, P., and Johnson, R. H. (1918). *Applied Eugenics.* New York: Macmillan.

Provine, W. B. (1973). Geneticists and the biology of race crossing. *Science* 182:790–6.

Provine, W. B. (1986). Geneticists and race. *Am. Zool.* 26:857–87.

Punnett, R. C. (1907). *Mendelism.* 2nd edition Cambridge: Cambridge University Press.

Punnett, R. C. (1917). Eliminating feeblemindedness. *J. Hered.* 8:464–65.

Stern, C., and Sherwood, E. R. (1966). *The Origin of Genetics: A Mendel Source Book.* Berkeley: University of California Press.

Yerkes, R. M. (ed.) (1921). *Psychological Examining in the United States Army.* Washington, DC: Memoirs of the National Academy of Sciences 15.

Index

Page numbers in bold denote figures and tables